# Springer Series in Chemical Physics

Volume 113

The purpose of this series is to provide comprehensive up-to-date monographs in both well established disciplines and emerging research areas within the broad fields of chemical physics and physical chemistry. The books deal with both fundamental science and applications, and may have either a theoretical or an experimental emphasis. They are aimed primarily at researchers and graduate students in chemical physics and related fields.

More information about this series at http://www.springer.com/series/676

Chang Q. Sun · Yi Sun

# The Attribute of Water

## Single Notion, Multiple Myths

Springer

Chang Q. Sun
School of Electrical and Electronic Engineering
Nanyang Technological University
Singapore
Singapore

Yi Sun
Xiangtan University
Changsha
China

and

China Jiliang University
Hangzhou
China

ISSN 0172-6218
Springer Series in Chemical Physics
ISBN 978-981-10-0178-9 ISBN 978-981-10-0180-2 (eBook)
DOI 10.1007/978-981-10-0180-2

Library of Congress Control Number: 2015960779



Printed on acid-free paper

This Springer imprint is published by SpringerNature
The registered company is Springer Science+Business Media Singapore Pte Ltd.

*The original version of the book frontmatter was revised:*
*For detailed information please see Erratum.*
*The Erratum to the book frontmatter is available at*
*DOI 10.1007/978-981-10-0180-2_19*

*Nothing is softer or more flexible than water, yet nothing can resist it.*

—Lao Tzu (571–470 B.C.)

*If there is magic on this planet, it is contained in water.*

—Loren Eiseley (1907–1977)

*The great seal of truth is simplicity.*

—Herman Boerhaave (1668–1738)

*Simplicity is about subtracting the obvious, and adding the meaningful.*

—John Maeda, *The Laws of Simplicity,* 2006

*The nature of the chemical bond bridges the structures and properties of crystals and molecules.*

—Linus Pauling, *The Nature of the Chemical Bond,* 1939

*Bond and nonbond relaxation and the associated energetics, localization, entrapment, and polarization of electrons mediate the performance of substance accordingly.*

—Chang Q. Sun, *Relaxation of the Chemical Bond,* 2014

*O:H–O bond segmental disparity and O–O repulsivity form the soul dictating the extraordinary adaptivity, cooperativity, recoverability, and sensitivity of water and ice*

*Dedicated to our beloved*

# Preface

Water is the source and central part of all life—without it, life cannot be sustained or evolve. Water is simple but magical, pure but elegant, perseverant but flexible. Water also symbolizes kindness, wisdom, wealth, and prosperity. Lao Tze, an ancient Chinese philosopher and thinker, wrote: *water represents the highest morality of human beings. It benefits all others, without expecting anything in return. It retains its ethical standard, but stays in where disdained* (上善若水，利万物而不争。处众人之所恶，故几于道).

Water is so magically sensitive to any perturbation of biological signals, radiations, and external constraints or stimuli that it lends itself to many romantic notions throughout history. Masaru Emoto described a variety of crystal patterns of ice grown from pure and polluted sources and influenced by emotions, thoughts, and voices in his book, *The Healing Power of Water*. James Brownridge dedicated himself for some 10 years to conducting over 20 experiments to identify the factors influencing the Mpemba effect, which describes the phenomenon where warmer water freezes more quickly than cold water, as first documented by Aristotle in 350 B.C. Gerald Pollack proposed *The Fourth Phase of Water* associated with the hydrophilic interface contact, which has explained numerous phenomena from the perspective of the exclusion zone of the layered, three-coordinated hydronium, $H_3O^+$, gel-like structure, capable of absorbing all types of energies, separating charges, and excluding microspores and organisms. Because of all these sensitive fascinations, water is described as having intelligence and spirit, and even as a messenger of God.

However, water is too strange, too anomalous, and too challenging, as noted by Philip Ball, a scientific writer and a former senior editor for *Nature*. Its versatile structural phases and strange behavior have fascinated inspiring minds such as Archimedes, Francis Bacon, René Descartes, Lord Kelvin, Isaac Newton, Siméon Denis Poisson, Thomas Young, Pierre-Simon Laplace, Carl Friedrich Gauss, Frantz Hofmeister, William Armstrong, Johann Gottlob Leidenfrost, Gilbert Newton Lewis, and Linus Pauling among many others. In 1611, Galileo Galilei and Ludovico delle Colombe ignited the debate on why ice floats, from the perspectives of buoyant force, surface tension, and mass density. In 1859, Michael Faraday, James Thomson, and James Forbes started the debate on ice regelation—the behavior of ice melting under pressure and refreezing when the pressure is relieved. Michael Faraday, James Thomson, William Thomson (Lord Kelvin), and Willard Gibbs had been involved in exploring the slipperiness of ice since 1850, based on the concepts of quasi-liquid skin, pressure melting, and friction heating.

In the 125th anniversary special issue, *Science* magazine listed *The Structure of Water* as one of the 125 big questions to mankind. In 2012, the Royal Society of Chemistry organized a competition offering a £1,000 award to the participant, out of 22,000 entries, who could provide the best possible explanation for the Mpemba effect. The International Union of Pure and Applied Chemistry (IUPAC) gathered over 30 experts in Pisa in 2005 to form a task force to revise the definition of the hydrogen bond, and the agreed-upon result was published in 2011. To commemorate the 400th anniversary of the debate between Galileo and Colombe on the mystery of floating ice, twenty-five bright minds met in Florence, Italy, in July 2013, for a week, to discuss unanswered questions on water. However, fierce debates regarding the structure and anomalies of water are still ongoing, *converging* Mark Twain's (1835–1910) saying that *whiskey is for drinking; water is for fighting over*. The current status is that each of the various anomalies of water and ice is debated by multiple theories. It has been a long-standing dream of the scientific society to develop one notion that reconciles as many anomalies as possible.

Alternative ways of thinking and approaching are necessary to resolve the anomalies of water and ice. Turning our perspectives from classical thermodynamics to hydrogen bond (O:H–O) cooperativity, from single snapshots to statistical means, from surface to skin, and from spectroscopy to spectrometrics would be essentially helpful. In place of the conventional approach in terms of dipole–dipole interaction in the "dipole sea" of water, we have focused on the performance of a representative O:H–O bond for all as an asymmetrical oscillator pair with short-range interactions and O–O Coulomb coupling. This premise enables us to examine the consequence of the O:H–O cooperativity and polarizability on the detectable properties of water and ice.

An oxygen atom always tends to hybridize its *sp* orbits when interacting with atoms of any arbitrary electropositive elements and therefore a water molecule takes the tetrahedral configuration not only in the gaseous phase but also in solid at temperatures of only several Kelvins, although the O:H–O bond containing angle and its segmental lengths are subject to relaxation under perturbation.

Rather than the O:H nonbond or the H–O polar-covalent bond alone, the O:H–O bond integrates both the O:H intermolecular and the H–O intramolecular asymmetrical, ultra-short-range interactions and the Coulomb repulsion between electron pairs on adjacent oxygen. Being conventionally overlooked, the Coulomb repulsion between oxygen anions and the segmental disparity of the O:H–O bond form the soul dictating the extraordinary adaptivity, cooperativity, recoverability, and sensitivity of water and ice when responding to perturbation at any level.

The segmental disparity and the strong H–O bond allow for molecular flipping vibration, but unlikely the "proton tunneling transition" (Bernal-Fowler 1933) or the "two-in two-out" proton frustration (Pauling 1935). Rather, the segmented O:H–O bond relaxes in a "master–slave" manner. One segment becomes stiffer if it turns to be shorter, and the other will become longer and softer. The O:H nonbond always relaxes more than the H–O bond. The flexible and polarizable O:H–O bond exists commonly to all phases irrespective of their geometries such as the superionic $OH_3^+$:$OH^-$ phase and the X phase of identical O:H and H–O distances. As uncovered by Yanming Ma at Jilin University, China, $(H_2O)_2$ only transits into the $OH_3^+$(hydronium):$OH^-$ (hydroxide) configuration under 2 TPa pressure and 2000 K temperature. The O:H–O bonding premise is more comprehensively appealing than the convention of intermolecular dipole–dipole interactions, and it is also general to situations with a presence of electron lone pairs.

In dealing with the strongly correlated and fluctuating water system, one should be more focused on the statistical mean of the collection of all correlated parameters than on the instantaneous accuracy of a certain isolated quantity under a specific condition at a certain point of time. One has to keep in mind the meaningful parameters and disregard those such as the long-range interactions and nuclear quantum dynamics as the common background of all to derive a simple solution for the seemingly unrelated phenomena.

The specific heat disparity between the O:H nonbond and the H–O bond discriminates the thermal dynamics of water and ice in different temperature regimes, which defines a quasisolid phase where the negative thermal expansion occurs. The rule of global bond contraction between undercoordinated atoms also applies to water molecules at the skins of bulk water, hydration shells, droplets, bubbles, and hydrophobically encapsulated water. Molecular undercoordination not only disperses the quasisolid phase outwardly to lower the freezing temperature and raise the melting point but also creates a supersolid skin phase that is less dense, ice like, elastic, and hydrophobic.

An HX-type acid dissolves into the $X^-$ anion and the $H^+$ that binds to an $H_2O$ to form the hydronium $H_3O^+$ tetrahedron with one lone pair and the $H_3O^+$ interacts with one of its four neighbors through the O–H↔H–O anti-hydrogen bond (anti-HB). The H↔H serves as a point breaker of the entire HB network, making the bulk water "fragile". A YOH-type base dissolves into the $Y^+$ cation and the $OH^-$ hydroxide with three lone pairs, and the $HO^-$ interacts with one of its four neighbors through the O:↔:O super-HB that serves as a point compressor, elongating the H–O bond and releasing heat when reaction rakes place. The $X^-$ and $Y^+$ solute ions create each an electric field to align, cluster, polarize, and stretch the $H_2O$ molecules in the hydration shells. Electrification of $H_2O$ molecular dipoles by the fields of artificially attached charge, or an externally applied potential will also align, stretch, and polarize the O:H–O bonds. Electrification of the O:H–O bond by opposite fields effects adversely than under either alone. The hydration shells of solute ions are essentially the same as the water skin exhibiting stronger supersolidity behavior. The macroscopic properties of water and ice depend discriminatively and functionally on the cooperative relaxation in length and energy of the segmented O:H–O bond and the associated polarization under excitation.

Phonon spectrometrics is a powerful tool that enables discovery of the molecular site, multifield driven O:H–O relaxation dynamics in terms of segmental lengths and stiffness, order of molecular fluctuation, and phonon abundance, which reveals sufficiently and exactly what is happening to water molecules of the same coordination environment under excitation. A combination of the micro-jet UPS and XPS collects molecular site-resolved information about electron polarization and the O 1s energy shift. Lagrangian mechanics is efficient in dealing with the O:H–O asymmetrical oscillator pair dynamics, which enables mapping the potential paths for the O:H–O bond at relaxation. Fourier thermo-fluid transport dynamics is essential for solving the heat conduction involved in the Mpemba paradox. The use of multiple strategies is necessary for unlocking the mysteries of water. Computations and spectrometrics serve as powerful tools for verifying the theoretical predictions that are key to solving the long-standing puzzles. These considerations have led to a set of experimental, numerical, and theoretical strategies that have enabled the presented efforts and progresses.

This volume deals with the scientific popularization, quantitative resolution, and insightful extension of the best known mysteries of water and ice. Consistent resolution to the noted mysteries verifies the validity of the O:H–O bond notion and the approaching strategies. This book also demonstrates how the segmented O:H–O bond responds adaptively and cooperatively to stimulus of chemical contamination, electrification, magnetization, mechanical compression, molecular undercoordination, thermal excitation, and their joint effect in a coordination-resolved manner, and how the bond relaxation changes the macroscopic properties of water and ice. This volume presents an effort to resolve, once and for all, the following systematic issues:

1. Crystallographic structure order (tetrahedrally-coordinated fluctuating monophase with a supersolid skin)
2. Density-geometry-size-separation correlation of molecules packed in water and ice
3. Bond-electron-phonon-property correlation of water and ice
4. Asymmetrical, short-range, and coupled potentials for the relaxed O:H–O bond
5. O:H–O bond relaxation kinetics crossing the phase diagram
6. Ice Regelation—compression lowers but tension raises the $T_m$ (O:H–O bond recoverability and quasisolid-phase boundary dispersivity)
7. Pressure-induced O:H–O bond proton centralization (O:H compression and H–O elongation)
8. Ice floating (specific heat disparity defined quasisolid phase that undergoes cooling expansion)
9. Mass density thermal oscillation of water and ice and coordination-resolved liquid O:H–O bond thermal dynamics (specific heat ratio entitled master–slave manner relaxation at different temperatures)
10. Unusual thermodynamics of skins, hydration shells, free and confined nanodroplets and nanobubbles (H–O contraction elongates and polarizes the O:H nonbond)

11. Hydrophobicity and hydrophilicity transition (dipole creation and annihilation)
12. Superlubricity of ice and quantum friction (electrostatic repulsivity and O:H phononic elasticity)
13. Supersolid solute hydration shells (elastic, polarized, hydrophobic, less dense, and thermally more stable)
14. Quasisolid phase boundary dispersivity (phonon frequency relaxation modulates the Debye temperatures)
15. Hofmeister effect—ions modulation of surface tension and DNA solubility (O:H–O bond relaxation and polarization)
16. Molecular bonding in Lewis solutions of acids, bases, and adducts or salts (H↔H anti-HB pointer breaker in acid solutions, O:↔:O super-HB point compressor in base solutions, and solute ionic polarizer in adduct solutions)
17. Discrimination of acid and salt solutions in stress and solubility (ionic electrification and discriminative polarization)
18. Hofmeister solution thermal stability—critical pressures, temperatures, and durations for phase transition (O:H–O bond deformation by the coupled fields)
19. Armstrong water floating bridge (long-range ordered electrification disperses the quasisolid phase boundary)
20. Electromelting (artificial electrification effect on quasisolidity)
21. Magnetization and electromagnetic radiation—(moving dipoles—Lorentz field—current induction—antiferromagnetism)
22. Soil wetting by aqueous solutions (electric fields superposition)
23. Correlation of H–O phonon frequency, lifetime, self-diffusion, skin stress, and solution viscosity
24. Mpemba paradox—warm water freezes quickly (O:H–O bond memory and water skin supersolidity)
25. Molecular-site-resolved O:H–O bonding dynamics in terms of segmental stiffness, structure order, phonon abundance, etc.

Water forms such a strongly correlated and fluctuating system that not only involves asymmetrical, ultra-short-range, and coupling interactions but also responds sensitively to any perturbation or radiation in an ultra-long range manner under a domino-like effect.

Water is much more interesting but less complicated than many of us could ever imagine. Nothing is more fun than playing with water and ice from the perspective of predictive bond-electron-phonon-property collaborative relaxation. It is really an enjoyable and fascinating experience to tackle these anomalies. It is our obligation and great pleasure to share these discoveries and progress, although some formulations and solutions might be subject to further improvement and refinement. Corrections, critiques, and better solutions are welcome and furthermore, much appreciated.

We hope that this volume, though it contributes a tiny drop to the ocean of water knowledge, could inspire fresh ways of thinking and approaching and stimulate more interest and activities toward uncovering the mysteries of water and ice, especially in the contexts of water being embedded in or interacting with other species. The strong correlation, fluctuation, localization, and polarization could be important ingredients in this understanding. Directing effort to interaction between water and soft matter and to water's role of messaging, regulating, repairing, and signaling in bioelectronics, food, drug, and life sciences could be even more challenging, fascinating, promising, and rewarding.

We express our sincere gratitude to friends and peers for their encouragement, invaluable input, and support, and to collaborators, particularly, research associates Dr. Xi Zhang, Dr. Yongli Huang, Dr. Zengsheng Ma, and Mr. Yong Zhou, for contributions. Last but not least, we thank our families, Mrs. Meng Chen, in particular, for assistance, patience, support, and understanding throughout this joyous and fruitful journey.

Singapore Chang Q. Sun
China Yi Sun

# Contents

# About the Authors

**Chang Q. Sun** received his B.Sc. degree in 1982 from Wuhan University of Science and Technology and the M.Sc. degree in 1987 from Tianjin University, China, and served on its faculty until 1992. He earned his Ph.D. in Surface Physics at Murdoch University, Australia in 1996.

Dr. Sun has been working on the Coordination Bond and Electronic Engineering since 1990 with creation of the unique Bond Relaxation Theory that has enabled: (1) invention of the Coordination-Resolved Electron and Phonon Spectrometrics; (2) reconciliation of the behavior of bonds and electrons associated with undercoordinated atoms of defects, skins, nanostructures of various shapes, and heterocoordinated atoms in chemisorption and interfaces; (3) formulation of the atomistic, multifield solid mechanics; (4) correlation of bond relaxation and polarization to detectable quantities of aqueous and solid specimens; and (5) resolution of multiple mysteries of water and ice.

His contribution has been featured in two monographs, the present *The Attribute of Water* and the *Relaxation of the Chemical Bond* (Springer, 2014; Chinese edition published by High Education Press, Beijing, 2015), and multiple treatises published in *Friction* (2015), *Chemical Reviews* (2015; 2012), *Coordination Chemistry Reviews* (2015), *Progress in Solid State Chemistry* (2015; 2007; 2006), *Surface Science Reports* (2013), *Progress in Materials Science* (2009; 2003), *Energy and Environment Science* (2011), *Science China* (2012), *Nanoscale* (2010), etc. Breakthroughs in water research have been featured by numerous media such as *Times, The Telegraph, Daily Mails, Physics Today, IOP News, Chemical World, Chem Views, Nature Chemistry, New Scientist,* etc.

Dr. Sun was conferred the First Laureate of the 25th Khwarizmi International Science Award in 2012 and the Inaugural Nanyang Award of Research in 2005, and finalist of Singapore Presidential/National Science Award in 2012 and 2003. He is currently on the Editorial Advisory Board of numerous scientific journals.

**Yi Sun** received her B.Sc. degree in Physics from University of Illinois at Urbana-Champaign (UIUC) in 2008, her M.A. in Economics and Ph.D. degree in Public Policy from the University of Chicago in 2016. Besides the current project, she has been devoted to the electron spectrometrics of binding energy shift of graphene, Ni/$TiO_2$, and Pt/CNT nanocrystals and interfaces and the superhydrophilicity, superfluidity, superlubricity, and supersolidity at the nanometric contacts from the perspective of undercoordination-induced quantum entrapment and polarization. She received the Commonwealth Edison/Bristow Award in Physics and was named a Chancellor's Scholar at UIUC in 2008.

# Nomenclature

| | |
|---|---|
| XPS | X-ray photoelectron spectroscopy |
| $\eta$ | Viscosity |
| $T_V/T_m$ | Evaporation/melting temperature |
| $\theta$ | Temperature |
| $T_C$ | Critical temperature for phase transition except for evaporation |
| STM/S | Scanning tunneling microscopy/spectroscopy |
| SFG | Sum frequency generation |
| RPS | Residual phonon spectroscopy |
| RLS | Residual length spectroscopy |
| NTE | Negative thermal expansion |
| NEP | Nonbonding electron polarization |
| $\rho$ | Mass density |
| H/LDL | High/low density liquid |
| FTIR | Fourier transformation infrared spectroscopy |
| $E_x$ | O:H–O segmental cohesive energy |
| $d_x$ | O:H–O segmental length |
| DOS | Density of states |
| CN(z) | Coordination number |
| BOLS | Bond order–length–strength correlation |
| $\eta_X$ | Specific heat of the xth segment |
| $\omega_X$ | O:H–O segmental phonon frequency |
| $\tau_H$ | H–O phonon lifetime |
| $\Delta E_{1s}$ | O 1s core—level shift |
| $\Theta_{DX}$ | O:H–O segmental Debye temperature (x = L for O:H; x = H for H–O) |
| $\mu$ | Self-diffusivity |
| $\gamma$ | Skin stress |

# Chapter 1
# Wonders of Water

- *Water and ice respond to stimuli or perturbations unexpectedly with derivatives of numerous anomalies.*
- *One phenomenon is often associated with multiple debating theories but one principle should reconcile all observations.*
- *Clarification, correlation, formulation, and quantification of hydrogen bond (O:H–O) relaxation and polarization dynamics and its consequence on detectable properties should be the focus tracking forward.*
- *Focusing on the statistical mean of all the correlated parameters simultaneously is more reliably revealing than on the instantaneous accuracy of a parameter at a given time for the strongly correlated and fluctuating system.*

**Abstract** Water and ice perform differently from other usual substance when subject to tiny perturbation but most of its mysteries remain unresolved up to date despite extensive dedications made by generations. The aim of this volume is to show that alternative ways of thinking and approaching could be efficient to making substantial and systematic progress towards consistent understanding of the performance of water and ice and quantitative information on the hydrogen bond (O:H–O) bond relaxation and polarization dynamics under various perturbations, as well as their consequences on the detestable and correlated properties of water and ice.

## 1.1 Significance of Water and Ice

Water covers 70 % of Earth's surface, makes up 60 % of the human body, and forms 90 % of human blood composition. Besides the scientific significance in subject areas such as astrophysics, agriculture, biology, climate, environment, galaxy, geology, and living cells, water is of paramount importance to civilization, economics, diplomacy, history, military, politics, and international relations of a nation. As noted by Vladyslav Goncharuk [1], a colloid and water chemist and Academician of the National Academy of Sciences of Ukraine, in his monograph *Drinking Water*: The intellect of any nation is determined by the quality of its drinking water and the progress of a civilization depends on the level of water supply and sewage systems. Mastering sufficient water resources, a nation will be privileged in dealing with diplomacy and international affairs. In another word, water resource is related to the peace of the world.



C.Q. Sun and Y. Sun, *The Attribute of Water*, Springer Series in Chemical Physics 113, DOI 10.1007/978-981-10-0180-2_1

Yet, each $H_2O$ molecule of this common, seemingly simple substance—one oxygen atom bound to two hydrogen atoms with $sp^3$-hybridized directional orbits—holds an ocean of full mystery. Water is simple yet magic, pure yet elegant, perseverant yet flexible. Water also symbolizes kindness, wisdom, wealthy, and prosperity. According to Lao Tze (571–531 B.C., an Ancient Chinese Philosopher and Thinker), water represents the highest morality of human beings, which is beneficial for all kinds of things without expecting for any return. It retains its ethical standard yet stays in where disdained.

As the key component of water and biomolecules, the hydrogen bond (O:H–O) bond determines ubiquitously the unusual behavior of water and ice and other molecules containing simultaneously such bonding "–" and nonbonding ":" lone pair interactions. The electron lone pair associated with atoms of electronegative elements such as oxygen, nitrogen, and fluorine. The O:H–O bond and its polarizability give water with unique properties, accelerates or slows reactions, and holds together the three-dimensional configurations of deoxyribonucleic acid (DNA), proteins, and other supramolecules [2–4]. Given its importance in nature [5–16] and in geochemical sciences [17, 18], and its role in DNA and protein folding [2, 19–21], gene delivering [22–24], cell culturing [25], drug target binding [26], ion channel activating and deactivating [27], regulating, signaling, messaging, etc., water demonstrates extraordinary high degree of adaptivity, cooperativity, sensitivity, and recoverability with memory in responding to constraints or perturbations.

$H_2O$ and the O:H–O bond between oxygen ions, being often confused with the H–O bond or the O:H nonbond, have been studied since the dawn of scientific thought. Approaches from the perspectives of classical thermodynamics and quantum mechanics have considerably advanced this subject field. Active areas regarding the intrinsic properties of water and ice include: (i) crystal structure optimization [28], phase formation and transition [29, 30]; (ii) reaction dynamics with other ingredients or at interfaces [31, 32]; (iii) O:H–O bond weak interactions [33, 34]; (iv) binding energy determination [35–41]; and, (v) phonon relaxation dynamics under various conditions [9, 32, 42–44].

Authoritative reviews and monographs on the advancement in this field have been focused on the following issues from various perspectives: phase and geometric structures [45–47], molecular clusters [48–51], ice nucleation and growth [16, 52], ice melting [53], slipperiness and friction of ice [54], and behavior of water ice subjected to positive pressure [55, 56] and negative pressure [48, 57]. Reviews have also covered topics on water surface charge density and polarization [58, 59], surface photoelectron emission [60, 61], phonon relaxation [18, 62, 63], water adsorption onto inorganic surfaces [64–68], imaging water molecules on different substrates [69], and ion effects on water properties and structures [70], among many other topics.

State-of-the-art probing techniques, such as neutron diffraction and X-ray diffraction (XRD) [71, 72], scanning tunneling microscopy/spectroscopy (STM/S) [73–75], O near k-edge X-ray absorption/emission fine structure spectroscopy (NEXFAS/NEXFES) [76, 77], microjet photoelectron emission spectroscopy (PES) [78], sum frequency generation (SFG) dielectric spectroscopy [59, 79], glancing-angle Raman spectroscopy [80, 81], etc., have propelled advancement in studying the droplets, bubblers, surface, and interface of water and ice [63, 82] with or without acid and salt solutions [83].

However, water and ice are too strange, too anomalous, and too challenging [5, 84, 85]. Of particular interest to scientists is what happens at the interfaces where water meets air or other substances such as cells, proteins, micro channels, hydrophobic and hydrophilic skins of other solids. Much has been theorized and calculated, despite insufficient corroborating experimental confirmation. A series lectures by renown experts in the Enrico Fermi summer school "Water: fundamentals as the basis for understanding the environment and promoting technology" held in Varenna, Italy, in July 2013 has addressed the up to date knowledge about water on questions that remain unanswered about this molecule, and topics covered included: water in relation to other liquids, biological water, local environment of water protons, atmospheric water, amorphous solid phases of water, NMR studies of water, spectroscopic studies of water, the structure of liquid water, and supercooled water, among others. A four-week symposium on the theme of "Water—the Most Anomalous Liquid" was held during 13 October and 07 November 2014 in Nordita, Stockholm, Sweden, which brought together experimentalists and theoreticians in strong synergy to explore interpretations and to provide a strong basis for further advances towards a unified picture of water.

The current situation is that each of the various anomalies of water and ice is associated with multiple debating theories yet one simple notion that reconciles as many as possible anomalies is lacking. Progress is below expectation regarding the structure order, local potentials, O:H–O bond relaxation dynamics, acid-base-adduct solutions, and the charge behavior in the skin and in solutions. Little has yet been known about the mechanism behind the weird properties of water and ice demonstrated under perturbation by fields such as coordination reduction, chemical reaction, compression, electrification, magnetization, radiation, thermal excitation, etc. Property variations rarely follow the rules that govern the performance of other "normal" substance. The ways of charge induction and energy absorption, energy retention, energy conduction and dissipation remain challenge. Difficulties remain in accurately determining the three-dimensional coordination order and thermodynamic behavior [86].

Insights into the unusual behavior of water and ice is still far from systematic and consistent. For instance, one often confused either the intramolecular H–O interaction or the intermolecular $H_2O$:$H_2O$ interaction as the hydrogen bond, O:H–O, between water molecule or between oxygen ions. The coordination number (CN) of a water molecule varies from 2 to 5, depending on experimental conditions.

## 1.2 Phase Diagram

Water and ice has insofar formed 17 types of structures in the P-V phase diagram, see Fig. 1.1 [87]. A water molecule, $H_2O$, forms in a V shape: two hydrogen atoms connected to a central oxygen atom, but water molecules are socialized like they are in a concert—their mobility is temperature and impurity concentration dependent. The

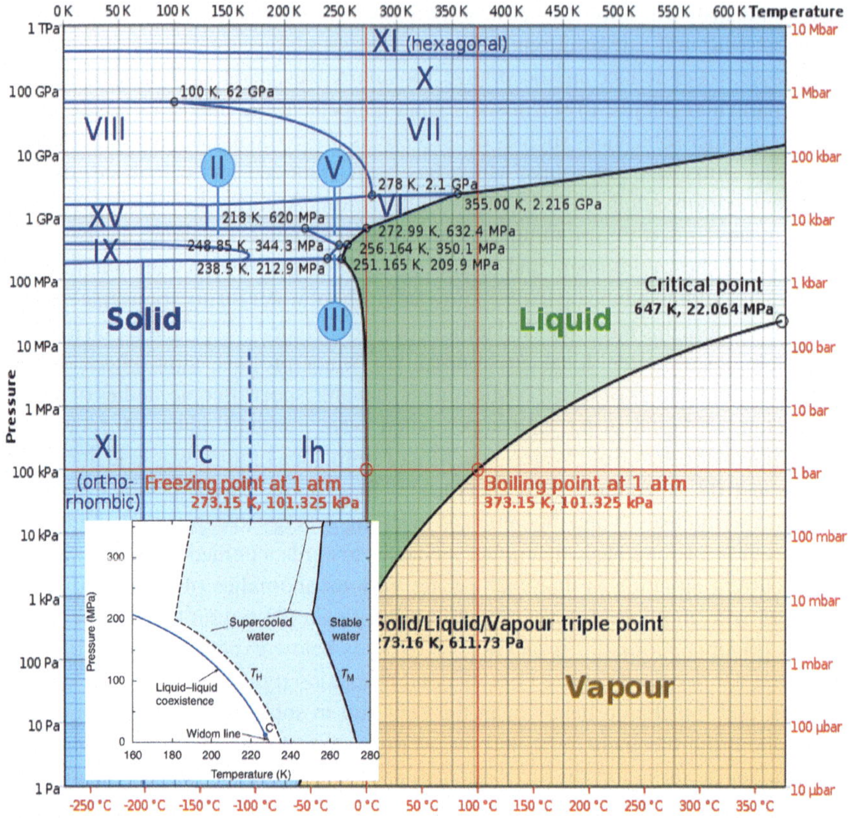


**Fig. 1.1** Phase diagram for water ice [87]. The phase boundaries can be categorized according to their slopes: $dT_C/dP < 0$ for II–V, Liquid—$I_h$, and VII–VIII transition; $dT_C/dP > 0$ for Liquid–Vapor, Liquid—(V, VI, and VII) transition. $dT_C/dP \cong \infty$ at the (VII and VIII)—X boundary occurring at 60 GPa; $dT_C/dP \cong 0$ at $I_h$ XI boundary in low temperatures. (Reprinted with permission from [87, 89, 90].) The *inset* shows the supercooling behavior of emulsified liquid water. The homogeneous ice nucleation temperature (second critical pint $T_H$) varies with pressure. (Reprinted with permission from [91, 92].) (The *thick lines* show the Raman probing paths for the hydrogen bonding dynamics, which will addressed in Chap. 4, The inset shows the Non Man' Land region of quasisolidity (often called supercooled water))

$H_2O$: $H_2O$ nonbond interaction in liquid is weaker than it is in solid; the O:H is even weakest in vapor. Classically, the concerted motion of water molecules is described by the interaction between molecular dipoles in a dipole sea, which overlooks the interaction between the intermolecular O:H nonbond and the intramolecular H–O bond. Coulomb repulsion between electron pairs on adjacent oxygen ions plays the key role in determining the concerted molecular association in water and ice [88].

Ice is the central to climate, geology, and life. Understanding its behavior is essential for predicting the future of our planet and unravelling the emergence of life in other planets of the Universe [93]. Water ice frosts planets, moons and comets in our Solar System. On Earth, white polar ice caps reflect up to 90 % of the

Sun's incoming radiation. On average, 7 % of the ocean's surface is frozen; sea ice alters ocean currents and limits the exchange of gases with seawater. Ice and snow coat 10 % of the land permanently and up to half of the Northern Hemisphere in midwinter. These blankets of frozen water insulate the ground and the oceans.

Ice clouds concentrate airborne chemicals and are sites for atmospheric chemistry. Above the poles, clouds of ice grains host ozone-depleting reactions, forming holes in the stratospheric ozone layer at high latitudes that expose millions of people to increased ultraviolet radiation. Chemical reactions in snow on the ground can produce ozone and other environmental pollutants. Organic toxins and mercury accumulate in snow can be released into rivers and oceans when the snow melts, where they enter the food web.

Ice behaves strangely at lower temperatures and higher pressures. In everyday ice cooled from liquid water, which scientists call Ice $I_h$, water molecules line up in a hexagonal pattern (the "h" stands for hexagonal); this is why snowflakes all have six-sided patterns. A variation called ice $I_c$ at lower temperatures, found in ice crystals floating high up in the atmosphere, forms cubic crystals, in fogs and clouds. The crystal structure of the ice is fairly loose—the reason that ice $I_h$ is less dense than liquid water—and the O:H nonbonds between hydrogen atoms and other water molecules are longer and weaker than the H–O atomic bonds. The cubic ice can also form artificially in graphene nanocapilaries at room temperature [94]. The square ice $I_c$ has a high packing density with a lattice constant of 0.283 nm and can assemble in bilayer and trilayer crystallites. Molecular dynamics (MD) simulations indicate that the square ice should be present inside hydrophobic nanochannels independently of the exact atomic nature of the channels.

At higher pressures, the usual hexagonal structure breaks down through bond angle and length relaxation [90], and the bonds rearrange themselves in more compact, denser crystal structures, neatly labeled with Roman numerals: Ice II, Ice III, Ice IV and so on. There are also several forms of ice in which the water molecules are arranged randomly, as in glass. At a pressure of about 200 MPa, Ice $I_h$ turns into a different type of crystalline ice, Ice II. Ice II does not occur naturally on Earth. Even at the bottom of the thickest portions of the Antarctic ice cap, the weight of three miles of ice pushes down at only one-quarter of the pressure necessary to make Ice II. But planetary scientists expected that Ice II, and possibly some other variations, like Ice VI (around 1 GPa $\sim 10^4$ atm pressure), exist inside icier bodies in the outer solar system, like the Jupiter moons Ganymede and Calisto.

With pressure high enough, the temperature needs not even be low for ice to form. Scientists considered what happens to tectonic plates after they are pushed back down into Earth's interior. At about 100 miles down, the temperature of these descending plates is 300–400 °C—well above the boiling point of water at the surface—but cool compared with that of surrounding rocks. The pressure of 2 GPa at this depth could be great enough to transform any water there into a solid phase known as Ice VII. Nobody knows whether ice can be found inside Earth, because no one has yet figured out a way to look 100 miles deep underground.

Just as salt melts ice at the surface, other molecules mixing could impede water freezing and ice melting. The critical temperatures for melting, freezing, evaporating and dew forming from gaseous phase to liquid are tunable not only by tuning the pressure but also by adding chemical impurities or by varying the confinement conditions. Water molecules at the hydrophobic interface perform differently from those at the hydrophilic interfaces.

Ice also changes its form with dropping temperatures. In hexagonal ice, the usual form, the oxygen atoms are fixed in position, but the O:H nonbond between water molecules are continually breaking and reattaching, tens of thousands of times a second. At temperatures cold enough—below −200 °C both H–O bond and O:H nonbond freeze without relaxation but bond angle undergoes cooling stretching [95], and normal ice starts changing into Ice XI (orthorhombic structure). Astronomers were probably already looking at Ice XI on the surface of Pluto and on the moons of Neptune and Uranus.

From ice XII to ice XVI, found just a decade ago, are also furnished with many wired features. For instance, the cadge-structured ice XVI has a density of 0.81 g/cm$^3$ as the stable low-temperature phase of water at negative pressures (that is, under tension) [96]. Composed of undercoordinated molecules, this hollow hydrate structure exhibits cooling expansion at temperatures about and below 55 K. This phase, as shown in Fig. 1.2, is also mechanically more stable and has at low temperatures larger lattice constants than the filled hydrate. When pressure is increased to 60 GPa, ice X phase forms with identical H–O and O:H length of 0.11 nm, reaching a density that is twice of ice $I_h$ [97, 98].

Ice can turn to be partially ionic: $2H_2O \rightarrow H_3O^{\delta+}{:}HO^{\delta-}$ ($\delta = 0.62$) by exchanging an H on a molecule with an electron on another at extremely high

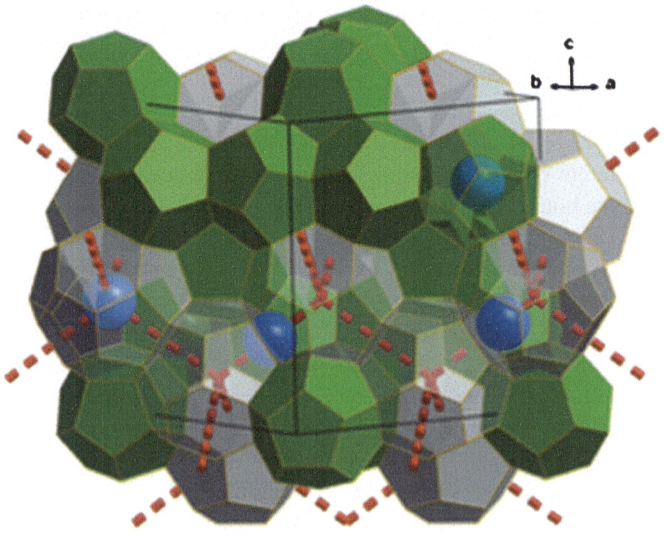


**Fig. 1.2** Cage structures of the XVI phase. Ne atoms (in *blue*) can easily travel between large cages (in *grey*) passing through six-membered rings of water molecules (*red dashed lines*). Removal of Ne atoms from the small cages (in *green*) requires the presence of a water vacancy in one of the five-membered rings [99] (Reprinted with permission from [96].) (color online)

pressure (2 TPa) and high temperature (2000 K) [38], optimized using the package of crystal structure analysis by particle swarm optimization (CALYPSO) [100]. The $H_3O^{\delta+}$ likes an $NH_3$ molecule that has one electron lone pair. The $HO^{\delta-}$ likes a HF having three lone pairs Fig. 1.3 illustrates the crystal structure of $P2_1$ viewed along a axis. The large spheres represent the oxygen atoms whereas the small ones are for hydrogen atoms. The labelled H–O distances vary from site to site in the range of 0.902 and 1.182 Å under 2 TPa pressure. The occurrence of this partial ionic phase transition takes place by breaking up the typical O–H covalently bonded tetrahedrons in the hydrogen symmetric atomic phases.

In addition to the critical point for liquid-vapor phase transition, a second critical point exists [91, 101] for supercooling water, in the range of 145 K < $T_{c2}$ < 175 K range and $P_{c2} \approx 200$ MPa pressure. This critical point, shown in Fig. 1.1 inset, was attributed to the homogeneous ice nucleation ($T_H$) or transition between the low-density and the high-density amorphous solid phase.

One may note that phase IV is absent in the phase diagram. Gerald Pollack [102], Professor of Biochemical Engineering at Washington University, suggests its presence at sites between water and hydrophilic interface. This phase is gel like, high density, positively charged $OH_3^+$ configuration), having the ability to exclude impurities and organisms.

In 2015, NASA announced the discovery of liquid water existence on Mars (see Fig. 1.4) as detected using infrared absorption spectroscopy with the characteristic phonon wave lengths of 1.4, 1.9 and 3.0 μm [103]. Spectrum analysis also found

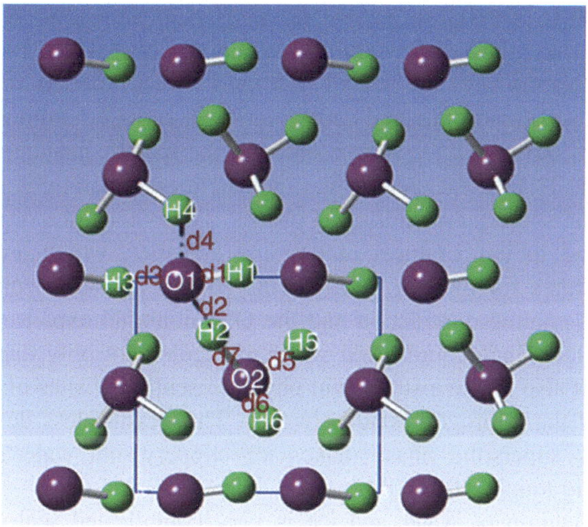


**Fig. 1.3** $2H_2O$ transition into the superionic $H_3O^{\delta+}$:$HO^{\delta-}$ ($\delta = 0.62$) under 2000 K temperatures and 2 TPa pressure. The *large spheres* represent the oxygen atoms whereas the *small ones* are for hydrogen atoms. The atoms and O–H distances are labelled in accordance (Reprinted with permission from [38]). The O:H–O bond remains but the $2H_2O$ evolves into respective $NH_3$ and HF like quasi-tetrahedron with different numbers of lone pairs

**Fig. 1.4** This NASA photo, taken by an instrument aboard the agency's Mars Reconnaissance Orbiter, shows dark narrow 100 m long streaks on the surface of Mars that scientists believe were caused by flowing streams of salty water (by David Templeton/Pittsburgh Post-Gazette)

the water to be a brine containing forms of perchlorate, a chlorine salt. The average temperature is −33 °C. Normally, only exists ice under these conditions but it is possible for salted solution in the quasisolid phase—lower freezing temperature and higher viscosity from the quasisolid [88].

Yet the molecular mechanisms underlying these processes remain largely unknown [93]. It is essential to understand how chemical reactions proceed in ice and snow, and where they occur within the grain and crystal structure for building snow or ice-cloud modules for atmospheric and climate models or extrapolating laboratory studies to environmental conditions with enough confidence.

Sun et al. [95] shows that the superposition of the specific heat curves of the O:H nonbond and the H–O bond creates two intersecting temperatures that define the boundaries of the solid/quasisolid/liquid phases. The boundaries of this quasisolid phase correspond to temperatures of extreme densities, which are close to the critical temperatures for melting and freezing. In the quasisolid phase, the H–O bond undergoes cooling contraction and the O:H nonbond expansion resulting in low density of ice. Undercoordinated water molecules not only disperse the quasisolid phase but also create a supersolid phase presenting at sites of skins, defects, and the edge of the hydrogen bond networks. This elastic, less dense, polarized, thermally stable supersolid phase makes ice slippery and water's skin elastic, hydrophobic, and tough [104, 105].

The phase diagram of water and ice is very complicated with boundaries of versatile slopes. One often connects the structure and property Q(PV, ST, …) of a substance directly to the external stimuli such as pressure, volume, and temperature, which is exactly what the classical thermodynamics deals with for large trunks of regular substance. Such treatment employs concepts of entropy, enthalpy, Gibbs and Helmholtz free energy, etc. These statistic quantities are associated with standard deviation δ depending on the sample size N in the form of $\sigma \propto N^{-1/2}$. This approach provides limited information on how the chemical bond responds to external stimulus in each phase and crossing the phase boundaries.

According to Pauling [106], the nature of the chemical bond bridges the structure and properties of crystals and molecules. Hence, formation and relaxation of the bond in angle and length and the associated energetics and dynamics of localization, entrapment, polarization, and transition of electrons mediate uniquely the structure and properties of a substance accordingly [107]. Therefore, the phase diagram is supposed to tell how the O:H–O bond relaxes over the entire phase map [90], as justified using phonon spectrometrics shortly.

## 1.3 Physical Anomalies

Water displays a surprising matrix of unusual properties. The origin of these anomalies is still a matter of debate, and so far a quantitative description of water's phase behavior starting from the molecular arrangements is still missing [108]. The following epitomizes some of the best-known mysteries desiring for consistent and quantitative resolution:

| Stimulus | No. | Mysteries |
|---|---|---|
| Intrinsic attributes | 1 | Packing order [109]. Water ice exhibit 17 phases with uncertainty of molecular CN, O–O distance, and electronic configuration [109] |
| | 2 | Hydrogen bond [88]. Discrepancy and confusion exist in hydrogen bond definition: O:H–O bond between oxygen, intramolecular H–O polar covalent bond, and intermolecular O:H nonbond |
| | 3 | Local potentials [110]. Determination of the intra- and intermolecular potentials was not possible using the state-of-the-art methods of diffraction, electron and phonon spectroscopy |
| | 4 | Structures versus properties [88]. Factors control the structures and properties and their interdependence remain unclear |
| Mechanical compressuion/tension | 5 | Ice regelation [111]. Ice melts under compression and freezes again when the pressure is relieved. A 210 MPa pressure lowers the $T_m$ from 273 to the limit of 251 K; a −95 MPa tension raises the $T_m$ by 6.5 K; 18.84 MPa compression lowers the temperature of maximum density from 277 to 273 K; −180 MPa raises this temperature to 320 K |
| | 6 | Proton symmetrization [97]. Both O:H and H–O lengths become 0.11 nm at 60 GPa in phase X; salt addition lengthens the critical the O-O distance |

(continued)

(continued)

| Stimulus | No. | Mysteries |
|---|---|---|
| | 8 | Phonon cooperative relaxation [112]. The low-frequency O:H phonon undergoes a redshift but the high-frequency H–O phonon undergoes a blue shift when the ice is compressed at ≤140 K up to a critical pressure ~3.3 GPa and then the trends transit under high pressures |
| | 9 | Compressibility anomaly [84]. Water has a compressibility of 0.46 $GPa^{-1}$ compared with 1.05 $GPa^{-1}$ for $CCl_4$ at 25 °C. The compressibility having a maximum at temperature just below the minimum density and then drops with the rise of temperature up to a minimum centered at 46.5 °C [113] |
| Thermal excitation | 10 | Ice floating [95]. Oxygen and Argon shrink their volumes by 19 % of their liquid upon freezing. The density drops by 8 % when water turns to be ice. The lower density of ice makes ice float on water |
| | 11 | Density oscillation [95]. The density of water and ice oscillates over the full temperature range in four regimes. Cooling contraction occurs in the liquid and in the ice $I_h$ and $I_c$ phases but cooling expansion happens at transition from liquid to ice. The density remains almost constant in phase XI (≤100 K) |
| | 12 | Second critical point [91, 101]. A second critical point exists, in addition to that for the normal liquid–vapor transition, in the 145 K < $T_{c2}$ < 175 K range and $P_{c2}$ ≈ 200 MPa pressure |
| | 13 | Extraordinary specific heat [95]. Water has extraordinarily high specific heat that changes with temperature abnormally and disobeys Debye approximation. It takes a lot of energy to heat water (about ten times as much as the same mass of iron), and it must lose a lot of energy to cool down. So the vast bodies of water on earth help keep the earth's temperature fairly steady. On the other hand, land masses heat up and cool down more quickly. Different parts of the atmosphere are heated differently, which generates wind |
| | 14 | Temperature and pressure resolved refractive index [88]. The index rises from 1.33026 at −30 °C to a maximum value of 1.33434 at just below 0 °C before falling increasingly to 1.31854 at 100 °C. Compression raises but heating lowers the refractive index |

(continued)

(continued)

| Stimulus | No. | Mysteries |
|---|---|---|
| Molecular undercoordination | 15 | Water's tough skin [114]. The surface tension of water is 72.75 $mJ/m^2$ compared with 26.6 $mJ/m^2$ for $CCl_4$ at 20 °C. The tension drops linearly with temperature |
| | 16 | Water skin density [105]. Classical thermodynamics predicts a denser skin but diffraction revealed a 5.9 % skin O–O elongation or 15.6 % density loss at 25 °C. In contrast, the skin O–O for liquid methanol contracts by 4.6 % associated with a 15 % density gain [115] |
| | 17 | Water skin refractive index [84]. The refractive index of water (measured at $\lambda$ = 589.2 nm) skin is higher than it is in the bulk |
| | 18 | Droplet floating and dancing [88, 116]. Water droplet does not emerge immediately into the water but it dances continuously at the skin of the bulk for rounds |
| | 19 | Slipperiness of ice [117]. Ice skin is one of the slipperiest of ever known. The slipperiness of wet surfaces is greatest for hydrophilic/hydrophobic contact but least for hydrophilic/hydrophilic interaction [118] |
| | 20 | Supercooling and superheating [119, 120]. Water nanodroplets or bubbles undergo superheating at melting and supercooling at freezing, whose extent is droplet size dependence. A 1.2 nm sized droplet freezes at temperature below 172 K and the monolayer skin melts at 320 K |
| | 21 | Identical H–O phonon frequency for water and ice skin [105]. The skin of −20 °C ice shares an identical H–O frequency 3450 $cm^{-1}$ with the skin of 25 °C water, in contrast to the bulk values of 3200(water)/3150(ice) $cm^{-1}$ and 3650 $cm^{-1}$ for the $H_2O$ monomer in gaseous phase |
| Impurity and hetero-coordination | 22 | Isotope effect on phonon relaxation [88]. Deuterium $^2H$ replacement of H attenuates the intensity and redshifts the frequencies of all Raman phonons |
| | 23 | Superhydrophobicity, superfluidity, superlubricity, and supersolidity at contacting interface [121] |
| | 24 | Hydrophobicity/hydrophilicity transition [88, 102]. An air gap presents between water and hydrophobic contacts. The curved water skin is thermally more stable. Hydrophilic contact creates exclusion zone that is gel-like, separating charges, and excludes organisms and impurities. Water–protein interface exhibits two characteristic life times |

(continued)

(continued)

| Stimulus | No. | Mysteries |
|---|---|---|
| Aqueous and capacitor electrification | 25 | Hofmeister series [122, 123]. Anions and cations change the surface tension of the solution and the solubility of proteins in a sequential order of salts. Salts also modulate the critical temperatures for water phase transition and the gelation time of solutions |
| | 26 | Solute phase transition—critical pressures, temperatures, and gelation times [124, 125]. Charge addition and electric bias modulate the temperatures of ice melting and water freezing |
| | 27 | Armstrong effect—Water Bridge [126]. Water bridge can form across two beakers under $10^6$ V/m field and sustain for hours under electric bias under the ambient temperature |
| Energy absorption, retention, transportation and dissipation | 28 | Mpemba paradox [127, 128]. Hotter water freezes faster than its cold, which is in counter-intuitive of Newton's law in thermal transportation |
| | 29 | Radiation absorption [129]. Water exchanges heat and gases with the environment and absorbs electromagnetic radiation and acoustic signals in a long-range manner |
| | 30 | Electromagnetic field modulated temperatures for melting and freezing [124, 130]. These critical temperatures vary with the intensity of the field. Superposition of fields generated by different sources may effect oppositely, such as solution Water Bridge [131] and soil wetting by aqueous solution [132] |

Water is also a "universal solvent". Many minerals and vitamins can be transported throughout the body after being dissolved. Dissolved sodium and potassium ions are essential for nerve impulses. Water also dissolves gases, such as oxygen from the air, enabling water-living animals to use oxygen. Water, a major component of blood, also dissolves carbon dioxide, a waste product from energy production in all cells, and transports it to the lungs, where it can be breathed out. However, water is repelled by oily compounds, so our cells have membranes made of these. Many of our proteins have partly oily regions, and they tend to fold together, repelled by the surrounding water. This is partly responsible for the many and varied shapes of proteins. These shapes are essential for carrying out functions vital to life.

## 1.4 Challenges and Objectives

Philip Ball announced in 2014 in a special issue of the *Europe Physics Journal* [129]: No one really understands water. It's embarrassing to admit it, but the stuff that covers two-thirds of our planet is still a mystery. Worse, the more we look, the more the problems accumulate: new techniques probing deeper into the molecular architecture of liquid water are throwing up more puzzles.

Although some of those puzzles—such as the tetrahedral coordination of a molecule in bulk water—seem now to be within reach of a resolution, other important questions, such as the nature of hydrophobic hydration and the existence of a second critical point in supercooled water [101], remain unanswered. More pertinent than the fact that some basic properties of water and hydration are still not understood, however, is the fact that many facets of water's behavior can now be seen to be inadequately accounted for by any picture that regards the liquid state as being the sum of its parts. That is to say, the phase, solvation and wetting behaviors of the liquid are not determined in a transparent way from the properties of the individual water molecules, but emerge from their collective interaction, over many length scales, in ways that can be highly contentious and context-dependent.

The difficulties in understanding liquid water are generally attributed to the cooperative hydrogen bonding, i.e., the binding energy of two H-bonded molecules is modified by the presence of a third molecule and nuclear quantum effects [133, 134]. Such effects occur because the proton is so light that classical mechanics can no longer adequately describe properties like spatial dispersion of the hydrogen positions, nuclear tunneling, zero-point energy and, naturally, quantization of nuclear motions.

According to Ball and Eshel [129], it is often insufficient to think only in terms of local interactions for understanding water's properties; one must take a systemic and holistic view, considering the ways in which water modifies its own tendencies in responding to its environment. Such a picture might be needed to explain multiple phenomena presented. We are at in the midst of a shift in our perception of water—a shift from the current molecular—level based approach (which focuses on the behavior of individual or small numbers of molecules) towards a new, systemic view of water. In this picture, water is perceived as an active substance that responds adaptively to external and internal constraints and signals. These responses can have profound effects on substances immersed in water, and in particular on the functioning of biological constituents, from molecules to living cells.

According to Pauling [106], the nature of the chemical bond bridges the structures and properties of crystals and molecules. Therefore, the interatomic bond and the performance of electrons are the starting point of consideration [107]. For other unusual substance, one bond represents all on average, but for water and ice, the represent hydrogen bond contains two parts—the O:H nonbond and the H–O polar-covalent bond. These two parts respond to perturbation cooperatively and differently associated with electron entrapment and polarization, which differs water and ice from other substance in physical properties.

Therefore, resolving mysteries of water ice regarding its structure order, O:H–O bond local potentials, and its physical anomalies and their interdependence under various stimuli is the key challenge. Bond relaxation in length and energy and the associated bonding charge entrapment and nonbonding electron polarization in the long-range responding to applied stimuli such as mechanical compression, molecular undercoordination (with fewer than four nearest neighbors as in the bulk), thermal, electric, and magnetic excitation and radiation, for instance, all of which change the structure and property of a substance in the respective manner [107].

In modeling consideration, it is proper to formulate the intramolecular interaction and the intermolecular interaction dependently. Long-range responding to external stimulus is necessary because of the high sensitivity and fluctuation of water liquid. Examination of the statistical mean of all the correlated quantities with certain rules and multiple means is more realistic and meaningful than focusing on the instantaneous accuracy of a certain quantity at a point of time for a strongly correlated and fluctuating system. Modeling hypotheses and expectations, numerical calculations, and experimental measurements should be consistent and correlated in addressing the change of multiple properties for the highly correlated and fluctuating water. Alternative ways of thinking and multiple means of approaching are necessary to deal with these difficult issues effectively.

This volume aims to deal with the correlation between the anomalous behavior of water ice and the relaxation dynamics of the O:H–O bond under afore-mentioned stimuli from the perspective of hydrogen bond cooperative relaxation. Extending the Ice Rule of Bernal and Fowler [135] suggests a tetrahedral block that contains two $H_2O$ molecules and four O:H–O bonds. This block unifies the length scale, geometric configuration, and mass density of molecular packing in water ice. This extension also clarifies the flexible and polarizable O:H–O bond that performs like a pair of asymmetric, coupled, H-bridged oscillator pair with ultra-short-range interactions and memory effect as well as extraordinary recoverability. Coulomb repulsion between electron pairs on adjacent oxygen anions and the disparity between the O:H and the H–O segments relax the O:H–O bond length and energy cooperatively when responds to perturbation.

The developed strategies have enabled clarification of origins of the following observations from the perspective of O:H–O segmental relaxation and associated bonding electron entrapment and nonbonding electron polarization:

(1) Pressure-induced proton centralization, ice regelation, and phase transition-temperature depression;
(2) Thermally-induced four-region oscillation of the mass density and the phonon frequency over the full temperature range;
(3) Molecular-undercoordination-induced supersolidity that is elastic, hydrophobic, thermally stable, with ultra-low density. The supersolid skin is responsible for the slipperiness of ice, the hydrophobicity and toughness of water skin, and the bi-phase structure of nanosized droplets and bubbles.
(4) Electrification of the O:H–O bond by the electric fields of solute point charge and a capacitor results in the Hofmeister series and the Armstrong water bridge.
(5) Magnetification of water dipoles by the Lorentz force: a magnetic field rotates the dipole at transitional motion and transits the rotating dipoles.
(6) O:H–O bond absorbs, retains, transmits, and dissipates all sorts of energy in a long-range manner with memory, which resolve the Mpemba paradox and the crystal pattern sensitivity of ice formation under perturbation.

However, there some limitations faced by conventional approaches. For instance, discriminating the cooperative relaxation of the O:H and the H–O segment is beyond the scope of neutron or X-ray diffraction that probes the O–O Radial Distribution Function (RDF) with structure factors. The relaxation of the stronger H–O bond dictates the electronic binding energy shift in all bands as the O:H energy is smaller than the tolerance level. The O K-edge absorption and emission spectroscopies collect information about the energy shift of the deeper O 1s core band and the upper valence band. The asymmetrical, ultra-short range interactions and the strong localization and polarization become dominance in quantum computations. London dispersion or the interaction between dipoles and the dipole-induced spontaneous dipoles also comes into play in determining the performance of water and ice.

Therefore, an interplay of density functional theory (DFT) and molecular dynamics (MD) calculations, Raman and IR phonon spectroscopy, XPS measurements has enabled clarification, correlation, formulation, and quantification of multiple best-known puzzles demonstrated by water ice. A Lagrangian solution has enabled mapping the landscape of the asymmetric potentials for the O:H–O bond at relaxation. The H–O bond relaxation shifts the melting point, O 1s binding energy, and high-frequency phonon whereas the O:H relaxation dominates polarization, viscoelasticity, and the O:H dissociation energy. Solving the Fourier thermal fluid transportation equation with adequate initial-and-boundary conditions would clarify the historical mystery of heating "emission–conduction–dissipation" in the Mpemba paradox. Attainments made in so far have thus verified our hypothesis and expectations consistently.

## 1.5 Scope of this Volume

This volume starts with a brief overview in Chap. 1 on the challenge, significance, and status in understanding the structure order, phase diagram, physical anomalies demonstrated by water and ice. Besides a brief introduction of the history background and the known structure models, Chap. 2 shows the essentiality of O $sp^3$-orobital hybridization that laid the foundation for the quasi-tetrahedron geometry for a water molecule interacting with hydrogen atoms through two bonding electron pairs and two nonbonding lone pairs. An extension of the quasi-tetrahedron geometry turns out an ideal tetrahedral block containing two $H_2O$ molecules and four O:H–O bonds. This building block unifies the size, separation, structure order and mass density of molecules packed in water and ice.

Chapter 3 is focused on the flexibility and polarizability of the O:H–O bond that performs like an asymmetric, coupled, H-bridged oscillator pair with ultra-short-range interactions. The O:H–O bond disparity and the O–O Coulomb repulsion discriminate water and ice from other usual materials in responding to stimulus. The O:H–O bond responds to various stimuli cooperatively associated with relaxation and polarization, which determines all detectable properties of water and ice including the phonon frequency, electron entrapment and polarization, reactivity, surface stress, solubility, thermal stability, viscoelasticity, etc.

Chapter 4 analyzes the phase diagram of water and ice from the perspective of O:H–O bond relaxation dynamics. The phase boundaries are categorized into four types according to their $T_C$–$P_C$ slopes. Reproducing the negatively sloped boundaries for the Liquid-Quasisolid and VII/VIII boundaries confirmed that the H–O bond energy dictates theses transition, which results in the 3.97 eV H–O bond energy; reproduction of the liquid-vapor phase boundary turns out the pressure dependence of the O:H length. The XI–$I_c$ phase boundary of constant $T_C$ indicates that the O:H–O bond does not respond to thermal excitation and the VII/VIII–XI boundary of constant $P_C$ suggests that the O:H and H–O segments relax identically under compression at this boundary. Seven paths of Raman probing confirmed the expected O:H–O bond relaxation dynamics crossing the phase diagram except for the "no man's land" regime that is subject to further verification.

Chapter 5 shows the mapping of the asymmetrical potential paths of the O:H–O bond at elongation and contraction, which further confirms the universality and adequateness of the O:H–O bond cooperativity notion. Lagrangian mechanics is hither to the most efficient means solving the O:H–O bond oscillator pair dynamics, which turns the known segmental length and vibration frequency into the respective force constant and cohesive energy, enabling the mapping of the potential paths.

Chapter 6 resolves the behavior of ice under compression, typically, the effect of regelation discovered by Michael Faraday in 1859. Compression shortens the O:H bond and stiffens its phonon while the H–O bond, as the slave, responds to compression oppositely, towards O:H–O length symmetrization. The softened H–O bond dictates the melting temperature of quasisolid ice so the regelation occurs. The resolution of regelation evidences the extraordinary recoverability of the O:H–O bond and the dispersivity of the quasisolid phase boundary.

Chapter 7 deals with O:H–O bond thermal relaxation over the full temperature range. Because of its disparity, each segment of the O:H–O bond has its own specific heat characterized by its Debye temperature and the thermal integral. The Debye temperature varies linearly with the characteristic phonon frequency and the integral is proportional to the segmental cohesive energy. The superposition of the specific heat curves defines two intersecting temperatures that divide the full temperature range into the liquid, quasisolid, solid $I_{h+c}$, and the XI phases with different specific heat ratios. The segment of relatively lower specific heat follows the regular rule of thermal expansion and the other part relaxes oppositely, so the mass density of water and ice oscillates over the full temperature range. This mechanism is completely different from the thermal expansion of other normal materials. The H–O cooling contraction and O:H expansion in the quasisolid phase elongate the O:H–O bond, which makes ice floats.

Chapter 8 resolves anomalies demonstrated by water molecules with fewer than four nearest neighbors presented in clusters, hydration shells, skins, and ultrathin films. Particular attention is given on verifying the undercoordination induced supersolidity and quasisolid phase dispersivity. The supersolid skin that is elastic, hydrophobic, less dense, and thermally more stable lubricates ice and toughens water skin. The boundary dispersion of the quasisolid phase results in phenomena of "supercooling" at freezing and "supercooling" at melting of nanodroplets and nanobubbles.

Chapter 9 extends the skin supersolidity to the superlubricity of ice, dry and wet contacts, in terms of phononic elasticity and electrostatic repulsivity. The supersolidity depresses the O:H phonon frequency and enhance its vibrational magnitude, which ensures the phononic elasticity; the nonbonding electron polarization due to H–O contraction ensures the electrostatic repulsion at the contacting interface. The electrostatic repulsivity and phononic elasticity claim the responsibility for the superlubricity of ice, wet and dry contacts with involvement of polarized lone electrons or electron lone pairs.

Chapter 10 extends the supersolidity to liquid skin, which also applies to the superhydrophobicity for water droplet on solids, superfluidity for microchanels, and supersolidity for solid $^4$He. The transition between superhydrophobicity and superhydrophilicity by UV radiation or plasma sputtering evidence the essentiality of the skin dipoles in the wetting/non-wetting transition, which also elaborates Wenzel-Cassie-Baster's notion for superhydrophobicity/superhydrophilicity enhancement by nanofabrication as the enhancement of energy quantum entrapment that raises the elasticity and the subjective polarization that enhances the repulsivity. Quantum entrapment dictates hydrophilicity but polarization dictates hydrophobicity.

Chapter 11 resolves the Mpemba paradox both experimentally and numerically, showing the essentiality of O:H–O bond memory and water skin supersolidity. Resolving the Fourier thermo-fluid dynamics revealed the essentiality of skin supersolidity that raises the local thermal conductivity favoing heat outward flow; experimental observations evidence the O:H–O bond memory that entitles the O:H–O bond emits energy at a rate proportional to its initial storage. The Mpemba effect only happens at the non-adiabatic source-drain interface in this "source-path-drain" cycling system, which is very sensitive to experimental conditions.

Chapter 12 deals with the O:H–O bond electrification by fields of solute ions in terms of molecular site, solute type, solute concentration, and temperature dependence of the O:H–O bond relaxation dynamics in aqueous solutions of salts, bases, and acids. Solute ions provide short-range electric fields that align, stretch, and polarize water molecular dipoles. Quantum fragilation due to the excessive $H^+$ on hydronium and quantum compression due to the additional lone pair on hydroxide formation resolve, respectively, the performance of acid and base solutions. Focus was given on the local bond stiffness, order of molecular fluctuation and the phonon abundance with multiple discoveries. For instance, electrification has the same effect of molecular undercoordination on relaxing and polarizing the O:H–O bond but the ionic effect proceeds throughout the solution. Molecules in the hydration shells are thermally more stable than those in the bulk or the H–O radical.

Chapter 13 discusses the interdependence of phonon frequencies, H–O phonon lifetime, solution viscosity, and skin stress, which forms the key component of

Hofmeister series—salt modulates the skin stress and the solubility of protein of the solution. The blueshift of the H–O phonon is associated with longer life, higher viscosity, and lower order of molecular fluctuation. Salts and bases raise but acids lower the skin stress of the solution.

Chapter 14 extends the salting effect to the conditions for phase transition of the solution—critical pressures, critical temperatures, and regelation times. Electrification stiffens the H-O phonon, lengthens the O:H-O bond, depresses the freezing temperature and elevates the melting point by dispersing the boundary of the quasisolid phase. High-pressure icing of the ambient solution conforms that compression recovers the deformed solvent O:H–O bond with excessive pressures transiting the solution into ice VI and then into ice VII. The additional pressure is solute concentration and type dependent, following the Hofmeister series. However, the NaI concentration increase has the same effect of heating on the phase transition under compression.

Chapter 15 extends the solute electrification to the long-range field of a capacitor with and without leaking and the combination of electric fields from different sources. The electrified quasisolidity and skin supersolidity enable the suspension of Armstrong's floating Water Bridge. Aqueous solutions can hardly form a Water Bridge because the cancellation of the electric fields of the solute ions and the capacitor. The cancellation effect of combined electric fields also promotes soil wetting by aqueous solution. O:H–O bond electrification by the opposite fields of the soil particles and the solutes reduces the viscosity of the solvent to an extent that is lower than it is under either field alone.

Chapter 16 deals with some miscellaneous issues on the basis of the established premises. These issues include the multiple field effects, droplet charging, energy exchange and absorption, and isotope effect on phonon relaxation, dielectric relaxation, negative thermal expansion, exclusion zone, magnetification, and the generality of the O:H–O bond in other lone-pair involved systems, etc.

Chapter 17 introduces strategies for probing and analyzing and their advantages and limitations. With the aid of X-ray and neutron diffractions, electron spectroscopies, quantum computations, phonon spectrometrics reveals the O:H–O bond relaxation dynamics in segmental length, energy, bond angle, H–O lifetime and viscosity, solubility, thermal stability, skin stress, of water and ice. This chapter also correlates these properties in terms of bond length, bond energy, and polarization.

This volume ends with Chap. 18 summarizing the gained understandings in terms of sixty laws for water and ice. Progresses evidence the efficiency and validity of ways of thinking, strategies of approaching, and the comprehensiveness of the present understandings—one notion for multiple myths.

In each chapter, we start with a digest of the challenge and clarification of a topic of specific concern, followed by historical background information for scientific popularization. We present then quantitative resolution based on experimental observations and numerical computations and followed by possible insight extension to related issues.

# References


1. V.V. Goncharuk, *Drinking Water: Physics, Chemistry and Biology* (Springer, 2014)
2. G.H. Zuo, J. Hu, H.P. Fang, Effect of the ordered water on protein folding: an off-lattice Go$ $($)over-bar-like model study. Phys. Rev. E **79**(3), 031925 (2009)
3. J.L. Kulp, D.L. Pompliano, F. Guarnieri, Diverse fragment clustering and water exclusion identify protein hot spots. J. Am. Chem. Soc. **133**(28), 10740–10743 (2011)
4. A. Twomey, R. Less, K. Kurata, H. Takamatsu, A. Aksan, In situ spectroscopic quantification of protein-ice interactions. J. Phys. Chem. B **117**(26), 7889–7897 (2013)
5. P. Ball, Water: water—an enduring mystery. Nature **452**(7185), 291–292 (2008)
6. I.V. Stiopkin, C. Weeraman, P.A. Pieniazek, F.Y. Shalhout, J.L. Skinner, A.V. Benderskii, Hydrogen bonding at the water surface revealed by isotopic dilution spectroscopy. Nature **474**(7350), 192–195 (2011)
7. D. Marx, M.E. Tuckerman, J. Hutter, M. Parrinello, The nature of the hydrated excess proton in water. Nature **397**(6720), 601–604 (1999)
8. Y. Yoshimura, S.T. Stewart, M. Somayazulu, H. Mao, R.J. Hemley, High-pressure x-ray diffraction and Raman spectroscopy of ice VIII. J. Chem. Phys. **124**(2), 024502 (2006)
9. D. Kang, J. Dai, Y. Hou, J. Yuan, Structure and vibrational spectra of small water clusters from first principles simulations. J. Chem. Phys. **133**(1), 014302 (2010)
10. J.W.M. Frenken, T.H. Oosterkamp, MICROSCOPY When mica and water meet. Nature **464** (7285), 38–39 (2010)
11. J.M. Headrick, E.G. Diken, R.S. Walters, N.I. Hammer, R.A. Christie, J. Cui, E.M. Myshakin, M.A. Duncan, M.A. Johnson, K.D. Jordan, Spectral signatures of hydrated proton vibrations in water clusters. Science **308**(5729), 1765–1769 (2005)
12. J.K. Gregory, D.C. Clary, K. Liu, M.G. Brown, R.J. Saykally, The water dipole moment in water clusters. Science **275**(5301), 814–817 (1997)
13. N. Bjerrum, Structure and properties of ice. Science **115**(2989), 385–390 (1952)
14. A.K. Soper, J. Teixeira, T. Head-Gordon, Is ambient water inhomogeneous on the nanometer-length scale? Proc. Nat. Acad. Sci. U.S.A. **107**(12), E44–081101 (2010)
15. C.S. Zha, R.J. Hemley, S.A. Gramsch, H.K. Mao, W.A. Bassett, Optical study of $H_2O$ ice to 120 GPa: dielectric function, molecular polarizability, and equation of state. J. Chem. Phys. **126**(7), 074506 (2007)
16. T. Bartels-Rausch, V. Bergeron, J.H. Cartwright, R. Escribano, J.L. Finney, H. Grothe, P.J. Gutiérrez, J. Haapala, W.F. Kuhs, J.B. Pettersson, Ice structures, patterns, and processes: a view across the icefields. Rev. Mod. Phys. **84**(2), 885 (2012)
17. R.J. Bakker, M. Baumgartner, Unexpected phase assemblages in inclusions with ternary $H_2O$-salt fluids at low temperatures. Cent. Eur. J. Geosci. **4**(2), 225–237 (2012)
18. R.J. Bakker, Raman spectra of fluid and crystal mixtures in the systems $H_2O$, $H_2O$-NaCl and $H_2O$-$MgCl_2$ at low temperatures: applications to fluid-inclusion research. Can. Miner. **42**, 1283–1314 (2004)
19. M. Smyth, J. Kohanoff, Excess electron localization in solvated DNA bases. Phys. Rev. Lett. **106**(23), 238108 (2011)
20. P. Baaske, S. Duhr, D. Braun, Melting curve analysis in a snapshot. Appl. Phys. Lett. **91**(13), 133901 (2007)
21. A. Kuffel, J. Zielkiewicz, Why the solvation water around proteins is more dense than bulk water. J. Phys. Chem. B (2012)
22. C. Castellano, J. Generosi, A. Congiu, R. Cantelli, Glass transition temperature of water confined in lipid membranes as determined by anelastic spectroscopy. Appl. Phys. Lett. **89** (23), 233905 (2006)
23. J.H. Park, N.R. Aluru, Water film thickness-dependent conformation and diffusion of single-strand DNA on poly(ethylene glycol)-silane surface. Appl. Phys. Lett. **96**(12), 123703 (2010)

24. F. Garczarek, K. Gerwert, Functional waters in intraprotein proton transfer monitored by FTIR difference spectroscopy. Nature **439**(7072), 109–112 (2006)
25. P. Ball, Water as an active constituent in cell biology. Chem. Rev. **108**(1), 74–108 (2008)
26. Y.B. Shan, E.T. Kim, M.P. Eastwood, R.O. Dror, M.A. Seeliger, D.E. Shaw, How does a drug molecule find its target binding site? J. Am. Chem. Soc. **133**(24), 9181–9183 (2011)
27. J. Ostmeyer, S. Chakrapani, A.C. Pan, E. Perozo, B. Roux, Recovery from slow inactivation in K channels is controlled by water molecules. Nature **501**(7465), 121–124 (2013)
28. G. Malenkov, Liquid water and ices: understanding the structure and physical properties. J. Phys.-Condens. Matter **21**(28), 283101 (2009)
29. H.M. Lee, S.B. Suh, J.Y. Lee, P. Tarakeshwar, K.S. Kim, Structures, energies, vibrational spectra, and electronic properties of water monomer to decamer. J. Chem. Phys. **112**(22), 9759 (2000)
30. H.G. Lu, Y.K. Wang, Y.B. Wu, P. Yang, L.M. Li, S.D. Li, Hydrogen-bond network and local structure of liquid water: an atoms-in-molecules perspective. J. Chem. Phys. **129**(12), 124512 (2008)
31. C.K. Lin, C.C. Wu, Y.S. Wang, Y.T. Lee, H.C. Chang, J.L. Kuo, M.L. Klein, Vibrational predissociation spectra and hydrogen-bond topologies of H + (H2O)(9–11). PCCP **7**(5), 938–944 (2005)
32. A. Lenz, L. Ojamae, A theoretical study of water equilibria: the cluster distribution versus temperature and pressure for $(H_2O)(n)$, n = 1–60, and ice. J. Chem. Phys. **131**(13), 134302 (2009)
33. S.O.N. Lill, Application of dispersion-corrected density functional theory. J. Phys. Chem. A **113**(38), 10321–10326 (2009)
34. S.N. Steinmann, C. Corminboeuf, Comprehensive bench marking of a density-dependent dispersion correction. J. Chem. Theory Comput. **7**(11), 3567–3577 (2011)
35. K. Kobayashi, M. Koshino, K. Suenaga, Atomically resolved images of I(h) ice single crystals in the solid phase. Phys. Rev. Lett. **106**(20), 206101 (2011)
36. A. Hermann, P. Schwerdtfeger, Blueshifting the onset of optical UV absorption for water under pressure. Phys. Rev. Lett. **106**(18), 187403 (2011)
37. W. Chen, X.F. Wu, R. Car, X-ray absorption signatures of the molecular environment in water and ice. Phys. Rev. Lett. **105**(1), 017802 (2010)
38. Y. Wang, H. Liu, J. Lv, L. Zhu, H. Wang, Y. Ma, High pressure partially ionic phase of water ice. Nat. commun. **2**, 563 (2011)
39. M. Abu-Samha, K.J. Borve, Surface relaxation in water clusters: evidence from theoretical analysis of the oxygen 1s photoelectron spectrum. J. Chem. Phys. **128**(15), 154710 (2008)
40. O. Bjorneholm, F. Federmann, S. Kakar, T. Moller, Between vapor and ice: free water clusters studied by core level spectroscopy. J. Chem. Phys. **111**(2), 546–550 (1999)
41. G. Ohrwall, R.F. Fink, M. Tchaplyguine, L. Ojamae, M. Lundwall, R.R.T. Marinho, A.N. de Brito, S.L. Sorensen, M. Gisselbrecht, R. Feifel, T. Rander, A. Lindblad, J. Schulz, L. J. Saethre, N. Martensson, S. Svensson, O. Bjorneholm, The electronic structure of free water clusters probed by Auger electron spectroscopy. J. Chem. Phys. **123**(5), 054310 (2005)
42. S. Hirabayashi, K.M.T. Yamada, Infrared spectra and structure of water clusters trapped in Argon and krypton matrices. J. Mol. Struct. **795**(1–3), 78–83 (2006)
43. P. Andersson, C. Steinbach, U. Buck, Vibrational spectroscopy of large water clusters of known size. Eur. Phys. J. D **24**(1–3), 53–56 (2003)
44. S. Maheshwary, N. Patel, N. Sathyamurthy, A.D. Kulkarni, S.R. Gadre, Structure and stability of water clusters $(H_2O)n$, n) 8–20: an ab initio investigation. J. Phys. Chem. A **105**, 10525–10537 (2001)
45. A. Nilsson, L.G.M. Pettersson, Perspective on the structure of liquid water. Chem. Phys. **389** (1–3), 1–34 (2011)
46. G.N.I. Clark, C.D. Cappa, J.D. Smith, R.J. Saykally, T. Head-Gordon, The structure of ambient water. Mol. Phys. **108**(11), 1415–1433 (2010)

47. C. Vega, J.L.F. Abascal, M.M. Conde, J.L. Aragones, What ice can teach us about water interactions: a critical comparison of the performance of different water models. Faraday Discuss. **141**, 251–276 (2009)
48. R. Ludwig, The importance of tetrahedrally coordinated molecules for the explanation of liquid water properties. Chem. Phys. Chem. **8**(6), 938–943 (2007)
49. B. Santra, A. Michaelides, M. Fuchs, A. Tkatchenko, C. Filippi, M. Scheffler, On the accuracy of density-functional theory exchange-correlation functionals for H bonds in small water clusters. II. The water hexamer and van der Waals interactions. J. Chem. Phys. **129** (19), 194111 (2008)
50. K. Liu, J.D. Cruzan, R.J. Saykally, Water clusters. Science **271**(5251), 929–933 (1996)
51. V. Buch, S. Bauerecker, J.P. Devlin, U. Buck, J.K. Kazimirski, Solid water clusters in the size range of tens-thousands of $H_2O$: a combined computational/spectroscopic outlook. Int. Rev. Phys. Chem. **23**(3), 375–433 (2004)
52. J.H. Cartwright, B. Escribano, C.I. Sainz-Diaz, The mesoscale morphologies of ice films: porous and biomorphic forms of ice under astrophysical conditions. Astrophys J **687**(2), 1406 (2008)
53. Y. Li, G.A. Somorjai, Surface premelting of ice. J. Phys. Chem. C **111**(27), 9631–9637 (2007)
54. A.-M. Kietzig, S.G. Hatzikiriakos, P. Englezos, Physics of ice friction. J. Appl. Phys. **107**(8), 081101–081115 (2010)
55. S.K. Sikka, S.M. Sharma, The hydrogen bond under pressure. Phase Transitions **81**(10), 907–934 (2008)
56. P. Pruzan, J.C. Chervin, E. Wolanin, B. Canny, M. Gauthier, M. Hanfland, Phase diagram of ice in the VII–VIII–X domain. Vibrational and structural data for strongly compressed ice VIII. J. Raman Spectrosc. **34**(7–8), 591–610 (2003)
57. K. Davitt, E. Rolley, F. Caupin, A. Arvengas, S. Balibar, Equation of state of water under negative pressure. J. Chem. Phys. **133**(17), 174507 (2010)
58. M. Chaplin, Theory versus experiment: what is the surface charge of water? Water **1**(1–28), (2009)
59. Y.R. Shen, V. Ostroverkhov, Sum-frequency vibrational spectroscopy on water interfaces: polar orientation of water molecules at interfaces. Chem. Rev. **106**(4), 1140–1154 (2006)
60. M. Faubel, K.R. Siefermann, Y. Liu, B. Abel, Ultrafast soft x-ray photoelectron spectroscopy at liquid water microjets. Acc. Chem. Res. **45**(1), 120–130 (2011)
61. A. Morita, T. Ishiyama, Recent progress in theoretical analysis of vibrational sum frequency generation spectroscopy. PCCP **10**(38), 5801–5816 (2008)
62. J.L. Skinner, P.A. Pieniazek, S.M. Gruenbaum, Vibrational spectroscopy of water at interfaces. Acc. Chem. Res. **45**(1), 93–100 (2012)
63. H.J. Bakker, J.L. Skinner, Vibrational spectroscopy as a probe of structure and dynamics in liquid water. Chem. Rev. **110**(3), 1498–1517 (2010)
64. C.H. Sun, L.M. Liu, A. Selloni, G.Q. Lu, S.C. Smith, Titania-water interactions: a review of theoretical studies. J. Mater. Chem. **20**(46), 10319–10334 (2010)
65. M.A. Henderson, The interaction of water with solid surfaces: fundamental aspects revisited. Surf. Sci. Rep. **46**(1–8), 5–308 (2002)
66. A. Hodgson, S. Haq, Water adsorption and the wetting of metal surfaces. Surf. Sci. Rep. **64** (9), 381–451 (2009)
67. A. Verdaguer, G.M. Sacha, H. Bluhm, M. Salmeron, Molecular structure of water at interfaces: wetting at the nanometer scale. Chem. Rev. **106**(4), 1478–1510 (2006)
68. J. Carrasco, A. Hodgson, A. Michaelides, A molecular perspective of water at metal interfaces. Nat. Mater. **11**(8), 667–674 (2012)
69. T. Kumagai, Direct observation and control of hydrogen-bond dynamics using low-temperature scanning tunneling microscopy. Prog. Surf. Sci. **90**(3), 239–291 (2015)
70. Y. Marcus, Effect of ions on the structure of water: structure making and breaking. Chem. Rev. **109**(3), 1346–1370 (2009)

71. J.A. Sellberg, C. Huang, T.A. McQueen, N.D. Loh, H. Laksmono, D. Schlesinger, R.G. Sierra, D. Nordlund, C.Y. Hampton, D. Starodub, D.P. DePonte, M. Beye, C. Chen, A.V. Martin, A. Barty, K.T. Wikfeldt, T.M. Weiss, C. Caronna, J. Feldkamp, L.B. Skinner, M.M. Seibert, M. Messerschmidt, G.J. Williams, S. Boutet, L.G. Pettersson, M.J. Bogan, A. Nilsson, Ultrafast x-ray probing of water structure below the homogeneous ice nucleation temperature. Nature **510**(7505), 381–384 (2014)
72. L.B. Skinner, C. Huang, D. Schlesinger, L.G. Pettersson, A. Nilsson, C.J. Benmore, Benchmark oxygen–oxygen pair-distribution function of ambient water from x-ray diffraction measurements with a wide Q-range. J. Chem. Phys. **138**(7), 074506 (2013)
73. C. Drechsel-Grau, D. Marx, Tunnelling in chiral water clusters: protons in concert. Nat. Phys. **11**(3), 216–218 (2015)
74. X. Meng, J. Guo, J. Peng, J. Chen, Z. Wang, J.-R. Shi, X.-Z. Li, E.-G. Wang, Y. Jiang, Direct visualization of concerted proton tunnelling in a water nanocluster. Nat. Phys. **11**(3), 235–239 (2015)
75. J. Guo, X. Meng, J. Chen, J. Peng, J. Sheng, X.Z. Li, L. Xu, J.R. Shi, E. Wang, Y. Jiang, support: real-space imaging of interfacial water with submolecular resolution-Supp. Nat. Mater. **13**, 184–189 (2014)
76. J. Meibohm, S. Schreck, P. Wernet, Temperature dependent soft x-ray absorption spectroscopy of liquids. Rev. Sci. Instrum. **85**(10), 103102 (2014)
77. H. Bluhm, D.F. Ogletree, C.S. Fadley, Z. Hussain, M. Salmeron, The premelting of ice studied with photoelectron spectroscopy. J. Phys.: Condens. Matter **14**(8), L227 (2002)
78. K.R. Wilson, B.S. Rude, T. Catalano, R.D. Schaller, J.G. Tobin, D.T. Co, R.J. Saykally, X-ray spectroscopy of liquid water microjets. J. Phys. Chem. B **105**(17), 3346–3349 (2001)
79. Y.R. Shen, Basic theory of surface sum-frequency generation. J. Phys. Chem. C **116**, 15505–15509 (2012)
80. S.N. Wren, D.J. Donaldson, Glancing-angle Raman spectroscopic probe for reaction kinetics at water surfaces. Phys. Chem. Chem. Phys. **12**(11), 2648–2654 (2010)
81. T.F. Kahan, J.P. Reid, D.J. Donaldson, Spectroscopic probes of the quasi-liquid layer on ice. J. Phys. Chem. A **111**(43), 11006–11012 (2007)
82. F. Zaera, Probing Liquid/solid interfaces at the molecular level. Chem. Rev. **112**, 2920–2986 (2012)
83. C.M. Johnson, S. Baldelli, Vibrational sum frequency spectroscopy studies of the influence of solutes and phospholipids at vapor/water interfaces relevant to biological and environmental systems. Chem. Rev. **114**(17), 8416–8446 (2014)
84. M. Chaplin. Water structure and science, http://www.lsbu.ac.uk/water/
85. Editorial, So much more to know. Science, **309**(5731), 78–102 (2005)
86. T. Head-Gordon, G. Hura, Water structure from scattering experiments and simulation. Chem. Rev. **102**(8), 2651–2669 (2002)
87. J. Baez, Phase diagram of water (2012) http://math.ucr.edu/home/baez/chemical/725px-Phase_diagram_of_water.svg.png
88. Y. Huang, X. Zhang, Z. Ma, Y. Zhou, W. Zheng, J. Zhou, C.Q. Sun, Hydrogen-bond relaxation dynamics: resolving mysteries of water ice. Coord. Chem. Rev. **285**, 109–165 (2015)
89. D.D. Kang, J. Dai, H. Sun, Y. Hou, J. Yuan, Quantum similation of thermally driven phase transition and O k-edge absorption of high-pressure ice. Sci. Rep. **3**, 3272 (2013)
90. X. Zhang, P. Sun, Y. Huang, T. Yan, Z. Ma, X. Liu, B. Zou, J. Zhou, W. Zheng, C.Q. Sun, Water's phase diagram: from the notion of thermodynamics to hydrogen-bond cooperativity. Prog. Solid State Chem. **43**, 71–81 (2015)
91. H. Kanno, K. Miyata, The location of the second critical point of water. Chem. Phys. Lett. **422**(4–6), 507–512 (2006)
92. V. Holten, M.A. Anisimov, Entropy-driven liquid-liquid separation in supercooled water. Sci. Rep. **2**, 713 (2012)
93. T. Bartels-Rausch, Chemistry: ten things we need to know about ice and snow. Nature **494** (7435), 27–29 (2013)

94. G. Algara-Siller, O. Lehtinen, F.C. Wang, R.R. Nair, U. Kaiser, H.A. Wu, A.K. Geim, I.V. Grigorieva, Square ice in graphene nanocapillaries. Nature **519**(7544), 443–445 (2015)
95. C.Q. Sun, X. Zhang, X. Fu, W. Zheng, J.-L. Kuo, Y. Zhou, Z. Shen, J. Zhou, Density and phonon-stiffness anomalies of water and ice in the full temperature range. J. Phys. Chem. Lett. **4**, 3238–3244 (2013)
96. A. Falenty, T.C. Hansen, W.F. Kuhs, Formation and properties of ice XVI obtained by emptying a type sII clathrate hydrate. Nature **516**(7530), 231–233 (2014)
97. C.Q. Sun, X. Zhang, W.T. Zheng, Hidden force opposing ice compression. Chem. Sci. **3**, 1455–1460 (2012)
98. J. Teixeira, High-pressure physics—the double identity of ice X. Nature **392**(6673), 232–233 (1998)
99. A. Falenty, A. Salamatin, W. Kuhs, Kinetics of $CO_2$-hydrate formation from ice powders: data summary and modeling extended to low temperatures. J. Phys. Chem. C **117**(16), 8443–8457 (2013)
100. Y. Wang, Y. Ma, Perspective: crystal structure prediction at high pressures. J. Chem. Phys. **140**(4), 040901 (2014)
101. P.H. Poole, F. Sciortino, U. Essmann, H.E. Stanley, Phase behaviour of metastable water. Nature **360**(6402), 324–328 (1992)
102. G.H. Pollack, *The Fourth Phase of Water: Beyond Solid, Liquid, and Vapor* (Ebner and Sons Seattle, USA, 2013)
103. L. Ojha, M.B. Wilhelm, S.L. Murchie, A.S. McEwen, J.J. Wray, J. Hanley, M. Masse, M. Chojnacki, Spectral evidence for hydrated salts in recurring slope lineae on Mars. Nat. Geosci. (2015). doi:10.1038/ngeo2546
104. C.Q. Sun, X. Zhang, J. Zhou, Y. Huang, Y. Zhou, W. Zheng, Density, elasticity, and stability anomalies of water molecules with fewer than four neighbors. J. Phys. Chem. Lett. **4**, 2565–2570 (2013)
105. X. Zhang, Y. Huang, Z. Ma, Y. Zhou, W. Zheng, J. Zhou, C.Q. Sun, A common supersolid skin covering both water and ice. PCCP **16**(42), 22987–22994 (2014)
106. L. Pauling, *The Nature of the Chemical Bond*, 3rd edn. (Cornell University press: Ithaca, 1960)
107. C.Q. Sun, *Relaxation of the Chemical Bond*. Springer Series in Chemical Physics 108, vol. 108 (Springer, Heidelberg, 2014), 807 pp
108. J. Russo, H. Tanaka, Understanding water's anomalies with locally favoured structures. Nat Commun. **5**, 3556 (2014)
109. Y. Huang, X. Zhang, Z. Ma, Y. Zhou, J. Zhou, W. Zheng, C.Q. Sun, Size, separation, structure order, and mass density of molecules packing in water and ice. Sci Rep. **3**, 3005 (2013)
110. Y. Huang, X. Zhang, Z. Ma, Y. Zhou, G. Zhou, C.Q. Sun, Hydrogen-bond asymmetric local potentials in compressed ice. J. Phys. Chem. B **117**(43), 13639–13645 (2013)
111. M. Faraday, Note on Regelation. Proc. R. Soc. Lon. **10**, 440–450 (1859)
112. Y. Yoshimura, S.T. Stewart, M. Somayazulu, H.K. Mao, R.J. Hemley, Convergent Raman features in high density amorphous ice, ice vii, and ice viii under pressure. J. Phys. Chem. B **115**(14), 3756–3760 (2011)
113. V. Holten, C. Bertrand, M. Anisimov, J. Sengers, Thermodynamics of supercooled water. J. Chem. Phys. **136**(9), 094507 (2012)
114. X. Xiao-Min, C. Lan, Z. Wen-Long, L. Long-Fei, Y. Yue-Bin, P. Zhi-Yong, Z. Jin-Xiu, Imaginary part of the surface tension of water. Chin. Phys. Lett. **31**(7), 076801 (2014)
115. K.R. Wilson, R.D. Schaller, D.T. Co, R.J. Saykally, B.S. Rude, T. Catalano, J.D. Bozek, Surface relaxation in liquid water and methanol studied by x-ray absorption spectroscopy. J. Chem. Phys. **117**(16), 7738–7744 (2002)
116. S. Trainoff, N. Philips. Water droplet dancing on water surfaces (2009), http://www.youtube.com/watch?v=pbGz1njqhxU
117. R. Rosenberg, Why is ice slippery? Phys. Today **58**(12), 50 (2005)

118. Z. Pawlak, W. Urbaniak, A. Oloyede, The relationship between friction and wettability in aqueous environment. Wear **271**(9), 1745–1749 (2011)
119. X. Zhang, P. Sun, Y. Huang, Z. Ma, X. Liu, J. Zhou, W. Zheng, C.Q. Sun, Water nanodroplet thermodynamics: quasi-solid phase-boundary dispersivity. J. Phys. Chem. B **119** (16), 5265–5269 (2015)
120. P.G. Debenedetti, H.E. Stanley, Supercooled and glassy water. Phys. Today **56**(6), 40–46 (2003)
121. C.Q. Sun, Y. Sun, Y.G. Ni, X. Zhang, J.S. Pan, X.H. Wang, J. Zhou, L.T. Li, W.T. Zheng, S.S. Yu, L.K. Pan, Z. Sun, Coulomb repulsion at the nanometer-sized contact: a force driving superhydrophobicity, superfluidity, superlubricity, and supersolidity. J. Phys. Chem. C **113** (46), 20009–20019 (2009)
122. F. Hofmeister, Concerning regularities in the protein-precipitating effects of salts and the relationship of these effects to the physiological behaviour of salts. Arch. Exp. Pathol. Pharmacol. **24**, 247–260 (1888)
123. X. Zhang, T. Yan, Y. Huang, Z. Ma, X. Liu, B. Zou, C.Q. Sun, Mediating relaxation and polarization of hydrogen-bonds in water by NaCl salting and heating. PCCP **16**(45), 24666–24671 (2014)
124. F. Mei, X. Zhou, J. Kou, F. Wu, C. Wang, H. Lu, A transition between bistable ice when coupling electric field and nanoconfinement. J. Chem. Phys. **142**(13), 134704 (2015)
125. D. Ehre, E. Lavert, M. Lahav, I. Lubomirsky, Water freezes differently on positively and negatively charged surfaces of pyroelectric materials. Science **327**(5966), 672–675 (2010)
126. W. Armstrong, Electrical phenomena the newcastle literary and philosophical society. Electr. Eng. **10**, 153 (1893)
127. X. Zhang, Y. Huang, Z. Ma, Y. Zhou, J. Zhou, W. Zheng, Q. Jiang, C.Q. Sun, Hydrogen-bond memory and water-skin supersolidity resolving the Mpemba paradox. PCCP **16**(42), 22995–23002 (2014)
128. E.B. Mpemba, D.G. Osborne, Cool? Phys. Educ. **14**: 410–413 (1979)
129. P. Ball, E. Ben-Jacob, Water as the fabric of life. Eur. Phys. J. Spec. Top. **223**(5), 849–852 (2014)
130. H. Qiu, W. Guo, Electromelting of Confined Monolayer Ice. Phys. Rev. Lett. **110**(19), 195701 (2013)
131. E.C. Fuchs, P. Baroni, B. Bitschnau, L. Noirez, Two-dimensional neutron scattering in a floating heavy water bridge. J. Phys. D Appl. Phys. **43**(10), 105502 (2010)
132. Z. Yu, H. Li, X. Liu, C. Xu, H. Xiong, Influence of soil electric field on water movement in soil. Soil Tillage Res. **155**, 263–270 (2016)
133. S.S. Xantheas, Cooperativity and hydrogen bonding network in water clusters. Chem. Phys. **258**(2), 225–231 (2000)
134. A.E. Reed, L.A. Curtiss, F. Weinhold, Intermolecular interactions from a natural bond orbital, donor-acceptor viewpoint. Chem. Rev. **88**(6), 899–926 (1988)
135. J.D. Bernal, R.H. Fowler, A theory of water and ionic solution, with particular reference to hydrogen and Hydroxyl ions. J. Chem. Phys. **1**(8), 515–548 (1933)

# Chapter 2
# Water Structure

- *Water prefers the fluctuating, tetrahedral geometry because O tends to hybridize its sp orbits.*
- *The tetrahedral geometry and mass density entitles the size and separation of molecules packed in water and ice.*
- *Presence of nonbonding lone pairs and bonding pairs determine the atomistic anisotropy of an $H_2O$.*
- *The O:H–O bond holds universally in all phases albeit its angle and segmental length relaxation.*

**Abstract** The geometry structure, mass density $\rho(g/cm^3)$, the size ($d_H$) and separation ($d_{OO} = d_L + d_H$) of molecules packed in water and ice are closely correlated, which reconciles: (i) the $d_x$ symmetrization in compressed ice, (ii) the $d_{OO}$ relaxation of cooling water and ice and, (iii) the $d_{OO}$ expansion between undercoordinated molecules. With any one of the $d_{OO}$, the density $\rho$, the $d_L$, and the $d_H$, as a known input, one can resolve the rest using this solution that is probing conditions or methods independent. Consistency between predictions and observations clarified that: (i) liquid water prefers *statistically* the mono-phase of tetrahedrally-coordinated structure with high fluctuation, (ii) the supersolid phase (strongly polarized and much less dense) exists only in regions consisting undercoordinated molecules and, (iii) repulsion between electron pairs on adjacent oxygen atoms dictates the cooperative relaxation of the segmented O:H–O bond.

## 2.1 Challenge: What Rules the Structure Order?

*Science* magazine enlisted the *Structure of Water* as one of the 125 big questions to the mankind in its 125 anniversary special issue published in 2005 (Fig. 2.1) [1] though the physical anomalies of ice and water are even more fascinating and challenging. Typical questions include:

(1) How many nearest neighbors does one molecule have?
(2) How do water molecules interact one another in water?
(3) Does water hold mono- or mixed phase structure?
(4) How are the molecular size, separation, geometry, and mass density correlated in water and ice?



C.Q. Sun and Y. Sun, *The Attribute of Water*, Springer Series in Chemical Physics 113, DOI 10.1007/978-981-10-0180-2_2

**Fig. 2.1** Water structure is one of the 125 big questions of mankind [1]

## 2.2 Clarification: O Hybridizes Its sp Orbits

Physical foundations [2, 3] define a $H_2O$ molecule in bulk water preferring the quasi tetrahedral structure. Molecules are associated with tetrahedrally-coordinated, strongly correlated, and fluctuating configuration, because of the anisotropy in the local charge and energy distribution surrounding the central oxygen of the tetrahedron, see Fig. 2.2[4]:

(1) Like nitrogen ($2s^2p^3$) and fluorine ($2s^2p^5$), the electronegative oxygen hybridizes its $2s^2p^4$ orbits upon interacting with atoms of electropositive elements to form a quasi $sp^3$ tetrahedron with creation of two bonding electron pairs, represented by "–", and two nonbonding electron lone pairs, denoted as ":", $2H^{\delta+}{:}O^{2\delta-}{-}2H^{\delta+}$ ($\delta < 1$ denotes the net charge, for simplicity in discussion, we let $\delta = 1$ from now onward), because of the O:H nonbond interaction, the molecule is often symbolled as the V-shaped $H_2O$ [5].
(2) The water molecule prefers the quasi tetrahedral structure over the temperature range from 5 K [6] to its gaseous phase under even extreme conditions of high temperature and pressure (2000 K and 2 TPa), at which the ionization takes place and the $(H_2O)_2$ transits into an $OH_3^+$ (hydronium, $NH_3$ like configuration with one lone pair) and an $OH^-$(hydroxide, HF like with three lone pairs) bridged also by the O:H–O interaction [7].
(3) The O:H–O bond between oxygen ions holds common disregarding the phase order or geometric structure. Cooperative relaxation of the O:H and the H–O bond and the ∠O:H–O containing angle alters water and ice from one phase to another, and determines the macroscopic properties of water and ice [8].
(4) Water molecules never stabilize with the tetrahedral geometry in liquid phase because of the atomistic anisotropy. As a statistic mean, the tetrahedron and

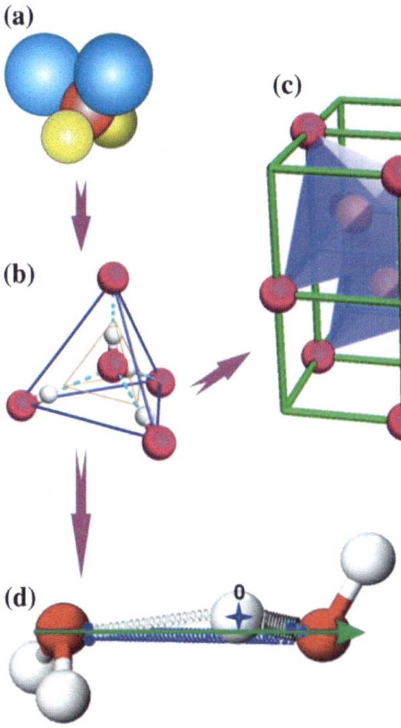


**Fig. 2.2** **a** An oxygen atom forms a quasi-tetrahedron with its neighboring H atoms through two bonding pairs (*yellow*, angled at 104.5° or less) and two nonbonding lone pairs (*blue*, angled at 109.5° or more) upon $sp^3$-orbit being hybridized [5]. An extension of this quasi-tetrahedron of $C_{2v}$ group symmetry yields **b** an ideal $C_{3v}$-symmetrical tetrahedron that contains two $H_2O$ molecules and four directional O:H–O bonds. Assembling such tetrahedral blocks around the central oxygen ion forms the ideal diamond structure in (**c**). This packing geometry and known mass density correlate the size and separation of molecular packed in water and in ice [4]. **d** The flexible and polarizable O:H–O bond forms a pair of asymmetrical, coupled, H-bridged oscillator pair whose relaxation in length and energy, and the associated electron entrapment and polarization, resolves mysteries of water ice. Bond angle relaxation/distortion discriminates geometrical topology such as 2D and caged ice [9] while the O:H–O bond retains. The $H^+$ is the coordination origin and the pairing dots denote electron pairs of oxygen. (Reprinted with permission from [4])

mass density ρ define the size $d_H$ (H–O length) and separation $d_L$ (O:H distance) of molecules packed in water and ice in the following manner (length unit in Å and density in g/cm$^3$; the $d_x$ (x = L and H) is the projection along the O:H–O direction [4]:

$$\begin{cases} d_{OO} = d_L + d_H = 2.6950\rho^{-1/3} & (O-O\ separation) \\ d_L = \dfrac{2d_{L0}}{1+exp[(d_H - d_{H0})/0.2428]}; & (d_{H0} = 1.0004, d_{L0} = 1.6946) \end{cases}. \tag{2.1}$$

## 2.3 History Background

Since 1809 when Gay-Lussac and Alexander von Humboldt firstly defined the essential nature of water (2 hydrogen + 1 oxygen = 2 water vapor in volume) [10], and in 1920 Peter Debye identified the intra- and intermolecular interference features of water using X-ray diffraction, there have been a huge number of models for water's structure. Here we list a few of them from the perspectives of intra- and intermolecular interactions, including Ice Rules of Pauling, Fowler, and Bernal and the one- and two-state models for molecular association, which remain as

| | |
|---|---|
|  | Louis Joseph Gay-Lussac (December 6, 1778–May 9, 1850) was a French chemist and physicist. He is known mostly for two laws related to gases, and for his work on alcohol-water mixtures, which led to the degrees of Gay-Lussac used to measure alcoholic beverages. |
|  | Alexander von Humboldt (September 14, 1769–May 6, 1859), German naturalist and explorer who was a major figure in the classical period of physical geography and biogeography. |
|  | Peter Joseph William Debye (March 24, 1884–November 2, 1966) was a Dutch–American physicist and physical chemist, and Nobel laureate in Chemistry in 1936 for his contributions to molecular structure, dipole moment relationship and for the diffraction of X-rays and electrons. |

highly-debated issues. Interested readers may be referred to the monograph *Biography of Water* authored by Philip Ball [11].

### 2.3.1 Typical Structural Models

| Models | | Contributors |
|---|---|---|
| Ice Rules for water molecules | | |
| Bernal–Fowler proton tunneling transition (Fig. 2.3a): $H_2O$ forms a V-shaped motif with an angle of 104.5 ± 1.5°, $d_H = 0.96$ Å, $d_{O\text{–}O} = 2.76$ Å. $H^+$ proton jumps from one molecule to next through $(H_2O)_2 \rightarrow OH^{\delta-} + OH_3^{\delta+}$ transition in a frequency of $f \sim 1.5$ THz | | Ralph H. Fowler, John Desmond Bernal (1933) [12] |
| Proton "two-in two-out" frustration (Fig. 2.3b): For each oxygen atom, two of the neighboring hydrogen atoms are near (forming the traditional $H_2O$ molecule), and two are further away (being the hydrogen atoms of neighboring water molecules). The H proton is frustrated in the two positions with identical probability. Even upon cooling to zero temperature, water ice has residual entropy, i.e., intrinsic randomness; the number of configurations, or the entropy, grows exponentially with the system size | | Linus Pauling (1935) [13] |
| Frustrationless and tunnelingless quasi-tetrahedron (Fig. 2.2):<br>(1) Oxygen hybridizes its sp-orbit intrinsically to form tetrahedral and directional bonds and nonbonds with atoms of electropositive elements disregarding structure phase;<br>(2) H–O is shorter and much stronger than the O:H nonbond; the tetrahedron is subjected to ceaseless rotation and vibration in liquid;<br>(3) H–O and O:H binding energy disparity prevents $H^+$ from tunneling across molecules or from "two-in two-out" frustrating transition | | Chang Q Sun (2013) [3, 5] |
| Models for water molecular association | | |
| Clustering | Icebergs in a fluid: Cold water contains microscopic 'icebergs' in a fluid of 'dipole sea'. Interstitial 'hydrone' ensures long-lived clusters of water molecules | Wilhelm Röntgen (1891); Henry Armstrong (1920); Oleg Samoilov (1940) [14] |
| | Flicking clusters: Clusters of water molecules built from surrounding water grow to a critical size and then spontaneously disperse. All of this happens repeatedly on a time scale of $10^{-10}$ to $10^{-11}$ s | Henry S. Frank and Wen-Yang Wen (1957) [15] |
| | Submicron-sized (millions molecules) coherent domains: Intermolecular bonds within those domains serve as antennae that receive electromagnetic energy from outside. With such energy, water molecules can release | Emilio del Giudice (2009) [16] |

(continued)

(continued)

| Models | | Contributors |
|---|---|---|
| | electrons, making them available for chemical reactions | |
| | Two types of intermixed clusters: One type is shell-like, and more-or-less collapsed, while the other is rather solid and more regularly structured. Molecules of water switch their allegiance rapidly between these two phases, but under a given set of conditions, the average number of molecules in each category remains constant | Martin Chaplin [17] |
| | Five-member shell model: The locally favored structures of water not only have translational order in the second shell but also contain five-membered rings of hydrogen-bonded molecules. The former promotes crystallization, whereas the latter causes frustration against crystallization | Russo and Tanaka (2014) [18] |
| Two-phase | Ice-like clumps in the disordered sea: Chains or rings containing up to about 100 molecules surrounded by a disordered organization | Anders Nilsson, Lars Petterson (2004) [19, 20] |
| | Low-density and high-density fragments: Water has two distinct states, low density and high density, particularly for supercooled water. Low-density water has an open tetrahedral structure, while high-density water has a more compact structure. The two states dynamically interchange with one another | Gene Stanley (1998) [21] |
| | Heterogeneity complexity model: The heterogeneity of water structure eases the water-molecule interchange | Rustum Roy (2005) [22] |
| | Homogeneous bulk with supersolid skin: Water molecules prefer the quasi-tetrahedron configuration with fluctuation but the skin molecular undercoordination relaxes the O:H–O bond substantially to lower the standard mass density by 25 %, whereas the geometry remains | Chang Q Sun (2013) [23] |
| Fluctuating mono-phase | Homogeneous tetrahedron with slight fluctuation: Water is homogeneous down to the molecular level where different water molecules form tetrahedral units of different perfection and/or participate in rings of different | Petkov et al. (2012) [24] |

(continued)

(continued)

| Models | | Contributors |
|---|---|---|
| | sizes. The local diversity of this tetrahedral network coupled to the flexibility of the hydrogen bonds | |
| | Homogeneous with slight distortion: The majority of molecules in liquid water retains the four-fold coordination with only moderately distorted tetrahedral configurations. For a significant fraction of molecules the hydrogen-bonding environments are highly asymmetric with extremely weak and distorted bonds | Kuhne and Khaliullin (2014) [25] |
| | Identical ice-like H-bond geometry: Water and ice (in the range of 251–288 K) share the same ice-like geometry according to the total energy yield near-edge X-ray absorption | Smith, Saykally et al.(2004) [26] |
| | Long-range order of tetrahedrally-coordinated signatures: Water prefers the tetrahedral hydrogen-bonded long range monophase though both one- and two-state models could fit the same set of X-ray absorption spectroscopic data | Head-Gordon (2006) [27] |
| Surface and interface water | Exclusion zones: Gel-like and $OH_3^+$ dominated zone presents at water/hydrophilic contacts | Gerald Pollack (2013) [28] |
| | Capillary polywater: Gel-like water forms in small capillary tubes | Boris Deryaguin (1960) [14] |
| | Supersolid phase associated with undercoordinated molecules:<br>(1) Supersolid phase forms at sites surrounding defects or free skins or hydrophobic capillary contacts because of molecular undercoordination that shortens and stiffens the H–O bond but lengthens and softens the O:H nonbond with dual polarization, resulting in this elastic, hydrophobic, less dense, and thermally stable phase.<br>(2) Supersolid skin of the same geometry but subject to strong O:H–O bond relaxation and polarization;<br>(3) Nanodroplets and bubbles prefer the two-phase structure in a core-shell configuration | Chang Q Sun et al. (2015) [3] |
| Quasisolid phase | (1) Quasisolid phase with negative thermal expansion coefficient presents at temperatures ranging from 258 to 277 K for density transition from minimum to maximum, or close to temperatures transiting from ice to liquid | Chang Q Sun et al. (2015) [3] |

(continued)

(continued)

| Models | | Contributors |
|---|---|---|
| | (2) Any stimulus that shifts the phonon frequencies disperses the boundaries of the quasisolid phase, resulting in supercooling/heating | |
|  | John Desmond Bernal, FRS, (10 May 1901–15 September 1971) was one of the best-known and most controversial UK scientists. Known as "Sage" to friends, Bernal is a pioneer in X-ray crystallography in molecular biology. | |
|  | Sir Ralph Howard Fowler, OBE FRS (17 January 1889–28 July 1944) was a British physicist and astronomer. Best known for Statistical physics and Fowler–Nordheim-type equations for the quantum mechanics of electron field emission. | |
|  | Linus Carl Pauling (February 28, 1901–August 19, 1994) was an American chemist, biochemist, peace activist, and educator. Nobel Laureates in Chemistry and peace. (public domain) | |

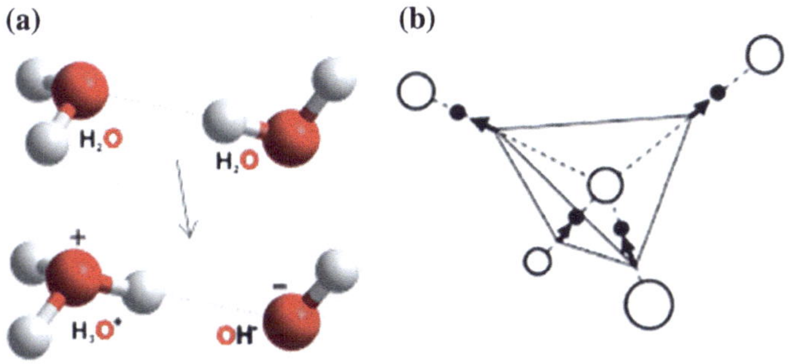


**Fig. 2.3** Ice Rules of **a** Bernal–Fowler for $(H_2O)_2 \leftrightarrow OH_3^+ : OH^-$ proton tunneling transition in 1.5 THz frequency [15] and **b** Pauling's "two-in two-out" $H^+$ proton frustration [13] in the arrangement of hydrogen atoms (*black circles*) about oxygen atoms (*open circles*) in ice. Hydrogen atoms frustrate in two positions off midway between two oxygen anions. Both tunneling and frustrating transition remain in favor of tetrahedron configuration

## 2.3.2 *Intermolecular Interactions*

Tremendous efforts have been made by generations in order to measure or calculate the $H_2O$ molecular structure [29, 30]. This is a difficult task due to the formation of a complex, dynamic, and fluctuating hydrogen bond network. From a theoretical point of view, the difficulties in understanding liquid water are often attributed to two entities:

(1) Cooperative hydrogen bonding suggests that the binding energy of two H-bonded molecules is modified by the presence of a third molecule and by the nuclear quantum effects.
(2) Such effects occur because the proton is so light that classical mechanics can no longer adequately describe properties such as the spatial dispersion of the hydrogen positions, nuclear tunneling, zero-point energy and naturally, quantization of nuclear motions.

Among various models for the structure of water ice, the rigid and non-polarizable TIPnP (n varies from 1 to 5) series [31, 32] and the polarizable models [33] are widely used in calculations. In the TIPnQ models, for instance, the V-shaped $H_2O$ geometry with a bond length of $r_{OH} = 0.9572$ Å and a bond angle of $\theta_{HOH} = 104.52°$ describes a water molecule in the gaseous phase. Figure 2.4 illustrates the TIP4Q/2005 structure model [34] that simplifies the $H_2O$ molecule as a dipole ($O^+$–$M^-$) with a fixed $H^+$ point charge. Lennard-Jones (L-J) potential represents interaction between such dipoles in the "dipole sea" bulk water.

Most available models consider inter- and intramolecular interactions independently with neglecting of their coupling or the possibility of its cooperative relaxation that induces charge entrapment and polarization. It is the right omitted context that dictates the adaptivity, cooperativity, sensitivity of water and ice when subject to perturbation. These rigid non-polarizable or polarizable models can

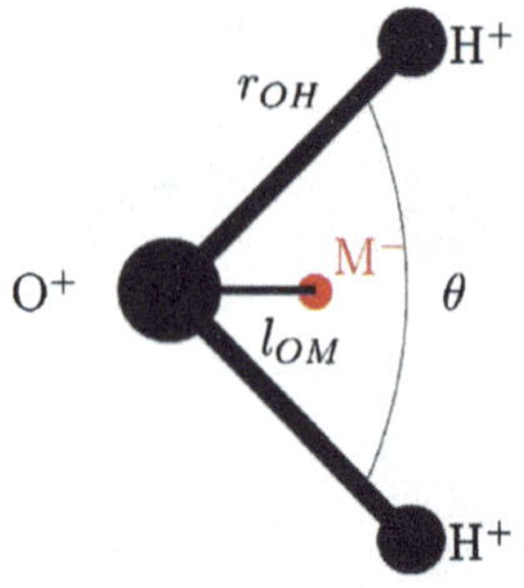


**Fig. 2.4** The V-shaped water molecular dipole is modeled as a rigid non-polarizable in TIP4Q/2005 model. There is a positive point charge $q_O$ on the oxygen atom, a positive point charge $q_H$ on each hydrogen atom and a negative point charge $q_M$ at site M located at a distance $l_{OM}$ from the oxygen atom along the bisector of angle ∠HOH. The molecule is electrically neutral, thus $q_M = -(2q_H + q_O)$, and $q_H$, $q_O$, $l_{OM}$, $\sigma_{OO}$, and $\varepsilon_{OO}$ are independent parameters to be determined, where $\sigma_{OO}$ and $\varepsilon_{OO}$ are the L-J (Lennard-Jones) potential parameters for the O–O interaction. $\delta < 1$ ($\sim$0.6 e) is the net fraction charge on ions. (reprinted with permission from [34]) (color online)

reproduce some anomalies of water ice but the outcome is often subject to satisfaction [34, 35].

The mixture models proposed by Röntgen [36] for water have been popular for many years. These models refer to at least two distinct species or domains: a low-density ice-like one, with nearly fully hydrogen bonded water molecules, and a high-density domain of individual or oligomeric water molecules with a few hydrogen bonds each. Robinson et al. [37] examined the volumetric properties of water at varying temperatures (−30 to +100 °C) [38] and pressures (0.1 MPa to 0.77 GPa) in terms of an explicit two-state mixture model. Such models depend on the consideration of the hydrogen bonding in liquid water. The description of water in terms of a multicomponent mixture model, though plausible, was challenged by Smith et al. [39] and Head-Gorden et al. [27] contended that the spectroscopic evidence for such models can equally well be explained in terms of a continuity of states with hydrogen bonds of varying strengths.

Although it is generally accepted that water is a highly structured liquid, there is little agreement on how to express the structure quantitatively and how to measure it experimentally. The experimental methods commonly employed to ascertain the molecular structure of liquids, namely X-ray and neutron diffraction, yield the structure factors and indirectly (after Fourier transformation) the total pair correlation function $g(r) = f(r/\sigma)$, where $r$ is the linear distance from the center of a given particle and $\sigma$ is the diameter of the particle. Application of these methods results in the structure of liquid water resembling that of liquid argon, a nonstructured liquid by all accounts [30]. There is, thus, more in the notion of the structure of water than what is measurable by $g(r)$, which is dominated by the strong repulsion of molecules that are too closely packed together. Partial pair correlation functions, such as all three $g_{OO}(r)$, $g_{OH}(r)$, and $g_{HH}(r)$ provide more information on the molecular structure of water.

The structure order and the length scale of molecular packed in water and ice remain open for discussion. Traditionally, independent and instantaneous accuracy is sought for one of the strongly correlated parameters, providing fuel for endless debating. For instance, the separation between adjacent oxygen atoms ($d_{OO}$) varies from 2.70 to 3.00 Å [40–52], and the molecular size ($d_H$) changes from 0.970 to 1.001 Å [53]. The molecular CN varies from two [20] to four or even greater [54], such as 4.3 in an occasion [55]. Liquid $H_2O$ is subject to the certainty of the homogeneous phase or the inhomogeneous low- and high-density biphase structural order [19, 56, 57] or the chain like icebergs [54].

As an important yet often-overlooked quantity for water ice, the mass density $\rho$, that can easily be determined, is able to unify the uncertainties of geometric configuration and length scale of ice and water. Based on the essential rules of $sp^3$-orbit hybridization of oxygen atoms [5, 13] and the O:H–O bond cooperativity [23, 58, 59], one should be able to resolve the structure configuration and these uncertainties without needing any assumptions or approximations.

## 2.4 Quantitative Resolution

### *2.4.1 Basic Rules*

#### 2.4.1.1 The $sp^3$ Orbital Hybridization

Atoms of electronegative elements A such as carbon, nitrogen, oxygen, and fluorine and their neighbors in the Periodic Table tend to hybridize their sp orbits to form four directional bonding orbits upon reacting with arbitrary atoms of electropositive elements B such as aliki and transition metals. The directional orbits can be occupied by sharing electron pairs to form the "–" covalent or polar covalent bond or by electron lone pairs ":" to form the nonbond in the following form [5]:

$$A + 4B \rightarrow (4\text{-}n)B^{\delta+} : A^{n\delta-} - nB^{\delta+}, \ (4\text{-}n \text{ rule}).$$

The number of lone pairs follows the 4-n rule with n being the valence value of the particular A atom. The AB combination also forms the hydrogen-bond like A:B–A and polar covalent B–A–B bond or their mixture. The covalency of the B–A bond and the net charge δ vary with the electronegativity difference between the bonding constituents. The difference in electronegativity between oxygen ($\eta_O$ = 3.5) and hydrogen ($\eta_H$ = 2.2) determines the covalency of the H–O bond ($\Delta\eta$ = 1.3). $\Delta\eta = 4.0 - 2.2 = 1.8$ for the HF is an ideal ionic bond and $\Delta\eta = 0$ an ideal covalent bond.

Methane $CH_4$, Ammonia $NH_3$, water $H_2O$, and fluoride HF share the same but slightly distorted tetrahedron with 4-n = 0, 1, 2, and 3 lone pairs, respectively. Hydrogen bond also forms in compounds with C, N, O, and F mixtures, such as N: B–O, F:B–O, and N:B–F, where the electron lone pair is the key element. B is an arbitrary element of electropositive than the acceptors. Therefore, the hydrogen bond is ubiquitous to form important functional groups in the organic and even the inorganic substance.

#### 2.4.1.2 Nonbonding Lone Pairs' Capabilities

The nonbonding lone pairs are associated with the process of tetrahedral-coordinated bond formation of O, N or F atoms in reactions [5, 60]. Replacing the $O^{2-}$ with $N^{3-}$ or $F^-$ and replacing the $H^{+/p}$ with any element $B^{+/p}$, an O:H–O like bond forms. The strength of the nonbonding lone pair also varies with the local environment [61] and applied stimulus. The O:H nonbond breaks when water molecules are disassociated at the evaporation point [62]. The $sp^3$ orbital dehybridization occurs at a certain temperature. For example, the O:Cu nonbond breaks at a temperature around 700 K, as probed using very-low-energy electron diffraction (VLEED) spectromentrics [63].

The weak O:H nonbonds, to quote Hoffmann [64], "… *are ubiquitous, and their prevalence gives them a power that belies their modest nature. In water, they influence the global geology and climate of the Earth. In biomolecules, they regulate the folding, signaling and messaging of proteins, and hold together the DNA's double helix*".

Because of its relatively weak interaction, O:H nonbond cohesive energy $E_L$ contributes insignificantly to the Hamiltonian and associated attributes [65]; however, these localized electrons add electronic states in the vicinity of Fermi energy ($E_F$), which is within the energy scope of an STM/S [65]. These nonbonding electrons neither follow usually observed dispersion relations nor occupy the allowed states of the valence band and below.

In addition to the weak interactions with energies around 50 meV, as detected using Raman and electron energy loss spectroscopy (EELS) [5], these lone pairs polarize their neighboring atoms into dipoles, which explains why the $CF_4$ serves as a medium for anticoagulant in artificial blood and for superhydrophobicity of Fluorides. As illustrated in Fig. 2.5, the central $C^{4+}$ ion is surrounded by $4F^-$ ions and each $F^-$ creates

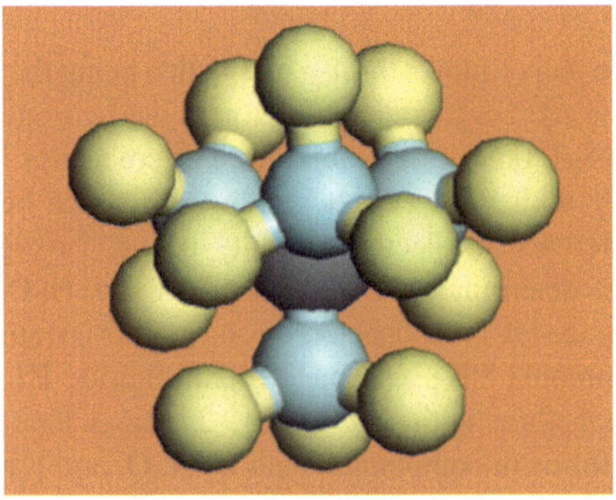

**Fig. 2.5** Hypothetic $CF_4$ [$C^{4+}$(*central black*) + $4F^-$(*yank*) + 4 × (3:, *yellow*)] with ":" being the nonbonding lone pair serves as anticoagulant in artificial blood [65] (color online)

3 pairs of nonbonding electrons. These 12 electron lone pairs form a function group that attracts and polarizes nearby atoms weakly, making it mobile in fluids. This cluster of centered electron lone pairs translates to an anticoagulating function. $NF_5$ and $SF_6$ also possess 15 and 18 lone pairs of electrons with structure being similar to $CF_4$, respectively, and hence are anticipated to display the same character as the $CF_4$.

The lone pair and dipole interactions not only act as the most important function groups in biological and organic molecules, but they also play important roles such as high-$T_C$ superconductivity and carriers in the topological insulators. An ultra-violet light irradiation or thermal excitation can dehybridize the $sp^3$-orbit, annihilating the lone pairs and dipoles and altering their functionalities. Proper aging recovers the $sp^3$ configuration of the central acceptor and the nonbonding lone pairs and antibonding dipoles appear again. Therefore, strong recoverablity of the tetrahedron structure is intrinsic to atoms of these electronegative elements.

#### 2.4.1.3 The $C_{2v}$ Symmetrical $H_2O$ Molecule

Specifically, an oxygen atom ($2s^22p^4$) catches one electron from each of its two nearest neighbors such as metal or hydrogen atoms and then hybridizes its *sp* orbits, creating four directional orbits [2]. Such a process of $sp^3$-orbital hybridization proceeds in four discrete steps, as revealed using STM and VLEED, when the oxygen reacts with atoms of any relatively less electronegative, irrespective of the structural phase of gaseous, liquid, or solid [5]. For the O–Cu(001) instance, the Cu–O–Cu angle is around 98 ± 5°, the Cu:O:Cu angle is 135 ± 5°, the Cu–O bond length varies from 1.63 to 1.75 Å and the Cu:O distance changes from 1.94 to 2.05 Å. The Cu–O bond contracts at the skin, which is associated with Cu:O expansion in the process of chemisorption under different oxygen exposures.

In the case of $H_2O$, one O forms two intramolecular H–O polar-covalent bonds with shared electron pairs of 4.0–5.1 eV binding energy [23, 66]. The $O^{2-}$ then fills up the rest two orbits with its nonbonding lone pairs to form the intermolecular O:H nonbond. The O:H nonbond energy is in the 0.1 eV order [67], which is only 2.5 % or less that of the H–O bond. The atomistic anisotropy of charge and energy surrounding the central oxygen ion permits a $H_2O$ molecule only $C_{v2}$ group symmetry. The energetic anisotropy drives the $H_2O$ molecules rotate and vibrate rapidly and ceaselessly in the network matrix. The $sp^3$-orbital hybridization happens at temperatures from 5 K [6] or even below in vacuum up to the hot gaseous phase under the ambient pressure.

#### 2.4.1.4 Water Structure: Correlation and Fluctuation

The packing order of $H_2O$ molecules follows the tetrahedral configuration even subjected to ultra-high temperature and high pressure [7]. Despite thermal fluctuation in the O:H length and in the ∠O:H–O containing angle, the average size and separation of molecules change when the $H_2O$ transits from the strongly ordered solid phase to the weakly ordered liquid and to the disordered amorphous or vapor state, as the tetrahedron configuration implicitly strongly retains.

An extension of the $C_{2V}$ quasi-tetrahedron configuration simplifies the situation substantially. This extension results in an ideal tetrahedron with higher $C_{3v}$ group symmetry and a flexible, polarizable O:H–O bond between oxygen ions, as shown in Fig. 2.2c, d. Containing two equivalent $H_2O$ molecules and four identical O:H–O bonds at different orientations, this symmetrical tetrahedron unifies the length scale and mass density in building up the bulk water and ice as a statistic expectation. In the hexagonal or cubic ice, $O^{2-}$ ions form each a tetrahedron with an O–O distance of 2.76 Å and an H–O bond 0.96 Å long in ice-VIII phase, compared with water at 4 °C, the O–O distance is 2.6950 Å and the H–O bond length is 1.0004 Å.

Relaxation of the segmented O:H–O bond and the associated electron entrapment and polarization govern the properties and performance of water and ice. Deformation of the bond angles create different geometers such as case structure [9, 23] but the O:H–O always retains.

#### 2.4.1.5 O:H–O Bond Coopcrativity

The building block in Fig. 2.2b results in two entities. One is the ideal geometry in Fig. 2.2c that water and ice prefer; the other is the O:H–O bond between oxygen ions shown in Fig. 2.2d with asymmetrical and ultra-short-range interactions, analogous to coupled springs. The $H^+$ proton being chosen as the coordinate origin donates its electron to the O shown on the right, to form the intramolecular H–O polar-covalent bond, whereas the electron lone pair ':' of the O shown on the left (blue pairing dots) polarizes the shared electron pair '–' and attracts the $H^+$ proton to form an intermolecular O:H nonbond without sharing any charge.

The key component is the Coulomb repulsion between electron pairs on adjacent oxygen ions, which has coventionally been overlooked. This O–O repulsion discriminates water and ice from other normal substance in responding to stimulus with strong adaptivity and recoverablity. As thus, the O:H and the H–O relax cooperatively and oppositely in direction and extent. Relaxation of either the H–O or the O:H segment plays a role in mediating separately the detectable properties of water and ice. Table 2.1 specifies the O:H–O bond identities compared to the C–C bond in a diamond.

#### 2.4.1.6 Proton Tunneling or Frustration Prohibition

According to Ice Rules of Bernal-Fowler-Pauling [12, 13], the minimum energy position of a $H^+$ proton in the O:H–O bond is not half-way between two adjacent oxygen ions. There are two equivalent positions that an $H^+$ proton may occupy on the O–O bond with the same probability, a far and a near position. $H^+$ proton jumps from one molecule to next through $(H_2O)_2 \rightarrow OH^{\delta-} + OH_3^{\delta+}$ transition in a frequency of $f \sim 1.5$ THz. Thus, the rule leads to the 'frustration' of the proton

**Table 2.1** Identities of the segmented O:H–O bond compared to the C–C bond in a diamond [59][a]

| | $d_x$ (Å) | $E_x$ (eV) | $\omega_x$ ($cm^{-1}$) | $\Theta_D$ (K) | $T_m$ (K) | Interaction | Properties |
|---|---|---|---|---|---|---|---|
| H–O(H) | 1.00 | 3.97–5.10 | >3000 | >3000 | ~5000 | Exchange | $T_C$, $O_{1s}$, $\omega_H$, $E_H$, etc. |
| O:H(L) | 1.70 | 0.05–0.10 | <300 | 198 | 273 | vdW-like | P, Y, $\omega_L$, $E_L$, etc. |
| O–O | – | – | – | – | – | Coulomb | ρ |
| C–C | 1.54 | 1.84 | 1331 | 2230 | 3800 | Exchange | – |

[a]The H–O cohesive energy $E_H$ determines H–O atomic dissociation, O 1s energy shift, H–O phonon frequency $\omega_H$, and the $T_C$ for phase transition, except for evaporation. O:H cohesive energy $E_L$ determines molecular dissociation, dipole moment P, elastic modulus Y, O:H phonon frequency $\omega_L$ etc

positions for the ground state configuration: for each oxygen ion, two of the neighboring protons must reside in the far positions and two of them in the near, and their transitions result in the 'two-in two-out' frustration or proton tunneling transition, see Fig. 2.3. The open tetrahedral structure of ice affords many equivalent states.

However, as it is presently understood, an $H_2O$ molecule undergoes only rotational and vibrational fluctuations without $H^+$ proton location frustrating or tunneling transition between molecules because of the stronger H–O bond and the weaker, fluctuating O:H interaction [3]. The O:H nonbond switches on and off ceaselessly with a sub-picosecond period [20, 24, 68, 69] or with a frequency of THz. However, transition from $(H_2O)_2$ to $OH^-$ : $OH_3^+$ through exchanging an electron of one water molecule with the $H^+$ proton of another frequently is unlike. Breaking an H–O bond requires at least 4.0 eV energy or 300 nm laser excitation, which is very unlikely under regular conditions, even with the aiding of catalysts. However, the basic tetrahedron and the O:H–O bond retain even in the $(H_2O)_2 \rightarrow OH^-$ : $OH_3^+$ transition that happened only under extremely high pressure and temperature 2 TPa and 2000 K [7].

An oxygen atom always tends to find four neighbors to form a stable tetrahedron; but the nonequivalent bond angles (∠H–O–H ≤ 104.5° and ∠H:O:H ≥ 109.5°) and the repulsion between the electron pairs on adjacent oxygen ions [23, 58] prevent the tetrahedron from being stay alone in the liquid phase but subject to ceaseless rotation and vibration. This anisotropy explains why water remains liquid at temperatures above the critical points of other liquids such as nitrogen. In contrast to $NH_3$ that retains liquid form only in the temperature range of 195–240 K, $H_2O$ forms liquid in 273–373 K. The strong fluctuation is analogous to the motion of a complex pendulum surrounded by four nonbonding interactions.

Nevertheless, the "two-in two-out" frustration is valid from the statistically macroscopic view in terms of free energy and entropy. One can observe statistically proton jumping back and forth because molecules rotate ceaselessly. Mechanisms of flicking fluctuation and proton frustration or tunneling transition are different though they can describe the same matter effectively at respective scale.

Its emphasized that it is more realistic and useful to consider the statistical mean of the structure order, length scale and mass density in a phase of question over a long time span at the molecular level rather than attempt to capture a snapshot of a quantity with instantaneous accuracy under a specific condition at the macroscopic scale [68].

## *2.4.2 Confirmation*

### 2.4.2.1 Molecular STM Images and Energies

Figure 2.6 shows the orbital images and the dI/dV spectra of a $H_2O$ monomer and a $(H_2O)_4$ tetramer deposited on a NaCl(001) surface probed using STM/S at 5 K [6].

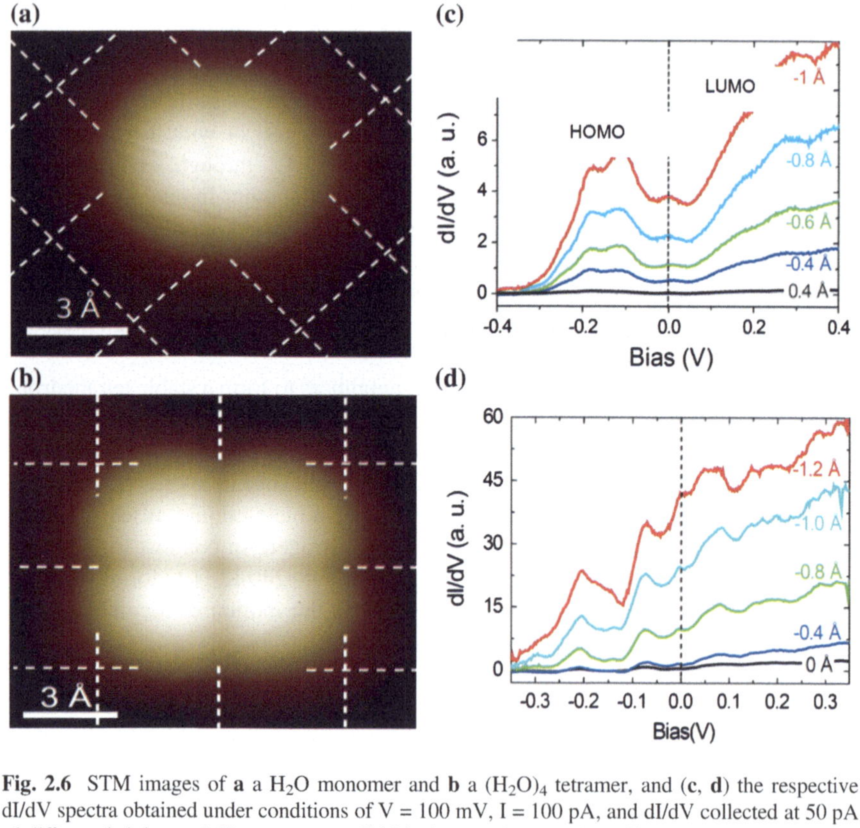


**Fig. 2.6** STM images of **a** a $H_2O$ monomer and **b** a $(H_2O)_4$ tetramer, and (**c**, **d**) the respective dI/dV spectra obtained under conditions of V = 100 mV, I = 100 pA, and dI/dV collected at 50 pA of different heights at 5 K temperature. Grid in images denotes the $Cl^-$ lattice of the NaCl(001) substrate (Reprinted with permission from [6].) LUMO ($>E_F$) and the HOMO ($<E_F$) indicated in (**b**) denote the orbital energy states

The highest occupied molecular orbital (HOMO) below the $E_F$ of the monomer appears as a double-lobe structure with a nodal plane in between, while the lowest unoccupied molecular orbital (LUMO) above $E_F$ appears as an ovate lobe developing between the two HOMO lobes. STS spectra at different depths discriminate the tetramer from the monomer in the density of states (DOS) crossing $E_F$.

These STM images [6] confirmed the occurrence of $sp^3$-orbit hybridization of oxygen in $H_2O$ monomer occurs at 5 K or lower and the intermolecular interaction involved in $(H_2O)_4$. According to the bond–band-barrier correlation notion [2, 5], the HOMO located below $E_F$ corresponds to the energy states occupied by electron lone pairs of oxygen, and the LUMO to states yet to be occupied by electrons of antibonding dipoles. The image of the monomer showing the directional lone pairs suggests that the lone pairs point into the open end of the surface. As the $H^+$ ion can only share its unpaired electron with oxygen, the $Cl^-$ ion in the NaCl substrate interacts with the $H^+$ only electrostatistically.

Figure 2.7 shows a real-space visualization of individual hydrogen nonbonding in 8-hydroxyquinoline (8-hq) molecular assemblies on a Cu(111) substrate, using noncontact atomic force microscopy, which resolves hydrogen nonbonding networks, sites, orientations, and lengths [70].

Single molecular imaging [6, 70] is indeed useful to examine the molecular attribute under specific conditions. However, solution or solid specimens comprise unaccountable number of $H_2O$ molecules in three-dimensional bonding networks, which evolves with the functioning environment or perturbation. The performance of the O:H–O bond in such a collection is greatly important in practical applications. Therefore, a set of theoretical, numerical, and detecting strategies for collecting and purifying comprehensive information on the O:H–O bond polarization and segmental cooperative relaxation in length and energy under various fields in different situations is highly challenging.

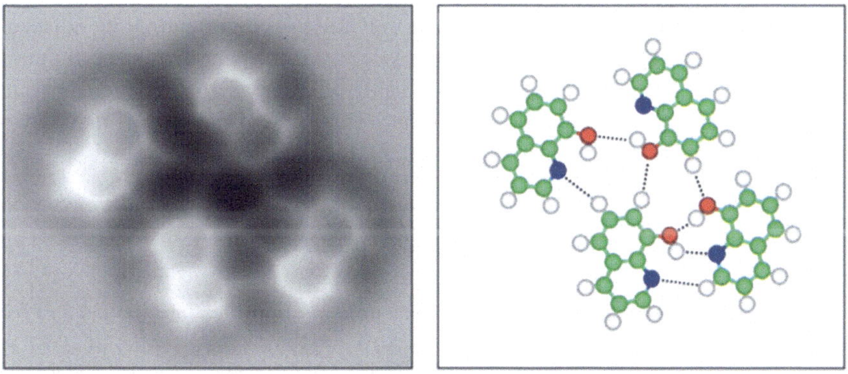

**Fig. 2.7** AFM image of hydrogen bond networking in 8-hydroxyquinoline (8-hq) molecular assemblies on a Cu(111) substrate revealed the nonbonding interaction with weak charge sharing [70]

**Table 2.2** Tetrahedron geometry and mass density ρ defined size $d_H$, separation $d_L$(or $d_{OO}$) and the lattice constant for molecules packed in water and ice (unit in Å unless indicated)

| ρ (g/cm$^3$) | $d_H$ | $d_L$ | $d_{OO}$ | a |
|---|---|---|---|---|
| 1.0 | 1.0004 | 1.6946 | 2.6950 | 6.2228 |
| 0.95 | 0.9813 | 1.7612 | 2.7425 | 6.3325 |
| 0.90 | 0.9612 | 1.8312 | 2.7924 | 6.4476 |
| 0.85 | 0.9393 | 1.9068 | 2.8461 | 6.5717 |
| 0.70 | 0.8577 | 2.1787 | 3.0364 | 7.0110 |
| 0.65 | 0.8221 | 2.2902 | 3.1123 | 7.1864 |
| 0.60 | 0.7792 | 2.4173 | 3.1965 | 7.3807 |
| 0.55 | 0.7243 | 2.5662 | 3.2905 | 7.5978 |
| 0.50 | 0.6457 | 2.7510 | 3.3967 | 7.8432 |

#### 2.4.2.2 Density–Geometry–Length Scale Correlation

The packing order in Fig. 2.2c defines that every other cube of $a^3$ volume accommodates two $H_2O$ molecules, which means each one cube has one on average. With the known mass of a $H_2O$ molecule consisting of 8 neutrons, 10 protons and 10 electrons, $M = (10 \times 1.672621 + 8 \times 1.674927 + 10 \times 9.11 \times 10^{-4}) \times 10^{-27}$ kg, and the known density $\rho = M/a^3 = 1$ (g/cm$^3$) at 4 °C and atmospheric pressure, this tetrahedrally-coordinated structure correlates unambiguously the molecular separation, $d_{OO}$ and mass density, as given (2.1).

Furthermore, formulating the pressure-resolved $d_L(P)$ and $d_H(P)$ [58], straightly yields the $d_x$ length cooperativity (unit in Å), free from probing conditions or methods. Equation (2.1) features the correlation among the $d_{OO}$, $d_L$, $d_H$ and $\rho$ [4]. Thus, given any one of these quantities for water and ice, one can determine all the rest. If the relaxation of $d_x$ matches the value of detection, the structure in Fig. 2.2c and the $d_x$ cooperativity derived herein are justified true and unique. Table 2.2 lists the interdependence of the mass density ρ, the size $d_H$ and separation $d_L$(or $d_{OO}$), and the lattice constant a for molecules packed in water and ice, on the basis of the tetrahedron bond geometry in Fig. 2.2c. Given any of the five quantities, one could be able to know the rest. For example, the unit cell size representing the adjacent two molecular layers' spacing in Fig. 2.2c is a = 6.2228 Å at 4 °C. The thinnest water film of 7.3 Å [71] corresponds to a value of $d_{OO}$ = 3.165 Å, giving rise to ρ = 0.6174 g/cm$^3$ for the hcp-structured water. The 3.7 Å step height of monolayer film corresponds to $d_{OO}$ = 3.205 Å and a 0.5946 g/cm$^3$ density.

#### 2.4.2.3 Uniqueness of Structure Solution

Figure 2.8a shows the conversion of the measured $\rho(T)$ profiles into the $d_{OO}(T)$ for water droplets of different sizes [72, 73]. Strikingly, the $d_{OO}$ values of 2.70 Å measured at 25 °C, and 2.71 Å measured at −16.8 °C [42] exactly match the derivative for the larger droplet, which testify to the truth of both (2.1) and the tetrahedral structure as representing the actually mean configuration of water and

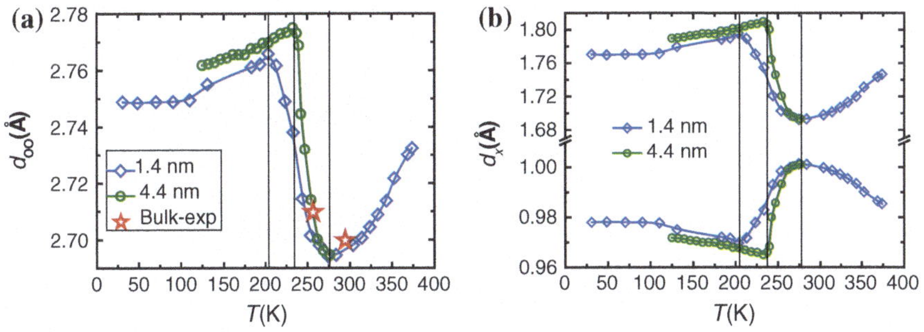


**Fig. 2.8** Conversion of **a** the $\rho(T)$ to the $d_{OO}(T)$; and from **b** $d_{OO}(T)$ to $d_x(T)$ for water droplets ($T < 273$ K) and water bulk ($T > 273$ K) [72, 73]. Matching the $d_{OO}(T)$ profile to the measured $d_{OO}$ at 25 °C and –16.8 °C in panel (**a**) [42] not only verifies the validity of the uniqueness of the tetrahedral configuration and the $d_x$ cooperativity, but also the accuracy and reliability of the data reported in [42] (Reprinted with permission from [4].)

ice. Furthermore, the data reported in [42] is essentially accurate and correct. The O–O neighboring distance of 2.80 Å [40] gives rise to $d_H = 0.9581$, $d_L = 1.8419$ Å, and $\rho = 0.8917$ g/cm$^3$ for the ambient water, according to (2.1).

One can obtain the $d_L$, $d_H$, $d_{OO}$ and $\rho$ by solving the following equation with any one of these parameters as known input [4]:

$$d_L - 2.5621 \times [1 - 0.0055 \times exp((d_{OO} - d_L)/0.2428)] = 0.$$

Matching observed $d_{OO}(T)$ [42] to that decomposed from the measured $\rho(T)$ for 4.4 and 1.4 nm sized water droplets [72, 73], shown in Fig. 2.8, confirms the following expectations and calculations [59]:

(1) Cooling shortens the O:H nonbond in the liquid ($T > 277$ K) and in the solid phase ($T < 202/242$ K), which lengthens the H–O bond slightly and raises the density at different rates.
(2) In the quasisolid transition phase, the relaxation process reverses, leading to the O–O length gain and density loss.
(3) At $T \leq 80$ K, $d_x$ remains almost constant because the specific heat $\eta_L \approx \eta_H \approx 0$ in this regime [59].
(4) Dispersion of the least-density temperature from 205 (for a 1.4 nm-sized droplet) to 242 K (4.4 nm-sized droplet) and to 258 K (for bulk water [59]) and the maximal-density temperature from the bulk value to higher takes pace as the droplet reduces its size [74].

Figure 2.9 shows the solution consistency that unifies the ($d_H$, $d_L$, $d_{OO}$, $\rho$) and the structural order pertaining to: (i) compressed ice [75], (ii) cooling water and ice [72, 73], and, (iii) water skin and molecular monomers [42, 51]. The currently

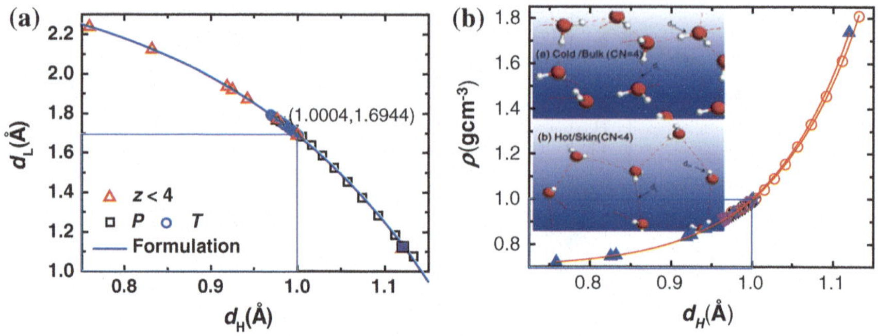


**Fig. 2.9** **a** $d_L$–$d_H$, and **b** $\rho$–$d_H$ correlation profiles for $H_2O$ molecules, which match (**a**) (i) ice under compression ($d_H > 1.00$ Å) [75]; (ii) water ice at cooling ($0.96 < d_H < 1.00$ Å) [72, 73]; and (iii) water skin and monomer ($d_H < 1.00$ Å) [40–42, 46–50]. The value $d_H = 1.0004$ Å is the standard at $\rho = 1$ g·cm$^{-3}$. For $d_H$ shorter than 0.96 Å, it matches (**b**) in the skin or clusters [23, 51, 52]. The *inset* in (**b**) illustrates a change of molecular size and separation with temperature and molecular CN (Reprinted with permission from [4].)

derived value $d_H = 1.0004$ Å at unit density lies within observed values ranging from 0.970 to 1.001 Å [53]. The $d_{OO}$ values [20, 69, 76] greater than the ideal value of 2.6950 Å at $\rho = 1$ (g·cm$^{-3}$) correspond to the skin that exists only in sites of water ice composed of molecules with fewer than four nearest neighbors [23].

## 2.5 Summary

(1) The basic rules of oxygen sp$^3$-orbital hybridization defines that water and ice prefer the fluctuating tetrahedrally-coordinated monophase with O:H–O segmental length relaxation being subject to coordination and thermal environment.

(2) The tetrahedral geometry and the known mass density unify the size and separation of molecules packed in water and ice.

(3) The skin low-density phase remains the same geometry but shorter and stiffer H–O bond and longer and softer O:H nonbond because of molecular under-coordination effect.

(4) In place of the classical proton tunneling transition and "two-in two-out" frustration mechanism in statistic thermodynamics, water molecules are subject to strong correlation and fluctuation in molecular rotation and vibration.

(5) It is much more realistic and revealing to consider the statistical mean of the structure order, length scale and mass density in a phase of question over a long time span rather than attempt to capture a snapshot of a quantity with instantaneous accuracy under a specific condition.

# References


1. Editorial, *So Much More to Know*. Science **309**(5731), 78–102 (2005)
2. C.Q. Sun, *Relaxation of the Chemical Bond*. Springer Series in Chemical Physics 108, vol. 108 (Springer, Heidelberg, 2014), 807 pp
3. Y. Huang, X. Zhang, Z. Ma, Y. Zhou, W. Zheng, J. Zhou, C.Q. Sun, Hydrogen-bond relaxation dynamics: resolving mysteries of water ice. Coord. Chem. Rev. **285**, 109–165 (2015)
4. Y. Huang, X. Zhang, Z. Ma, Y. Zhou, J. Zhou, W. Zheng, C.Q. Sun, Size, separation, structure order, and mass density of molecules packing in water and ice. Sci. Rep. **3**, 3005 (2013)
5. C.Q. Sun, Oxidation electronics: bond-band-barrier correlation and its applications. Prog. Mater. Sci. **48**(6), 521–685 (2003)
6. J. Guo, X. Meng, J. Chen, J. Peng, J. Sheng, X.Z. Li, L. Xu, J.R. Shi, E. Wang, Y. Jiang, support: real-space imaging of interfacial water with submolecular resolution-Supp. Nat. Mater. **13**, 184–189 (2014)
7. Y. Wang, H. Liu, J. Lv, L. Zhu, H. Wang, Y. Ma, High pressure partially ionic phase of water ice. Nat. Commun. **2**, 563 (2011)
8. X. Zhang, P. Sun, Y. Huang, T. Yan, Z. Ma, X. Liu, B. Zou, J. Zhou, W. Zheng, C.Q. Sun, Water's phase diagram: from the notion of thermodynamics to hydrogen-bond cooperativity. Prog. Solid State Chem. **43**, 71–81 (2015)
9. A. Falenty, T.C. Hansen, W.F. Kuhs, Formation and properties of ice XVI obtained by emptying a type sII clathrate hydrate. Nature **516**(7530), 231–233 (2014)
10. C.H. Langford, R.A. Beebe, *The Development of Chemical Principles* (Courier Corporation, 1995)
11. P. Ball, The hidden structure of liquids. Nat. Mater. **13**(8), 758–759 (2014)
12. J.D. Bernal, R.H. Fowler, A theory of water and ionic solution, with particular reference to hydrogen and hydroxyl Ions. J. Chem. Phys. **1**(8), 515–548 (1933)
13. L. Pauling, The structure and entropy of ice and of other crystals with some randomness of atomic arrangement. J. Am. Chem. Soc. **57**, 2680–2684 (1935)
14. P. Ball, Water: water—an enduring mystery. Nature **452**(7185), 291–292 (2008)
15. H.S. Frank, W.-Y. Wen, Ion-solvent interaction. Structural aspects of ion-solvent interaction in aqueous solutions: a suggested picture of water structure. Discuss. Faraday Soc. **24**, 133–140 (1957)
16. E.D. Giudice, A. Tedeschi, Water and autocatalysis in living matter. Electromagn. Biol. Med. **28**(1), 46–52 (2009)
17. M. Chaplin. *Water Structure and Science:* http://www.lsbu.ac.uk/water/
18. J. Russo, H. Tanaka, Understanding water's anomalies with locally favoured structures. Nat. Commun. **5**, 3556 (2014)
19. J.A. Sellberg, C. Huang, T.A. McQueen, N.D. Loh, H. Laksmono, D. Schlesinger, R.G. Sierra, D. Nordlund, C.Y. Hampton, D. Starodub, D.P. DePonte, M. Beye, C. Chen, A.V. Martin, A. Barty, K.T. Wikfeldt, T.M. Weiss, C. Caronna, J. Feldkamp, L.B. Skinner, M.M. Seibert, M. Messerschmidt, G.J. Williams, S. Boutet, L.G. Pettersson, M.J. Bogan, A. Nilsson, Ultrafast X-ray probing of water structure below the homogeneous ice nucleation temperature. Nature **510**(7505), 381–384 (2014)
20. P. Wernet, D. Nordlund, U. Bergmann, M. Cavalleri, M. Odelius, H. Ogasawara, L.A. Naslund, T.K. Hirsch, L. Ojamae, P. Glatzel, L.G.M. Pettersson, A. Nilsson, The structure of the first coordination shell in liquid water. Science **304**(5673), 995–999 (2004)
21. O. Mishima, H.E. Stanley, The relationship between liquid, supercooled and glassy water. Nature **396**(6709), 329–335 (1998)
22. R. Roy, W.A. Tiller, I. Bell, M.R. Hoover, The structure of liquid water; novel insights from materials research; potential relevance to homeopathy. Mater. Res. Innovations **9**(4), 98–102 (2005)

23. C.Q. Sun, X. Zhang, J. Zhou, Y. Huang, Y. Zhou, W. Zheng, Density, elasticity, and stability anomalies of water molecules with fewer than four neighbors. J. Phys. Chem. Lett. **4**, 2565–2570 (2013)
24. V. Petkov, Y. Ren, M. Suchomel, Molecular arrangement in water: random but not quite. J. Phys.: Condens. Matter **24**(15), 155102 (2012)
25. T.D. Kuhne, R.Z. Khaliullin, Nature of the asymmetry in the hydrogen-bond networks of hexagonal ice and liquid water. J. Am. Chem. Soc. **136**(9), 3395–3399 (2014)
26. J.D. Smith, C.D. Cappa, K.R. Wilson, B.M. Messer, R.C. Cohen, R.J. Saykally, Energetics of hydrogen bond network rearrangements in liquid water. Science **306**(5697), 851–853 (2004)
27. T. Head-Gordon, M.E. Johnson, Tetrahedral structure or chains for liquid water. PNAS **103** (21), 7973–7977 (2006)
28. G.H. Pollack, *The Fourth Phase of Water: Beyond Solid, Liquid, and Vapor* (Ebner & Sons, Seattle, 2013)
29. E.C. Fuchs, A.D. Wexler, A.H. Paulitsch-Fuchs, L.L.F. Agostinho, D. Yntema, J. Woisetschlager, The Armstrong experiment revisited. Eur. Phys. J. Spec. Top. **223**(5), 959–977 (2014)
30. Y. Marcus, Effect of ions on the structure of water: structure making and breaking. Chem. Rev. **109**(3), 1346–1370 (2009)
31. C. Vega, J.L.F. Abascal, M.M. Conde, J.L. Aragones, What ice can teach us about water interactions: a critical comparison of the performance of different water models. Faraday Discuss. **141**, 251–276 (2009)
32. V. Molinero, E.B. Moore, Water modeled as an intermediate element between carbon and silicon. J. Phys. Chem. B **113**(13), 4008–4016 (2009)
33. P.T. Kiss, A. Baranyai, Density maximum and polarizable models of water. J. Chem. Phys. **137**(8), 084506–084508 (2012)
34. J. Alejandre, G.A. Chapela, H. Saint-Martin, N. Mendoza, A non-polarizable model of water that yields the dielectric constant and the density anomalies of the liquid: TIP4Q. Phys. Chem. Chem. Phys. **13**, 19728–19740 (2011)
35. C. Vega, J.L.F. Abascal, Simulating water with rigid non-polarizable models: a general perspective. Phys. Chem. Chem. Phys. **13**(44), 19663–19688 (2011)
36. W.K. Röntgen, Ueber die constitution des flüssigen wassers Ann. Phys. U. Chim. (Wied) **45**, 91–97 (1892)
37. C.H. Cho, J. Urquidi, S. Singh, S.C. Park, G.W. Robinson, Pressure effect on the density of water. J. Phys. Chem. A **106**(33), 7557–7561 (2002)
38. M. Vedamuthu, S. Singh, G.W. Robinson, Properties of liquid water: origin of the density anomalies. J. Phys. Chem. **98**(9), 2222–2230 (1994)
39. J.D. Smith, C.D. Cappa, K.R. Wilson, R.C. Cohen, P.L. Geissler, R.J. Saykally, Unified description of temperature-dependent hydrogen-bond rearrangements in liquid water. PNAS **102**(40), 14171–14174 (2005)
40. L.B. Skinner, C. Huang, D. Schlesinger, L.G. Pettersson, A. Nilsson, C.J. Benmore, Benchmark oxygen-oxygen pair-distribution function of ambient water from x-ray diffraction measurements with a wide Q-range. J. Chem. Phys. **138**(7), 074506 (2013)
41. K.T. Wikfeldt, M. Leetmaa, A. Mace, A. Nilsson, L.G.M. Pettersson, Oxygen-oxygen correlations in liquid water: addressing the discrepancy between diffraction and extended x-ray absorption fine-structure using a novel multiple-data set fitting technique. J. Chem. Phys. **132** (10), 104513 (2010)
42. U. Bergmann, A. Di Cicco, P. Wernet, E. Principi, P. Glatzel, A. Nilsson, Nearest-neighbor oxygen distances in liquid water and ice observed by x-ray Raman based extended x-ray absorption fine structure. J. Chem. Phys. **127**(17), 174504 (2007)
43. J. Morgan, B.E. Warren, X-ray analysis of the structure of water. J. Chem. Phys. **6**(11), 666–673 (1938)
44. L.A. Naslund, D.C. Edwards, P. Wernet, U. Bergmann, H. Ogasawara, L.G.M. Pettersson, S. Myneni, A. Nilsson, X-ray absorption spectroscopy study of the hydrogen bond network in the bulk water of aqueous solutions. J. Phys. Chem. A **109**(27), 5995–6002 (2005)

45. L. Orgel, The hydrogen bond. Rev. Mod. Phys. **31**(1), 100–102 (1959)
46. K.R. Wilson, B.S. Rude, T. Catalano, R.D. Schaller, J.G. Tobin, D.T. Co, R.J. Saykally, X-ray spectroscopy of liquid water microjets. J. Phys. Chem. B **105**(17), 3346–3349 (2001)
47. A. Narten, W. Thiessen, L. Blum, Atom pair distribution functions of liquid water at 25 C from neutron diffraction. Science **217**(4564), 1033–1034 (1982)
48. L. Fu, A. Bienenstock, S. Brennan, X-ray study of the structure of liquid water. J. Chem. Phys. **131**(23), 234702 (2009)
49. J.-L. Kuo, M.L. Klein, W.F. Kuhs, The effect of proton disorder on the structure of ice-Ih: a theoretical study. J. Chem. Phys. **123**(13), 134505 (2005)
50. A. Soper, Joint structure refinement of x-ray and neutron diffraction data on disordered materials: application to liquid water. J. Phys.: Condens. Matter **19**(33), 335206 (2007)
51. K.R. Wilson, R.D. Schaller, D.T. Co, R.J. Saykally, B.S. Rude, T. Catalano, J.D. Bozek, Surface relaxation in liquid water and methanol studied by x-ray absorption spectroscopy. J. Chem. Phys. **117**(16), 7738–7744 (2002)
52. K. Liu, J.D. Cruzan, R.J. Saykally, Water clusters. Science **271**(5251), 929–933 (1996)
53. M. Hakala, K. Nygård, S. Manninen, L.G.M. Pettersson, K. Hämäläinen, Intra- and intermolecular effects in the Compton profile of water. Phys. Rev. B **73**(3), 035432 (2006)
54. A.K. Soper, Recent water myths. Pure Appl. Chem. **82**(10), 1855–1867 (2010)
55. L.B. Skinner, C.J. Benmore, J.C. Neuefeind, J.B. Parise, The structure of water around the compressibility minimum. J. Chem. Phys. **141**(21) (2014)
56. A.K. Soper, Supercooled water continuous trends. Nat. Mater. **13**(7), 671–673 (2014)
57. D. Paschek, R. Ludwig, Advancing into water's "no man's land": two liquid states? Angew. Chem. Int. Ed. Engl. **53**(44), 11699–11701 (2014)
58. C.Q. Sun, X. Zhang, W.T. Zheng, Hidden force opposing ice compression. Chem. Sci. **3**, 1455–1460 (2012)
59. C.Q. Sun, X. Zhang, X. Fu, W. Zheng, J.-L. Kuo, Y. Zhou, Z. Shen, J. Zhou, Density and phonon-stiffness anomalies of water and ice in the full temperature range. J. Phys. Chem. Lett. **4**, 3238–3244 (2013)
60. W.T. Zheng, C.Q. Sun, Electronic process of nitriding: mechanism and applications. Prog. Solid State Chem. **34**(1), 1–20 (2006)
61. M. Hus, T. Urbic, Strength of hydrogen bonds of water depends on local environment. J Chem Phys **136**(14), 144305 (2012)
62. P.C. Cross, J. Burnham, P.A. Leighton, The Raman spectrum and the structure of water. J. Am. Chem. Soc. **59**, 1134–1147 (1937)
63. C.Q. Sun, Time-resolved VLEED from the O-Cu(001): atomic processes of oxidation. Vacuum **48**(6), 525–530 (1997)
64. R. Hoffmann, Little interactions mean a lot. Am. Sci. **102**(2), 94 (2014)
65. C.Q. Sun, Dominance of broken bonds and nonbonding electrons at the nanoscale. Nanoscale **2**(10), 1930–1961 (2010)
66. S.A. Harich, D.W.H. Hwang, X. Yang, J.J. Lin, X. Yang, R.N. Dixon, Photodissociation of H2O at 121.6 nm: a state-to-state dynamical picture. J. Chem. Phys. **113**(22), 10073–10090 (2000)
67. M. Zhao, W.T. Zheng, J.C. Li, Z. Wen, M.X. Gu, C.Q. Sun, Atomistic origin, temperature dependence, and responsibilities of surface energetics: an extended broken-bond rule. Phys. Rev. B **75**(8), 085427 (2007)
68. T.D. Kuhne, R.Z. Khaliullin, Electronic signature of the instantaneous asymmetry in the first coordination shell of liquid water. Nat. Commun. **4**, 1450 (2013)
69. A. Nilsson, C. Huang, L.G.M. Pettersson, Fluctuations in ambient water. J. Mol. Liq. **176**, 2–16 (2012)
70. J. Zhang, P. Chen, B. Yuan, W. Ji, Z. Cheng, X. Qiu, Real-space identification of intermolecular bonding with atomic force microscopy. Science **342**(6158), 611–614 (2013)
71. G. Zhao, Q. Tan, L. Xiang, D. Cai, H. Zeng, H. Yi, Z. Ni, Y. Chen, Structure and properties of water film adsorbed on mica surfaces. J. Chem. Phys. **143**(10), 104705 (2015)

72. F. Mallamace, C. Branca, M. Broccio, C. Corsaro, C.Y. Mou, S.H. Chen, The anomalous behavior of the density of water in the range 30 K < T < 373 K. PNAS **104**(47), 18387–18391 (2007)
73. M. Erko, D. Wallacher, A. Hoell, T. Hauss, I. Zizak, O. Paris, Density minimum of confined water at low temperatures: a combined study by small-angle scattering of X-rays and neutrons. PCCP **14**(11), 3852–3858 (2012)
74. X. Zhang, P. Sun, Y. Huang, Z. Ma, X. Liu, J. Zhou, W. Zheng, C.Q. Sun, Water nanodroplet thermodynamics: quasi-solid phase-boundary dispersivity. J. Phys. Chem. B **119**(16), 5265–5269 (2015)
75. Y. Yoshimura, S.T. Stewart, M. Somayazulu, H. Mao, R.J. Hemley, High-pressure x-ray diffraction and Raman spectroscopy of ice VIII. J. Chem. Phys. **124**(2), 024502 (2006)
76. C. Huang, K.T. Wikfeldt, T. Tokushima, D. Nordlund, Y. Harada, U. Bergmann, M. Niebuhr, T.M. Weiss, Y. Horikawa, M. Leetmaa, M.P. Ljungberg, O. Takahashi, A. Lenz, L. Ojamäe, A.P. Lyubartsev, S. Shin, L.G.M. Pettersson, A. Nilsson, The inhomogeneous structure of water at ambient conditions. Proc. Natl. Acad. Sci. **106**(36), 15214–15218 (2009)

# Chapter 3
# O:H–O Bond Cooperativity

- *The segmented O:H–O bond approximates an asymmetrical, H-bridged oscillator pair coupled with O–O repulsion, which extends to X:B–A short-range interaction in general.*
- *O:H–O segmental disparity and O–O repulsivity discriminate water and ice from other "normal" substance in responding to perturbation.*
- *When stimulated, $O^{2-}$ anions relax in the same direction but by different amounts along the O:H–O bond.*
- *O:H–O bond cooperative relaxation dictates the adaptivity, recoverability, and memory-ability of water and ice.*

**Abstract** As the basic structure element, hydrogen bond (O:H–O) is universal to all phases of water and ice irrespective geometric configuration or fluctuation order. The O:H–O bond integrates the asymmetric, coupled, short-range intermolecular and intramolecular interactions, whose segmental length and energy respond to stimulations sensitively in a "mater-slave" manner. If one segment shortens and turns to be stiffer, the other will expand and become softer. The O:H nonbond always relaxes more than the H–O bond in length. Such a manner of segmental cooperative relaxation and the associated polarization and bond angle relaxation discriminates ice and water from other substance in responding to stimuli of chemical, electrical, mechanical, thermal, and undercoordination effect, which reconcile almost all detectable properties of water and ice.

## 3.1 Challenge: How Does a Hydrogen Bond Work?

One often refers the hydrogen bond to either the intramolecular H–O polar-covalent bond or to the intermolecular O:H nonbond alone, as epitomized in Fig. 3.1a, with specification of the donor and accepter, which overlook the Coulomb repulsion between electron pairs on adjacent $O^{2-}$ anions. Questions remain on the following:

(1) Does a hydrogen bond bridge adjacent water molecules or pair neighboring $O^{2-}$ anions?
(2) What is the consequence of molecular separation ($d_{OO}$) change on the molecular size ($d_{H\text{–}O}$)?
(3) How are the intermolecular and intramolecular interactions correlated to anomalies of water and ice?
(4) How does an external stimulus mediate the hydrogen bond length and energy and properties of water and ice?


C.Q. Sun and Y. Sun, *The Attribute of Water*, Springer Series in Chemical Physics 113, DOI 10.1007/978-981-10-0180-2_3

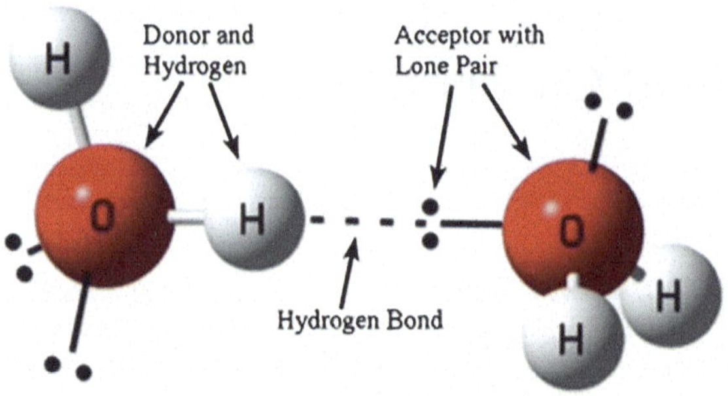


**Fig. 3.1** Confused definition of a hydrogen bond (O:H) between water molecules. A hydrogen atom bonded to the donor (generally N, O, or F) connects to a *lone electron pair* of an "acceptor" (again N, O, or F). The donor's hydrogen receives a partial positive charge, which is then attracted to the negatively charged lone pair on the acceptor (Free Wikipedia)

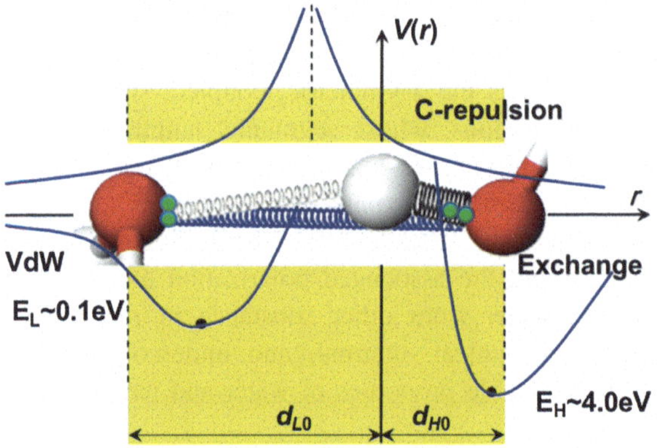


**Fig. 3.2** Segmentation and short-range interactions in an O:H–O bond. A hydrogen atom donates partially of its electron to the oxygen in the right and interacts with the electron lone pair in the left to form the H–O bond and O:H nonbond, respectively [1]. The segmented O:H–O bond forms an asymmetrical, H-bridged, coupled oscillator pair with the O:H vdW-like nonbond interaction (*left-hand side*, superposition of the Coulomb attract between the lone pair and the $H^+$ proton and the van der walls interaction between dipoles) of 0.1 eV order, the H–O bond exchange (*right-hand side*) interaction of 4.0 eV scale, and the O–O Coulomb repulsion. One switches off a particular potential and the other on at an atomic site, when it moves across the boundary. No dispersion of a particular potential is allowed cross the boundary. The $O^{2-}$ anions always dislocate in the same direction by different amounts along the O:H–O bond when stimulated. The $d_{L0}$ and $d_{H0}$ are the references under standard conditions and springs represent respective short-range interactions (Reprinted with permission from [2].)

## 3.2 Clarification: O:H–O Bond Oscillator Pair

Figure 3.2 illustrates the segmented O:H–O bond and its asymmetric and short-range potentials with the following attributes:

(1) O:H–O bond consists of both the intramolecular H–O polar-covalent bond and the intermolecular O:H nonbond rather than either of them alone.
(2) Besides the asymmetrical and short-range interactions of intra- and intermolecular bonding, Coulomb coupling between electron pairs on adjacent oxygen is vital to the adaptivity, cooperativity, flexibility, recoverability, and memory of the O:H–O bond.
(3) The specific heats of the O:H and H–O bond are different. A superposition of the specific heat curves yields two critical temperatures of extreme densities, which divides water and ice into the liquid/quasisolid/solid phases.
(4) The quasisolid phase boundaries disperse when the frequencies of the O:H and H–O phonon shift, which results in "supercooling" in freezing and "superheating" in melting, for nanodroplets instance.

## 3.3 History Background

### *3.3.1 O:H Nonbond or Hydrogen Bond?*

We have covalent bonds, ionic bonds, metallic bonds, vdW bonds, and hydrogen bonds. These bonds bring together atoms, molecules, or ions in their respective chemically acceptable and meaningful manner. A hydrogen bond is an interaction wherein a hydrogen atom is attracted to two atoms unevenly, rather than just one, and acts like a bridge between them [3, 4].

In the classical view, the hydrogen bond is often referred to O:H interaction that is highly electrostatic and sometimes even partly covalent. Gradually, the concept of a hydrogen bond has become more relaxed to include weaker and more dispersive interactions, provided some electrostatic characteristics. A great variety of very strong, strong, moderately strong, weak, and very weak hydrogen bonds are mentioned conventionally. Weak hydrogen bonds are now invoked in several matters in structural chemistry and biology. While strong hydrogen bonds are easily covered by all existing definitions of the phenomenon, the weaker ones may pose a challenge with regard to nomenclatures and definitions.

The O:H nonbond is a unique phenomenon in structural chemistry and biology [3]. Its fundamental importance lies in its role in molecular association. Its functional importance stems from both thermodynamic and kinetic reasons. In supramolecular chemistry, the hydrogen bond is able to control and direct the structures of molecular assemblies because it is sufficiently strong and sufficiently directional. This control is both reliable and reproducible and extends to the most delicate of architectures. In mechanistic biology, it is of vital importance because it lies in an energy range intermediate between vdW interactions and covalent bonds. This energy range is one that permits hydrogen bonds (meant O:H nonbonds) to both associate and dissociate quickly at ambient temperatures. This twin ability renders the interaction well suited to achieving specificity of recognition within short time spans, a necessary condition for biological reactions that must take place at room temperature or around. Hydrogen bond (lone pair) forms a key functional group in the ion channel signalling regulating; messaging, and DNA folding and unfolding.

Desiraju and Thomas have discussed the history of the hydrogen bond in *The Weak Hydrogen Bond* published in 1999 [3]. The term of *hydrogen bonds* was firstly proposed in 1902 by Werner [5] and 1910 by Hantzsch [6] who employed the concept *Nebenvalenz* (secondary valence) to describe the binding situation in ammonia salts. Pfeiffer et al. [7] in 1913 formulated the reduced reactivity of compounds with C=O and OH groups placed adjacently, with amines and hydroxides. This report could be the first hydrogen bond in organic chemistry. Moore and Winmill [8] employed the term *weak union* to describe the weaker basic properties of trimethylammonium hydroxide relative to tetramethylammonium hydroxide. Latimer and Rodebush [9] suggested that

> a free pair of electrons on one water molecule might be able to exert sufficient force on a hydrogen held by a pair of electrons on another water molecule to bind the two molecules together' and that 'the hydrogen nucleus held between two octets constitutes a weak bond.

Alfred Werner (12 December 1866–15 November 1919) was a Swiss chemist who won the Nobel Prize in Chemistry in 1913 for proposing the octahedral configuration of transition metal complexes. Werner developed the basis for modern coordination chemistry

Arthur Rudolf Hantzsch (7 March 1857–14 March 1935), was a German chemist. The Hantzsch Pyrrole Synthesis, named for him, is the reaction of β-ketoesters with ammonia and α-haloketones to give substituted Pyrroles that are found in a variety of natural products with biological activity, so the synthesis of substituted pyrroles has important applications in medicinal chemistry

### 3.3.2 Pauling's Notion

It is Pauling [10] who firstly used the term 'hydrogen bond' to account for the residual entropy of ice in 1935. Other work on diketopiperazine by Corey [11] in 1938 and on glycine by Albrecht and Corey [12] in 1939 mentioned 'hydrogen bonds' while the paper by Senti and Harker (1940) [13] on acetamide speaks of 'N–H–O bridges'. Pauling [14] drew the subject of hydrogen bonding into the chemical mainstream in 1939. Pauling was clear and unambiguous in the use of the word *bond* when he stated that

*'under certain conditions an atom of hydrogen is attracted by rather strong forces to two atoms, instead of only one, so that it may be considered to be acting as a bond between them'*, with noting, at that point of time, neither the asymmetrical interactions between hydrogen and the bridged oxygen anions nor the Coulomb repulsion between electron pairs on O anions were considered.

In a configuration such as X–H···A, it is the H atom that is considered to be the seat of bonding and not the H···A entity. Given such an interpretation, the use of the word *bridge* is hardly objectionable, and the sometimes heated discussions as to whether or not an interaction of a particular geometry is a hydrogen *bond* are perhaps unnecessary. After all, if the H atom is accepted as a bridging or bonding agent between the elements X and A, then this should suffice for an operational definition of the hydrogen bond, according to Desiraju and Thomas.

The second core idea of Pauling is that the hydrogen bond is an *electrostatic interaction*. He stated thus:

> …the hydrogen atom, with only one stable 1 s orbital, can form only one covalent bond, that the hydrogen bond is largely ionic in character; and that it is formed only between the most electronegative atoms.

The electrostatic nature of the hydrogen bond and indeed the unique ability of the H atom to form these associations arise from the fact that the solitary electron on the H atom is on time-average situated between H and X, and that with increasing electronegativity of X, the H atom is increasingly deshielded in the forward direction.

Pauling assumed that only if X and A are very electronegative, would the deshielding of H and in turn the electrostatic attraction between H and A be sufficiently high to term the interaction of a hydrogen bond. In practical terms, this means that the hydrogen bond phenomenon would be restricted to interactions X–H···A, where X and A can be any of the electronegative elements [3] such as N, O, F, Cl, Br, and I that undergo sp-orbital hybridization with creation of electron lone pairs when they react with electropositive elements. The number of lone pairs follows the 4-n rule where n is the valance value of the element A [15].

Both these ideas were developed and refined further culminating in the definition of a hydrogen bond by Pimentel and McClellan in 1960 [16]. This is the first of the modern definitions of the phenomenon:

> A hydrogen bond is said to exist when (i) there is evidence of a bond, and (ii) there is evidence that this bond sterically involves a hydrogen atom already bonded to another atom.

This definition makes no assumptions about the nature of X and A and that it enables an evaluation of the hydrogen bonding potential of groups like C–H, P–H, and As–H among others, and of π-acceptors. Because its single electron is involved in the covalent bond X–H, the H atom is always deshielded in the forward direction. This deshielding occurs irrespective of the nature of the X atom. Does this mean then that an X–H group is always a potential hydrogen bond donor, even if there is no accumulation of electron density on the X–atom?

Jeffrey and Saenger [17] pose the question in 1991:

> Should the C–H···O = C interaction be referred to as a hydrogen bond, even though there is every reason to suspect that the carbon atom is not electronegative and may even carry a positive charge?

By Pauling's definition, the answer is no. By Pimentel and McClellan's definition, the answer is yes. Refinement of the latter definition led to a quantification by Steiner and Saenger [18] who consider a hydrogen bond as

> any cohesive interaction X–H···A where H carries a positive and A carries a negative (partial or full) charge and the charge on X is more negative than on H.

A positive charge on the atom X is not precluded. This definition is incomplete in that it highlights only the electrostatic character of hydrogen bonds and is restrictive with respect to border-line cases [3].

The two central ideas of Pauling on hydrogen bonds are that they are related through the concept of strength. Bonding would seem to imply strength and unless the electrostatic nature of the association X–H···A was pronounced, it would not seem to be particularly strong. Desiraju and Steiner [3] showed that while the most familiar properties of hydrogen bonds depend on their electrostatic character, it is not necessary for a hydrogen bond to be strong to retain many of these characteristics. A hydrogen bond, in keeping with Pimenel and McClellan, is defined then on phenomenological rather than energetic grounds [3].

### *3.3.3 IUPAC Definition*

The International Union of Pure and Applied Chemistry (IUPAC) defined in 1997 the hydrogen bond as [19]

> … a form of association between an electronegative atom and a hydrogen atom attached to a second, relatively electronegative atom. It is best considered as an electrostatic interaction, heightened by the small size of hydrogen, which permits proximity of the interacting dipoles or charges. Both electronegative atoms are usually (but not necessarily) from the first row of the Periodic Table, i.e., N, O, or F. Hydrogen bonds may be intermolecular or intramolecular. With a few exceptions, usually involving fluorine, the associated energies are less than 20–25 $kJmol^{-1}$ ….

A recommendation was made to the IUPAC in 2005 to update the definition of the hydrogen bond [4]. According to new definition [4], a typical hydrogen bond is depicted as

> X–H···Y–Z, where the three dots denote the bond. X–H represents the hydrogen-bond donor. The acceptor may be an atom or an anion Y, or a fragment or a molecule Y–Z, where Y is bonded to Z. In specific cases, X and Y can be the same with both X–H and Y–H bonds being equal. In any event, the acceptor is an electron-rich region such as, but not limited to, a lone pair in Y or a π-bonded pair in Y–Z.

A hydrogen bond is thought a complex interaction with electrostatic, dispersive, covalency, and polarization components. It is misleading to think that only the H···Y part constitutes the hydrogen bond. A hydrogen bond is not a very weak covalent bond. It is also not a very strong van der Waals interaction. It is not even a particularly strong type of directional dipole–dipole interaction, according to Desiraju [4].

## 3.4 Quantitative Resolution

### *3.4.1 Hydrogen Bond Generality*

#### 3.4.1.1 Extended (X:B–Y) Hydrogen Bond

In place of over complicated situations consisting versatile long- and short-range, dispersion, classical and quantum interactions, this volume takes the O:H–O bond only with the short-range inter- and intramolecular interactions and the interoxygen repulsion. The presence of the electron lone pairs ":" associated with N, O, F is the key ingradient. This O:H–O bond with asymmetrical, short-range, and coupling interactions are shown indeed most revealing.

The classical hydrogen bond ($O^{-}$:$H^{+/p}$–$O^{-}$) for water and ice, known for over 100 years, plays an essential role in the structure and function of biological molecules. The '–' represents the bonding pair shared by O and H and ':' is the unshared nonbonding lone pair of O. Hydrogen bonds are responsible for the strength and elasticity of materials such as wood or a spider's web, molecular binding, as well as base pairing and folding in DNA. Hydrogen bonds are also responsible for the synthesis and transferring of protein signaling, messaging, and regulating [20, 21].

The formation of the hydrogen bond is not due to the presence of hydrogen or oxygen atoms but a consequence of the nonbonding lone pairs [22] or π-bonding electrons. If the lone-pair-induced $B^{\delta+/p}$ bonds further to an electronegative element *A*, then an H-bond like ($X^{-}$:$B^{+/p}$–$A^{-}$) forms. H-bond like differs from the classical hydrogen bond simply in that, the $B^{+/p}$ replaces the $H^{+/p}$ and the $A^{-}$ or the $X^{-}$ replaces the $O^{-}$. If an atom of another electronegative element, such as C, replaces one of the oxygen ions then the ($O^{2-}$:$B^{+/p}$–$C^{4-}$) bond forms, which was ever specified as the anti-hydrogen bond [23]. This is also an H-bond like.

Formation of such an H-bond like depends merely on the existence of the lone pair rather than the particular O or H atoms involvement. Hence, the H-bond like is more generally applicable though it is often referred to no as such. The same is true for the hydrocarbon bond like. The hydrocarbon bond is polar covalent in nature. The naked $H^+$ also polarizes and attracts electrons of its neighboring atoms. Hydrocarbon bond like can form by replacing the $H^+$ with a less electronegative $B^+$ ion than carbon. The O:H–O bond in water and ice is only one among many types.

The production of nonbonding lone pairs, anti-bonding dipoles, H-bond likes and the hydrocarbon–bond like are ubiquitous but they are often overlooked in reality. However, these interaction events indeed play crucial roles in determining the physical properties of a system that involves electronegative additives. Quite often, a system contains several kinds of chemical bonds, such as in graphite and in an oxide. Because of the $sp^2$-orbital hybridization of carbon, the vdW bond is dominant between the [0001] layers through $\pi$–$\pi$ interaction while the stronger covalent bond is dominant in the (0001) plane of the graphite. The intralayer covalent bond is even shorter (1.42 Å) and stronger than that (1.54 Λ) in a diamond. $O^-$:$B^{+/p}$–$O^-$ bond formation involves sharing pairs of bonding electrons, non-bonding lone pairs, and anti-bonding dipoles.

From an energy point of view, bond formation lowers the system energy and stabilizes the system. Anti-bond dipole formation requires additional energy. Although it is energetically less favorable, the anti-bond can still form by polarization as a by-product of bonding and non-bonding. Occupation of the orbits by non-bonding electron lone pairs of an electronegative element, in principle, neither raises nor lowers the system energy with respect to the initially specific energy level of the isolated atoms of the electronegative element [24–26].

From the band structure point of view, the anti-bond derived density-of-state (DOS) locate at energy above $E_F$ or near to it due to the energy rise of the polarized electrons. Figure 3.3 illustrates the residual energy states upon hydrogen bond formation in conductors and semiconductors. The DOS features for bonding are located below the originally occupied levels of the electronegative element while the DOS features of non-bonding lone pairs are located between that of the bond and that of the anti-bond. Hydrogen-bond like forms between a dipole and an electronegative atom by transferring the dipole electrons from the high-energy anti-bonding states to the lower bonding states of the atom, which stabilizes the system. Bond and anti-bond formation will produce holes below the $E_F$ of the host material [27], which is responsible for the transition from metal to semiconductor when a compound forms.

#### 3.4.1.2 Electron Localization and Polarization

Figure 3.4 shows the DFT-derived trajectory of the strongly localized electrons (in red) of an ice $I_h$ unit cell with gridded reference. As expected, both bonding and

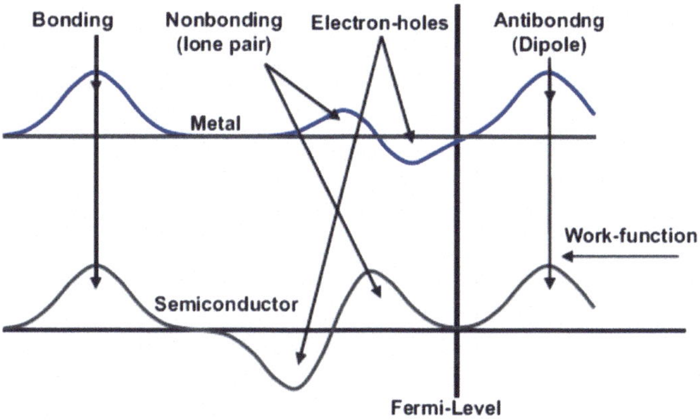


**Fig. 3.3** $NB_3$ and $B_2O$ and BF tetrahedron coordination induced valence DOS for metals and semiconductors with four excessive DOS features: bonding ($\ll E_F$), lone pairs ($<E_F$), electron holes ($<E_F$), and dipoles ($>E_F$). These DOS features are crucial to the performance of a compound (Reprinted with permission from [28].)

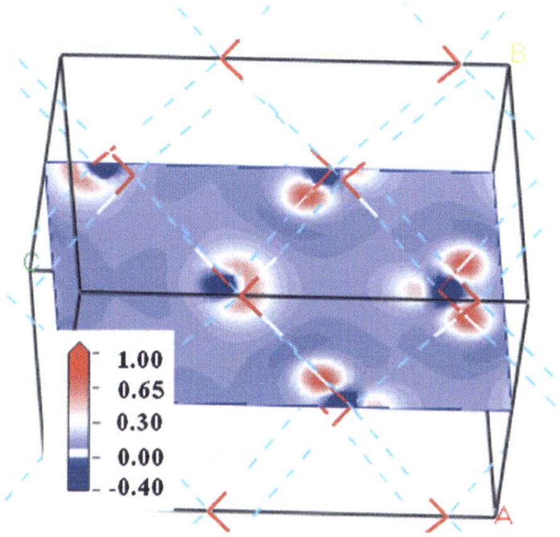


**Fig. 3.4** **a** The residual charge density of an ice-VIII unit cell shows strong localization. The residual charge density is the difference between the charge of an $H_2O$ molecule and that of an isolated O atom. The positive regions (*red*) correspond to the gain of electrons. The negative regions (*blue*) correspond to the charge loss (Reprinted with permission from [1].) (colour online)

nonbonding electron pairs are strongly localized at sites close to oxygen anions (in blue). The localization of the electron pairs lays the foundation for Coulomb repulsion between adjacent oxygen anions introduced in this present way [1]. Such often-overlooked repulsion forms the soul dictating the unusual relaxation dynamics of the O:H–O bond and the anomalies of water and ice.

### 3.4.1.3 Asymmetrically Short-Range Interactions

Segmentation of the O:H–O bond is necessary into a shorter and stiffer H–O covalent bond with a stronger exchange interaction and a longer and softer O:H nonbond with a weaker vdW-like interaction, as illusrated in Fig. 3.2 [29, 30]. The vdW-like interaction contains electrostatic interaction between the lone-pair of $O^{2-}$ and the $H^+$ proton in addition to the Londen and Gesson dispersion interactions (instantaneous dipole–induced dipole forces), so the nonbond interaction is slightly stronger than the ideal vdW bond that denotes purely dipole–dipole interaction.

The $H^+$ proton always remains closer to the O (right-hand side in Fig. 3.2), and keeps away from the other nearest O atom because of the disparity in the H–O bond and the O:H nonbond interactions. Proton frustration or tunneling transition, jumping back and forward between two locations along the O–O is strictly forbidden though it may happen from the macroscopical and statistical point of view. The O:H–O bond links the O–O in all phases even the X phase, regardless of geometric configurations or phase structures [2]. The $sp^3$ dehybridization of an oxygen in $H_2O$, or breaking the H–O bond requires at least 4.0 eV energy.

The following potentials describe the O:H–O bond asymmetrical short-range interactions [31–33]:

$$\begin{cases} V_L(r_L) = V_{L0}\left[\left(\frac{d_{L0}}{r_L}\right)^{12} - 2\left(\frac{d_{L0}}{r_L}\right)^{6}\right] & (\text{O:H vdW-like L–J potential } (V_{L0},\ d_{L0})) \\ V_H(r_H) = V_{H0}\left[e^{-2\alpha(r_H - d_{H0})} - 2e^{-\alpha(r_H - d_{H0})}\right] & (\text{H–O Morse interaction } (\alpha, V_{H0}, d_{H0})) \\ V_C(r_C) = \frac{q_O^2}{4\pi\varepsilon_r\varepsilon_0 r_C} & (\text{O–O Coulomb repulsion } (q_O, \varepsilon_r)) \end{cases}, \quad (3.1)$$

where $V_{L0}$ ($E_{L0}$) is the potential well depth for the O:H nonbond; $V_{H0}$ ($E_{H0}$) is the H–O bond energy; $r_x$ (x = L, H) are the interatomic distance and $r_C = r_L + r_H$ is the O–O separation (the lengths of springs); and $d_{x0}$ is the length at equilibrium. The parameter $\alpha$ determines the width of the Morse potential; $q_O$ (around 0.6 e) denotes the net charge on an $O^{2-}$; $\varepsilon_r = 3.2$ is the relative dielectric constant of ice, and $\varepsilon_0 = 8.85 \times 10^{-12}$ F/m is the dielectric constant of the vacuum. The $q_O$ and $\varepsilon_r$ are subject to variation with external excitations such as coordination and chemical conditions. A sum of all long-range interactions and proton interactions is treated as the background, which is responsible for the flicking of the $H^+$ proton coordination origin that is fixed in the present iteration.

Having the smallest number of adjustable parameters, the Morse potential suffices for the exchange interaction. The L–J potential approximates the O:H nonbond interaction with slight electrostatic nature since no charge is shared within the O:H segment, which also sums all possible interactions between O:H including dispersion, polarization, and Coulomb attraction between lone pair and proton. It is meaningless to say whether one potential is better than the other when considering O:H and H–O interactions; however, the equilibrium coordinates of the potentials as to the bond length and bond energy are of major concern, the shape of a potential curve does not come into play when dealing with situation of quasi-equilibrium.

Because of the asymmetrical and short-range nature of these interactions, the solid lines in Fig. 3.2 are valid only within the respectively shaded range for the basic O:H–O unit. A particular potential must be switched off and the other switched on immediately when moving to the boundary of the region, or to any atomic site. No spatial decay of any potential exists, irrespective regime. Any dispersion of the H–O bond energy in the O:H region by an exponential cut off is still greater than the O:H cohesive energy that is 0.1 eV order or less than 3 % of the H–O bond energy. Therefore, it is essential to consider the O:H–O bond disparity, asymmetry, and short-range interactions, in dealing with water ice or specimens with lone pair involvement.

## *3.4.2 O:H–O Bond Segmentation*

### 3.4.2.1 Lengths and Containing Angle

In place of the convention of donor and acceptor, the O:H–O bond is segmented into two parts. One is the intermolecular O:H nonbond linking the proton and oxygen by the lone pair. The other is the intramolecular H–O polar-covalent bond with shared electron pairs. The H–O bond length also represents the molecular size that will change with stimulus.

Three variables describe the O:H–O bond relaxation dynamics. They are the ∠O:H–O angle $\theta$ and the segmental lengths $d_x$ ($x$ = L and H). The $d_{OO}$ is an auxiliary parameter that sums the $d_H$ and $d_L$. Relaxation of the angle $\theta$ contributes little to the physical properties, except for mass density and crystal geometry. Response of $d_x$ and its bond energy $E_x$ to a stimulus changes inclusively the physical properties of water and ice such as phonon relaxation, density variation, dielectrics, $T_C$ change, O 1 s energy shift, viscoelasticity, etc.

Another dimension is the electron polarization or depolarization for the physical properties such as chemical reactivity, hydrophobicity, radiation absorptivity, and solubility of water or aqueous solutions when interacting with other subjects.

### 3.4.2.2 Coordination Origin

In the present iteration, rapid H proton is taken as the coordination origin fixed in a position of the network matrix. Quantum calculations decompose the measured V–P equation of states into the $d_H$ and $d_L$ serving as reference in subsequent investigations [29, 34]. Accordingly, the cohesive energy of $E_H$ and $E_L$ are of concern while the angle contributes insignificantly to the energies. Therefore, segmentation of the X–H(B):A bond with H(B) being the coordination origin is necessary for simplifying the hydrogen bond relaxation in length and energy.

## 3.4.3 O:H–O Bond Relaxation

### 3.4.3.1 Local Bond Average

Fourrier transformation transfers the coupling arguments in the wave function, exp[i (k·r−ωt)], or between the real and the reciprocal (momentum) spaces, and between the time and frequency (energy) domains. This principle govens the spectroscopy techniques that collect information from real space and time domain and then convert the information into characteristic peaks in the spectrum in the reciprocal domains. For example, the X-ray diffraction collects and sorts the interlayer spacing in the reciprocal space, disregarding the number and manner of distribution. A certain spectral peak corresponds to all such oriented interlayers of the same registry disregarding their locations in specimen. A phonon spectral peak collects vibrations of the same frequency without needing to known where and how the oscillators are distributed in the specimen.

In practice, one can take a representative bond for the average of all bonds involved in a substance to examine spectrometrically the relaxation dynamics of the representative. Such an iteration forms the local bond average (LBA) approach indicating that the number and nature of the bonds for a substance do not change unless phase transition takes place [35]. One can then focus on the representative bond and monitor its length and vibrational frequency change.

Figure 3.5 shows a pairing potential u(r) for the dimer bond in a regular substance. The coordinates ($d$, $E_b$) at equilibrium correspond to the bond length and bond energy, which relax under an external stimulus, regardless of the shape of the particular u(r). Therefore, the change of ($d$, $E_b$) under excitation means the bond relaxation, which is of immediate concern [22]. The difference between water ice

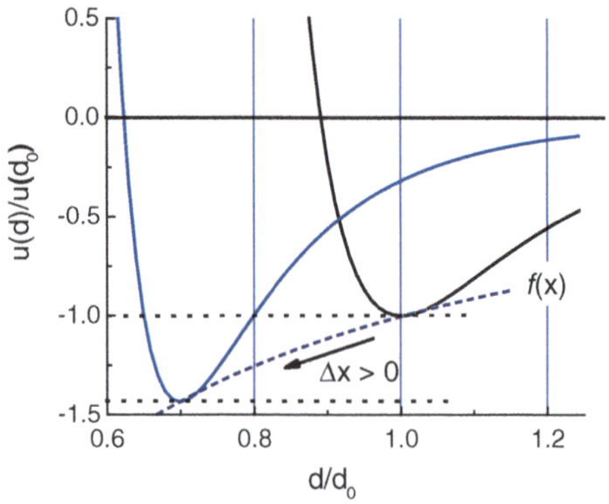


**Fig. 3.5** The long-range, mono-well potential for paring atoms in a "normal" substance [36]. Compression stores energy by shortening and stiffening the bond whereas tension does the opposite, along an $f$(x = P) path (Reprinted with permission from [1].)

and other normal substance is that the representative O:H–O bond consists two asymmetrical segments but the latter has only one segment.

Generally, external stimuli such as compression (P) and thermal excitation (T) modulate the length *d* and energy *E* of the representative bond along a path of *f* (*x*) in the potential curve [35]. For instance, compression stores energy into a substance by shortening and stiffening all bonds with possible plastic deformation when the compression is high when relieved. Tension does the opposite [36]. The following formulates bond relaxation in length and energy under stimuli (P, T) [22]:

$$\begin{cases} d(P,T) = d_b\left(1 + \int_{T_0}^{T} \alpha(t)dt\right)\left(1 - \int_{P_0}^{P} \beta(p)dp\right) \\ E(P,T) = E_b\left(1 - \dfrac{\int_{T_0}^{T} \eta(t)dt + \int_{V_0}^{V} p(v)dv}{E_b}\right) \end{cases} \tag{3.2}$$

where $T_0$ and $P_0$ are the ambient referential conditions. The α(t) is the thermal expansion coefficient. $\beta = -\partial v/(v\partial p)$ is the compressibility ($p < 0$, compressive stress) or extensibility ($p > 0$ tensile stress). The $v = sd$ is the volume of a bond (the product of length d and its cross-sectional area s). The $\eta(t)$ is the specific heat of the representative bond in Debye approximation. The Debye temperature $\Theta_D$ determines the rate of the $\eta(t)$ curve to it saturation. The integration of the $\eta(t)$ from 0 K to the $T_m$ equals the bond energy [1]. Conventional text books often constrain activities such as thermal vibration and compression deformation with the frame of the u(r) without such relaxation that offsets the potential curve.

#### 3.4.3.2 Driving Forces

Unlike normal substance, only one part of the O:H–O bond follows the regular rule of stimulated relaxation, but the other part relaxes oppositely in length and energy because of O–O Coulomb repulsion. Figure 3.6 illustrates forces acting on the electron pairs of oxygen ions. Background long-range interactions due to other $H_2O$ molecules or protons [37] and the nucleus quantum effect on fluctuations [38] are taken as the background of common. The forces include,

(1) Coulomb repulsion between electron pairs on adjacent $O^{2-}$ ions is the first-order differentiation of the Coulomb potential, $f_q = -\partial V_C(r)/\partial r$, out of equilibrium. Replacing one $O^{2-}$ ion with an ion of acid, salt, sugar, protein, or a biomolecule or a cell, or simply adding guest ions, mediates the Coulomb repulsion $f_q$ by varying the ionic size and charge quantity as well as the dielectrics. This replacement may mediate the O:H dissociation energy and the functionalities of the O:H–O bond and the solubility of aqueous solutions [39].

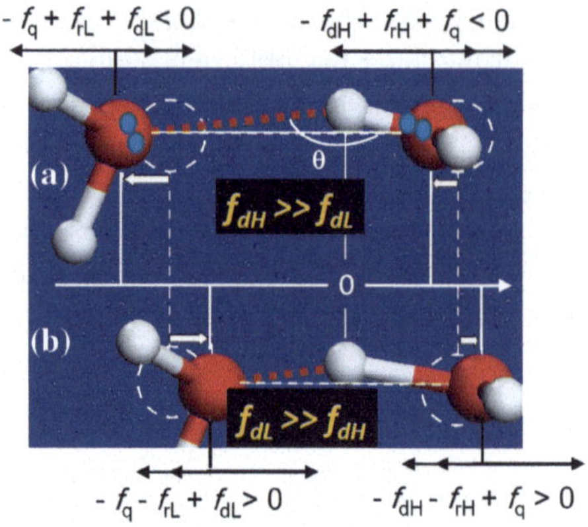


**Fig. 3.6** Forces and relaxation dynamics of the segmented O:H–O bond with $H^+$ as the coordinate origin. Forces include the Coulomb repulsion $f_q$, deformation recovery $f_{rx}$, and the external force driving relaxation $f_{dx}$. External stimulus dislocates O atoms in the same direction but by different amounts because of the Coulomb repulsion and O:H–O segmental disparity. The softer O:H always relaxes always more than H–O. **a** Cooling in the quasi-solid phase [40], electrification, molecular undercoordination, [30], and tension elongate the O:H–O bond. **b** Compression [29], cooling in the *liquid* and *solid phases* [40] and base hydration shorten the O:H–O bond. Relaxation in the specified direction takes place when the $f_{dX}$ competition meets the condition $f_{dH} \gg f_{dL}$ or $f_{dH} \ll f_{dL}$

(2) Pointing either towards or away from the coordinate origin, the driving force dislocates oxygen atoms $f_{dx}$. An applied stimulus (e.g. mechanical compression, molecular undercoordination, thermal excitation, chemical reaction, electrification, etc.) provides this driving force.
(3) The resistant force of deformation recovery, $f_{rx} = -\partial V_x(r)/\partial r$, approximates to the first-order differentiation of the respective $V_x(r)$ at equilibrium. The $f_{rx}$ always points opposite to the direction of O ion dislocation.

The following relationships define the 'master' segment that drives the O:H–O relaxation and determines the O–O length gain or loss under applied stimulus (see Fig. 3.6):

| (a) (freezing, undercoordination, salting) | (b) (compression, liquid and solid cooling) |
|---|---|
| $\begin{cases} -f_q + f_{rL} + f_{dL} < 0 & (1) \\ -f_{dH} + f_{rH} + f_q < 0 & (2) \end{cases}$ | $\begin{cases} -f_q - f_{rL} + f_{dL} > 0 & (3) \\ -f_{dH} - f_{rH} + f_q > 0 & (4) \end{cases}$ |
| $or\ (1) + (2),$ | $or\ (3) + (4),$ |
| $f_{rL} + f_{dL} - f_{dH} + f_{rH} < 0$ | $-f_{rL} + f_{dL} - f_{dH} - f_{rH} > 0$ |
| $or,$ | $or,$ |
| $f_{dH} > (f_{rL} + f_{dL} + f_{rH})$ | $f_{dL} > (f_{rL} + f_{dL} + f_{rH})$ |

$$\left.\begin{array}{r} f_{dH} > (f_{dL} + f_{rL} + f_{rH}) \\ f_{dL} > (f_{dH} + f_{rL} + f_{rH}) \\ f_{dH} = f_{dL} \end{array}\right\} \Rightarrow \Delta d_{O-O} \left\{ \begin{array}{c} > \\ < \\ = \end{array} \right\} 0 \qquad (3.3)$$

Segments of the O:H–O bond relax cooperatively because of the joint effect of O–O Coulomb coupling and applied stimulus. One segment dominates the relaxation and the other follows under a certain stimulus. The segment that drives relaxation is assigned to be the 'master', and the other as the 'slave'. When stimulated, the master segment relaxes and pushes or pulls the electron pair of the slave $O^{2-}$ through the repulsion; once the stimulus is released, deformation recovers and ensures the adaptivity and recoverability. Meanwhile, the repulsion widens the $\angle$O:H–O angle θ and polarizes the electron pairs during relaxation.

The H–O bond serves as the master to drive the relaxation, if the driving forces meet the criterion, $f_{dH} \gg f_{dL}$. In this situation, the master H–O relaxes less than the slave O:H, resulting in a net O–O length gain or loss and an accompanying volume variation. If $f_{dL} \gg f_{dH}$, the master and the slave exchange roles and the situation reverses. At $f_{dH} = f_{dL}$, there is a transition between O–O expansion and contraction, as both segments change their signs of relaxation, which correspond to the density extremes.

Interoxygen repulsion has an influence in not only O:H–O relaxation but also in the bonding dynamics of oxygen chemisorption. An STM and VLEED study revealed that the $O^{2-}$–$Cu^{+}$ bond and the $O^{2-}$:$Cu^{P}$ nonbond relax cooperatively and oppositely in lengths during oxygen chemisorption onto the Cu(001) skin. The $O^{2-}$–$Cu^{+}$ contracts to 0.163 nm while the $O^{2-}$:$Cu^{P}$ expands to 0.195 nm in the $Cu^{P}$:$O^{2-}$–$Cu^{+}$ configuration, with the creation of the $Cu^{P}$ dipoles and the missing Cu atoms [15].

### 3.4.4 O:H–O Bond Mechanical Disparity

One may analogue the O:H and H–O segmental interactions as each a rod of different cross-sectional areas $s_x$ of the same substance. A compression of $P$ pressure will insert different forces on these segments along the rods. Letting the compression force be $f_{dx} \approx P/s_x$, the mechanical disparity of the O:H–O bond may be derived under the quasi-equilibrium condition (compression shortens the O:H so $f_{dL} - f_{dH} > 0$, see (3.3b):

$$\begin{array}{l} f_{dL} - f_{dH} = P(1/s_L - 1/s_H) = (f_{rL} + f_{rH}) > 0 \\ or \\ s_H - s_L > s_L s_H > 0 \end{array} . \qquad (3.4)$$

This relationship indicates that the effective cross-sectional area of the H–O bond $s_H$ is greater than that of the O:H nonbond, which explains why the O:H 'masters' the relaxation dynamics of water ice under compression and why the O:H nonbond always relaxes more than the H–O bond [29]. Hence, compression shortens and stiffens the O:H nonbond and spontaneously lengthens and softens the H–O bond; negative compression (tension) will effect reversely [29].

## *3.4.5 Molecular Undercoordination*

### 3.4.5.1 General BOLS-NEP Notion

According to the bond order-length-strength correlation and nonbonding electron polarization (BOLS-NEP) notion, see (3.5) [22, 41], atomic undercoordination shortens and stiffens the remaining bonds between undercoordinated atoms spontaneously with an association of local densification and quantum entrapment of the bonding and core electrons. This process occurs regardless of the nature of the bond or the structure phase unless coupling correlation exists between a segmented bond. Furthermore, the locally and densely entrapped bonding charge in turn polarizes the nonbonding electrons pertained to lone pairs, dangling bonds, or conduction electrons at the upper edge of the conduction band [28].

Unifying the unusual performance of adatoms, defects, terrace edges, grain boundaries, and solid skins of various curvatures, the BOLS-NEP is responsible not only for the size dependence of the known bulk properties but also for the emergence of anomalies of materials at the nanometer scale. Size emergence means properties that the bulk parent does not display such as catalytic enhancement, toxicity, and dilute magnetism of the noble metals [42] and ZnO [43]. The polarization creates Dirac-Fermi polarons at graphene zigzag edges and graphite point defects [44], serving as carriers for topological insulators.

The following formulates the BOLS notion in terms of coefficient for bond length ($d_z$) contraction $C(z)$, bond energy $E_z$ gain, and atomic cohesive energy change $E_{B,\ z}$, where subscript $z$ denotes an atom with $z$-coordination number (CN), and subscript $b$ denoting a fully coordinated atom in the bulk ($z$ = 12 for the fcc structure as a standard). The bond nature index $m$ correlates the bond length and energy, which keeps a constant for a specific substance:

$$\begin{cases} C(z) & = d_z/d_0 = 2/\{1 + \exp[(12 - z)/(8z)]\} & (\textit{BOLS-coefficient}) \\ E_z & = C_z^{-m} E_b & (\textit{Single-bond-energy}) \\ E_{B,z} & = zE_z & (\textit{Atomic-coherency}) \end{cases} \quad (3.5)$$

Figure 3.7 formulates the BOLS-NEP notion in comparison with the measured data for atomic chains, liquid and solid skins, Au nanoparticles, graphite and carbon nanotubes, etc. [41].

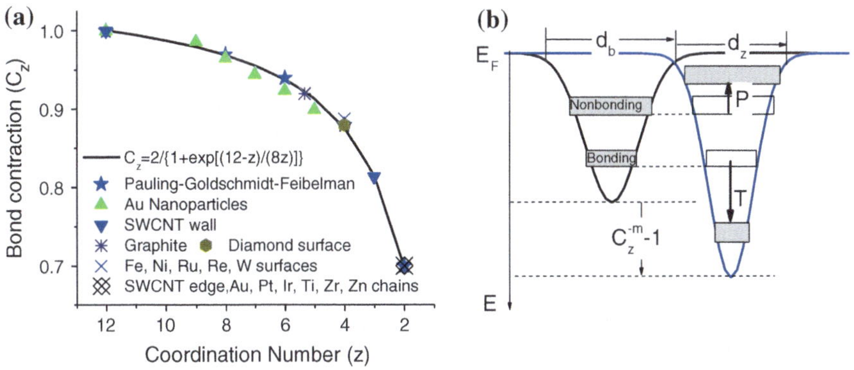


**Fig. 3.7** Schematic illustration of the BOLS-NEP notion [28]. **a** Atomic undercoordination shortens and stiffens the local bond ($d_z/d_0 = z < 1$; $E_z/E_b = C_z^{-m} > 1$). Bond contraction raises the local density of the bonding charge and binding energy. Bond stiffening deepens the local potential well, which **b** entraps (T) the core and the bonding electrons accordingly. The densely entrapped electrons in turn polarize (P) the nonbonding electrons shifting up in energy. The T and P evolution dynamics modulates the Hamiltonian by crystal potential screening and splitting and charge distribution in all energy bands of a substance. Scattered symbols in (**a**) represent observations by Goldschmidt [45] Feibelman [46] and Huang et al. [47] from gold clusters

### 3.4.5.2 H–O Contraction and Dual Polarization

Water molecules with fewer than the ideal four nearest neighbours in the bulk should follow the BOLS-NEP notion. However, the involvement of the lone-pair interaction and O–O repulsion prevents the O:H and the H–O from following the BOLS-NEP notion simultaneously because the ':' lone pairs screen an $H_2O$ molecule becoming nearly isolated. The binding energy disparity means that the stronger H–O bond serves as the 'master' to contract by a different amount from what the BOLS notion predicts. The contraction of the H–O bond is associated with lengtheing of the 'slave' O:H nonbond by Coulomb repulsion. In another word, molecular undercoordinattion shrinks the molecular size but enlarge their separations.

It is universally true that one segment of the O:H–O bond will be stiffer if it becomes shorter; it will be softer if it becomes longer [29]. The stiffness of the segment is characterized by the respective phonon frequency $\omega_x$ ($x$ = L for the O:H nonbond; $x$ = H for the H–O bond). Therefore, the phonon frequency shift $\Delta\omega_x$ tells directly the variation in length, strength, stiffness of the particular segment subjected to an applied stimulus. Because of the Coulomb repulsion, $\omega_L$ and $\omega_H$ shift such that if one becomes stiffer, the other will become softer. Polarization enhancement will offset up the $\omega_L$ slightly. Therefore, the contraction of the H–O bond is associated with its stretching phonon stiffening from the value for bulk water 3200–3450 $cm^{-1}$ for the skin. The O:H nonbond elongation softens its phonon from 200 to lower frequencies for the skin by Coulomb repulsion.

There are dual processes of nonbonding electron polarization pertaining to undercoordinated water molecules [1]. The shorter H–O covalent bond results in local densification of its charge and energy. The stiffer H–O bond deepens the segmental potential to entrap the core and bond electrons. These densely entrapped electrons of an oxygen atom and the H–O bond polarize the lone pairs of its own. On the other hand, the polarised lone pairs on the adjacent oxygen atoms polarize and repel one another, resulting in a second round of polarization. This explains why the surface of water ice is so strongly polarized, and why it is elastic, hydrophobic, viscoelastic and slippery, in the case of ice. Furthermore, the polarization slows the molecular dynamics or reduces the degree of fluctuation, which prolongs the H–O phonon lifetime and enhance the macroscopical viscoelaticity of liquid water.

## 3.4.6 Thermodynamics: Specific-Heat Disparity

### 3.4.6.1 Segmental Specific Heat

Generally, the specific heat of a substance is regarded as a macroscopic quantity integrated over all bonds of the specimen, which is also the amount of energy required to raise the temperature of the substance by 1 K degree. The thermal response of the specimen follows the specific heat of Debye description. However, in dealing with the representative for all bonds of the entire specimen, it is necessary to consider the specific heat per bond that is obtained by dividing the bulk specific heat by the total number of its bonds [35].

For a specimen of other usual materials, one bond represents all on average; therefore the thermal response is the same for all the bonds, without any differences in cooling contraction or thermal expansion [48]. For water and ice, however, the representative O:H–O bond is composed of two strongly coupled segments with strong disparity in cohesive energy that determines the specific heat of the Debye approximation, $\eta_x$. These two segments response independently to a thermal excitation in their respective ways.

### 3.4.6.2 Specific Heat-Phonon Frequency-Cohesive Energy

Two parameters characterize one specific heat curve. One is the Debye temperature $\Theta_{\mathrm{Dx}}$, and the other is the integration of the $\eta_x$ curve from 0 K to $T_m$ for a solid. The liquid and gaseous phases have their own specific heat and latent heat at transitions. These specific heats continue smoothly at the transition temperatures. The $\Theta_{Dx}$ determines the rate at which the specific-heat curve reaches saturation in the solid phase. The specific-heat curve of a segment with a relatively low $\Theta_{\mathrm{Dx}}$ value reaches saturation more rapidly than the other segment, since $\Theta_{Dx}$, which is lower than $T_{mx}$, is proportional to the characteristic vibration frequency $\omega_x$ of the respective segment in Einstein's relationship $k\Theta_{Dx} = \hbar\omega$ with $\hbar$ and $k$ being the Planck and Boltzmann constant, respectively.

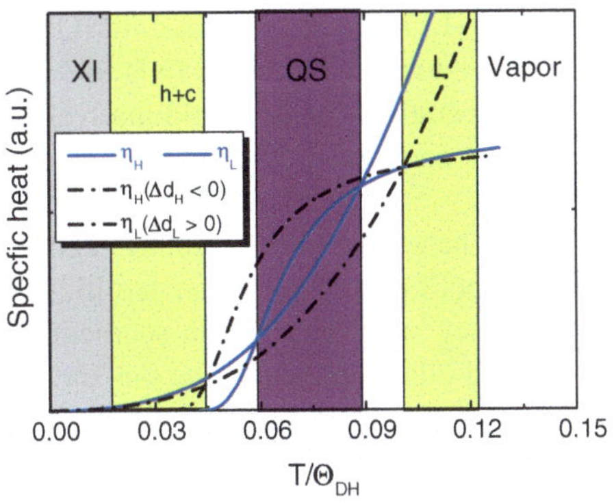


**Fig. 3.8** A superposition of the two specific-heat curves yields two intersecting temperatures that divide the full temperature range into four regimes of different $\eta_L/\eta_H$ ratios in addition to the gaseous phase ($\eta_L = 0$), which correspond to phases of liquid (L: $\eta_L/\eta_H < 1$), quasisolid (QS: $\eta_L/\eta_H > 1$), solid ($I_{h+c}$: $\eta_L/\eta_H < 1$) and XI ($\eta_L \approx \eta_H \approx 0$). The $\eta_L$ in the solid phase differs from the $\eta_L$ in liquid, which does not influence the validity of the hypothesis (Reprinted with permission from [40].)

Conversely, the integral of the specific-heat curve from 0 K to the melting point $T_{mx}$ determines the cohesive energy per segment $E_x$ [35]. The $T_{mx}$ is the temperature at which the vibration amplitude of an atom or a molecule expands abruptly to more than 3 % of its diameter irrespective of the environment or the size of a molecular cluster [49, 50].

Thus, see Fig. 3.8:

$$\begin{cases} \Theta_{DL}/\Theta_{DH} \approx 198/\Theta_{DH} \approx \omega_L/\omega_H \approx 200/3200 \sim 1/16 \\ \left(\int_0^{T_{mH}} \eta_H dt\right)/\left(\int_0^{T_{mL}} \eta_L dt\right) \approx E_H/\mathrm{E_L} \approx 4.0/0.1 \sim 40 \end{cases} . \quad (3.6)$$

Analysis of the temperature-dependence of water surface tension yielded $\Theta_{DL} = 198$ K < 273 K ($T_{mL}$) and $E_L = 0.095$ eV [51]. Hence, $\Theta_{DH} \approx 16 \times \Theta_{DL} \approx 3200$ K. The O:H specific heat $\eta_L$ ends at 273 K and the H–O specific heat $\eta_H$ ends at T ≥ 3200 K. Numerical duplication of the compression-induced $T_C$ change for the ice VII–VIII and the $T_{mL}$ for the liquid/quasisolid phase transition results in an $E_H$ value of 3.97 eV [30]. That is, the area covered by the $\eta_H$ curve is 40 times greater that covered by the $\eta_L$ curve.

#### 3.4.6.3 O:H–O Oscillating Relaxation

The superposition of these two $\eta_x$ curves results in the heat capacity of water ice that differs from the specific heat of other, 'normal', materials. Such a specific-heat superposition yields two intersecting temperatures that divide the full temperature range into four regimes with different $\eta_L/\eta_H$ ratios; see Fig. 3.8. These regions correspond to phases of liquid (L: $\eta_L/\eta_H < 1$), quasisolid (QS: $\eta_L/\eta_H > 1$), and solid ($I_{h+c}$: $\eta_L/\eta_H < 1$). At extremely low temperatures (XI) $\eta_L \approx \eta_H \approx 0$. O:H–O bond segmental length relaxes cooperatively and discriminatively in different regimes. The intersecting temperatures correspond to extreme densities at boundaries of quasisolid phase.

The number of the specific-heat defined temperature regimes coincides with that demonstrated by the density ρ(T) profiles of water ice [52–54] over the full temperature range. This consistency suggests that the segment of lower specific heat serves as the master while the other part serves as the slave during thermal excitation. The master segment is more active than the other when the O:H–O bond is thermally invoked. The thermal expansion or cooling contraction of the master segment drives an asymmetrical and cooperative relaxation of the entire O:H–O bond.

According to the specification, the O:H nonbond serves as the master to contract/expand in the liquid (L) and in the solid ($I_{h+c}$) phases, and the slave H–O bond relaxes oppositely and slightly, leading to the seemingly 'normal' process of O:H–O cooling contraction and water volume densification, but by a completely different and unaware mechanism.

In the quasisolid phase, the master-slave roles interchange and the water volume expands, resulting in floating of ice. The intersection points correspond to a maximum density at 4 °C and a minimum density below the freezing point for bulk water ice [52, 55]. At extremely low temperatures, both $\eta_L$ and $\eta_H$ approach zero, which means that neither segment is active in responding to thermal excitation in this regime except for the ∠O:H–O containing angle.

The thermodynamic disparity of the O:H–O bond indicates that the H–O bond, rather than the O:H nonbond networks, dictates the extremely high heat capacity of water and ice [56]. According to the current notion, cooling drives the oscillation of the O–O length and mass density in these four regions, $\rho \propto (d_{OO})^{-3}$:

$$\left.\begin{array}{lll} QS & (\eta_H < \eta_L): & f_{dH} > (f_{dL} + f_{rL} + f_{rH}) \\ I, L & (\eta_L < \eta_H): & f_{dL} > (f_{dH} + f_{rL} + f_{rH}) \\ XI & (\eta_L \cong \eta_H \cong 0): & \Delta\theta > 0; \quad \Delta d_x = 0 \\ \text{QS boundary} & (\eta_L = \eta_H): & f_{dH} = f_{dL} \end{array}\right\} \Rightarrow \Delta d_{OO} \left\{\begin{array}{c} > \\ < \\ = \\ = \end{array}\right\} 0. \tag{3.7}$$

### 3.4.6.4 Supercooling or Superheating?

Most strikingly, the quasisolid phase boundary is dispersive. The direction and extent of the dispersion depends on the phonon frequency shift $\Delta\omega_x$. Molecular undercoordination ($z < 4$), negative compression (tension), or electrification stretches the quasisolid phase boundaries by modulating the respective $\Theta_{Dx}$ through stiffening the $\omega_H$ and softening the $\omega_L$ phonon. Therefore, nanodroplets and nanobubbles with high-fraction of undercoordinated molecules and ice under tension demonstrate the melting point elevation and freezing depression [57]. However, ice under compression demonstrates inversely-ice regelation [58]. This dispersion also happens when the water molecular dipole is stretched by any external fields, such as electrification and magnetization.

Nanodroplets and nanobubbles of water, which have a higher proportion of undercoordinated skin molecules, should follow this trend. It is true that the least-density temperature drops from 258 to 205 K when the bulk water is divided into droplets of 1.4 nm size [40, 52–54], and that the $T_m$ increases from 273 K at the bulk centre to 310 K at the skin of water [59]. The $T_m$ for a monolayer water film is even higher [60, 61]. Compression effects oppositely to molecular undercoordination on the $\omega_x$ and $\Theta_{Dx}$ relaxation. Observations indicate that ice melts at +6.5 °C under −95 MPa negative compression and melts at −22 °C under 210 MPa compression [62]. Electrification by the short-range fields of ions in aqueous solutions or by the long-range field of a capacitor has the same effect of molecular undercoordination on the phase boundary dispersion.

Supercooling, also known as undercooling [63], is the process of lowering the temperature of a liquid or a gas below its freezing point without it becoming a solid. Superheating is otherwise. Supercooled water occurs in the form of small droplets in clouds and plays a key role in the processing of solar and terrestrial radiative energy fluxes. Supercooled water is also important for life at subfreezing conditions for the commercial preservation of proteins and cells, and for the prevention of hydrate formation in nature gas pipelines.

Supercooling/heating is often confused with freezing/melting point depression/elevation. Freezing point depression occurs when a solution can be cooled below the freezing point of the corresponding pure liquid due to the presence of the solute; an example of this is the freezing point depression that occurs when salt is added to pure water.

## 3.4.7 Electromagnetic Activation and Isotope Effect

### 3.4.7.1 Electrification: Water Bridge and Hofmeister Effect

Molecular dipoles of liquid water are rotating and moving ceaselessly approaching their optimal geometries without any external field perturbation. However, an electric field will align, stretch, and polarize the molecular dipole. One can imagine what will happen to the dipole when it subject to any of the electric fields:

$$E(x) = \begin{cases} E_0 & \text{(homogeneous)} \\ Ax^{-2} & \text{(point charge)} \\ A\left(x^{-2} - (x-L)^{-2}\right) & \text{(mirror charges)} \end{cases}$$

where $A = q/4\pi\varepsilon_0$ is a constant. The first expression is an uniform field between two plates of a capacitor, which approaches the Armstrong effect [64]—a quasisolid bridge forms over two beakers under the applied DC or AC bias of $10^6$ volt/m with or without a current flow. The current flow generates a magnetic field circling the current I, following the right-hand rule. This magnetic field induces another electric field mainly in the skin of the bridge column to against the current flow, according to Faraday's law of induction, but this amount is negligibly small. Water bridges demonstrate indeed density gradients (7 % edge to core) observed using optical techniques, structural anisotropy observed in neutron scattering, optical birefringence from polarized light scattering and changes in the OH stretch vibration observed using Raman and infrared measurements [65, 66].

The rest two describe the electric field surrounding an ionic pint charge, or a pair of anion and cation, for sugar or salt solutes at different concentrations that the Hofmeister effect is involved [66, 67]. An addition of aqueous anions or cations modulates the surface tension and the solubility of proteins [68] of the solution. Ions addition also lowers the freezing temperature of the liquid water, associated with $\omega_H$ stiffening and $\omega_L$ softening [39].

Under the applied electric field, a water molecular dipole will firstly align along the electric field, forming a hydration shell with dipoles heading or tailing toward the central ions. The electric field will stretch the dipole by lengthening the O:H nonbond and softening its phonon. The H–O bond will become shorter and stiffer because the O:H–O bond elongation weakens the O–O Coulomb repulsion, being the same effect to liquid heating and molecular undercoordination. The O:H phonon softening lowers the freezing point $T_N$, while the H–O bond stiffening raises the melting temperature $T_m$ [39].

In the field of mirror charges, an anion (−q) and a cation (+q), as shown in Fig. 3.9, the molecular dipole is subject to alignment, relaxation, and polarization

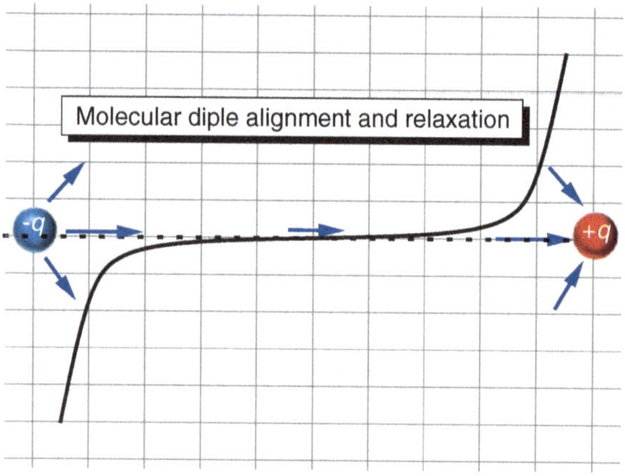


**Fig. 3.9** Molecular dipoles (*blue arrows*) tend to align along the field lines, which lengthens and polarizes the O:H–O bond in an electric field of an ion or a pair of mirror ions. The extent of relaxation and polarization varies with the intensity of electric field. Dipole stretching results in lengthening and softening of the O:H nonbond and shortening and stiffening of the H–O bond, which effects the same to molecular undercoordination and liquid heating [39] (colour online)

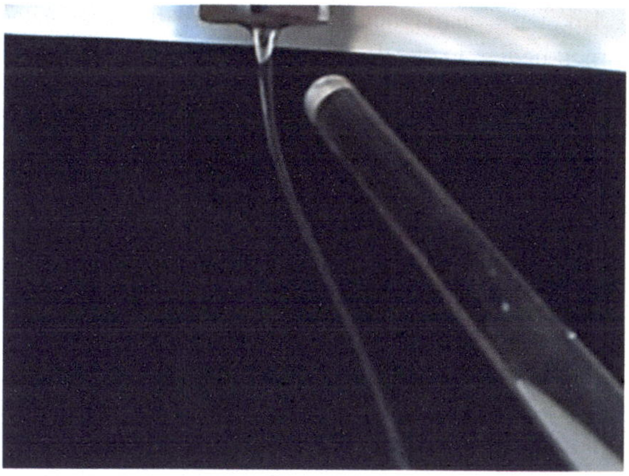

**Fig. 3.10** A thin stream of water is deflected by an electric field of the charged rod (https://www.youtube.com/watch?v=7b-w0oWttN0)

whose extent is more sensitive to ion-dipole distance than the dipole in the field created by single charge center. As the field is inhomogeneous, the extent of O:H–O bond relaxation and polarization varies. The length scale of interaction depends on the separation and charge quantity between the pairing ions. The inner field of the dipole screens the ionic field. Because of the O:H weak interaction, the perturbation effects in a long distance manner.

One can test the polarizability of water molecules by bending a thin water stream using a charged object such as a balloon or a glass rod rubbed with hairs, see Fig. 3.10. The water steam will deviate from its initial path towards the charged object. This simple observation testifies to that the polarized molecular dipoles are subject to electrostatic attraction by the charged object through the graded field. An electrostatic charged bar attracts and deviates strongly the stream of water that has electric dipole momentum. The charged rod always induces charge in opposite sign to the water molecules [69].

#### 3.4.7.2 Magnetization: Moving Dipole in the Lorentz Field

A moving charge in velocity $v$ is subject to the Lorentz force in an electromagnetic field, which ensures the charge to do circular motion in an orbital of R radius:

$$\overrightarrow{F} = m\overrightarrow{a} = q\left(\overrightarrow{E} + \overrightarrow{v} \times \overrightarrow{B}\right) = mv^2/R$$

Meanwhile, a molecular dipole will rotate around an axis along the $B$ field. Under an inhomogeneous electromagnetic field, water molecules will go both transitional motion and angular rotation. The complex motion and rotation also relax the O:H–O bond to a certain extent and hence modulate the critical temperatures and the vibrational frequencies though this effect may not be so significant as the electric field.

Likewise, a magnet can also bend the thin water stream slightly in direction perpendicular to the B field. With a strong spherical Neodymium magnet, water shows its diamagnetism: when the magnet is really close, the point where water-fall moves just a few millimeters away from the magnet.

#### 3.4.7.3 Energy Absorption, Emission, Conduction, and Dissipation

Water absorbs all sorts of energies from bioelectronic signals, sound waves, and electromagnetic radiations by exciting fluctuations, dipole rotations, phonon vibrations, electron polarizations (transition from ground to excited states), bond relaxation, and bond dissociation at energies up to 4.0 eV. The response of liquid water to external stimulus is expected to be ultra-long-range order exhibiting a domino manner because of the extremely weak O:H interaction [1].

The O:H–O bond segmental disparity and the O–O Coulomb coupling ensures the adaptivity, cooperativity, recoverability, sensitivity, memory of water and ice when subject to perturbation. The O:H–O bond memory ensure its emitting energy at a rate of initial storage dependent when subject to cooling in the Mpemba paradox-warmer water freezes faster [70].

The crystal patterns of ice grown from pure and polluted sources could be influenced by emotions, thoughts, and voices [71]. The exclusion zone [72] associated with hydrophilic interface extends to depth of micrometers, which absorbs all sorts of energy, separates charges, and excludes microspores and organisms. The O:H–O bond segmental disparity and O–O Coulomb coupling dictate its adaptivity, cooperativity, memory-ability, recoverability and sensitivity, which stems anomalies of water and ice. Water is thus attributed to having intelligence and spirit or being the messenger of the God.

### 3.4.8 O:H–O Bond Cooperative Relaxation

#### 3.4.8.1 Segmental Length Cooperativity

Letting $k_x$ be the force constant and $\delta d_x$ the extent of relaxation for the respective segment, the $f_{rx,}$ $k_x$, and $\delta d_x$ follow the relationship at equilibrium $f_{dH} = f_{dL}$:

$$f_{rH} + f_{rL} = k_H \delta d_H + k_L \delta d_L = 0,$$

which yields,

$$k_L / k_H = -\frac{\delta d_H / \delta t}{\delta d_L / \delta t} = -\frac{\delta^2 d_H / \delta t^2}{\delta^2 d_L / \delta t^2}. \tag{3.8}$$

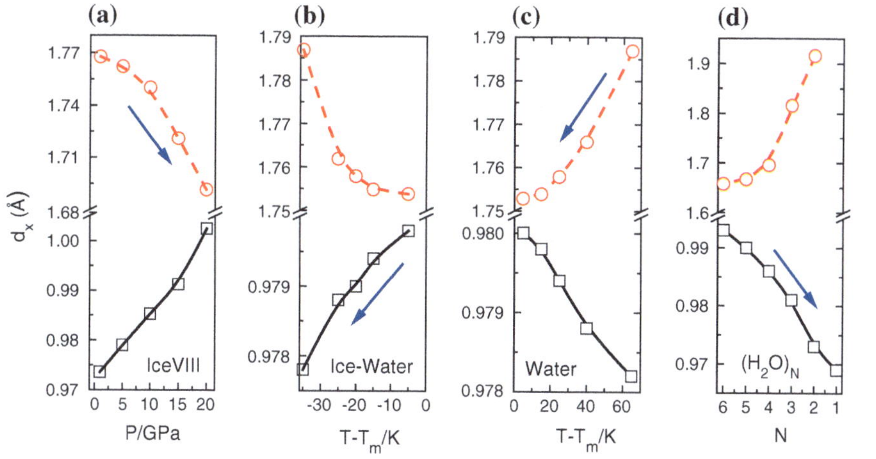


**Fig. 3.11** Universality of the O:H–O bond cooperative relaxation under stimulation of: **a** mechanical compression; **b** quasisolid phase thermal excitation; **c** liquid phase thermal excitation; and **d** cluster size reduction (molecular undercoordination). If either of the O:H or the H–O shrinks, the other one expands, regardless of the applied stimulus or structural phases, because of the interoxygen Coulomb coupling. Arrows indicate the master segments and their relaxation directions. The arrows in (**a**) and (**c**) show the direction of density gain as $d_{O\text{–}O}$ shortens; the arrows in (**b**) and (**d**) show the direction of density loss. All processes are reversible [1]

The variable $t$ represents any stimulus of $T$, $P$ or $N$ for $(H_2O)_N$ clusters, or beyond. This relationship indicates that the slopes and curvatures of the O:H and H–O relaxation curves are inversely negatively proportional to each other.

Figure 3.11 confirms the universality of the O:H–O bond cooperative relaxation under various stimuli obtained from MD calculations using the force field code of Sun [73]. The slopes and curvatures of the $d_x$–$t$ curves are indeed asymmetrical and cooperative, as (3.8) predicts. If one segment is shortened, the other one in the same panel is lengthened; the O:H segment always relaxes more than the H–O. The two curves in one panel relax in a manner either 'face to face' (in a) or 'back to back' (in b, c, d), due to the curvature correlation. The O:H serves as the master and the H–O as slave under mechanical compression (a) and thermal excitation in the region of $T > T_m$ (c); the H–O bond serves as the master in the quasisolid phase (b) at $T < T_m$ and subjected to molecular undercoordination (d). The arrows next to the 'master' segment point in the direction of density gain (a, c) due to O:H–O shortening or density loss (b, d).

The O–O distance dominates the mass density of water ice in the manner of $\rho \propto (d_{O\text{–}O})^{-3} \propto (d_H + d_L)^{-3}$. The $d_x$ is the projection along the O–O without contribution from the ∠O:H–O angle contribution, which remains >160° in all phases at the ambient pressure [40]. The angle difference between 160° and 180° deviates the length scale by only 3 % or less [40].

When the phase structures are different, there are other possible volume changes. For example, ice VII has a smaller volume and longer intermolecular distance than ice Ic because the former has double the network of the latter. Ice VII and VIII have similar network connectivities but different crystal symmetries [74]. The transition between phase VII and VIII is of the first order [75, 76]. However, volume change by such structure variation contributes insignificantly to the O:H–O bond relaxation that dictates the detectable qualities and the anomalous behavior of water ice.

Accurate measurement of the H–O or O:H length relaxation dynamics is not frequently as one often measures the O–O distance using neutron or X-ray diffraction. However, the extended tetrahedron allows one to determine the ρ, $d_H$, $d_L$, $d_{O\text{–}O}$ given any one of them as known input [2].

### 3.4.8.2 Cooperative Relaxation of Characteristic Phonons

Figure 3.12 shows typical phonon spectra for ambient water probed using Fourier transform infrared absorption (FTIR), Raman reflection, and neutron diffraction [77]. Neutron scattering gives more comprehensive information of the phonon states compared to FTIR and Raman as the latter two are subject to the selection rule for phonon excitation [78–80].

The characteristic frequencies (or energies) correspond to vibration modes of segmental stretching and bond bending. Features centered at $\omega_H \approx 3450$ and

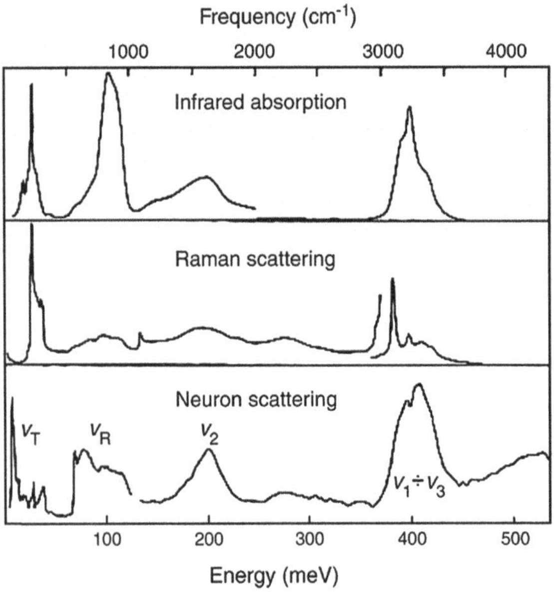


**Fig. 3.12** Typical vibrational spectra measured from bulk water at the ambient temperature. Peaks at $\omega_L \approx 200$ and 75 $cm^{-1}$ correspond to stretching of the O:H in the bulk and skin; the peak at $\omega_{B1} \approx 500$–700 $cm^{-1}$ is from ∠O:H–O bending; features $\omega_{B2} \approx 1600$–1750 $cm^{-1}$ are the ∠H–O–H bending libration mode; features centered at $\omega_H \approx 3200$ and 3450 $cm^{-1}$ are the H–O stretching mode in the bulk and in the skin of water (Reprinted with permission from [77].)

3200 cm$^{-1}$ corresponds to the H–O stretching phonons in the skin and in the bulk, respectively. Peaks at $\omega_L \approx 75$ and 200 cm$^{-1}$ arise from stretching of the O:H in the skin and in the bulk. The peak at $\omega_{B1}$ at 500–700 cm$^{-1}$ is the ∠O:H–O bending mode. The peak of $\omega_{B2}$ at 1600–1750 cm$^{-1}$ is the librational mode of ∠H–O–H bending. Water molecules in the gaseous phase exhibit a peak in the vicinity of $\omega_H \approx 3650$ cm$^{-1}$ and the ∠O:H:O bending mode is at 50 cm$^{-1}$, which are absent from Fig. 3.12. The intensity of the low-frequency modes is so low that they are often overlooked. A residual phonon spectrometrics (RPS), subtracting the spectrum collected at smaller incident beam angle by the reference collected at larger beam angles, can discriminate the skin feature from the bulk.

The libration mode is insensitive to experimental conditions. The intensity of the peaks below 500 cm$^{-1}$ is rather weak, but it is very sensitive to a stimulus; any perturbation, even sunlight irradiation, changes the spectra of liquid water [81] because of the high sensitivity of the O:H nonbond. Monitoring the cooperative relaxation of high-frequency $\omega_H$ and low-frequency $\omega_L$ phonons would suffice to examine the cooperativity of the O:H–O bond under excitation. Figure 3.13 shows

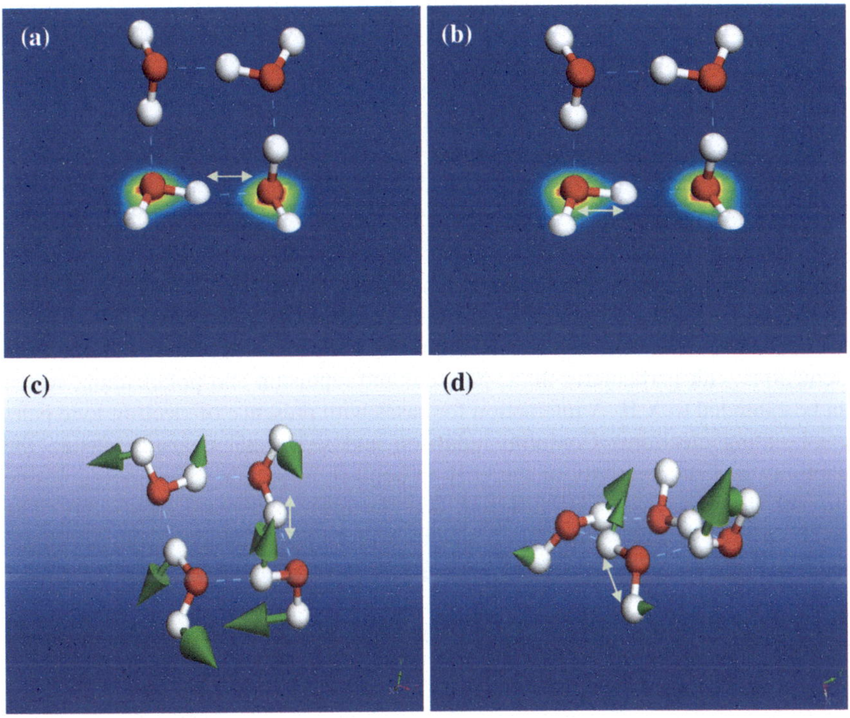


**Fig. 3.13** $(H_2O)_4$ tetramer vibration modes. **a** O:H stretching, **b** H–O stretching, **c** O:H–O bending, and **d** H–O–H bending with characteristic frequencies that change with the number of $(H_2O)_N$ size, as listed in Table 3.1

**Table 3.1** $(H_2O)_N$ size and mode resolved phonon frequencies of water derived from MD calculations, compared with measurements for skin and bulk water

| | $H_2O$ monomer | $(H_2O)_2$ dimer | $(H_2O)_3$ trimer | $(H_2O)_4$ tetramer | $(H_2O)_5$ penamer | Water skin | Bulk water |
|---|---|---|---|---|---|---|---|
| O:H stretching | – | 184 | 251 | 229 | 198 | 75 | 180 |
| ∠O:H–O bending | – | 414 | 409 | 431 | 447 | – | 500 |
| ∠H-O-H bending | – | 1638 | 1644 | 1654 | 1646 | 1600 | 1600 |
| H–O stretching | – | 3565 | 3387 | 3194 | 3122 | 3450 | 3200 |
| H–O stretching[50–52,55] | 3650 | 3575 | 3520 | 3353 | 3326 | | |

[a] $10^{-12}$ s = 1 THz = 33 $cm^{-1}$. ∠O:H–O bending is also called libration mode

the vibration modes and Table 3.1 shows the $(H_2O)_N$ size and mode resolved phonon frequencies in comparison with measurements.

The advantage of phonon spectrometrics is the Fourier transformation between the real and the energy spaces. Each spectral feature represents all bonds with the same vibrational attributes, regardless of their number or location in real space and irrespective of stimulus type. The $\omega_L$ and the $\omega_H$ undergo spontaneously cooperative relaxation. If one is subject to redshift the other will be shift blue, because of the effect of Coulomb coupling.

## 3.5 Summary

The representative O:H–O bond bridging $O^{2-}$ anions represents all essential interactions with particularly the O–O Coulomb repulsion and polarization, which can be extended to X:B–A interaction for cases with presence of electron lone pairs or π-bonding electrons. Coulomb repulsion and segmental disparity ensure the adaptivity and cooperativity of the O:H–O bond in responding to stimulus. If one segment shortens, it will be stiffened, and the other part reacts oppositely. The specific heat disparity defines the boundaries of the solid/quasisolid/liquid phase, which will disperse when subject to stimulus that varies the $\omega_x$ and the respective Debye temperature $\Theta_{Dx}$. O:H–O bond absorbs all sorts of energies in a long-range domino manner because of the wide-range energy of phonon frequencies and bond and nonbond relaxation.

## References


1. Y. Huang, X. Zhang, Z. Ma, Y. Zhou, W. Zheng, J. Zhou, C.Q. Sun, Hydrogen-bond relaxation dynamics: resolving mysteries of water ice. Coord. Chem. Rev. **285**, 109–165 (2015)
2. Y. Huang, X. Zhang, Z. Ma, Y. Zhou, J. Zhou, W. Zheng, C.Q. Sun, Size, separation, structure order, and mass density of molecules packing in water and ice. Sci. Rep. **3**, 3005 (2013)
3. G.R. Desiraju, T. Steiner, *The Weak Hydrogen Bond: in Structural Chemistry and Biology*, vol 9 (Oxford University Press, 2001)
4. G.R. Desiraju, A bond by any other name. Angew. Chem. Int. Ed. **50**, 52–59 (2011)
5. A. Werner, Ueber Haupt-und Nebenvalenzen und die Constitution der Ammoniumverbindungen. Justus Liebigs Annalen der Chemie **322**(3), 261–296 (1902)
6. A. Hantzsch, Über die Isomerie-Gleichgewichte des Acetessigesters und die sogen. Isorrhopesis seiner Salze. Ber. Dtsch. Chem. Ges. **43**(3), 3049–3076 (1910)
7. P. Pfeiffer, P. Fischer, J. Kuntner, P. Monti, Z. Pros, Zur Theorie der Farblacke. II. Justus Liebigs Annalen der Chemie **398**(2), 137–196 (1913)
8. T.S. Moore, T.F. Winmill, The state of amines in aqueous solution. J. Chem. Soc. Trans. **101**, 1635–1676 (1912)
9. W.M. Latimer, W.H. Rodebush, Polarity and ionization from the standpoint of the Lewis theory of valence. J. Am. Chem. Soc. **42**, 1419–1433 (1920)
10. L. Pauling, The structure and entropy of ice and of other crystals with some randomness of atomic arrangement. J. Am. Chem. Soc. **57**, 2680–2684 (1935)
11. R.B. Corey, The crystal structure of diketopiperazine. J. Am. Chem. Soc. **60**(7), 1598–1604 (1938)
12. G. Albrecht, R.B. Corey, The crystal structure of glycine. J. Am. Chem. Soc. **61**(5), 1087–1103 (1939)
13. F. Senti, D. Harker, The crystal structure of rhombohedral acetamide. J. Am. Chem. Soc. **62** (8), 2008–2019 (1940)
14. L. Pauling, *The Nature of the Chemical Bond*, 3rd edn. (Cornell University Press, Ithaca, 1960)
15. C.Q. Sun, Oxidation electronics: bond-band-barrier correlation and its applications. Prog. Mater Sci. **48**(6), 521–685 (2003)
16. G.C. Pimentel, A. McClellan, Hydrogen bonding. Annu. Rev. Phys. Chem. **22**(1), 347–385 (1971)
17. G.A. Jeffrey, W. Saenger, *Hydrogen Bonding in Biological Structures* (Springer Science & Business Media, 2012)
18. T. Steiner, W. Saenger, Role of CH. cntdot. cntdot. cntdot. O hydrogen bonds in the coordination of water molecules. Analysis of neutron diffraction data. J. Am. Chem. Soc. **115** (11), 4540–4547 (1993)
19. A.D. McNaught, A.D. McNaught, *Compendium of Chemical Terminology*, vol 1669 (Blackwell Science Oxford, 1997)
20. W.G. Han, C.T. Zhang, A theory of nonlinear stretch vibrations of hydrogen-bonds. J. Phys. Condens. Matter **3**(1), 27–35 (1991)
21. R.H. Crabtree, Chemistry—a new type of hydrogen bond. Science **282**(5396), 2000–2001 (1998)
22. C.Q. Sun, in *Relaxation of the Chemical Bond*. Springer Series in Chemical Physics 108, vol 108 (Springer, Heidelberg, 2014), 807 pp
23. P. Hobza, Z. Havlas, The fluoroform center dot center dot center dot ethylene oxide complex exhibits a C–H center dot center dot center dot O anti-hydrogen bond. Chem. Phys. Lett. **303** (3–4), 447–452 (1999)
24. J. Coey, H. Sun, Improved magnetic properties by treatment of iron-based rare earth intermetallic compounds in anmonia. J. Magn. Magn. Mater. **87**(3), L251–L254 (1990)
25. P.W. Atkins, *Physical Chemistry*, 4th edn. (Oxford University Press, 1990)

26. S.R. Morrison, *The Chemical Physics of Surfaces* (Plenum Press, London, 1977)
27. C.Q. Sun, C.L. Bai, A model of bonding between oxygen and metal surfaces. J. Phys. Chem. Solids **58**(6), 903–912 (1997)
28. C.Q. Sun, Dominance of broken bonds and nonbonding electrons at the nanoscale. Nanoscale **2**(10), 1930–1961 (2010)
29. C.Q. Sun, X. Zhang, W.T. Zheng, Hidden force opposing ice compression. Chem. Sci. **3**, 1455–1460 (2012)
30. C.Q. Sun, X. Zhang, J. Zhou, Y. Huang, Y. Zhou, W. Zheng, Density, elasticity, and stability anomalies of water molecules with fewer than four neighbors. J. Phys. Chem. Lett. **4**, 2565–2570 (2013)
31. R.F. McGuire, F.A. Momany, H.A. Scheraga, Energy parameters in polypeptides. V. An empirical hydrogen bond potential function based on molecular orbital calculations. J. Phys. Chem. **76**, 375–393 (1972)
32. N. Kumagai, K. Kawamura, T. Yokokawa, An interatomic potential model for $H_2O$: applications to water and ice polymorphs. Mol. Simul. **12**(3–6), 177–186 (1994)
33. Y. Huang, X. Zhang, Z. Ma, Y. Zhou, G. Zhou, C.Q. Sun, Hydrogen-bond asymmetric local potentials in compressed ice. J. Phys. Chem. B **117**(43), 13639–13645 (2013)
34. Y. Huang, X. Zhang, Z. Ma, W. Li, Y. Zhou, J. Zhou, W. Zheng, C.Q. Sun, Size, separation, structural order, and mass density of molecules packing in water and ice. Sci. Rep. **3**, (2013)
35. C.Q. Sun, Thermo-mechanical behavior of low-dimensional systems: the local bond average approach. Prog. Mater Sci. **54**(2), 179–307 (2009)
36. X.J. Liu, M.L. Bo, X. Zhang, L.T. Li, Y.G. Nie, H. TIan, Y. Sun, S. Xu, Y. Wang, W. Zheng, C.Q. Sun, coordination-resolved electron spectrometrics. Chem. Rev. **115**(14), 6746–6810 (2015)
37. Y. Liu, J. Wu, Communication: long-range angular correlations in liquid water. J. Chem. Phys. **139**(4), 041103 (2013)
38. X.Z. Li, B. Walker, A. Michaelides, Quantum nature of the hydrogen bond. Proc. Natl. Acad. Sci. U.S.A. **108**(16), 6369–6373 (2011)
39. X. Zhang, T. Yan, Y. Huang, Z. Ma, X. Liu, B. Zou, C.Q. Sun, Mediating relaxation and polarization of hydrogen-bonds in water by NaCl salting and heating. PCCP **16**(45), 24666–24671 (2014)
40. C.Q. Sun, X. Zhang, X. Fu, W. Zheng, J.-L. Kuo, Y. Zhou, Z. Shen, J. Zhou, Density and phonon-stiffness anomalies of water and ice in the full temperature range. J. Phys. Chem. Lett. **4**, 3238–3244 (2013)
41. C.Q. Sun, Size dependence of nanostructures: impact of bond order deficiency. Prog. Solid State Chem. **35**(1), 1–159 (2007)
42. E. Roduner, Size matters: why nanomaterials are different. Chem. Soc. Rev. **35**(7), 583–592 (2006)
43. J.W. Li, S.Z. Ma, X.J. Liu, Z.F. Zhou, C.Q. Sun, ZnO meso-mechano-thermo physical chemistry. Chem. Rev. **112**(5), 2833–2852 (2012)
44. W.T. Zheng, C.Q. Sun, Underneath the fascinations of carbon nanotubes and graphene nanoribbons. Energy Environ. Sci. **4**(3), 627–655 (2011)
45. V.M. Goldschmidt, Crystal structure and chemical correlation. Ber. Dtsch. Chem. Ges. **60**, 1263–1296 (1927)
46. P.J. Feibelman, Relaxation of hcp(0001) surfaces: a chemical view. Phys. Rev. B **53**(20), 13740–13746 (1996)
47. W.J. Huang, R. Sun, J. Tao, L.D. Menard, R.G. Nuzzo, J.M. Zuo, Coordination-dependent surface atomic contraction in nanocrystals revealed by coherent diffraction. Nat. Mater. **7**(4), 308–313 (2008)
48. M.X. Gu, Y.C. Zhou, C.Q. Sun, Local bond average for the thermally induced lattice expansion. J. Phys. Chem. B **112**(27), 7992–7995 (2008)
49. M.A. Omar, *Elementary Solid State Physics: Principles and Applications* (Addison-Wesley, New York, 1993)

50. F.A. Lindemann, The calculation of molecular natural frequencies. Phys. Z. **11**, 609–612 (1910)
51. M. Zhao, W.T. Zheng, J.C. Li, Z. Wen, M.X. Gu, C.Q. Sun, Atomistic origin, temperature dependence, and responsibilities of surface energetics: an extended broken-bond rule. Phys. Rev. B **75**(8), 085427 (2007)
52. F. Mallamace, C. Branca, M. Broccio, C. Corsaro, C.Y. Mou, S.H. Chen, The anomalous behavior of the density of water in the range 30 K < T < 373 K. Proc. Natl. Acad. Sci. U.S.A. **104**(47), 18387–18391 (2007)
53. M. Erko, D. Wallacher, A. Hoell, T. Hauss, I. Zizak, O. Paris, Density minimum of confined water at low temperatures: a combined study by small-angle scattering of X-rays and neutrons. PCCP **14**(11), 3852–3858 (2012)
54. K. Rottger, A. Endriss, J. Ihringer, S. Doyle, W.F. Kuhs, Lattice-constants and thermal-expansion of $H_2O$ and $D_2O$ Ice ih between 10 and 265 K. Acta Crystallographica B **50**, 644–648 (1994)
55. E.B. Moore, V. Molinero, Structural transformation in supercooled water controls the crystallization rate of ice. Nature **479**(7374), 506–508 (2011)
56. S.V. Lishchuk, N.P. Malomuzh, P.V. Makhlaichuk, Contribution of H-bond vibrations to heat capacity of water. Phys. Lett. A **375**(27), 2656–2660 (2011)
57. X. Zhang, P. Sun, Y. Huang, Z. Ma, X. Liu, J. Zhou, W. Zheng, C.Q. Sun, Water nanodroplet thermodynamics: quasi-solid phase-boundary dispersivity. J. Phys. Chem. B **119**(16), 5265–5269 (2015)
58. X. Zhang, Y. Huang, P. Sun, X. Liu, Z. Ma, Y. Zhou, J. Zhou, W. Zheng, C.Q. Sun, Ice regelation: Hydrogen-bond extraordinary recoverability and water quasisolid-phase-boundary dispersivity. Sci. Rep. **5**, 13655 (2015)
59. X. Zhang, Y. Huang, Z. Ma, Y. Zhou, W. Zheng, J. Zhou, C.Q. Sun, A common supersolid skin covering both water and ice. PCCP **16**(42), 22987–22994 (2014)
60. H. Qiu, W. Guo, Electromelting of confined monolayer ice. Phys. Rev. Lett. **110**(19), 195701 (2013)
61. C. Wang, H. Lu, Z. Wang, P. Xiu, B. Zhou, G. Zuo, R. Wan, J. Hu, H. Fang, Stable liquid water droplet on a water monolayer formed at room temperature on ionic model substrates. Phys. Rev. Lett. **103**(13), 137801–137804 (2009)
62. G. Malenkov, Liquid water and ices: understanding the structure and physical properties. J. Phys. Condens. Matter **21**(28), 283101 (2009)
63. P.G. Debenedetti, H.E. Stanley, Supercooled and glassy water. Phys. Today **56**(6), 40–46 (2003)
64. W. Armstrong, Electrical phenomena the newcastle literary and philosophical society. Electr. Eng. **10**, 153 (1893)
65. L.B. Skinner, C.J. Benmore, B. Shyam, J. Weber, J.B. Parise, Structure of the floating water bridge and water in an electric field. Proc. Natl. Acad. Sci. **109**(41), 16463–16468 (2012)
66. E.C. Fuchs, J. Woisetschlager, K. Gatterer, E. Maier, R. Pecnik, G. Holler, H. Eisenkolbl, The floating water bridge. J. Phys. D-Appl. Phys. **40**(19), 6112–6114 (2007)
67. F. Hofmeister, Concerning regularities in the protein-precipitating effects of salts and the relationship of these effects to the physiological behaviour of salts. Arch. Exp. Pathol. Pharmacol **24**, 247–260 (1888)
68. P. Lo Nostro, B.W. Ninham, Hofmeister phenomena: an update on ion specificity in biology. Chem. Rev. **112**(4), 2286–322 (2012)
69. X. Meng, J. Guo, J. Peng, J. Chen, Z. Wang, J.-R. Shi, X.-Z. Li, E.-G. Wang, Y. Jiang, Direct visualization of concerted proton tunnelling in a water nanocluster. Nat. Phys. **11**(3), 235–239 (2015)
70. X. Zhang, Y. Huang, Z. Ma, Y. Zhou, J. Zhou, W. Zheng, Q. Jiang, C.Q. Sun, Hydrogen-bond memory and water-skin supersolidity resolving the Mpemba paradox. PCCP **16**(42), 22995–23002 (2014)
71. M. Emoto, E. Puttick, *The Healing Power of Water*. (Hay House, Incorporated, 2007)

72. G.H. Pollack, *The Fourth Phase of Water: Beyond Solid, Liquid, and Vapor* (Ebner & Sons Seattle, USA, 2013)
73. H. Sun, COMPASS: an ab initio force-field optimized for condensed-phase applications overview with details on alkane and benzene compounds. J. Phys. Chem. B **102**(38), 7338–7364 (1998)
74. V.F. Petrenko, R.W. Whitworth, *Physics of Ice* (Clarendon Press, 1999)
75. P. Pruzan, J.C. Chervin, B. Canny, Determination of the d2o ice vii-viii transition line by raman-scattering up to 51 gpa. J. Chem. Phys. **97**(1), 718–721 (1992)
76. P. Pruzan, J.C. Chervin, B. Canny, Stability domain of the ice-VIII proton-ordered phase at very high-pressure and low-temperature. J. Chem. Phys. **99**(12), 9842–9846 (1993)
77. R. Aswani, J.C. Li, A new approach to pairwise potentials for water–water interactions. J. Mol. Liq. **134**(1–3), 120–128 (2007)
78. J. Li, Inelastic neutron scattering studies of hydrogen bonding in ices. J. Chem. Phys. **105**(16), 6733–6755 (1996)
79. A.I. Kolesnikovm, J. Li, S.F. Parker, R.S. Eccleston, C.-K. Loong, Vibrational dynamics of amorphous ice. Phys. Rev. B **59**(5), 3569–3578 (1999)
80. J. Li, D. Ross, Evidence for two kinds of hydrogen bond in ice. Nature **365**, 327–329 (1993)
81. T. Yokono, S. Shimokawa, M. Yokono, H. Hattori, Infra-red spectroscopic study of structural change of liquid water induced by sunlight irradiation. Water **1**, 29–34 (2009)

# Chapter 4
# Phase Diagram: Bonding Dynamics

- *Raman spectroscopy visualizes directly O:H–O bonding dynamics across the phase diagram.*
- *Slopes of the $T_C(P_C)$ boundaries categorize the phase boundaries into four groups.*
- *The elongation/compression of a certain segment dictates the negatively/positively-sloped $T_C(P_C)$ boundaries; ∠O:H–O relaxation with equal $\Delta d_x$ dictate the zero or the $\delta(P_C)$-sloped boundaries.*
- *Reproduction of the VII/VIII and the Liquid/Vapor boundaries result in the $E_H$ and the $d_L(P)$ function, respectively.*

**Abstract** Phonon spectrometric mapping of the O:H–O bond relaxation dynamics across the phase diagram along the following paths confirmed the reality of the O:H–O cooperativity mechanism: (i) liquid water at 300 K and ice at 80 K as a function of pressure, (ii) liquid water cooling from 350 to 80 K under the ambient pressure, (iii) mechanical freezing of the ambient water under compression up to 4.0 GPa, and, (iv) liquid water heating from 253 to 753 K under 30 MPa pressure. Observations classify the $T_C(P)$ phase boundaries of water and ice into four types according to their slopes. O:H compression dictates the positively-sloped such as Vapor/Liquid boundaries; the H–O elongation dictates the negatively-sloped such as VII/VIII boundaries, while O:H–O frozen dictates the XI/$I_c$ constant $T_C$ boundary and the symmetrical relaxation governs the X/(VII, VIII) constant $P_C$ boundaries.

## 4.1 Challenge: What Is Behind the Phase Diagram?

Water and ice exhibits at least 17 phases with four types $T_C(P_C)$ boundaries according to their slopes (see Fig. 1.1 in Chap. 1). Classical thermodynamics, such as Clausius–Clapeyron equation [1] and August-Roche-Magnus formula [2], mainly describe the T-P relationship along the boundary between Liquid and Vapor phase in terms of entropy, latent heat, Gibbs free energy, etc. Typical questions regarding the phase diagram include:

1. How to correlate the O:H–O bonding kinetics to the phase diagram?
2. What is the O:H–O bond going on when transits from one phase to another?



C.Q. Sun and Y. Sun, *The Attribute of Water*, Springer Series in Chemical Physics 113, DOI 10.1007/978-981-10-0180-2_4

3. How does the O:H–O bond respond to mechanical compression and thermal excitation?
4. What determines the phase boundaries of different slope and what the hidden information is?

The phase diagram of water and ice in Chap. 1 shows four vertical and three lateral thick lines depict paths along which the Raman spectroscopy probed.

## 4.2 Clarification: O:H–O Bonding Dynamics

Raman spectroscopy probed the H–O and O:H phonon relaxation dynamics along seven different paths as the thick lines indicated:

(1) under constant pressures of 0.1 and 30 MPa (lateral red thick lines) and
(2) constant temperatures of 10, 80, 140 and 298 K (vertical dark thick lines)

These paths cross most phases and boundaries that are categorized according to their slopes [3]:

$$T_C(P_C) = \begin{pmatrix} \delta(P_C) \\ Const \\ f_1(P_C) \\ f_2(P_C) \end{pmatrix} \Rightarrow \frac{dT_C(P_C)}{dp} \begin{cases} \cong \delta(P_C) & (X/(Xi, VII, VIII)) \\ \cong 0 & (IC/XI, XV/VI, etc) \\ > 0 & (Liquid/(Vapor, III, IV, V, VII)) \\ < 0 & (VII/VIII, T_m;\ T_H) \end{cases}$$

The O:H–O bond remains common to all phases and boundaries albeit segmental length and the containing angle relaxation, irrespective of geometries including the $OH_3^+$:$OH^-$ superionic phase obtained at extremely high pressure (2 TPa) and temperature (2000 K) [4]. Raman probing under constant pressures and constant temperatures crosses most phases and boundaries revealed the following:

(1) Within a certain phase, the O:H and the H–O segment relax cooperatively—if one segment becomes stiffer/shorter the other will be softer/longer—resulting from the O–O Coulomb repulsion.
(2) Both segments may change abruptly irregularly when it crosses a phase boundary because of a transition of the ∠O:H–O containing angle and the O–O Coulomb repulsion.
(3) Elongated/compressed segment dominates boundaries of negative/positive slopes; Identical length relaxation/frozen determine boundaries of $\delta(P_c)$/zero slopes.
(4) The abnormal homogeneous freezing (including Widom) line may result from the O:H nonbond elongated by molecular undercoordination and mechanical compression in the particular regime.

## 4.3 History Background

Few records are available on the development of the phase diagram, particularly, on formulating the boundaries in general cases. Available descriptions are mainly on the $T_C(P_C)$ phase boundaries between Liquid and Vapor phase in terms of classical thermodynamics. For instances, Clausius–Clapeyron equation [1] describes water vapor under typical atmospheric conditions (near standard temperature and pressure) and August–Roche–Magnus formula [2] approximates the temperature dependence of the saturation vapor pressure $P_S$:

$$\begin{cases} \frac{dP_s}{dT} = \frac{L_v(T)P_s}{R_v T^2} & (Clausius-Clapeyron) \\ P_s(T) = 6.1094 \exp\left(\frac{17.625T}{T+243.04}\right) & (August-Roche-Magnus) \end{cases} \quad (4.1)$$

where $L_v$ is the specific latent heat of evaporation of water and $R_v$ is the gas constant of vapor.

Rudolf Julius Emanuel Clausius (2 January 1822–24 August 1888), was a German physicist and mathematician and one of the central founders of the science of thermodynamics. His most important paper, *On the Moving Force of Heat*, published in 1850, first stated the basic ideas of the second law of thermodynamics. In 1865 he introduced the concept of entropy

Benoît Paul Émile Clapeyron (26 February 1799–28 January 1864) was a French engineer and physicist, one of the founders of thermodynamics

According to Pauling [5], the nature of the chemical bond bridges the structure and properties of a substance, and therefore, the structure and properties of a substance varies with bond relaxation that can be realized by controlling the external stimuli such as pressure, temperature, coordination environment, chemical composition, electric and magnetic field [6]. This bonding premise considers only the change of the order, length, and energy of the interatomic bond under given excitations. Pauling's premise and the bond relaxation theory [6] inspired us to revisit the physical nature of a complex system by decomposing it into a simple representative of all bonds involved. Therefore, understanding how the water molecule works at transition from the perspective of O:H–O bond relaxation in segmental length and energy, and the containing angle is critical to understanding how water interacts with all the biological molecules in living organisms and how it respond to constraints and excitations.

## 4.4 Quantitative Resolution

### *4.4.1 Energy Required for Phase Transition*

The atomic cohesive energy that is the sum of bond energy over all coordinates of a specific atom determines the thermal stability at the specific atomic site. If the atomic cohesive energy is higher than it is in the bulk, the critical temperature $T_C$ for the local phase transition becomes higher; otherwise, the $T_C$ is lower.

For a given specimen, the bond nature and the total number of bonds do not change before a phase transition taking place. However, the bond length and bond energy will respond to the external stimulus such as coordination environment, temperature, pressure, and electromagnetic radiation field. Therefore, one can focus on the response of the representative for all bonds to approximate the thermodynamic behavior of the entire specimen. Using this approach of local bond average (LBA), one will be free from considering the concepts such as surface stress, surface energy, latent heat, and the entropy, as implemented in the classical thermodynamic theories.

On the other hand, if the applied pressure is increased, all bonds become shorter and stronger because of the volume shrinkage and deformation energy storage. The pressure-induced energy storage is equally distributed to all bonds without needing discrimination of bond distribution in real space. We can then focus on the relaxation of the representative bond, which is the foundation of Fourier transformation in spectrometrics of electrons and phonons. Taking the average over all bonds of the entire specimen renders no physical indication of the approach but a substantial simplification.

Compression stores energy into the representative bond in the way,

$$\int_{P_0}^{P_C} dE = -\int_{V_0}^{V} p dv = \int_{P_0}^{P_C} v dp - \int_{P_0 V_0}^{P_C V_C} d(vp)$$

If this amount of energy is high enough, phase transition takes place at the phase boundary corresponding the critical temperature $T_C$ and pressure $P_C$. For water and ice, the representative O:H–O bond comprises two segments that respond to compression oppositely,

$$\frac{dd_L}{dp} < 0, \frac{dd_L}{dp} > 0$$

The $T_C$ and $P_C$ are thus correlated by ($v_x = S_x d_x$),

$$T_{xC} \propto \sum_{H,L} E_{xC} = \sum_{H,L} \left( E_{x0} - \int_{P_0}^{P_{xC}} p dv_x \right)$$

The $E_{x0}$ is the segmental cohesive energy required for segmental dissociation. The sum of the normalized integrals, as Fig. 4.1 illustrates, determines the pressure trend of the $\Delta T_{xC}$:

$$\frac{\Delta T_C(P_C)}{T_C(P_0)} = -\sum_{H,L} \frac{s_x \int_{P_0}^{P_C} p \frac{dd_x}{dp} dp}{E_{x0}}$$

Therefore, the $T_C$, $P_C$, and the pressure trend of $d_x$ determines the thermodynamic behavior of water and ice in phase relaxation and phase transition. Figure 4.1 shows the energy stored into the O:H and the H–O segment by compression mathematically,

$$-s_x p dv_x = -s_x p \frac{dd_x}{dp} dp = v_x dp - d(pv_x).$$

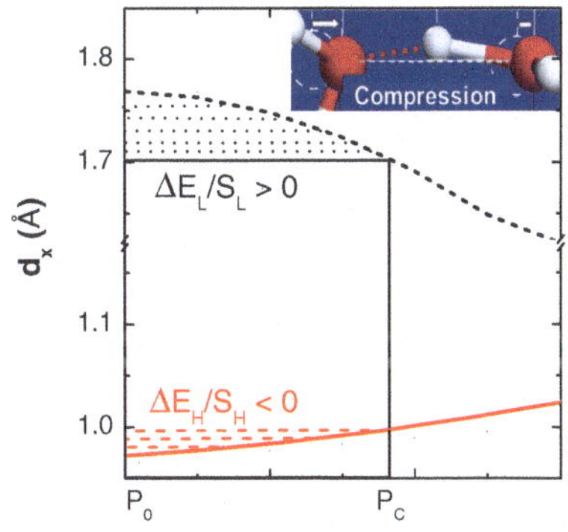


**Fig. 4.1** Compression stores energy into the O:H nonbond and evacuates energy from the H–O bond. The *red shaded area* $\Delta E_H/S_H$ is the H–O energy loss and the *dark shaded area* $\Delta E_L/S_L$ is the O:H nonbond energy gain. $S_L \ll S_H$ is the respective segmental cross-sectional area (see Sect. 4.3). The $\Delta E_H$ or the $\Delta E_L$ determines discriminatively the $T_C$-$P_C$ for a specific phase transition

The black shaded area is energy stored into the O:H and the red energy loss by the H–O segment. The $\Delta E_L > 0$ and the $\Delta E_H < 0$ but $\Delta E_L \ll -\Delta E_H$ because of the O:H shortens longer than H–O elongates and the $S_H \gg S_L$. Generally, the sum of the two parts determines the thermodynamics of water and ice but one segment dominates one occasion for water and ice.

### 4.4.2 *Compression Induced O:H–O Bonding Kinetics*

The Raman frequency shift $\Delta\omega_x$ is proportional to the segmental stiffness that depends on the length and energy in the form $\Delta\omega_x \propto \sqrt{E_x}/d_x \propto \sqrt{Y_x d_x}$. The segment cohesive energy $E_x$ is inversely proportional to its length $d_x$ in a certain power. Phonon blue shift happens if the specific segment becomes shorter.

According to the principle of Fourier transformation, the characteristic phonon spectral peak represents all segments of the same kind disregarding their locations and numbers. Therefore, the Raman phonon spectroscopy straightforwardly tells how the segmented O:H–O bond changes across the entire T–P phase diagram. The following show four paths of Raman probing in the phase diagram under constant temperatures.

#### 4.4.2.1 O:H–O Bond Relaxation at 10 K

The in situ Raman spectroscopy shown in Fig. 4.2 revealed the following [7]:

(1) Both $\omega_H$ and $\omega_L$ undergo a redshift at 1.4 GPa pressure and below, which covers the XI, IX, II, and XV phase and boundaries. The simultaneous $\omega_x$ redshift indicates that both segments undergo compression expansion, which could result from polarization and containing angle bending in these phases. Compression enhances polarization [8].
(2) Increasing pressure from 1.4 to 3.7 GPa, the $\omega_H$ continues its redshift but the $\omega_L$ starts the blue shift, agreeing with expectation of O:H compression and H–O elongation. This transition happens at the boundary between the XV and the VIII phase.
(3) A sudden blueshift happens to both segments at 3.7–5.0 GPa, which suggests the degeneration of the O–O repulsion. This transition pressure corresponds to a certain yet unclear boundary that could not be found from the phase diagram.
(4) A further abrupt $\omega_H$ red and $\omega_L$ blue shift takes place at 5.0–13.0 GPa and then is followed continuously by the expected $\omega_H$ redshift and $\omega_L$ blueshift.

#### 4.4.2.2 80 and 140 K Temperatures

The Raman spectra probed at 80 and 140 K shown in Fig. 4.3 [7] revealed the following:

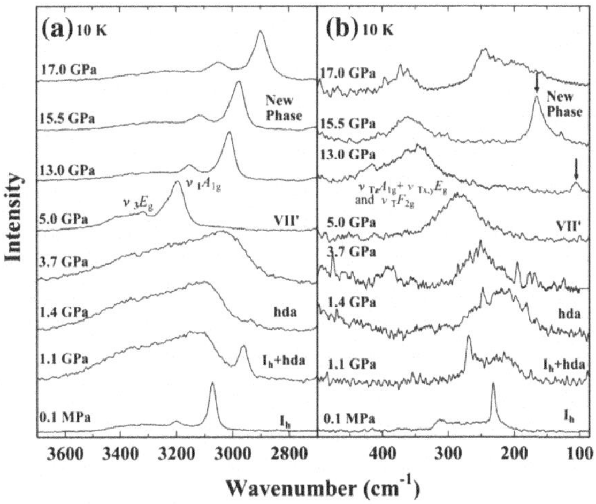


**Fig. 4.2** In situ Raman probing the phase diagram at 10 K revealed versatile cooperative shift of **a** the $\omega_H$ and **b** the $\omega_L$ across the XI, IX, II, XV and VIII phases (Reprinted with permission from [7].)

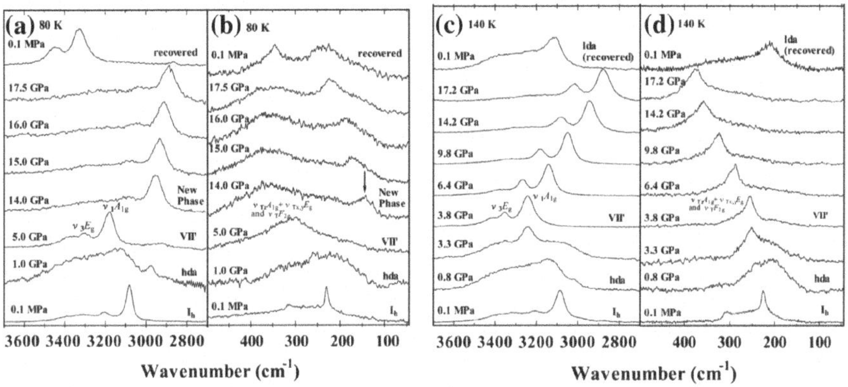


**Fig. 4.3** In situ Raman phonon $\omega_x$ shift at **a**, **b** 80 K and at **c**, **d** 140 K (Reprinted with permission from [7].)

(1) The expected $\omega_H$ redshift and $\omega_L$ blueshift happen continuously at 14.0/3.8 GPa and above for the O:H–O bond at 80/140 K.
(2) Unexpected $\omega_L$ redshift accompanied with $\omega_H$ blueshift happen at pressure up to 1.0/0.8 GPa for O:H–O bond at 80/140 K.
(3) Blueshift happens to both $\omega_x$ at 5.0/3.3 GPa pressure and followed by a sudden $\omega_L$ blueshift and $\omega_H$ redshift transition when pressure increases from 5.0 to 14 GPa at 80 K.
(4) The phonon frequencies remain stable when transited from 3.3 to 3.8 GPa at 140 K temperature.

#### 4.4.2.3 Mechanical Icing of 298 K Water

When subjected to compression, liquid water turns into ice-VI and then ice-VII at room temperature [9]. Figure 4.4 shows the pressure-dependent Raman spectra of water at 298 K. There are two components in each branch, the peaks centered at 75/3450 $cm^{-1}$ correspond to the skin and the ones centered at 200/3200 $cm^{-1}$ to the bulk liquid [10]. The constant 3450 $cm^{-1}$ frequency indicates the hydrophobic contact between water ice and the wall of the diamond cell that contains the testing sample.

Within the Liquid (unapparent), ice VI and VII phases, the $\omega_x$ shifts cooperatively, following the expected trend of compression—$\omega_L$ blueshift accompanied with $\omega_H$ redshift, or O:H compression and H–O elongation.

During the Liquid-VI phase transition, the pressure suddenly drops from 1.33 to 1.14 GPa though the volume of the diamond compression cell containing the water sample shrinks continually and slowly [11]. The sharp spectral features indicate ice formation. At the boundary between ice-VI and ice-VII, the pressure drops from 2.23 to 2.17 GPA. At both pressures of phase transition, the $\omega_H$ and the $\omega_L$ simultaneously

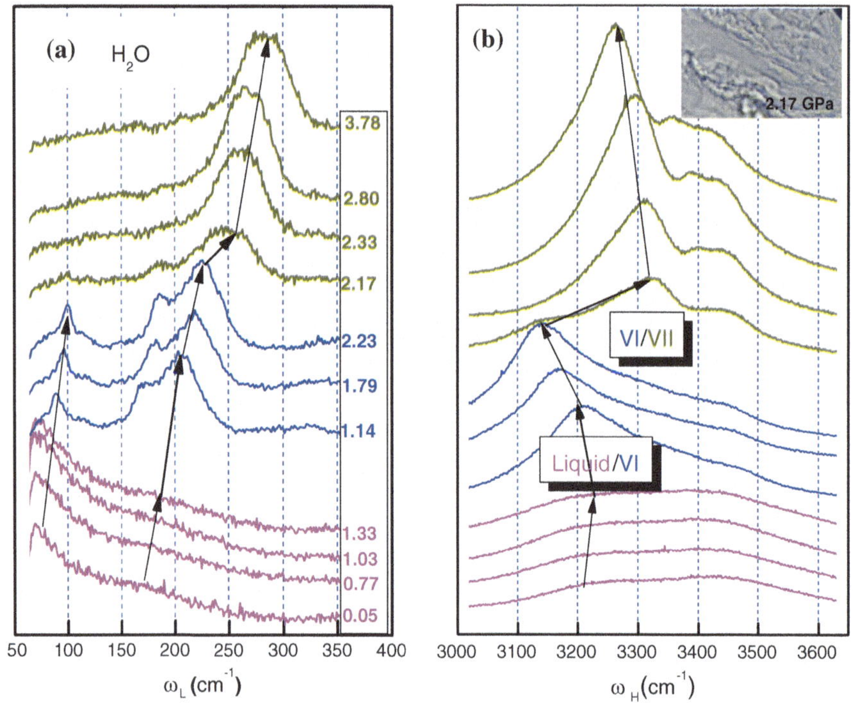


**Fig. 4.4** Raman spectra for mechanical freezing of the ambient deionized water (298 K) reveals abrupt drop in pressure at Liquid/VI (from 1.33 to 1.14 GPa) and at VI/VII (2.23–2.17 GPa) transition. *Inset* (**c**) is the optical image of ice at 2.17 GPa. The simultaneous *blueshift* of the $\omega_L$ and $\omega_H$ indicates that the O:H nonbond and the H–O bond undergo contraction because of degeneration of O–O repulsion (Reprinted with permission from [11].)

**Table 4.1** O:H–O bond compressing relaxation at different phases and boundaries

| T (K) | | XI(o) | XI/IX | IX | IX/II | II | II/XV | XV | XV/VIII | VIII |
|---|---|---|---|---|---|---|---|---|---|---|
| 10 | $\Delta\omega_L$ | <0(till 3.7 GPa) | | | | | | | >0(17.0 GPa) | |
| | $\Delta\omega_H$ | <0 | | | | | | | <0 | |
| | | Ic | Ic/IX | IX | IX/II | II | II/XV | XV/VIII | VIII | |
| 80 | $\Delta\omega_L$ | <0(till 5.0 GPa) | | | | | | | >0(17.5 GPa) | |
| | $\Delta\omega_H$ | >0 | | | | | | | <0 | |
| | | Ic | Ic/IX | IX | IX/II | II | II/XV | XV/VIII | XV/VIII | VIII |
| 140 | $\Delta\omega_L$ | <0 (till 3.8 GPa) | | | | | | | >0(17.2 GPa) | |
| | $\Delta\omega_H$ | >0 | | | | | | | <0 | |
| | | Liquid | L/VI | VI | VI/VII | VII | – | | | |
| P (GPa) | | | 1.33 → 1.14 | | 2.23 → 2.17 | | | | | |
| 298 K | $\Delta\omega_L$ | >0 | ≫0 | >0 | ≫0 | >0 | | | | |
| | $\Delta\omega_H$ | <0 | ≫0 | <0 | ≫0 | <0 | | | | |

The no man's land region shows abnormal $\omega_L$ softening and $\omega_H$ stiffening. Both $\omega_L$ and $\omega_H$ are subject to blue shift at Liquid/VI and VI/VII phase boundaries, indicating Coulomb repulsion weakening. O:H–O bond segmental lengths always relax oppositely to their respective phonon frequencies in sigh (after Figs. 4.2, 4.3 and 4.4 [3, 7])

undergo a blueshift. The $\omega_H$ shifts from 3150 to 3330 at VI/VII phase transition indicates its spontaneous contraction.

The sudden drop in pressure and the simultaneous blueshift of both $\omega_x$ at transition indicate the degeneration of the O–O repulsion and polarization associated with possible bond angle relaxation. Both O:H and H–O segments undergo contraction because of the degenerated Coulomb repulsion.

Table 4.1 summarizes the O:H–O bonding kinetics under compression at these temperatures. In the temperature 140 K and below and in the phases of XI, IX, II, XV and VIII, the O:H–O bond relaxes unexpectedly, indicating much more complicated mechanisms to be explored using the Raman spectroscopy at reduced pressure steps and using computations to gain comprehensive information. Sufficiently small pressure step ensures fine details of the O:H–O bonding in the phases and boundaries across, particular in the No Man's Land regime.

## 4.4.3 *Thermally Stimulated O:H–O Bonding Dynamics*

### 4.4.3.1 Ambient-Pressure Freezing: Quasisolid Phase

Generally, the characteristic $\omega_L$ and $\omega_H$ shift in opposite direction because of the Coulomb repulsion. If one undergoes blue shift, the other does red with limited exception shown in Sect. 4.2. This cooperative phonon relaxation forms the straightforward yet simple way advancing the understanding of multifield phonon spectrometrics. Figure 4.5 shows the Raman spectra of water droplet of millimetre size cooled from 298 to 98 K using programmed liquid nitrogen. The spectra show

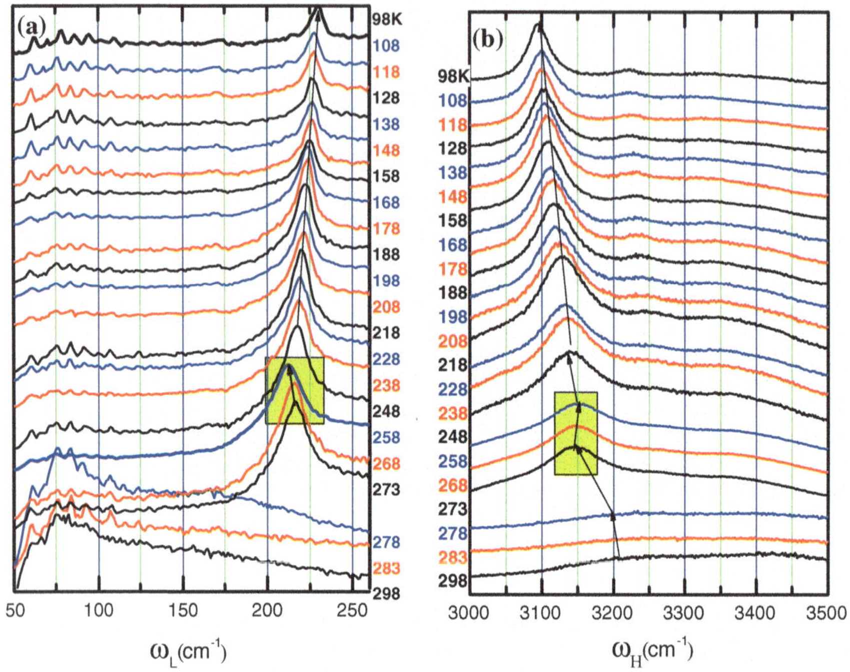


**Fig. 4.5** Temperature-dependent Raman shifts of **a** $\omega_L < 300\ \text{cm}^{-1}$ and **b** $\omega_H > 3000\ \text{cm}^{-1}$ show regions of $T > 273$ K, $273 \geq T \geq 258$ K, and $T < 258$ K (reprinted with permission from [12]). The thermal path crosses the liquid, quasisolid (*shaded*), $I_h$ and $I_c$ phases in the phase diagram. The frequency oscillation follows the relationship defined by the segmental specific heat disparity notion [19]

characteristic peaks of 75/3450 for the skin and 175/3200 $\text{cm}^{-1}$ for the bulk at the ambient temperature, respectively.

Agreeing with the expected O:H–O length, mass density, and stiffness cooperative oscillation over the full temperature range, the spectra show expected three regimes transiting at the liquid/quasisolid/solid phase boundaries at 273 and 258 K, respectively [12]:

(1) In the liquid phase, $T \geq 273$ K, cooling stiffens $\omega_L$ abruptly from 175 to 220 $\text{cm}^{-1}$ and softens $\omega_H$ from 3200 to 3140 $\text{cm}^{-1}$ with indication of ice forming at 273 K. The cooperative $\omega_x$ shift indicates that cooling shortens and stiffens the O:H bond but lengthens and softens the H–O bond in the liquid phase, which confirms again that the O:H bond cooling contraction dominates O:H–O relaxation in liquid phase.

(2) In the quasisolid phase, $273 \geq T \geq 258$ K, the situation reverses. Cooling stiffens $\omega_H$ from 3140 to 3150 $\text{cm}^{-1}$ and softens $\omega_L$ from 220 to 215 $\text{cm}^{-1}$ (see the shaded areas). Consistent with the Raman $\omega_H$ shift measured at temperatures around 273 K [13, 14], the cooperative shift of $\omega_x$ confirms the

switching of the master and the slave roles of the O:H and H–O during freezing; H–O contraction dominates in this quasi-solid phase.

(3) In the solid $I_{h+c}$ phase, T ≤ 258 K, the master–slave role reverts to its behavior in the liquid region, albeit with a different relaxation rate. Cooling from 258 to 98 K stiffens $\omega_L$ from 215 to 230 $cm^{-1}$ and softens $\omega_H$ from 3150 to 3100 $cm^{-1}$ as it cools. Earlier Raman spectroscopy revealed that the $\omega_L$ for bulk ice and $D_2O$ drops monotonically with the rise of temperature and the data fluctuates at 260 ± 10 K [15]. The supplementary peaks at about 300 and 3450 $cm^{-1}$ change insignificantly with temperature; the skin $\omega_H$ of about 3450 $cm^{-1}$ in water and ice is thermally insensitive [16]. The cooling softening of $\omega_H$ agrees with that measured using IR spectroscopy of ice clusters of 8–150 nm size [17]. When the temperature drops from 209 to 30 K, $\omega_H$ shifts from 3253 to 3218 $cm^{-1}$.

(4) Figure 4.6 shows that both the $\omega_H$ and the $\omega_L$ remain almost constant at $T < 60$ K [17]. Using IR spectroscopy, Medcraft et al. [18] measured the size- and temperature-dependence of $\omega_L$ in the temperature range 4–190 K. They found that heating softens the $\omega_L$ at $T > 80$ K but the $\omega_L$ remains almost unchanged below 60 K. This observation evidences that neither the length nor the stiffness or energy of these two segments change in this temperature regime IV because of their extremely low specific heat ($\eta_x \approx 0$).

#### 4.4.3.2 Phase XI O:H–O Bond Frozen

Figure 4.6 shows the $\omega_H$ peak shift with temperature and typical $\omega_L$ spectra for nanodroplet [17, 18]. At temperatures below 80 K, the $\omega_x$ remain almost constant,

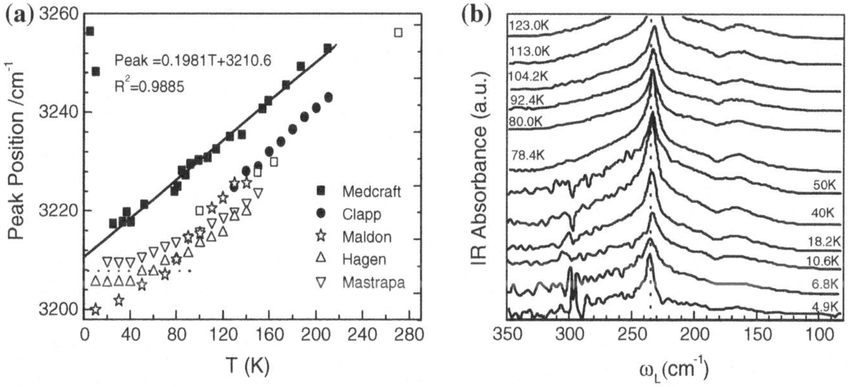


**Fig. 4.6** Insignificant shift of **a** the $\omega_H$ and **b** the $\omega_L$ at $T \leq 60$ K within phase XI for nanodroplet. This indicates that $\eta_x \cong 0$ almost silences the O:H–O bond length and stiffness in this temperature regime [12]. *Broken lines* guide viewing (Reprinted with permission from [17, 18] and references therein.)

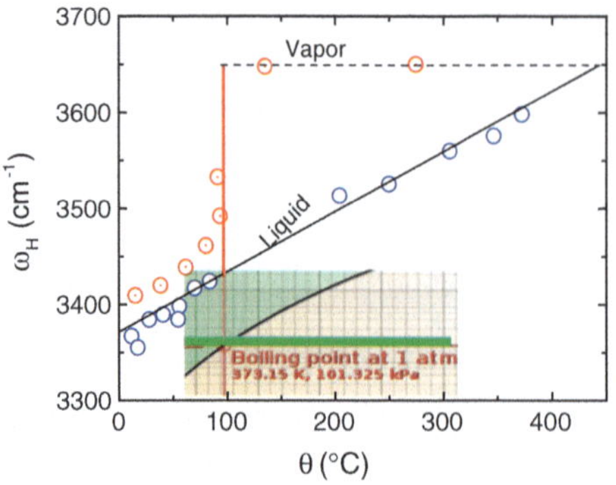


**Fig. 4.7** Infrared absorption $\omega_H$ peak shift of liquid water under ambient pressure and turning to valor phase (*red circles*) and Raman reflection $\omega_H$ peak shift of liquid water to superheated state (blue circles). *Inset* shows the probing path crossing the Liquid-Vapor transition in the phase diagram (Reprinted with permission from [20].)

which indicates that neither the segmental length nor the segmental energy change with temperature within phase XI. This observation is right the prediction of the specific heat disparity notion. At extremely low temperatures, the specific heat $\eta_x \cong 0$, and neither the O:H nor the H–O responds to temperature change except for cooling stretching of ∠O:H–O the containing angle [19].

#### 4.4.3.3 Ambient-Pressure Vaporization

Using IR absorption spectroscopy, Cross et al. [20] probed the $\omega_H$ shift of liquid water at heating under the ambient pressure. As shown in Fig. 4.7, the $\omega_H$ increases monotonically from 3350 to 3650 cm$^{-1}$ for a monomer in two paths. For superheating liquid, the $\omega_H$ increases monotonically but at Liquid/Vapour transition, the $\omega_H$ increases abruptly from 3450 to 3650 cm$^{-1}$. This observation evidences that the H–O bond is subject to heating contraction that is accompanied by O:H heating expansion [12]. The Vapour phase consists of monomers only.

#### 4.4.3.4 30 MPa Liquid Heating

According to the phase diagram in Sect. 4.1, water under 30 MPa pressure remains its liquid state in the temperature range from 253 to 753 K. Figure 4.8 shows the Raman spectra of the H–O stretching vibration, $\omega_H$, of water under such conditions, as probed by Hu et al. [21]. Initially, the spectra are dominated by both the 3200 cm$^{-1}$ bulk mode and the 3450 cm$^{-1}$ skin mode. The bulk mode corresponds to H–O length of 1.00 Å and O:H length of 1.68 Å while the skin mode to the H–O length of 0.95 Å and O:H length of 1.90 Å [22] because of molecular undercoordination induced H–O contraction and O:H elongation [10].

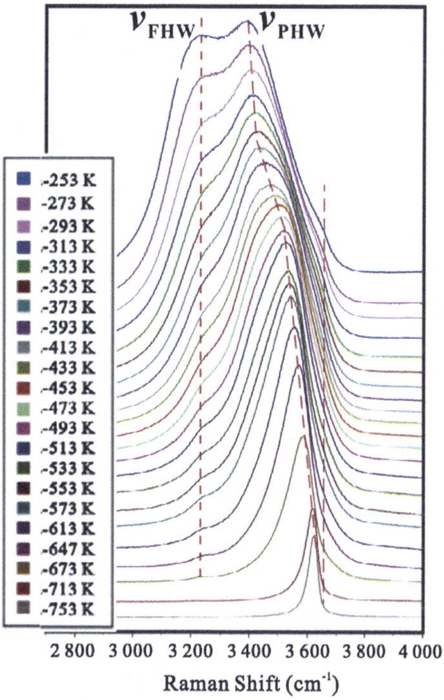


**Fig. 4.8** Raman probing of liquid water along 30 MPa pressure from 253 to 753 K temperature (Reprinted with permission from [21].)

The presence of this skin mode in the present case indicates that the water/cell contacting interface is hydrophobic. As temperature rises, the 3450 cm$^{-1}$ mode shifts upward and increases its intensity rendering the intensity loss of the 3200 cm$^{-1}$ mode. The significant blue shift of the 3450 cm$^{-1}$ up to 3620 cm$^{-1}$ indicates that the H–O bond becomes even shorter and stiffer, which is the same to liquid water heating under the ambient condition [20]. The $\omega_H$ for a monomer is 3650 cm$^{-1}$ and for a dimer it is 3575 cm$^{-1}$. The liquid water under 30 MPa approaches dimers at 673 K and monomers to 753 K.

Raman spectroscopy crossing the Vapor, Liquid, Quasisolid, Ih, IC, and XI phases at the ambient pressure revealed the oscillation of the phonon frequencies and segmental lengths, which agree with the expectation based on segmental specific heat disparity, as discussed in Sect. 4.3. A superposition of the $\eta_x(\Theta_{Dx})$ curves defines two intersecting temperatures that correspond to extreme densities. The extreme densities satisfy the relationships defined by the specific heat disparity principle, as summarized in Table 4.2.

**Table 4.2** O:H–O bond heating relaxation at 0.1 and 30 MPa

| P (MPa) | | XI | $XI/I_c$ | $I_{h+c}$ | $I_h/IV$ | IV(QS) | IV/L | L | L/Vapor | Vapor |
|---|---|---|---|---|---|---|---|---|---|---|
| 0.1 (atom) | $\Delta\omega_L$ | ≅0 | | <0 | ≪0 | >0 | ≪0 | <0 | ≫0 | =0 |
| | $\Delta\omega_H$ | ≅0 | | >0 | ≫0 | <0 | ≪0 | >0 | ≪0 | =0 |
| $\eta_x$ ratio | $\eta_H/\eta_L$ | 1 | | >1 | 1 | <1 | 1 | >1 | | |
| 30 | $\Delta\omega_L$ | Liquid phase at 30 MPa and 253–653 K range | | | | | | <0 | | |
| | $\Delta\omega_H$ | | | | | | | >0 | | |

The phonon frequencies shift cooperatively and oscillates in the full temperature range, follow the rule of O:H–O bond cooperativity and segmental specific heat disparity [12]. At boundaries, the specific heat ratio, $\eta_H/\eta_L = 1$. Within a certain phase, the segment of lower specific heat serves as the master to drive thermal expansion and the other serves as the slave to relax in opposite direction (Sect. 4.3) (after Figs. 4.5, 4.6, 4.7 and 4.8.)

## 4.4.4 *Phase Boundaries: O:H–O Bond Relaxation*

### 4.4.4.1 Formulation

For other 'normal' substance, the $T_C$ is proportional to the atomic cohesive energy, $T_C \propto zE_z$, where $z$ is the effective atomic CN and $E_z$ is the cohesive energy of a bond between the $z$-coordinated atoms [23]. However, for water molecules, the $T_C$ is proportional to the $E_x$ where the subscript x needs to be certain. Molecular CN contributes indirectly to the bond length and bond energy but not to the $T_C \propto E_x$, because of the 'isolation' of the $H_2O$ molecule by its surrounding four lone pairs.

Usually, compression shortens the softer O:H nonbond more than the stiffer H–O bond elongates because of the repulsion between electron pairs on adjacent O ions. The shortened O:H nonbond is associated with energy and vibration frequency gain but the lengthened H–O bond is suffered frequency and energy loss. However, as reveled in **Error! Reference source not found.** and Fig. 4.2, situation may reverse at the No Man's Land regime, where the O:H nonbond dis subject to compression elongation.

The following generalizes the $T_C(P)$ as a function of the segmental volume change $v_x = s_x d_x$ under pressure (x = H for the H–O bond and x = L for the O:H nonbond) with $s_x$ being the cross-sectional area of the specific segment,

$$\frac{\Delta T_C(P_C)}{T_C(P_0)} = -\sum_{H,L} \frac{s_x \int_{P_{x0}}^{P_{xC}} p \frac{dd_x}{dp} dp}{E_{x0}} = \begin{cases} >0 & (dd_x/dp < 0) \\ <0 & (dd_x/dp > 0) \\ \cong 0 & (dd_x/dp \cong 0) \\ \cong \delta(P_{xC}) & (dd_H/dp + dd_L/dp \cong 0) \end{cases} \tag{4.2}$$

The $E_{x0}$ is the reference of the segmental cohesive energy at the ambient conditions. The bonding parameters $d_L$, $d_H$, and $\theta$ dominate the respective $T_C(P_C)$ phase boundary function.

The O:H–O segmental length and stiffness is subject to abruption at boundaries because the relaxation of the network geometry. The relaxation may degenerate the

O–O Coulomb repulsion that cause the abnormal $\omega_H$ blue shift, particularly, for the 298 K water transiting from the VI to the VII phase for instance.

#### 4.4.4.2 H–O Bond Elongation: $dT_C/dP < 0$

The $T_C(P_c)$ profiles for the Liquid/quasisolid and the Ice VII/VIII phase transition epitomize the situation of negative slopes, which requires,

$$\begin{cases} \frac{\Delta T_C(P)}{T_C(P_0)} = -\sum_{H,L} \frac{s_x \int_{Px_0}^{Pxc} p \frac{dd_x}{dp} dp}{E_{x0}} < 0 \\ dd_x/dp > 0 \end{cases}. \quad (4.3)$$

With the known pressure dependence of the $d_x(P)$ for ice [8], one can find that only the $d_H$ satisfy this criterion—compression elongates it. The measured $T_C(P_C)$ for the VII–VIII phase transition [24–26], and the $T_m(P_C)$ for ice melting (−22°C at 210 MPa; +6.5°C at −95 MPa) [27, 28] can therefore be reproduced.

Matching the $T_C(P_C)$ profiles for VII/VIII phase transition yields an $E_H$ value of 3.97 eV by taking the H atomic diameter of 0.106 nm as the H–O bond diameter [29]. This $E_H$ value agrees with the energy of 4.66 eV for dissociating the H–O bond of water molecules deposited on a $TiO_2$ substrate with less than a monolayer coverage, and 5.10 eV for dissociating water monomers in the gaseous phase [30]. Molecular undercoordination differentiates values of 5.10, 4.66 and 3.97 eV for the H–O bond in various coordination environments [10]. Reproduction of both the $T_C(P_C)$ for the VII–VIII and the $T_m(P_C)$ indicates that the $d_x(P)$ follows the same quantitative relationship in these two situations (Fig. 4.9).

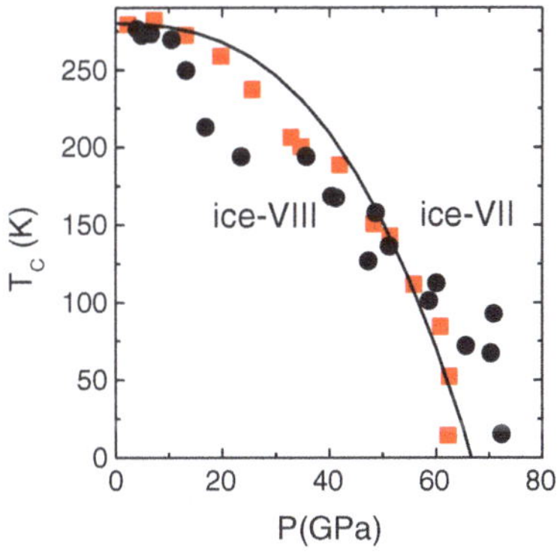


**Fig. 4.9** Theoretical reproduction of the measured $T_C(P_C)$ for the VII/VIII phase transition confirms that the $E_H$ dictates the $T_C(P_C)$ with derivative of $E_H$ = 3.97 eV for bulk water and ice (Reprinted with permission from [8, 9].)

#### 4.4.4.3 No Man's Land O:H Elongation: $dT_C/dP < 0$

The inset in the phase diagram in Chap. 1 shows the $T_H$ for homogeneous freezing and the Widom line nearby the supercooled liquid (quasisolid) region, which satisfy (4.3) but this time the $T_H(P_C)$ is dominated by $d_x = d_L$ lengthening because of the following:

(1) Water nanodroplet is subject to $\omega_H$ stiffening and $\omega_L$ softening, which offsets the $\Theta_{Dx}$ temperatures and disperses the quasisolid phase boundaries or the critical temperature of extreme densities—raising the $T_m$ and lowering the $T_H$ that are a few degrees away from the quasisolid phase boundaries [31].
(2) As shown in the phase diagram and Fig. 4.2, compression softens the O:H nonbond and lowers the $\Theta_{DL}$ in some occasions, which further depresses the critical temperature for the least density, nearby the $T_N$.
(3) These two effects cause the depression of the $T_N$ as pressure increases up to 200 MPa. The region between $T_m$ and $T_N$ is the quasisolid phase, often called supercooled state. This understanding may help in explaining the shape of the Widom line, which may correspond to the least density temperature. However, this suggestion is subject to further phonon spectrometrics verification.

#### 4.4.4.4 O:H Nonbond Contraction: $dT_C/dP > 0$

The positively-sloped $T_C(P_C)$ boundary for Liquid–Vapor phase transition (vaporization or dew formation) meeting the criteria,

$$\begin{cases} \dfrac{\Delta T_C(P)}{T_C(P_0)} = -\sum\limits_{H,L} \dfrac{s_x \int_{P_{x0}}^{P_{xc}} p \frac{dd_x}{dp} dp}{E_{x0}} > 0 \\ dd_x/dp < 0 \end{cases} \quad (4.4)$$

Such boundaries result from O:H nonbond shortening. With the known $T_C(P_C)$ profile one can resolve the $d_x(P)$ relationship. A numerical fitting to the $T_C(P)$ profile for the Liquid-Vapor transition [32, 33], see Fig. 4.10a, for instance, yields the following,

$$\frac{T_C(LnP)}{225.337} = 1 + 0.067757 \times \exp\left(\frac{LnP}{5.10507}\right) = 1 + A\exp\left(\frac{LnP}{B}\right) = 1 + AP^{\frac{1}{B}} \quad (4.5)$$

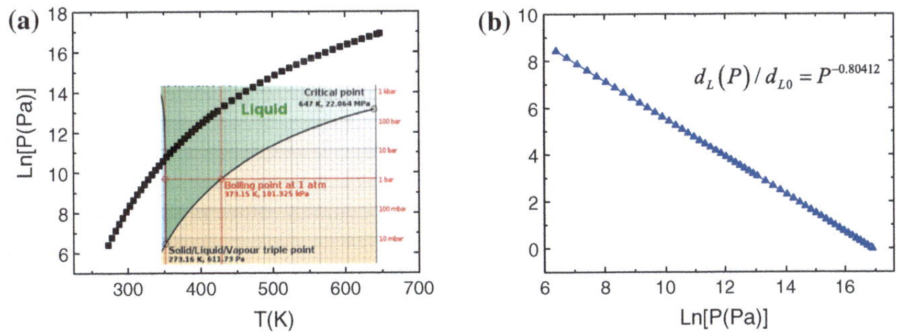


**Fig. 4.10** **a** Fitting the $T_C(P)$ boundary for Liquid-Vapor phase transition (sourced from [32, 33]) leads to **b** the pressure trend of the $d_L(P)$ along the Liquid-Vapor phase boundary (in Ln scales) (Reprinted with permission from [3].)

Equaling (4.4) and (4.5),

$$AP^{\frac{1}{B}} = -\frac{s_L}{E_{L0}}\int_{V_0}^{V} p\frac{dd_L}{dp}dp \tag{4.6}$$

yields the pressure dependent,

$$d_L(P) = d_L(P_0, T_0) \times \exp\left(\frac{-LnP}{1.2436}\right) = d_L(P_0, T_0) \times P^{-0.80412} \tag{4.7}$$

Figure 4.10b plots the pressure trend of the O:H (approaches to O–O in vapor) length at the Liquid-Vapor phase boundary obtained by taking the $d_L$(0.1 MPa, 373 K) as unity of standard. Indeed, the slope of the $d_L(P)$ is negative. Likewise, one can obtain the pressure trend of the O:H distance iterating the same for a specific phase boundary of $dT_C/dP > 0$.

#### 4.4.4.5 O:H–O Bond Frozen: $dT_C/dP \cong 0$

The zero-sloped $T_C(P_C)$ boundary for XI-$I_C$ phase transition means that the $T_C(P_C)$ boundary is energy independent, neither the O:H nor the H–O undergoes relaxation,

$$\begin{cases} \frac{\Delta T_C(P)}{T_C(P_0)} = -\sum_{H,L} \frac{s_x \int_{P_{x0}}^{P_{xC}} p\frac{dd_x}{dp}dp}{E_{x0}} = 0 \\ dd_x/dp = 0 \end{cases} \tag{4.8}$$

$T_C(P) \approx 70$ K (this TC vary with droplet size) at the $I_c$-XI phase boundary is within the regime where the specific heat $\eta_x \approx 0$ [12]. Neither the O:H nor the H–O bond is subject to length and energy change but the ∠O:H–O containing angle become dominance in the structure relaxation [12].

#### 4.4.4.6 O:H–O Bond Symmetrization: $dT_C/dp \cong \delta(P_C)$

Boundaries such as X-(XI, VII, VIII) transition occurs at rather high pressures and the $T_C$ is insensitive to temperature,

$$\begin{cases} \frac{\Delta T_C(P)}{T_C(P_0)} = -\sum_{H,L} \frac{s_x \int_{P_{X0}}^{P_{XC}} p \frac{dd_x}{dp} dp}{E_{x0}} = \delta(P_C) = \begin{cases} 1, & P = P_C \\ 0, & else \end{cases} \\ \frac{dd_L}{E_{L0}dp} + \frac{dd_H}{E_{H0}dp} = 0 \end{cases} \quad (4.9)$$

Insignificant length relaxation happens to both segments at these boundaries but geometrical evolution takes place. Figure 4.11 shows the (VII, VIII)/X phase boundaries obtained using path-integral MD calculations [34] and IR and Raman spectroscopic measurements [35–39]. The ice-X phase boundary at pressures about 60 GPa changes insignificantly with temperature. Heating O:H elongation and H–O contraction compensated for mechanical compression on the O:H–O bond relaxation [34, 40]. Indeed, the O:H and the H–O are identical in length of 0.11 Å at the X/(VII, VIII) boundary under 60 GPa compression [8].

Table 4.3 summarizes O:H–O bond relaxation at different boundaries. In place of classical thermodynamics, the notion of O:H–O bond relaxation dynamics is more revealing.

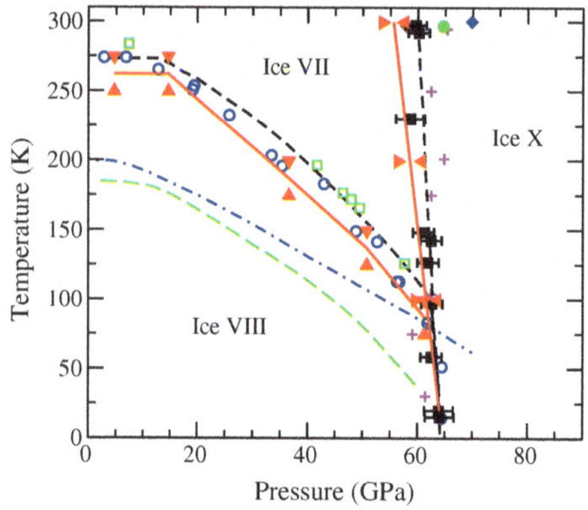


**Fig. 4.11** Phase diagram for ice VII, VIII, and X transition with the X phase boundary at 60 GPa varying insignificantly with temperature. *Lines* denote calculated values [34, 41]; *scattered symbols* are experimental results [35–39] (Reprinted with permission from [34].)

**Table 4.3** O:H–O bond relaxation origin for the slopes of phase boundaries [3]

| | | Typical boundaries | Origin |
|---|---|---|---|
| $dT_c/dP$ | <0 | VII/VIII, IX/II, II/V, $I_h$/Liquid | $\Delta d_H > 0$: H–O elongation |
| | | Widom line, $T_H$ (homogeneous freezing) | $\Delta d_L > 0$: O:H elongation (Table 4.1) |
| | >0 | II/III, liquid/(III, IV, V, VII), liquid/vapor | $\Delta d_L < 0$: O:H compression |
| | =0 | Ic/XI, XV/V, etc. | $\Delta d_H = \Delta d_L = 0$ ($\eta_x \cong 0$) |
| | ∞ | X/XI, X/(VII, VIII), IX/(XI, Ic, Ih), II/XV/VIII | $\Delta d_H = \Delta d_L < 0$ (O:H and H–O length symmetrization) |

## 4.5 Summary

Raman probing the phase diagram revealed the following:

(1) O:H compression and H–O elongation dominate the O:H–O bond relaxation in the pressure domain except for at very low temperature though the latter remains to be refined.
(2) O:H–O bond specific heat disparity governs its thermally oscillating relaxation in regimes of liquid, quasisolid, $I_h$ and $I_C$, and XI phases.
(3) H–O bond relaxation dictates the negatively-sloped $T_C(P)$ boundaries for the VII/VIII and the Liquid/Quasisolid phase transition.
(4) O:H nonbond relaxation dominates not only the positively-sloped $T_C(P)$ profile for the Liquid-Vapor phase transition but also the negatively-sloped $T_C(P)$ profile for quasisolid–solid homogeneous freezing. However, the latter is subject to further experimental and computational verification.
(5) O:H–O containing angle relaxation governs those of zero-sloped ($T_C$ = constant at the Ic/XI boundary) or infinitely-sloped ($P_C$ = constant at the (XII, XIII)-X boundary) $T_C(P)$ profiles. Numerical reproduction of the negatively-sloped $T_C(P)$ curves results in the H–O cohesive energy 3.97 eV for water and ice and duplication of the $T_C(P)$ boundary for Liquid-Vapor transition turns out the pressure trend of the O:H length change. Raman examination revealed that the O:H and H–O contract simultaneously when turning the ambient water into ice VI by compression.

## References

1. K. Wark, *Generalized Thermodynamic Relationships in the Thermodynamics*, 5th edn. (McGraw-Hill, Inc, New York, 1988)
2. O. Alduchov, R. Eskridge, *Improved Magnus' Form Approximation of Saturation Vapor Pressure* (Department of Commerce, Asheville, NC, 1997)

3. X. Zhang, P. Sun, Y. Huang, T. Yan, Z. Ma, X. Liu, B. Zou, J. Zhou, W. Zheng, C.Q. Sun, Water's phase diagram: from the notion of thermodynamics to hydrogen-bond cooperativity. Prog. Solid State Chem. **43**, 71–81 (2015)
4. Y. Wang, H. Liu, J. Lv, L. Zhu, H. Wang, Y. Ma, High pressure partially ionic phase of water ice. Nat. Commun. **2**, 563 (2011)
5. L. Pauling, *The Nature of the Chemical Bond*, 3rd edn. (Cornell University press, Ithaca, 1960)
6. C.Q. Sun, *Relaxation of the Chemical Bond*. Springer Series in Chemical Physics, Vol. 108 (Springer, Heidelberg, 2014), 807 pp
7. Y. Yoshimura, S.T. Stewart, M. Somayazulu, H.K. Mao, R.J. Hemley, Convergent Raman features in high density amorphous ice, ice VII, and ice VIII under pressure. J. Phys. Chem. B **115**(14), 3756–3760 (2011)
8. C.Q. Sun, X. Zhang, W.T. Zheng, Hidden force opposing ice compression. Chem. Sci. **3**, 1455–1460 (2012)
9. M. Chaplin, *Water structure and science,* http://www.lsbu.ac.uk/water/
10. C.Q. Sun, X. Zhang, J. Zhou, Y. Huang, Y. Zhou, W. Zheng, Density, elasticity, and stability anomalies of water molecules with fewer than four neighbors. J. Phys. Chem. Lett. **4**, 2565–2570 (2013)
11. X. Zhang, T. Yan, B. Zou, C.Q. Sun, *Mechano-freezing of the ambient water,* http://arxiv.org/abs/1310.1441
12. C.Q. Sun, X. Zhang, X. Fu, W. Zheng, J.-L. Kuo, Y. Zhou, Z. Shen, J. Zhou, Density and phonon-stiffness anomalies of water and ice in the full temperature range. J. Phys. Chem. Lett. **4**, 3238–3244 (2013)
13. I. Durickovic, R. Claverie, P. Bourson, M. Marchetti, J.M. Chassot, M.D. Fontana, Water-ice phase transition probed by Raman spectroscopy. J. Raman Spectrosc. **42**(6), 1408–1412 (2011)
14. X. Xue, Z.-Z. He, J. Liu, Detection of water-ice phase transition based on Raman spectrum. J. Raman Spectrosc. **44**(7), 1045–1048 (2013)
15. G.P. Johari, H.A.M. Chew, T.C. Sivakumar, Effect of temperature and pressure on translational lattice vibrations and permittivity of ice. J. Chem. Phys. **80**(10), 5163 (1984)
16. T.F. Kahan, J.P. Reid, D.J. Donaldson, Spectroscopic probes of the quasi-liquid layer on ice. J. Phys. Chem. A **111**(43), 11006–11012 (2007)
17. C. Medcraft, D. McNaughton, C.D. Thompson, D.R.T. Appadoo, S. Bauerecker, E.G. Robertson, Water ice nanoparticles: size and temperature effects on the mid-infrared spectrum. PCCP **15**(10), 3630–3639 (2013)
18. C. Medcraft, D. McNaughton, C.D. Thompson, D. Appadoo, S. Bauerecker, E.G. Robertson, Size and temperature dependence in the far-IR spectra of water ice particles. Astrophys. J. **758** (1), 17 (2012)
19. Y. Huang, X. Zhang, Z. Ma, Y. Zhou, W. Zheng, J. Zhou, C.Q. Sun, Hydrogen-bond relaxation dynamics: resolving mysteries of water ice. Coord. Chem. Rev. **285**, 109–165 (2015)
20. P.C. Cross, J. Burnham, P.A. Leighton, The Raman spectrum and the structure of water. J. Am. Chem. Soc. **59**, 1134–1147 (1937)
21. Q. Hu, X. Lü, W. Lu, Y. Chen, H. Liu, An extensive study on Raman spectra of water from 253 to 753 K at 300 MPa: a new insight into structure of water. J. Mol. Spectrosc. **292**, 23–27 (2013)
22. X. Zhang, Y. Huang, Z. Ma, Y. Zhou, W. Zheng, J. Zhou, C.Q. Sun, A common supersolid skin covering both water and ice. PCCP **16**(42), 22987–22994 (2014)
23. C.Q. Sun, Size dependence of nanostructures: impact of bond order deficiency. Prog. Solid State Chem. **35**(1), 1–159 (2007)
24. K. Aoki, H. Yamawaki, M. Sakashita, Observation of Fano interference in high-pressure ice VII. Phys. Rev. Lett. **76**(5), 784–786 (1996)

25. M. Song, H. Yamawaki, H. Fujihisa, M. Sakashita, K. Aoki, Infrared absorption study of Fermi resonance and hydrogen-bond symmetrization of ice up to 141 GPa. Phys. Rev. B **60** (18), 12644 (1999)
26. P. Pruzan, J.C. Chervin, E. Wolanin, B. Canny, M. Gauthier, M. Hanfland, Phase diagram of ice in the VII-VIII-X domain. Vibrational and structural data for strongly compressed ice VIII. J. Raman Spectrosc. **34**(7–8), 591–610 (2003)
27. G. Malenkov, Liquid water and ices: understanding the structure and physical properties. J. Phys.-Condens. Matter **21**(28), 283101 (2009)
28. J.L. Green, D.J. Durben, G.H. Wolf, C.A. Angell, Water and solutions at negative pressure: Raman spectroscopic study to −80 Megapascals. Science **249**(4969), 649–652 (1990)
29. C.Q. Sun, H.L. Bai, B.K. Tay, S. Li, E.Y. Jiang, Dimension, strength, and chemical and thermal stability of a single C-C bond in carbon nanotubes. J. Phys. Chem. B **107**(31), 7544–7546 (2003)
30. S.A. Harich, D.W.H. Hwang, X. Yang, J.J. Lin, X. Yang, R.N. Dixon, Photodissociation of $H_2O$ at 121.6 nm: a state-to-state dynamical picture. J. Chem. Phys. **113**(22), 10073–10090 (2000)
31. X. Zhang, P. Sun, Y. Huang, Z. Ma, X. Liu, J. Zhou, W. Zheng, C.Q. Sun, Water nanodroplet thermodynamics: quasi-solid phase-boundary dispersivity. J. Phys. Chem. B **119**(16), 5265–5269 (2015)
32. J. Baez, *Phase diagram of water* (2012), http://math.ucr.edu/home/baez/chemical/725px-Phase_diagram_of_water.svg.png
33. Chemical data, http://chemlab.qdu.edu.cn/html/other/bhszqb.htm
34. D.D. Kang, J. Dai, H. Sun, Y. Hou, J. Yuan, Quantum similation of thermally driven phase transition and O k-edge absorption of high-pressure ice. Sci. Rep. **3**, 3272 (2013)
35. M. Song, H. Yamawaki, H. Fujihisa, M. Sakashita, K. Aoki, Infrared investigation on ice VIII and the phase diagram of dense ices. Phys. Rev. B **68**(1), 014106 (2003)
36. P. Pruzan, J.C. Chervin, B. Canny, Stability domain of the ice-VIII proton-ordered phase at very high-pressure and low-temperature. J. Chem. Phys. **99**(12), 9842–9846 (1993)
37. A.F. Goncharov, V.V. Struzhkin, H.-K. Mao, R.J. Hemley, Raman spectroscopy of dense $H_2O$ and the transition to symmetric hydrogen bonds. Phys. Rev. Lett. **83**(10), 1998–2001 (1999)
38. K. Aoki, H. Yamawaki, M. Sakashita, H. Fujihisa, Infrared absorption study of the hydrogen-bond symmetrization in ice to 110 GPa. Phys. Rev. B **54**(22), 15673–15677 (1996)
39. V.V. Struzhkin, A.F. Goncharov, R.J. Hemley, H.K. Mao, Cascading Fermi resonances and the soft mode in dense ice. Phys. Rev. Lett. **78**(23), 4446–4449 (1997)
40. D.D. Kang, J.Y. Dai, J.M. Yuan, Changes of structure and dipole moment of water with temperature and pressure: a first principles study. J. Chem. Phys. **135**(2), 024505 (2011)
41. K. Umemoto, R.M. Wentzcovitch, S. de Gironcoli, S. Baroni, Order-disorder phase boundary between ice VII and VIII obtained by first principles. Chem. Phys. Lett. **499**(4–6), 236–240 (2010)

# Chapter 5
# O:H–O Bond Asymmetrical Potentials

- *O:H–O bond persists in all phases irrespective of crystal geometry or structural fluctuation.*
- *O:H–O approximates an asymmetrical oscillator pair coupled by O–O Coulomb repulsion.*
- *Lagrangian solution transforms the segmental length and vibration frequency into the respective force constant and cohesive energy, which maps the potential paths of the O:H–O bond at relaxation.*
- *One can calibrate the O:H–O bond segmental length, vibration frequency, cohesive energy, and the mass density of water ice with any one of them as a known input.*

**Abstract** Lagrangian solution of oscillator dynamics transforms the observed H–O bond and O:H nonbond lengths and their characteristic phonon frequencies ($d_x$, $\omega_x$) into their respective force constants and cohesive energies ($k_x$, $E_x$), which results in mapping of the potential paths for the O:H–O bond cooperative relaxation under stimulus. Results show that molecular undercoordination not only reduces its size ($d_H$) with enhanced H–O energy from the bulk value of 3.97 to 5.10 eV for a $H_2O$ monomer but also enlarges their separation ($d_L$) with O:H energy reduction from 95 to 35 meV for a dimer. The H–O energy gain raises the melting point of water skin from the bulk value 273 to 310 K, and the O:H energy loss lowers the freezing temperature of a 1.4 nm sized droplet from the bulk value 258 to 202 K. However, compression does the opposite to molecular undercoordination on bond relaxation but the same on polarization.

## 5.1 Challenge: Symmetrical or Asymmetrical?

Hydrogen bond interaction potential is the key for its responding to external stimulus. Figure 5.1 shows the typical double-well potentials of the symmetrical [1] and the asymmetrical [2] forms. These potentials facilitate the Ice Rules of Bernal-Fowler-Pauling [3, 4]. Debating remains on the following issues:

(1) Is the potential a long- or short-range order, symmetric or asymmetric?
(2) How does the potential evolve with external stimuli?
(3) Does the $H^+$ proton undergoes "two-in two-out" frustration or intermolecular "tunneling" transition?



C.Q. Sun and Y. Sun, *The Attribute of Water*, Springer Series in Chemical Physics 113, DOI 10.1007/978-981-10-0180-2_5

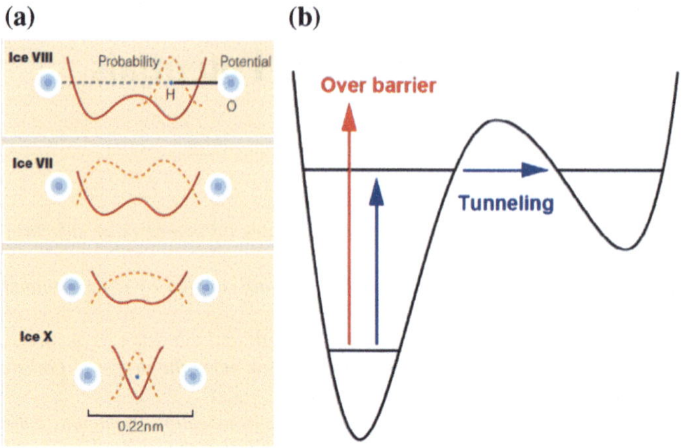


**Fig. 5.1** **a** The symmetrical double-well potential [1] describes the $H^+$ frustration in two identical sites between adjacent oxygen anions, which reduce into a single well when subjected to compression towards the ice-X phase with identical O:H and H–O length. **b** The asymmetrical double-well potential [2] facilitates $H^+$ tunneling from one location to the next (*lifted*) and reversely. Transition requires excitations from the reactant ground state to the higher excited state and proceeds via a process of over barrier (*red arrow*) or a vibration—assisted barrier tunneling (*blue arrow*). These potentials facilitate the Ice Rules of Bernal-Fowler-Pauling [3, 4] (Reprinted with permission from [1, 2].) (color online)

(4) How can one quantify the potential using existing experimental and theoretical approaches?

The left symmetrical potentials describe the "two-in two-out" proton frustration of a water molecule [4]. The right asymmetrical potentials represent the $H^+$ proton tunneling between two molecules [3]. The $H^+$ and one electron changes their location to turn the lone pair of one oxygen into H–O bond and the H–O into lone pair of the other neighboring oxygen, which only alters the O:H–O bond direction without rendering the $sp^3$ configuration. However, no means is available yet to certain or probe the shape of the O:H–O bond potentials, particularly, the potential paths for the O:H–O bond relaxing under stimulus.

## 5.2 Clarification: Asymmetrical, Coupled, and Short Range

Figure 5.2 shows the potential paths for the O:H–O bond contraction by compression [5] and elongation by molecular undercoordination [6], derived from the measured segmental length and phonon frequency ($d_x$, $\omega_x$) at each point of equilibrium under excitation [7, 8] using Lagrangian transformation. The apparently

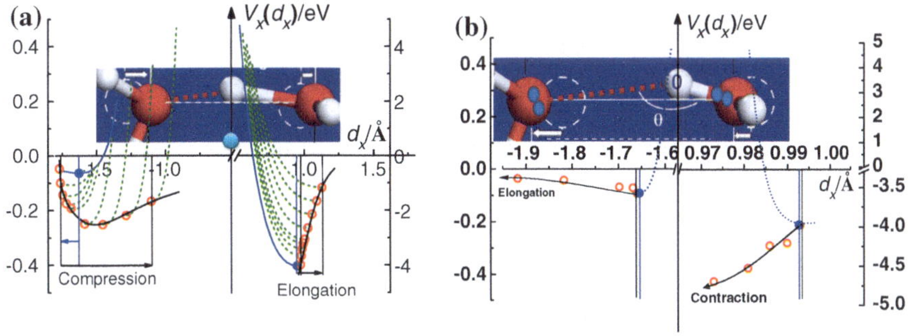


**Fig. 5.2** Potential paths (*red circles*) for the O:H–O bond **a** contraction and **b** elongation. Plot (**a**) is derived from mechanical compression of ice at 80 K (r. to l.: $P = 0, 5, 10, 15, 20, 30, 40, 50, 60$ GPa) [5] and (**b**) from molecular undercoordination in $(H_2O)_N$ clusters at the ambient (r. to l.: $N = 6, 5, 4, 3, 2$) [6]. *Blue dots* are states at the $V'_x = 0$ equilibrium without involvement of O–O repulsion. *Red circles* in the leftmost (**a**) and in the rightmost (**b**) are states at $V'_x + V'_C = 0$ equilibrium with O–O repulsion being involved. Rest *red circles* are states subjecting to the $V'_x + V'_C + f_{ex} = 0$. The $f_{ex}$ is the hidden non-conservative force due to stimulus, $V'_x$ is the gradient of the inter- and intra-molecular potential and $V'_C$ is the gradient of O–O Coulomb potential (Reprinted with permission from [5, 6].) (color online)

mobile $H^+$ proton is fixed presently as the coordination origin for the O:H–O bond that has the following attributes:

(1) O:H–O segmental disparity and O–O Coulomb repulsion define the segmental potentials and their adaptive and cooperative relaxation in responding to perturbation.
(2) O:H–O bond elongation occurs under electrification, hydrophobic capillary confinement, liquid and solid heating, molecular undercoordination, quasisolid cooling, or tension by lengthening the O:H nonbond and shortening the H–O bond.
(3) Mechanical compression, base hydration, liquid and solid cooling, quasisolid heating shortens the O:H–O bond in the opposte manner, elongating O:H more than H–O shortening.
(4) O:H–O bond compression enlarges the $H_2O$ molecular size ($d_H$) but reduces their separations ($d_L$) associated with H–O bond stretching phonon softening and O:H phonon stiffening. O:H–O bond elongation has the opposite effect on segmental lengths and phonon frequencies.

## 5.3 Historical Background

The accurate description of hydrogen-bond interactions is of vital importance not only to computational modeling of biological systems but also to understanding and manipulating O:H–O bonding dynamics. The intra- and intermolecular interactions determine the geometry and physical anomalies of water and ice. However, little effort has been made in probing the potentials because of the limitation of theoretical approaches and experimental methods, preventing one from judging the actual situation of the potentials experimentally [9] or computationally [10].

According to the work of Frank Stillinger published in 1980 [11],

> A comprehensive molecular theory for water is needed for two reasons: first, this substance is a major chemical constituent of our planet's surface and as such it may have been indispensable for the genesis of life. Second, it exhibits a fascinating array of unusual properties both in pure form and as a solvent. Physical scientists have attempted to respond to this need for nearly a century, but a satisfactory molecular theory has only begun to emerge in 1970s.

Stillinger referred to the progress made in understanding the nature and geometry of the non-covalent interactions between water molecules via hydrogen bonding or lone pair nonbonding interaction referred herewith. He focused on the spontaneous creation and annihilation of local clusters of several water molecules, characterized by strong hydrogen bonds and nearly tetrahedral angles. He presented a molecular-level view of water's structure, saying that

> All the properties of water and aqueous solutions ultimately must be explained in terms of intermolecular forces that are present.

John Finney [12] commented in his monograph entitled "*Water? What's so special about it?*" published in 2004, that:

> Although we pay lip service to the biological importance of water, we do not understand what it is about the molecule that makes it a particularly 'fit' molecule to form the 'matrix of life'.

Finney proposed that water's anomalous properties (such as its density maximum and singularities in its thermodynamic response functions) should not be perceived as "mysteries", as they can be explained on the basis of water molecule interactions at the molecular level. He claimed that

> Some demystification of water might help us to understand more clearly its role in the molecular-level processes that are important for maintaining life.

The focus in his article was also on the development in understanding the role of tetrahedral geometry in governing the local order of water, and the central role of the understanding in explaining water anomalies. However, Finney suggested that such understanding "may be incidental to the molecular-level biological fitness of water", admitting that the molecular-level picture falls short of providing a comprehensive theory of water's role in the genesis of life.

On the basis of neutron and X-ray diffraction, Alan Soper [13] proposed a softer core potential as he believes that many of the classical water potentials might have a core which is strongly repulsive at short distances producing too sharp a peak in the real-space at too short a distance. Leetmaa et al. [14] reported on the consistency of such potential models with infrared/Raman and X-ray absorption spectra. However, Leetmaa claimed that the overall agreement of calculated spectra based upon the established models is still unsatisfactory, and that no water model is available to describe equally well the IR/Raman phonon and electron (X-ray absorption) spectroscopy, and neutron and X-ray diffraction observations.

They furthermore showed that there is no strict proof of tetrahedral water based on diffraction and IR/Raman data, and that the tetrahedral structure model must, to fit diffraction data, be less structured than most models obtained from molecular dynamics simulations [15]—which was also strongly criticized, since it had been demonstrated that by using computer simulation fitting procedures. As commented in the *Discussion and Debate: Water Complexity—More than a Myth*?, perfectly acceptable fits to scattering data are always attainable even if they were physically meaningless [13].

Chumaevskii and Rodnikova [16] noted that the inconsistency of Raman spectra of water reveals a non-homogeneity of the H-bond network related to a (bi-, tri-) furcation of hydrogen bonds as due to the presence of defects in the hydrogen bond network, which correspond to different conformations of fully hydrogen-bonded five-molecule structural units.

Teixeira [1] suggested in 1998 that a symmetrical double-well potential exists between neighboring O ions to accommodate the $H^+$ proton "frustrating" [4] between these two wells, see Fig. 5.1a. When the O ions are forced closer, the pairing potential wells turn into single located midway between O ions. Figure 5.1b shows an asymmetrical potentials represent the $H^+$ proton tunneling between two molecules [3]. The $H^+$ and one electron changes their location to turn the lone pair of one oxygen into H–O bond and the H–O into lone pair of the other neighboring oxygen, which only alters the O:H–O bond direction without rendering the $sp^3$ configuration.

In 2004, Wernet et al. [17] hypothesized instead the existence of an asymmetrical H-bonding potential which Soper [9] further investigated by assuming different charges on hydrogen protons in order to create this asymmetry and examine whether that could be supported by diffraction data. Considerable efforts have also been made since then by Wikfeldt et al. [18], Leetmaa et al. [15], Nilsson and Pettersson [19], and Kuhne and Khaliullin [20]. Kumagai [2] proposed in 2015 an asymmetrical "double-well" potential for the proton tunneling transition in the process of $(H_2O)_2 \rightarrow OH_3^+ : OH^-$ transition noted by Bernal and Fowler [3] to explain his observations using low-temperature STM.

However, the state-of-the-art techniques, including X-ray/neutron diffraction, electron spectroscopy, and IR/Raman phonon spectroscopy, could hardly certain this assumption until recently [5, 6] when the presently confirmed potentials of asymmetrical, coupled, short-range nature.

Setting the mobile $H^+$ as the coordination origin and considering the O:H–O disparity with O–O repulsion make the things real yet simple. A Lagrangian-Laplace transformation of the observed segmental length and phonon frequency of the O:H–O bond oscillator pair into the respective force constant and cohesive energy, which enable probing the potential paths for the O:H–O bond undergoing relaxation under various stimuli.

## 5.4 Quantitative Resolution

### 5.4.1 Lagrangian Oscillating Dynamics

The segmented O:H–O bond performs as an asymmetric oscillator pair coupled by the Coulomb interaction and bridged by the H atom at the fixed coordination origin [21]. The reduced mass of the $(H_2O)$:$(H_2O)$ oscillator is $m_L = 18 \times 18/(18 + 18) m_0 = 9m_0$ and that of the H–O oscillator is $m_H = 1 \times 16/(1 + 16)m_0 = 16/17m_0$ with $m_0$ being the unit proton mass of $1.66 \times 10^{-27}$ kg. The motion of the coupled O:H–O oscillator pair follows Lagrangian equation [5]:

$$\frac{\mathrm{d}}{\mathrm{d}t}\left(\frac{\partial L}{\partial(\mathrm{d}q_i/\mathrm{d}t)}\right) - \frac{\partial L}{\partial q_i} = Q_i \tag{5.1}$$

The Lagrangian $L = T - V$ consists of the total kinetic energy $T$ and the total potential energy $V$ of the three-body system. $Q_i$ is the generalized non-conservative forces. Here, it is the external forces driving relaxations including mechanical compression, molecular undercoordination, electrification, thermal excitation, and any radiation absorption [8]. The time-dependent $q_i(t) = u_x$, represents the generalized variables, denoting the coordinates of O atoms in the O:H–O bond represented by the x springs. The kinetic energy $T$ consists of two terms, as the H is fixed as the coordination origin without motion,

$$T = \frac{1}{2}\left[m_L\left(\frac{\mathrm{d}u_L}{\mathrm{d}t}\right)^2 + m_H\left(\frac{\mathrm{d}u_H}{\mathrm{d}t}\right)^2\right] \tag{5.2}$$

The potential energy $V$ is composed of three terms: the vdW-like L–J interaction $V_L(u_L)$, the exchange interaction $V_H(u_H)$, and the Coulomb repulsion $V_C(u_C) = V_C(u_H - u_L)$. The $u_x$ is the coordinate and $d_x$ is the respective segmental length. Therefore, $d_{C0} = u_{H0} - u_{L0}$ is the nearest O–O distance at equilibrium without involvement of the Coulomb repulsion. The $d_C = u_H - u_L$ is the O–O distance with Coulomb repulsion inclusion. The O atom dislocates from the equilibrium to another equilibrium by $\Delta_x = u_x - u_{x0}$ upon the Coulomb repulsion being involved. A harmonic approximation of these potentials at each equilibrium by omitting the higher-order terms in their Taylor's series yields,

$$\begin{aligned} V &= V_L(u_L) + V_H(u_H) + V_C(u_H - u_L) \\ &\approx [V_L(u_{L0}) + V_H(u_{H0}) + V_C(u_C)] - V_C'\Delta u_C + \frac{1}{2}\left[k_L\Delta u_L^2 + k_H\Delta u_H^2 + k_C\Delta u_C^2\right] \end{aligned} \tag{5.3}$$

where $V_x(u_{x0})$ is the potential well depths ($n = 0$ terms) of the respective segment. $\Delta u_x$ is the amplitude of vibration. Noting that the Coulomb potential is out of equilibrium and that the repulsion force is always positive, one can then expand these potentials at their joint equilibrium based on harmonic approximation, which ensures sufficient accuracy of the elucidated potential paths [5].

In the Taylor series, the terms of $n = 1$ equal zero at equilibrium, $V'_x(u_{x0}) = 0$, without Coulomb repulsion; $V'_x(u_x) + V'_C(u_C) = 0$, with Coulomb repulsion; $V'_x(u_x) + V'_C(u_C) + f_z = 0$, under the non-conservative force of external excitation. Here $V'_C \neq 0$ denotes the first order derivative of the Coulomb potential at the joint equilibrium of $V'_x(u_x) + V'_C(u_C) = 0$. Terms of $n = 2$, or the potential curvatures, denote the force constants, i.e., $k_x = V'' = \mathrm{d}^2V_x/\mathrm{d}u_x^2|_{u_{x0}}$ for harmonic oscillators. Terms of $n \geq 3$ are insignificant and negligible, which is adequate for seeking the nature and trend of the potential paths.

The addition of the Coulomb repulsion dislocates both O ions slightly outwardly by $\Delta_x$ from their respective initial equilibrium, shifting the atomic distance from $d_{x0}$ to $d_x = d_{x0} + \Delta_x$. The Coulomb repulsion raises the respective potential well depths from $E_{x0}$ to $E_x$ by the same amount as both oxygen ions are subject to the same repulsive force. The non-conservative force dislocates $O^{2-}$ anions in the same direction because of Coulomb repulsion. The O in the O:H dislocates more than that of the H–O because of the potential energy disparity.

Substituting (5.2) and (5.3) into (5.1) yields the motion equations for the coupled O:H–O oscillator pair,

$$\begin{cases} m_L \dfrac{\mathrm{d}^2u_L}{\mathrm{d}t^2} + (k_L + k_C)u_L - k_C u_H + k_C(\Delta_L - \Delta_H) - V'_C - f_z = 0 \\ m_H \dfrac{\mathrm{d}^2u_H}{\mathrm{d}t^2} + (k_H + k_C)u_H - k_C u_L - k_C(\Delta_L - \Delta_H) + V'_C + f_z = 0 \end{cases} \tag{5.4}$$

This coupled motion equation could be readily decoupled using Laplace transformation and reverted by using the inverse Laplace transformation upon gained the solution.

## 5.4.2 Analytical Solutions

### 5.4.2.1 General Solution

A Laplace transformation of the Lagrangian equation turns out the following,

$$\begin{cases} u_L = \dfrac{A_L}{\omega_L}\sin\omega_L t + \dfrac{B_L}{\omega_H}\sin\omega_H t \\ u_H = \dfrac{A_H}{\omega_L}\sin\omega_L t + \dfrac{B_H}{\omega_H}\sin\omega_H t. \end{cases} \tag{5.5}$$

The coefficient denotes the vibrational amplitude of the respective segment. The $\omega_x$ is the angular frequency of the x oscillator. The O coordinates of the O:H and the H–O segments share the same form of eigen values of stretching vibration. The following correlates the $k_x$ and the $\omega_x$,

$$k_{H,L} = 2\pi^2\mu_{H,L}c^2\left(\omega_L^2 + \omega_H^2\right) - k_C \pm \sqrt{\left[2\pi^2\mu_{H,L}c^2(\omega_L^2 - \omega_H^2)\right]^2 - \mu_{H,L}k_C^2/\mu_{L,H}} \quad (5.6)$$

where $c$ is the velocity of light travelling in vacuum. Omitting the Coulomb repulsion will decouple the coupled oscillators into the isolated $(H_2O)$:$(H_2O)$ and H–O dimer oscillators with respective vibration frequency of $\omega_x = (2\pi c)^{-1}\sqrt{k_x/\mu_x}$, which is the same as empirically derived from the Taylor series of the respective interaction potential.

#### 5.4.2.2 Specific Solution

Given the frequency $\omega_x$ and the force constant $k_C$ of Coulomb repulsion, one can obtain the force constant $k_x$, the potential well depth $E_{x0}$, and the binding energy $E_x$, at each equilibrium of these two segments during relaxation under stimulation. The force constant due to Coulomb repulsion is $k_C = q_O^2/\left(2\pi\varepsilon_r\varepsilon_0 d_C^3\right) = 0.17$ eV/Å$^2$ at equilibrium by taking $\varepsilon_0 = 8.85 \times 10^{-12}$ F/m for vacuum, $\varepsilon_r = 3.2$ for ice, and $q_O = 0.652$ e for the undercoordinated skin molecules, as optimized using DFT calculations [22].

Calculations [5] resulted in that the $k_x(\omega_x)$ changes with the respective segmental vibration frequency $\omega_x$. The terms of the $k_L(\omega_H)$ and the $k_H(\omega_L)$ remain, however, almost constant. Therefore, (5.6) simplifies into the expression for the coupled oscillators,

$$\omega_x = (2\pi c)^{-1}\sqrt{\frac{k_x + k_C}{m_x}} \quad (5.7)$$

Table 5.1 shows the procedure and the outcome of derivatives with known ($d_x$, $\omega_x$) as input. For instance, at equilibrium, $V_x'(u_{x0}) = 0$ defines the $E_{x0}$ and $d_{x0}$; $V_x'(u_x) + V_c'(u_C) = 0$ defines the $E_x$. The difference between $E_x$ and $E_{x0}$ is the Coulomb repulsion energy $E_C$. Likewise, $V_x'' = k_x$, $V_x'' + V_C'' = k_x + k_C$, see (5.7). The known $k_C$ and $\omega_x$ define the $V_x''$ value and parameters involved in the respective potential function.

With the derived values of $k_L = 2.39$ eV/Å$^2$, $k_H = 36.09$ eV/Å$^2$, and the known $E_H = 3.97$ eV and $E_L = 0.095$ eV [21], one can determine all the parameters in the L–J and the Morse potentials, as well as the force fields of the O:H–O bond under the ambient conditions,

**Table 5.1** Derivatives of the L–J and Morse potentials at each equilibrium positions, which transits the known ($d_x$, $\omega_x$) into the ($k_x$, $E_x$)

| Derivatives | O:H potential | H–O potential | Outcome |
|---|---|---|---|
| $V_x'(u_{x0}) = 0$ | – | – | $E_{x0}$ |
| $V_x'(u_x) + V_c'(u_C) = 0$ | – | – | $E_x$, $u_x$ |
| $V_x'' = k_x(k_C, \omega_x)$ | $72E_{L0}/d_{L0}^2$ | $2\alpha^2 E_{H0}$ | $\alpha$ |
| $V_x'''$ | $-1512E_{L0}/d_{L0}^3$ | $-6\alpha^3 E_{H0}$ | – |

$$\begin{cases} k_L = 72E_{L0}/d_{L0}^2 = 2.39\,\mathrm{eV/A^2} \\ k_H = 2\alpha^2 E_{H0} = 36.09\,\mathrm{eV/A^2} \end{cases}$$
or
$$\begin{cases} E_{L0} = 2.39 \times 1.654^2/72 = 0.091\,\mathrm{eV} \\ \alpha = (36.09/3.97/2)^{1/2} = 2.13\,\mathrm{A}^{-1} \end{cases} \quad (5.8)$$

With the known Coulomb potential, the computed segmental $d_x$ and $\omega_x$ functions for $(H_2O)_N$ clusters, [8] one can obtain parameters in the *L–J* ($E_{L0}$, $d_{L0}$) and the Morse ($E_{H0}$, $\alpha$) potentials, as illustrated in Table 5.1.

The $E_L$ may have different values subjecting to experimental conditions or approaching methods. For instance, the $E_L$ varies from 0.05 eV for ice at zero pressure to 0.25 eV at 40 GPa and turns to 0. 16 eV at 60 GPa pressure [5]. It is therefore meaningful to consider the $E_x$ values associated with experimental conditions. The currently used $E_H = 3.97$ eV was obtained by fitting the $T_C - P$ profiles for both ice VII–VIII phase transition and ice melting and the $E_L = 0.095$ eV by fitting to the temperature dependence of the water skin stress [21].

## 5.5 Potential Paths for the Relaxed O:H–O Bond

### *5.5.1 O:H–O Bond Compression*

#### 5.5.1.1 Force Constant Versus Phonon Frequency

If the $\omega_x$ and the $k_C$ are given, one can obtain the force constants $k_x$, the potential well depths $E_{x0}$, and the binding energy $E_x$, at each equilibrium state of the two segments. The force constant due to Coulomb repulsion is $k_C = q_O^2/\left(2\pi\varepsilon_r\varepsilon_0 d_C^3\right)$. The $q_O = 0.62$ e. In this situation, the $k_C$ equals to 0.17 eV/Å$^2$ at 0 GPa.

Figure 5.3 shows the functional dependence of the $k_x$ on the $\omega_x$. The $k_L$ increases from 1.44 to 5.70 eV/Å$^2$ while the $k_H$ increases from 21.60 to 42.51 eV/Å$^2$ with their respective frequency. The cross of $k_L(\omega_H)$ and the $k_H(\omega_L)$ remains, however, almost constant. Therefore, (5.6) can be simplified as,

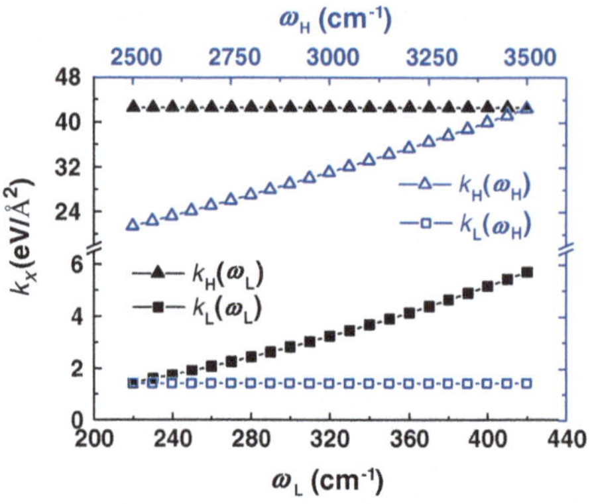


**Fig. 5.3** Functional dependence of the force constants $k_x$ on the vibration frequencies of $\omega_x$ with $k_C = 0.17$ eV/Å$^2$. The $k_L$ and the $k_H$ are much more sensitive to their respective frequency than the across $k_L(\omega_H)$ and the $k_H(\omega_L)$ that remain almost constant

$$k_{H,L} = 4\pi^2 c^2 m_{H,L} \omega_{H,L}^2 - k_C$$

or,

$$\omega_{H,L} = (2\pi c)^{-1} \sqrt{\frac{k_{H,L} + k_C}{m_{H,L}}} \tag{5.9}$$

With the measured $\omega_L = 237.42$ cm$^{-1}$ and $\omega_H = 3326.14$ cm$^{-1}$ for the ice-VIII phase under the atmospheric pressure [23–25], (5.7) yields $k_L = 1.70$ eV/Å$^2$ and $k_H = 38.22$ eV/Å$^2$. With the known $d_L = 0.1768$ nm and $d_H = 0.0975$ nm under Coulomb repulsion [7], one can obtain the free length $d_{L0}$ of 0.1628 nm, and the $d_{H0}$ of 0.0969 nm without involvement of the Coulomb repulsion. Coulomb repulsion lengthens the O–O distance from 0.2597 to 0.2733 nm by 0.0136 nm at ambient pressure.

With the derived values of $k_L = 1.70$ eV/Å$^2$, $k_H = 38.22$ eV/Å$^2$, and $E_{H0} = 3.97$ eV, one can determine all the parameters in the vdW-like and the Morse potentials, as well as the force fields of the O:H–O bond at the ambient pressure,

$$\begin{cases} k_L = 72 E_{L0}/d_{L0}^2 = 1.70\,\text{eV}/\text{A}^2 \\ k_H = 2\alpha^2 E_{H0} = 38.22\,\text{eV}/\text{A}^2 \end{cases}$$

or

$$\begin{cases} E_{L0} = 1.70 \times 1.628^2/72 = 0.062\,\text{eV} \\ \alpha = (38.22/3.97/2)^{1/2} = 2.19\,\text{A}^{-1} \end{cases} \tag{5.10}$$

#### 5.5.1.2 Segmental Length-Stiffness-Energy Correlation

MD decomposition of the measured V(P) profile of compressed ice [26] results in the $d_x(P)$ curves [7] that meet at $d_L = d_H = 0.11$ nm under 59 ~ 60 GPa pressure, which matches exactly the measured proton symmetrization of ice [27, 28]. This matching indicates that the MD derived $d_x(P)$ relation represents the true cooperativity of the segmental length $d_x$. Equation (5.11) and Fig. 5.4a, b show the numerical generalization of the pressure dependent $d_x(P)$ and the measured phonon relaxation dynamics $\omega_x(P)$ [23–26]. Lagrangian solution transforms the measured $\omega_x$ and $d_x$ into the force constant $k_x$ and bond energy $E_x$, at the equilibrium states. Table 5.1 and Fig. 5.3 feature the outcomes:

$$\begin{pmatrix} d_H/0.9754 \\ d_L/1.7687 \\ \omega_H/3326.140 \\ \omega_L/237.422 \end{pmatrix} = \begin{pmatrix} 1 & 9.510 \times 10^{-2} & 0.2893 \\ 1 & -3.477 \times 10^{-2} & -1.0280 \\ 1 & -0.905 & 1.438 \\ 1 & 5.288 & -9.672 \end{pmatrix} \begin{pmatrix} P^0 \\ 10^{-2}P^1 \\ 10^{-4}P^2 \end{pmatrix} \quad (5.11)$$

The Coulomb interaction offsets the intrinsic force constant only of the oscillators by the $k_x$. The measured $d_x$ and the $k_x$ that is available based on the known $m_x$, $k_C$, and the measured $\omega_x$, determine the other parameters involved in the respective potentials. The force constants and the bond energies can also be formulated as functional dependent on the pressure:

$$\begin{pmatrix} k_H/38.223 \\ k_L/1.697 \\ E_H/3.970 \\ E_L/0.046 \end{pmatrix} = \begin{pmatrix} 1 & -1.784 & 3.113 \\ 1 & 13.045 & -15.258 \\ 1 & -1.784 & 3.124 \\ 1 & 25.789 & -49.206 \end{pmatrix} \begin{pmatrix} P^0 \\ 10^{-2}P^1 \\ 10^{-4}P^2 \end{pmatrix} \quad (5.12)$$

Results shown in Fig. 5.4c indicate that the $k_C$ keeps almost constant when the O:H–O contracts. The $k_L$ increases more rapidly than the $k_H$ reduces because of the potential disparity. Figure 5.5d shows that increasing the pressure from 0 to 20 GPa stiffens the O:H bond from 0.046 to 0.190 eV while softens the H–O bond from 3.97 to 3.04 eV, as a result of O:H–O bond cooperativity. As featured in Table 5.1, when the pressure is increased to 60 GPa, the $k_L = 10.03$ eV/Å$^2$ and $k_H = 11.16$ eV/Å$^2$, the $E_L$ recovers slightly.

An extrapolation of the $d_x(P)$ in Fig. 5.4 matches the measured length symmetry of H-bond at 60 GPa [27]. The $k_C$ couples the two segments in a manner of almost constant. Agreement between the scattered data of harmonic approximation at each equilibrium site and the continuum functions $V_x(d_x)$ in (d) verifies the validity of the on-site harmonic approximation. Equations (5.11) and (5.12) generalize the pressure dependence of these parameters.

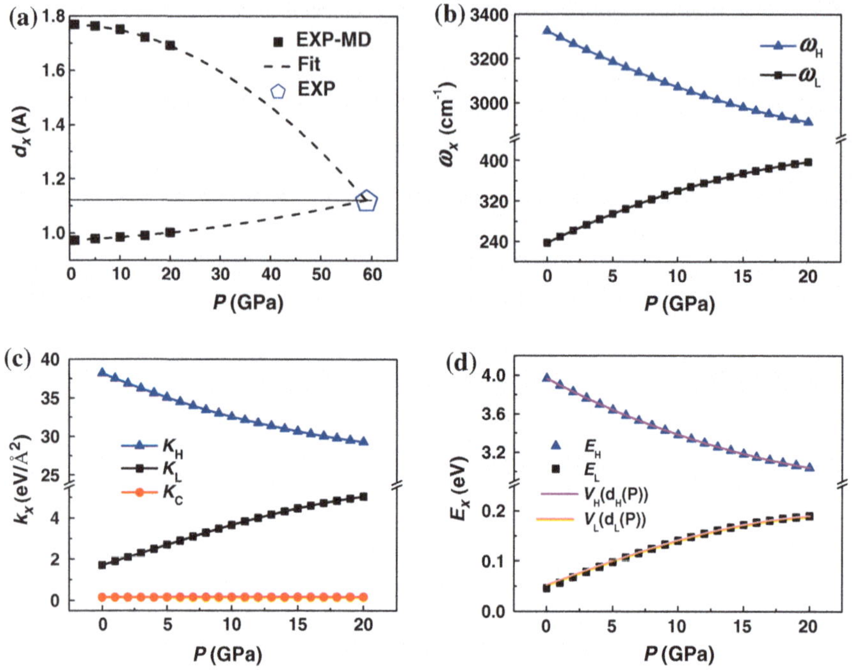


**Fig. 5.4** Cooperative relaxation of the segmental **a** length $d_x(P)$ [7, 26] (EXP–MD denotes MD derivatives from matching to the measured V–P profile of ice), **b** phonon frequency $\omega_x(P)$ [23–26], **c** force constant $k_x(P)$, and (**d**) cohesive energy $E_x(P)$ of the compressed O:H–O bond

### 5.5.1.3 Potential Paths

Table 5.1 also shows that compression shortens and stiffens the softer O:H bond, which lengthens and softens the H–O bond through the Coulomb repulsion, which compress the O–O distance towards O:H and H–O length symmetrization [7, 23–27, 29]. As the $d_L$ is shortened by 4.3 % from 1.768 to 1.692 Å, the $d_H$ lengthens by 2.8 % from 0.975 to 1.003 Å when the pressure is increased from 0 to 20 GPa [7]. When the pressure goes up to 60 GPa, the O:H bond almost equals to the elongated H–O bond in length of about 1.11 Å, forming a symmetric O:H–O bond without altering the nature of it though the segmental length and force constant approaches to equality. This observation clarifies that the $sp^3$-hybridized oxygen could hardly be de-hybridized by compression in the ice X phase. The asymmetrical, local, short-range potentials pertaining to the segmented O:H–O bond have thus been resolved and clarified, irrespective of the manner or the driving force for relaxation.

Figure 5.2b shows the $V_x$(r) paths for the O:H–O bond compression. Both oxygen ions moves firstly outward from their ideal equilibrium because of Coulomb repulsion. Upon being compressed, both O ions move to the right along the O:H–O potential paths with respect to the H coordination origin. The intrinsic equilibrium

**Table 5.2** Pressure dependence of the O:H–O segmental bond energy ($E_x$), force constant ($k_x$), and the deviated displacement ($\Delta_x$) from the equilibrium position [5]

| $P$ (GPa) | $-E_L$ (eV) | $-E_H$ (eV) | $k_L$ (eV/Å$^2$) | $k_H$ (eV/Å$^2$) | $-\Delta_L$ ($10^{-2}$ nm) | $\Delta_H$ ($10^{-4}$ nm) |
|---|---|---|---|---|---|---|
| 0 | 0.046 | 3.97 | 1.70 | 38.22 | 1.41 | 6.25 |
| 5 | 0.098 | 3.64 | 2.70 | 35.09 | 0.78 | 6.03 |
| 10 | 0.141 | 3.39 | 3.66 | 32.60 | 0.51 | 5.70 |
| 15 | 0.173 | 3.19 | 4.47 | 30.69 | 0.36 | 5.26 |
| 20 | 0.190 | 3.04 | 5.04 | 29.32 | 0.27 | 4.72 |
| 30 | 0.247 | 2.63 | 7.21 | 25.31 | 0.14 | 3.85 |
| 40 | 0.250 | 2.13 | 8.61 | 20.49 | 0.08 | 3.16 |
| 50 | 0.216 | 1.65 | 9.54 | 15.85 | 0.05 | 2.71 |
| 60 | 0.160 | 1.16 | 10.03 | 11.16 | 0.04 | 3.35 |

Subscript $x$ denotes L and H. The measured $d_x(P)$ and $\omega_x(P)$ [7, 23–26] are used as input in calculations

position of the oxygen in the H–O bond almost superposes on its equilibrium position under Coulomb repulsion, with a displacement of only $6.25 \times 10^{-4}$ nm at first. However, in the O:H, the repulsion induced displacement of O is $1.41 \times 10^{-2}$ nm, evidencing that the O:H nonbond is much softer than the H–O bond.

The combination of the mechanical compression, interoxygen Coulomb repulsion, and the segmental disparity drive both $O^{2-}$ relax along the potential paths. The dislocation steps ($10^{-4}$ nm for H–O and $10^{-2}$ nm for O:H) of O atoms are within the limit of fluctuation [30] and they are too small to be detected in reality. However, an accumulation of these mall steps turned out the final proton symmetrization of ice under compression (Table 5.2).

### 5.5.2 *O:H–O Bond Elongation*

Given the $\omega_L = 218$ cm$^{-1}$ and $\omega_H = 3225$ cm$^{-1}$ for the $(H_2O)_6$ cluster [8] and the known $k_C$, (5.7) yields $k_L = 2.39$ eV/Å$^2$ and $k_H = 36.09$ eV/Å$^2$. With the known $d_L = 1.659$ Å and $d_H = 0.993$ Å, under Coulomb repulsion, one can obtain the length $u_{L0}$ of $-1.654$ Å, and the $u_{H0}$ of 0.993 Å at equilibrium. Coulomb repulsion lengthens the O–O distance by 0.0053 Å from 2.6467 to 2.6520 Å, under the equilibrium conditions with and without Coulomb repulsion involvement. These values may be subject to accuracy due to the artifact in calculation algorithm and limitation of available probing techniques. However, these values do reflect the nature origin and the general trend of the segmental relaxation- under Coulomb repulsion and molecular undercoordination that dominates the performance of molecules involved in the skins, capillary confined and free nanodroplets and nanobubbles.

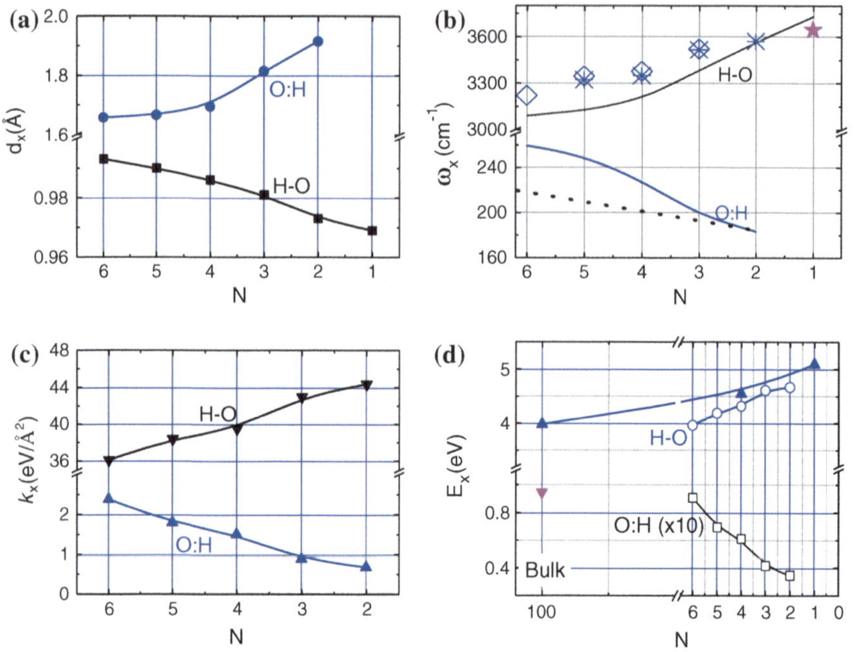


**Fig. 5.5** $(H_2O)_N$ cluster size (N) dependence of the segmental **a** length $d_x$, **b** phonon frequency $\omega_x$, **c** force constant $k_x$, and **d** cohesive energy $E_x$ of the O:H–O bond [8]. Scattered data in (**b**) are measurements [39–42]. The broken line in (**b**) modifies the calculation $\omega_L$ with respect to that 220 cm$^{-1}$ for bulk water and to that the calculated $\omega_H$ matches the measured value at N = 2 (calculations often over estimate $\omega_L$). Solid triangles in (**d**) are measured $E_L$ for the bulk (0.095 eV) [32] and $E_H$ are measurements from bulk (3.97 eV) [22], skin (4.66 eV) [21], and gaseous monomers (5.10 eV) [43]

#### 5.5.2.1 Segmental Length-Stiffness-Energy Correlation

Figure 5.5a, b and Table 5.3 featured the $N$ dependence of the $d_x$ and $\omega_x$ values obtained using DFT computations [8]. The present analytical solution transforms these eluded $d_x$ and $\omega_x$ into the force constant and bond energy ($k_x$, $E_x$) of the respective O:H–O segment, from one equilibrium to another, as the N is reduced. This transformation maps the potential paths for the O:H–O bond at elongation, see results featured in Table 5.4 and Fig. 5.5c, d. The counters of the equilibrium points follow the forms of the respective potentials. Therefore, one has to consider the relaxation of the bond potentials at equilibrium when the object is subject to stimulus, instead of keeping the potentials at fixed equilibria.

It is noted that the potential paths derived from cluster size reduction reconcile the paths due to O:H–O bond elongation without discrimination of the stimulus such as electrification, tension, solid and liquid heating, despite accuracy refinement in a specific situation.

**Table 5.3** DFT-derived segmental length $d_x$, ∠O:H–O containing angle θ, and phonon frequency $\omega_x$ for $(H_2O)_N$ clusters [8]

| | Monomer | Dimer | Trimer | Tetramer | Pentamer | Hexamer | Bulk [31] |
|---|---|---|---|---|---|---|---|
| N | 1 | 2 | 3 | 4 | 5 | 6 | Ih |
| $d_H$(Å) | 0.969 | 0.973 | 0.981 | 0.986 | 0.987 | 0.993 | 1.010 |
| $d_L$(Å) | – | 1.917 | 1.817 | 1.697 | 1.668 | 1.659 | 1.742 |
| θ (°) | – | 163.6 | 153.4 | 169.3 | 177.3 | 168.6 | 170.0 |
| $\omega_L$ ($cm^{-1}$) | – | 184 | 198 | 229 | 251 | 260 | – |
| $\omega_L$ ($cm^{-1}$) | – | 184 | 190 | 200 | 210 | 218 | 220 |
| $\omega_H$($cm^{-1}$) | 3732 | 3565 | 3387 | 3194 | 3122 | 3091 | – |
| $\omega_H$($cm^{-1}$)[50–52,55] | 3650 | 3575 | 3525 | 3380 | 3350 | 3225 | 3150 |
| $\Theta_{DL}$(K) | – | 167 | 171 | 180 | 189 | 196 | 198 [32] |
| $\Theta_{DH}$(K) | 3650 | 3575 | 3525 | 3380 | 3350 | 3225 | 3150 |
| $T_N$(K) | – | 94 | 110 | 180 | 188 | 246 | 258 |
| $T_m$(K) | – | 322 | 318 | 291 | 289 | 273 | 273 |

Presented are also N dependence of the Debye temperatures $\Theta_{Dx}$, freezing temperature $T_N$, and melting point $T_m$ estimated herewith, indicates the corrected $\omega_L$ as captioned in Fig. 5.5[a]

[a]Experimentally observed $T_m$ elevation and $T_N$ depression

$T_m$ = 325 K (monolayer) [33]; 310 K (skin of bulk) [22]

$T_N$ = 242 K (4.4 nm droplet) [34]; 220 K (3.4 nm droplet) [34]; 205 K (1.4 nm droplet) [35]; 172 K (1.2 nm droplet) [36]; <120 K (1–18 molecules) [37]

**Table 5.4** N dependence of the ($k_x$, $E_x$, $\Delta_x$) for the O:H–O bond in $(H_2O)_N$ clusters

| $N$ | $E_L$ (meV) | $E_H$ (eV) | $k_L$ (eV/Å$^2$) | $k_H$ (eV/Å$^2$) | $\Delta_L$ ($10^{-3}$ Å) | $\Delta_H$ ($10^{-4}$ Å) |
|---|---|---|---|---|---|---|
| 6 | 90.70 | 3.97 | 2.39 | 36.09 | 4.53 | 2.99 |
| 5 | 69.39 | 4.20 | 1.81 | 38.39 | 5.95 | 2.80 |
| 4 | 66.13 | 4.23 | 1.67 | 39.01 | 6.34 | 2.71 |
| 3 | 40.54 | 4.62 | 0.90 | 42.99 | 10.84 | 2.26 |
| 2 | 34.60 | 4.68 | 0.69 | 44.35 | 13.23 | 2.05 |

The O:H relaxes ($\Delta_L$) is more significant that the H–O bond does ($\Delta_H$) from one N value to the next

MD calculations [22] suggested that the H–O bond contracts from the bulk value of ~1.00 to ~0.95 at the skin, which is associated with O:H elongation from ~1.68 to ~1.95 Å with high fluctuation. This occurrence results in a 6.8 % $d_{OO}$ elongation or a 13 % density loss of the skin. The $d_{OO}$ of 2.965 Å [38] yields the lengths of $d_H$ = 0.8406 Å and $d_L$ = 2.1126 Å, which turns out a 0.75 g $cm^{-3}$ skin mass density. Consistency between these observations and the Lagrangian transformation reveals the true situation of the O:H–O bond relaxation and the associated mass density and their correlation with the O:H–O bond potential paths varying with molecular undercoordination.

**Table 5.5** $(H_2O)_N$ cluster size N dependent energetics at different order of differentials

| $N$ | $E_x$ (eV) | | | | | |
|---|---|---|---|---|---|---|
| | O:H potential $V_L(r)$ | | | H–O potential $V_H(r)$ | | |
| | 0th ($10^{-3}$) | 2nd ($10^{-5}$) | 3rd ($10^{-6}$) | 0th | 2nd ($10^{-6}$) | 3rd ($10^{-10}$) |
| 6 | 90.73 | 2.45 | 0.47 | 3.97 | 1.62 | 0.10 |
| 5 | 69.42 | 3.20 | 0.80 | 4.19 | 1.51 | 9.03 |
| 4 | 66.16 | 3.35 | 0.88 | 4.25 | 1.43 | 8.27 |
| 3 | 40.59 | 5.27 | 2.21 | 4.65 | 1.10 | 5.29 |
| 2 | 34.66 | 6.02 | 2.93 | 4.79 | 0.94 | 4.09 |

Both the O:H nonbond and H–O bond are very insensitive to the higher-order derivatives

#### 5.5.2.2 Precision and Reliability

Table 5.5 lists the zeroth, the first, the second, and the third derivatives of the Taylor series as a function of $(H_2O)_N$ cluster size. For both the O:H nonbond and the H–O bond, the contribution of the high order terms tò the respective binding energy is indeed negligibly small. Therefore, the harmonious approximation at each equilibrium site is valid in the first order approximation for the nature origin and the trend of O:H–O bond relaxation in length and energy.

#### 5.5.2.3 Potential Paths

Figure 5.2b shows the evolution of the O:H–O bond potential paths with $(H_2O)_N$ size, which agrees with experimental results (Fig. 5.5d). At $N = 6$, O atoms dislocate from the equilibrium (blue dots) outwardly by different amounts with identical energy elevation once the Coulomb repulsion is involved. When the N reduces from 6 to 2, the H–O bond contracts from 0.993 to 0.973 Å and its cohesive energy shifts from 3.97 to 4.68 eV, agreeing with the changing trend from bulk (3.97 eV) to the skin (4.66 eV) and to the monomer in gaseous phase (5.10 eV) [43]. The O:H expands from 1.659 to 1.917 Å and the energy shifts from 90.70 to 34.60 meV, compared with the bulk value of 95 meV [32]. These values may be subject to accuracy, as the DFT calculations are algorithm sensitive [8]. Compared is the O:H–O potential path for compressed ice, which takes the opposite trend of cluster size reduction [5].

## 5.6 Generality of $\rho$–$d_x$–$\omega_x$–$E_x$ Correlation

With the afore derived segmental length $d_x$, phonon frequency $\omega_x$, cohesive energy $E_x$, and the mass density $\rho$, one can unify and calibrate them with one known as input, regardless of the stimulus or probing method.

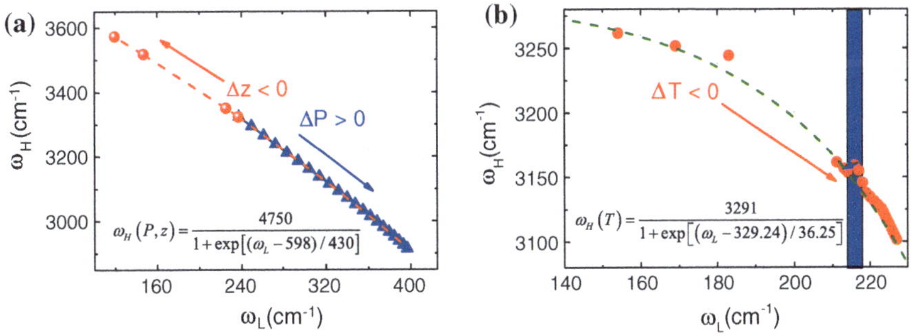


**Fig. 5.6** Segmental phonon frequency correlation under **a** pressure and undercoordination and **b** thermal excitation

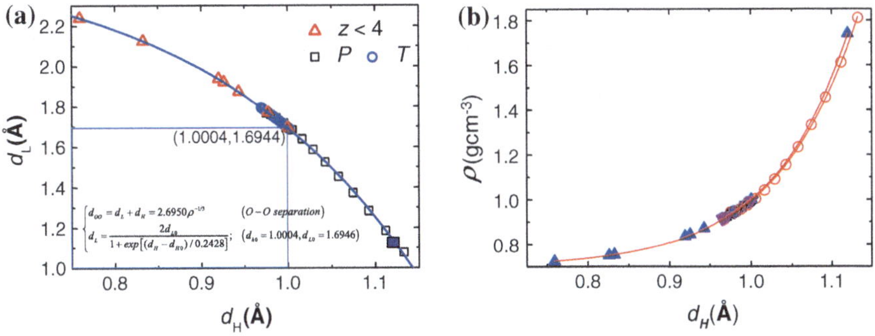


**Fig. 5.7** The molecular size $d_H$ dependence of **a** molecular separation $d_L$(or $d_{OO} = d_H + d_L$), and **b** mass density ρ, and the tetrahedral packing order of $H_2O$ molecules in the situations: (i) ice under compression ($d_H > 1.00$ Å) [26], (ii) water ice at cooling ($0.96 < d_H < 1.00$ Å) [34, 35], and (iii) liquid surface and dimer ($d_H < 1.00$ Å) [30, 44–50]. The derived $d_H = 1.0004$ Å at $\rho = 1$ is within the measurements ranging from 0.970 to 1.001 Å [51]. The $d_H$ shorter than 0.96 Å corresponds to the supersolid phase in regions of molecules having fewer than four neighbors [8, 38, 52]. In such regions, a $H_2O$ molecule shrinks in size and expands in separation because of inter electron-pair repulsion [8] (Reprinted with permission from [53].)

Figure 5.6 correlates the $\omega_H$ to the $\omega_L$ for the relaxed O:H–O bond in the form of:

$$\omega_H(P, z) = \frac{4750}{1 + \exp[(\omega_L - 598)/430]}$$

Figure 5.7 correlates the molecular separation $d_L$ and the mass density ρ of water and ice to the $d_H$ relaxation, which matches observations from compressed ice, cooling water and ice, skins, and droplets. Figure 5.8 features the $d_H$ dependence of

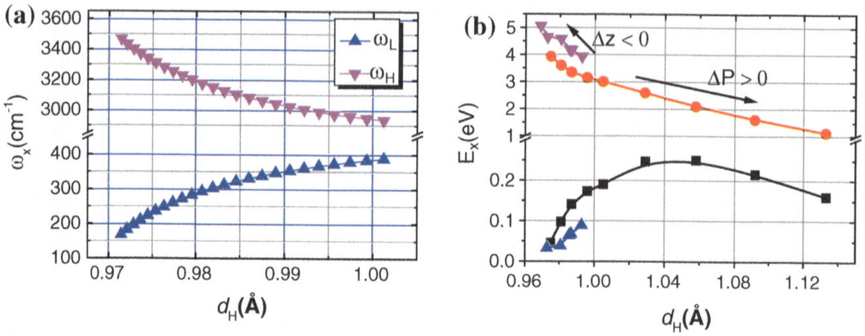


**Fig. 5.8** The $d_H$ dependence of segmental **a** phonon frequency $\omega_x$ and (**b**) cohesive energy $E_x$. Mismatch in **b** suggests parameterization revision in solving the Lagrangian motion equations

the $\omega_x$ and $E_x$ in the processes of O:H–O bond relaxation. One can estimate the segmental lengths, vibration frequencies, cohesive energy and the mass density with one of them as known input.

## 5.7 Summary

A combination of the Lagrangian mechanics of oscillator vibration, MD decomposition of volume evolution, and Raman spectroscopy of phonon relaxation has transformed the observed ($d_x$, $\omega_x$) into the ($k_x$, $E_x$) and thus enabled probing the potential paths for the relaxed O:H–O bond. This analytical solution has derived bond energy, force constant, potential field of each segment and their stimulus dependence based on the measurements.

Consistency between calculations and experimental observations on the O:H–O energy relaxation evidences consistently the persistence and significance of the asymmetrical, short-range interactions and Coulomb repulsion in the flexile, polarizable O:H–O bond. The relaxation induced by molecular undercoordination and mechanical compression could be general to other situations containing bonding and nonbonding interactions despite accuracy in a specific situation.

## References

1. J. Teixeira, High-pressure physics - the double identity of ice X. Nature **392**(6673), 232–233 (1998)
2. T. Kumagai, Direct observation and control of hydrogen-bond dynamics using low-temperature scanning tunneling microscopy. Prog. Surf. Sci. **90**(3), 239–291 (2015)
3. J.D. Bernal, R.H. Fowler, A theory of water and ionic solution, with particular reference to hydrogen and hydroxyl ions. J. Chem. Phys. **1**(8), 515–548 (1933)

4. L. Pauling, The structure and entropy of ice and of other crystals with some randomness of atomic arrangement. J. Am. Chem. Soc. **57**, 2680–2684 (1935)
5. Y. Huang, X. Zhang, Z. Ma, Y. Zhou, G. Zhou, C.Q. Sun, Hydrogen-bond asymmetric local potentials in compressed ice. J. Phys. Chem. B **117**(43), 13639–13645 (2013)
6. Y. Huang, X. Zhang, Z. Ma, G. Zhou, Y. Gong, C.Q. Sun, Potential paths for the hydrogen-bond relaxing with (H2O)N cluster size. J. Phys. Chem. C **119**(29), 16962–16971 (2015)
7. C.Q. Sun, X. Zhang, W.T. Zheng, Hidden force opposing ice compression. Chem. Sci. **3**, 1455–1460 (2012)
8. C.Q. Sun, X. Zhang, J. Zhou, Y. Huang, Y. Zhou, W. Zheng, Density, elasticity, and stability anomalies of water molecules with fewer than four neighbors. J. Phys. Chem. Lett. **4**, 2565–2570 (2013)
9. A.K. Soper, An asymmetric model for water structure. J. Phys.-Condens. Matter **17**(45), S3273–S3282 (2005)
10. M. Korth, Empirical Hydrogen-bond Potential FunctionsuAn Old Hat Reconditioned. ChemPhysChem **12**(17), 3131–3142 (2011)
11. F.H. Stillinger, Water revisited. Science **209**(4455), 451–457 (1980)
12. J.L. Finney, Water? what's so special about it? Philos. Trans. Roy. Soc. B: Biol. Sci. **359** (1448), 1145–1165 (2004)
13. F. Bruni, M.A. Ricci, A.K. Soper, Obtaining distribution functions for water from diffraction data, in *Francesco Paolo Ricci: His Legacy and Future Perspectives of Neutron Scattering*, ed. by M. Nardone, M.A. Ricci (Societ'a Italiana di Fisica, Bologna, Italy, 2001)
14. M. Leetmaa, M. Ljungberg, H. Ogasawara, M. Odelius, L.-Å. Näslund, A. Nilsson, L.G. Pettersson, Are recent water models obtained by fitting diffraction data consistent with infrared/Raman and x-ray absorption spectra? J. Chem. Phys. **125**(24), 244510 (2006)
15. M. Leetmaa, K.T. Wikfeldt, M.P. Ljungberg, M. Odelius, J. Swenson, A. Nilsson, L.G.M. Pettersson, Diffraction and IR/Raman data do not prove tetrahedral water. J. Chem. Phys. **129** (8), 084502 (2008)
16. N. Chumaevskii, M. Rodnikova, Some peculiarities of liquid water structure. J. Mol. Liq. **106** (2–3), 167–177 (2003)
17. P. Wernet, D. Nordlund, U. Bergmann, M. Cavalleri, M. Odelius, H. Ogasawara, L.A. Naslund, T.K. Hirsch, L. Ojamae, P. Glatzel, L.G.M. Pettersson, A. Nilsson, The structure of the first coordination shell in liquid water. Science **304**(5673), 995–999 (2004)
18. K.T. Wikfeldt, M. Leetmaa, M.P. Ljungberg, A. Nilsson, L.G.M. Pettersson, On the range of water structure models compatible with X-ray and neutron diffraction data. J. Phys. Chem. B **113**(18), 6246–6255 (2009)
19. A. Nilsson, L.G.M. Pettersson, Perspective on the structure of liquid water. Chem. Phys. **389** (1–3), 1–34 (2011)
20. T.D. Kuhne, R.Z. Khaliullin, Electronic signature of the instantaneous asymmetry in the first coordination shell of liquid water. Nat. Commun. **4**, 1450 (2013)
21. Y. Huang, X. Zhang, Z. Ma, Y. Zhou, W. Zheng, J. Zhou, C.Q. Sun, Hydrogen-bond relaxation dynamics: resolving mysteries of water ice. Coord. Chem. Rev. **285**, 109–165 (2015)
22. X. Zhang, Y. Huang, Z. Ma, Y. Zhou, W. Zheng, J. Zhou, C.Q. Sun, A common supersolid skin covering both water and ice. PCCP **16**(42), 22987–22994 (2014)
23. P. Pruzan, J.C. Chervin, E. Wolanin, B. Canny, M. Gauthier, M. Hanfland, Phase diagram of ice in the VII-VIII-X domain. Vibrational and structural data for strongly compressed ice VIII. J. Raman Spectrosc. **34**(7–8), 591–610 (2003)
24. M. Song, H. Yamawaki, H. Fujihisa, M. Sakashita, K. Aoki, Infrared absorption study of Fermi resonance and hydrogen-bond symmetrization of ice up to 141 GPa. Phys. Rev. B **60** (18), 12644 (1999)

25. Y. Yoshimura, S.T. Stewart, M. Somayazulu, H.K. Mao, R.J. Hemley, Convergent Raman features in high density amorphous ice, ice VII, and ice VIII under pressure. J. Phys. Chem. B **115**(14), 3756–3760 (2011)
26. Y. Yoshimura, S.T. Stewart, M. Somayazulu, H. Mao, R.J. Hemley, High-pressure x-ray diffraction and Raman spectroscopy of ice VIII. J. Chem. Phys. **124**(2), 024502 (2006)
27. M. Benoit, D. Marx, M. Parrinello, Tunnelling and zero-point motion in high-pressure ice. Nature **392**(6673), 258–261 (1998)
28. A.F. Goncharov, V.V. Struzhkin, H.-K. Mao, R.J. Hemley, Raman spectroscopy of dense $H_2O$ and the transition to symmetric hydrogen bonds. Phys. Rev. Lett. **83**(10), 1998–2001 (1999)
29. P. Loubeyre, R. LeToullec, E. Wolanin, M. Hanfland, D. Husermann, Modulated phases and proton centring in ice observed by X-ray diffraction up to 170 GPa. Nature **397**(6719), 503–506 (1999)
30. L.B. Skinner, C. Huang, D. Schlesinger, L.G. Pettersson, A. Nilsson, C.J. Benmore, Benchmark oxygen-oxygen pair-distribution function of ambient water from x-ray diffraction measurements with a wide Q-range. J. Chem. Phys. **138**(7), 074506 (2013)
31. C.Q. Sun, X. Zhang, X. Fu, W. Zheng, J.-L. Kuo, Y. Zhou, Z. Shen, J. Zhou, Density and phonon-stiffness anomalies of water and ice in the full temperature range. J. Phys. Chem. Lett. **4**, 3238–3244 (2013)
32. M. Zhao, W.T. Zheng, J.C. Li, Z. Wen, M.X. Gu, C.Q. Sun, Atomistic origin, temperature dependence, and responsibilities of surface energetics: an extended broken-bond rule. Phys. Rev. B **75**(8), 085427 (2007)
33. H. Qiu, W. Guo, Electromelting of confined monolayer ice. Phys. Rev. Lett. **110**(19), 195701 (2013)
34. M. Erko, D. Wallacher, A. Hoell, T. Hauss, I. Zizak, O. Paris, Density minimum of confined water at low temperatures: a combined study by small-angle scattering of X-rays and neutrons. PCCP **14**(11), 3852–3858 (2012)
35. F. Mallamace, C. Branca, M. Broccio, C. Corsaro, C.Y. Mou, S.H. Chen, The anomalous behavior of the density of water in the range 30 K < T < 373 K. Proc. Natl. Acad. Sci. U.S.A. **104**(47), 18387–18391 (2007)
36. F.G. Alabarse, J. Haines, O. Cambon, C. Levelut, D. Bourgogne, A. Haidoux, D. Granier, B. Coasne, Freezing of water confined at the nanoscale. Phys. Rev. Lett. **109**(3), 035701 (2012)
37. R. Moro, R. Rabinovitch, C. Xia, V.V. Kresin, Electric dipole moments of water clusters from a beam deflection measurement. Phys. Rev. Lett. **97**(12), 123401 (2006)
38. K.R. Wilson, R.D. Schaller, D.T. Co, R.J. Saykally, B.S. Rude, T. Catalano, J.D. Bozek, Surface relaxation in liquid water and methanol studied by x-ray absorption spectroscopy. J. Chem. Phys. **117**(16), 7738–7744 (2002)
39. P.C. Cross, J. Burnham, P.A. Leighton, The Raman spectrum and the structure of water. J. Am. Chem. Soc. **59**, 1134–1147 (1937)
40. Q. Sun, The Raman OH stretching bands of liquid water. Vib. Spectrosc. **51**(2), 213–217 (2009)
41. J. Ceponkus, P. Uvdal, B. Nelander, Water tetramer, pentamer, and hexamer in inert matrices. J. Phys. Chem. A **116**(20), 4842–4850 (2012)
42. S. Hirabayashi, K.M.T. Yamada, Infrared spectra and structure of water clusters trapped in argon and krypton matrices. J. Mol. Struct. **795**(1–3), 78–83 (2006)
43. S.A. Harich, D.W.H. Hwang, X. Yang, J.J. Lin, X. Yang, R.N. Dixon, Photodissociation of $H_2O$ at 121.6 nm: a state-to-state dynamical picture. J. Chem. Phys. **113**(22), 10073–10090 (2000)
44. U. Bergmann, A. Di Cicco, P. Wernet, E. Principi, P. Glatzel, A. Nilsson, Nearest-neighbor oxygen distances in liquid water and ice observed by x-ray Raman based extended x-ray absorption fine structure. J. Chem. Phys. **127**(17), 174504 (2007)
45. K.R. Wilson, B.S. Rude, T. Catalano, R.D. Schaller, J.G. Tobin, D.T. Co, R.J. Saykally, X-ray spectroscopy of liquid water microjets. J. Phys. Chem. B **105**(17), 3346–3349 (2001)
46. A. Narten, W. Thiessen, L. Blum, Atom pair distribution functions of liquid water at 25 C from neutron diffraction. Science **217**(4564), 1033–1034 (1982)

47. L. Fu, A. Bienenstock, S. Brennan, X-ray study of the structure of liquid water. J. Chem. Phys. **131**(23), 234702 (2009)
48. J.-L. Kuo, M.L. Klein, W.F. Kuhs, The effect of proton disorder on the structure of ice-Ih: a theoretical study. J. Chem. Phys. **123**(13), 134505 (2005)
49. A. Soper, Joint structure refinement of x-ray and neutron diffraction data on disordered materials: application to liquid water. J. Phys.: Condens. Matter **19**(33), 335206 (2007)
50. K.T. Wikfeldt, M. Leetmaa, A. Mace, A. Nilsson, L.G.M. Pettersson, Oxygen-oxygen correlations in liquid water: addressing the discrepancy between diffraction and extended x-ray absorption fine-structure using a novel multiple-data set fitting technique. J. Chem. Phys. **132** (10), 104513 (2010)
51. M. Hakala, K. Nygård, S. Manninen, L.G.M. Pettersson, K. Hämäläinen, Intra- and intermolecular effects in the Compton profile of water. Phys. Rev. B **73**(3), 035432 (2006)
52. K. Liu, J.D. Cruzan, R.J. Saykally, Water clusters. Science **271**(5251), 929–933 (1996)
53. Y. Huang, X. Zhang, Z. Ma, W. Li, Y. Zhou, J. Zhou, W. Zheng, C.Q. Sun, Size, separation, structural order, and mass density of molecules packing in water and ice. Sci. Rep. **3** (2013)

# Chapter 6
# Mechanical Compression

- *Compression shortens the O:H nonbond and lengthens the H–O bond towards $H^+$ centralization with strong polarization that enlarges the band gap.*
- *Compression depresses the $T_m$ by dispersing the quasisolid-phase boundary through $\omega_x(\Theta_{Dx})$ relaxation.*
- *O:H–O bond recovers when the mechanical compression, molecular undercoordination, or thermal excitation is relieved.*
- *Persistence of $sp^3$—orbital hybridization of $O^{2-}$ entitles O:H–O extraordinary recoverability from deformation and dissociation.*

**Abstract** Water and ice respond to mechanical compression unusually with numerous anomalies. Regelation, i.e., ice melts under compression and freezes again when the pressure is relieved, evidences that the O:H–O bond extraordinary recoverability and that quasisolid phase boundary dispersivity. An oxygen atom always finds bonding partners to retain its $sp^3$-orbital hybridization once the O:H breaks, which ensures O:H–O bond recoverability to its original state once the pressure is removed. On the other hand, mechanical compression shortens the O:H nonbond and soften its phonon but the H–O bond responds to compression oppositely, lowering the H–O phonon frequency, which offsets the Debye temperature and the boundaries of the quasisolid phase outwardly, which elevates the freezing point and depresses the melting point, so regelation takes place. Reproduction of the $T_m(P)$ profile clarifies that the H–O bond energy $E_H$ determines the $T_m$ with derivative of $E_H$ = 3.97 eV for bulk water and ice.

## 6.1 Challenge: Why Does Pressure Melt Ice?

Compression results in multiple physical anomalies of water and ice as exampled as follows

(1) Ice regelation–ice melts under compression and freezes again when the pressure is relieved [1–4], as illustrated in Fig. 6.1a.
(2) The $T_m$ drops by a maximal amount of −22 °C under 210 MPa pressure but a −95 MPa pressure (tension) raises it up to +6.5 °C [5, 6].
(3) Compression shortens the O:H nonbond and stiffens its phonon but the H–O bond responds to compression oppositely towards $H^+$ proton centralization at 60 GPa, see Fig. 6.1b.



C.Q. Sun and Y. Sun, *The Attribute of Water*, Springer Series in Chemical Physics 113, DOI 10.1007/978-981-10-0180-2_6

(a)

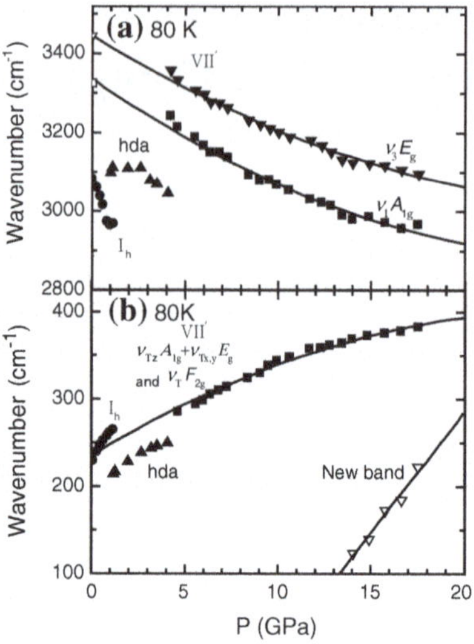


**Fig. 6.1** **a** A weighted wire cuts a block of ice through without severing it [7] (Reprinted with permission from [5, 6].) **b** Compression shifts the O:H (<400 $cm^{-1}$) and H–O phonon (>2900 $cm^{-1}$) at 80 K and pressure higher than 5 GPa but relaxes the O:H bond abnormally when cross the $I_C$, IX, II phases below 5 GPa (Reprinted with permission from [8].)

(4) Compression also enlarges the band gap of ice and offsets the maximal density temperature from 277 to 270 K.

## 6.2 Clarification: H–O Bond Elongation Depresses the $T_m$

Figure 6.2 illustrates the mechanisms for anomalies pertained ice compression:

(1) Phonon relaxation $\Delta\omega_x$ offsets the specific heat and hence it disperses the quasisolid-phase boundaries cooperatively according to $\hbar\omega_x \cong k\Theta_{Dx}$ relationship [9], see Fig. 6.2a.
(2) Quasisolid phase dispersion offsets the upper boundary closes to the $T_m$ and the lower to temperature of homogenous freezing $T_N$ in opposite directions [10].
(3) The H–O bond energy dictates the $T_m$ as Fig. 6.2b formulated [11].
(4) An oxygen atom always seeks for partners to recover its $sp^3$-orbital configuration once the O:H nonbond breaks [12], which entitles O:H–O bond to recover completely from its relaxation or dissociation once the excitation is relieved [13].

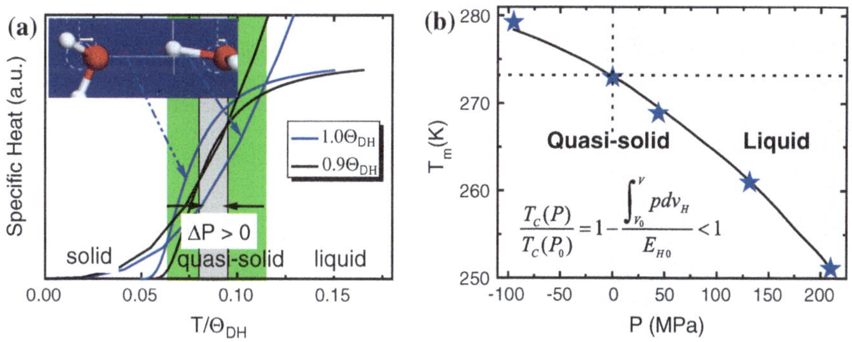


$$\frac{T_C(P)}{T_C(P_0)} = 1 - \frac{\int_{V_0}^{V} p dv_H}{E_{H0}} < 1$$

**Fig. 6.2** **a** Pressure squeezed inwardly quasisolid phase boundaries and hence lowers the $T_m$ for melting and raises the $T_N$ for homogeneous ice nucleation. **b** Theoretical reproduction [9] of the pressure dependence of the $T_m(P)$ in the phase diagram indicates the H–O bond recoverable relaxation dominance of the Regelation (Reprinted with permission from [5, 6].)

## 6.3 Historical Background

### *6.3.1 Discovery of Ice Regelation*

Discovered by Faraday [4], Thomson [14], and Forbes [1, 2] in 1850s, regelation is the phenomenon of ice melting under pressure and freezing again when the pressure is relieved at temperatures around −10 °C. Faraday firstly noted that [4],

> Two pieces of thawing ice, if put together, adhere and become one; at a place where liquefaction was proceeding, congelation suddenly occurs. The effect will take place in air, in water, or in vacuo. It will occur at every point where the two pieces of ice touch; but not with ice below the freezing point, i.e., with dry ice, or ice so cold as to be everywhere in the solid state.

According to the literature record [3, 14]:

> The phenomenon (Regelation) is a consequence of the properties, announced from theory by James Thomson, and then exemplified by an experiment; and the explanation depends on the theories put forward by him—the first (1857) founded on the lowering of the freezing point of water by pressure, and the second (1861) founded on the tendency to melt given by the application to the solid ice of forces whose nature is to produce change of form as distinguished from forces applied alike to the liquid and solid. The stress upon the ice, due to its pressure on the network, gives it a tendency to melt at the point in contact with the wire, and the ice, in the form of water intermixed with fragments and new crystals, moves so as to relieve itself of pressure.

The generally accepted explanation for this phenomenon is that sufficient compressive stresses exist at the contact area to cause melting when the pieces of ice are brought together, and when this stress is released solidification occurs. This was originally proposed by James Thomson and endorsed experimentally by his brother, Lord Kelvin (William Thomson).

| | |
|---|---|
|   | Michael Faraday [15] FRS (22 September 1791–25 August 1867) was an English scientist who contributed to the fields of electromagnetism and electrochemistry. He discovered the Ice Regelation and explained the slipperiness of ice in terms of liquid like skin that retains on the surface but the liquid like skin fuse two pieces of ice when they contact each other under compression |
|  | James David Forbes FRS FRSE FGS (20 April 1809–31 December 1868) was a Scottish physicist and glaciologist who worked extensively on the conduction of heat and seismology |
|  | James Thomson (16 February 1822–8 May 1892) was an engineer and physicist whose reputation is substantial though it is overshadowed by that of his younger brother (Right) William Thomson (Lord Kelvin) |
|  | Lord Kelvin (26 June 1824–17 December 1907) was a British mathematical physicist and engineer. He conducted the mathematical analysis of electricity and formulation of the first and second laws of thermodynamics. For his work on the transatlantic telegraph project he was knighted by Queen Victoria, becoming Sir William Thomson (Free Wikimedia) |

### 6.3.2 *Liquid Fusion Mechanism*

However, consistent understanding or numerical reproduction of regelation has yet been achieved since 1850s when Faraday, Thomson, and Forbes debated on possible mechanisms from the classical thermal dynamics and quasi-liquid skin viewpoints [3, 4]. Faraday supposed that a particle of water, which could retain the liquid state whilst touching ice only on one side, could not retain the liquid state if it were touched by ice on both sides; but became solid, the general temperature remaining the same.

Thomson who discovered that pressure lowered the freezing-point of water and attributed the regelation to the fact that two pieces of ice could not be made to bear on each other without pressure; and that the pressure, however slight, would cause *fusion* at the place where the particles touched, accompanied by relief of the pressure and resolidification of the water at the place of contact. Forbes assented to neither of these views; but due to the gradual liquefaction of ice, and assuming that ice is essentially colder than ice—cold water, i.e., the water in contact with it, he concluded that two wet pieces of ice will have the water between them frozen at the place where they come into contact.

It might be true that regelation can occur for substances with the property of expanding upon freezing, but mechanisms of the freezing expansion and for the regelation remain yet unclear [10]. Faraday concluded that this phenomenon occurs only to ice after conducting numerous experiments in 1859 [4]:

> Many salts were tried (without much or any expectation),—crystals of them being brought to bear against each other by torsional force, in their saturated solutions at common temperatures. In this way the following bodies were experimented with:—Nitrates of lead, potassa, soda; sulphates of soda, magnesia, copper, Zinc; alum; borax; chloride of ammonium; ferro—prussiate of potassa; carbonlate of soda; acetate of lead; and tartrate of potassa and soda; but the results with all were negative. My present conclusion therefore is that the property is special for water; and that the view I have taken of its physical cause does not appear to be less likely now than at the beginning of this short investigation, and therefore has not sunk in value among the three explanations given.

Unfortunately, after one and a half century long debating, quantitative resolution of ice regelation has yet been reached. This section shows that the present theory of O:H–O bond cooperativity under compression resolves this mystery quantitatively.

### 6.3.3 *Proton Centralization via Quantum Tunneling*

In 1972, Holzapfel [16] firstly predicted that, under compression, an O:H–O bond might be transformed from the highly asymmetrical O:H–O configuration to a symmetrical state in which the H proton lies midway between two $O^{2-}$ ions, leading to a non-molecular symmetrical phase of ice-X. Goncharov et al. [17] confirmed this prediction in 1998 using in situ high-Pressure Raman spectroscopy. The proton centralization in the O:H–O bond of ice-VIII occurred at about 60 GPa and 100 K, and no further shift of phonon frequency was observed when they increased the pressure, since the O:H and H–O had both reached identical lengths (0.11 nm) [18, 19]. Proton centralization also occurs in liquid $H_2O$ at 60 GPa and 85 K, and to liquid $D_2O$ at 70 GPa and 300 K [20].

Compression-induced proton centralization evolves the pairing potential wells into a single well located midway between $O^{2-}$ ions [21] which was attributed to "translational proton quantum tunneling" [18, 22, 23] or to the extraordinary, yet unclear, behavior of the inter- and intra-molecular bonds [24].

## 6.4 Quantitative Resolution

### *6.4.1 O:H–O Bond Symmetrization*

Generally, compression shortens all bonds of a normal substance. However, compression shortens and stiffens the softer O:H nonbond and meanwhile lengthens and softens the stiffer H–O bond by different amounts. The O:H shortens more than the H–O elongation through interoxygen repulsion. The cohesive energy of the H–O bond dictates the $T_m$ and the cohesive energy of the O:H nonbond decides the temperatures for freezing $T_N$ and boiling $T_V$.

Figure 6.3a shows the $V/V_0(P)$ profiles of water (300 K) and ice (77 K) measured using in situ high-pressure and low-temperature synchrotron XRD and Raman spectroscopy [25]. Molecular dynamics calculations converted the $V/V_0 - P$ profiles into the $d_x/d_{x0} - P$ curves [11]. As shown in Fig. 6.3b, compression shortens the O:H nonbond from 1.767 to 1.692 Å, and meanwhile, the compression lengthens the

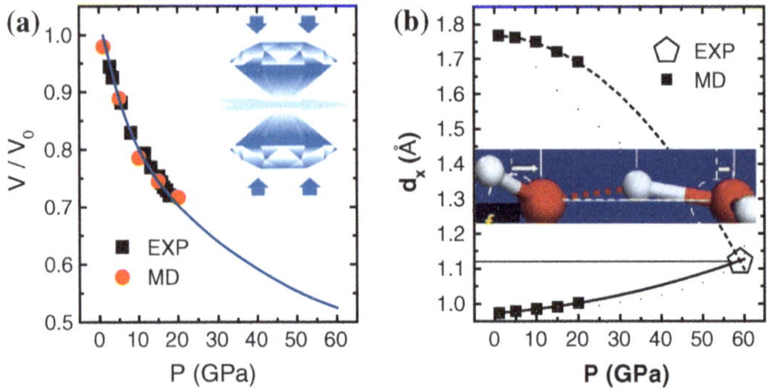


**Fig. 6.3** MD conversion of the measured **a** $V/V_0$–$P$ profiles [25] (*inset* illustrates experimental set up [29]) into **b** the $d_x$ –P curves meeting at the point of proton centralization occurring at 59 GPa and 2.20 Å [17, 18]. Hiding the variable P in (**b**), the $d_L$–$d_H$ correlation being independent of experimental conditions or probing method in the temperature range of 77–300 K (Reprinted with permission from [11].)

H–O bond from 0.974 to 1.003 Å when the pressure is increased from 1 to 20 GPa [26–28]. An extrapolation of the $d_x/d_{x0} - P$ curves joins the two segments at the exact point of proton centralization occurred in phase X [17, 18, 25] at 58.6–59.0 GPa with an O–O distance of 2.20–2.21 Å [18]. Observations evidence sufficiently the O–O repulsion that couples the O:H–O bond relaxing cooperatively.

Therefore, the proton centralization arises from pressure-induced O:H–O bond asymmetrical relaxation rather than the transitional $H^+$ proton quantum tunneling that is unlikely because of the strong H–O bond of 3.97 eV energy. Constrained by measured proton centralization [18] and the $V/V_0 - P$ profiles [25], the $d_x/d_{x0} - P$ curves represent the true situation in a broad temperature range (77–300 K), irrespective of the probing conditions or techniques. Table 6.1 features the DFT–MD derivatives of the O:H–O bond segmental relaxation dynamics and the band gap change as a function of pressure.

The following polynomials formulate the pressure dependence of the O:H–O bond length and volume relaxations. The referential atmospheric pressure $P_0 = 10^{-4}$ GPa approximates to zero:

$$\begin{pmatrix} d_H/0.975 \\ d_L/1.768 \\ V/1.060 \end{pmatrix} = 1 + \begin{pmatrix} 9.510 & 2.893 \\ -3.477 & -10.280 \\ -238.000 & 47.000 \end{pmatrix} \begin{pmatrix} 10^{-4}P^1 \\ 10^{-5}P^2 \end{pmatrix}. \quad (6.1)$$

Furthermore, first principle and Quantum Monte Caro calculations [30] suggested that the contribution to the lattice energy from the O:H intermolecular

**Table 6.1** Pressure dependence of the mass density $\rho$, segmental lengths $d_x$, O–O distance $d_{OO}$, and band gap $E_G$ of ice [11]

| P (GPa) | DFT | | | | MD | | |
|---|---|---|---|---|---|---|---|
| | ρ (g/cm³) | $d_H$ (Å) | $d_L$ (Å) | $E_G$ (eV) | $d_H$ (Å) | $d_L$ (Å) | $d_{O\text{-}O}$ (Å) |
| 1 | 1.659 | 0.966 | 1.897 | 4.531 | 0.974 | 1.767 | 2.741 |
| 5 | 1.886 | 0.972 | 1.768 | 4.819 | 0.979 | 1.763 | 2.742 |
| 10 | 2.080 | 0.978 | 1.676 | 5.097 | 0.985 | 1.750 | 2.736 |
| 15 | 2.231 | 0.984 | 1.610 | 5.353 | 0.991 | 1.721 | 2.713 |
| 20 | 2.360 | 0.990 | 1.556 | 5.572 | 1.003 | 1.692 | 2.694 |
| 25 | 2.479 | 0.996 | 1.507 | 5.778 | – | – | – |
| 30 | 2.596 | 1.005 | 1.460 | 5.981 | – | – | – |
| 35 | 2.699 | 1.014 | 1.419 | 6.157 | – | – | – |
| 40 | 2.801 | 1.026 | 1.377 | 6.276 | – | – | – |
| 45 | 2.900 | 1.041 | 1.334 | 6.375 | – | – | – |
| 50 | 2.995 | 1.061 | 1.289 | 6.459 | – | – | – |
| 55 | 3.084 | 1.090 | 1.237 | 6.524 | – | – | – |
| 60 | 3.158 | 1.144 | 1.164 | 6.590 | – | – | – |

Discrepancy in DFT and MD derivatives exist due to artifacts in the algorithms so one needs to focus on the trend of change instead of the accuracy

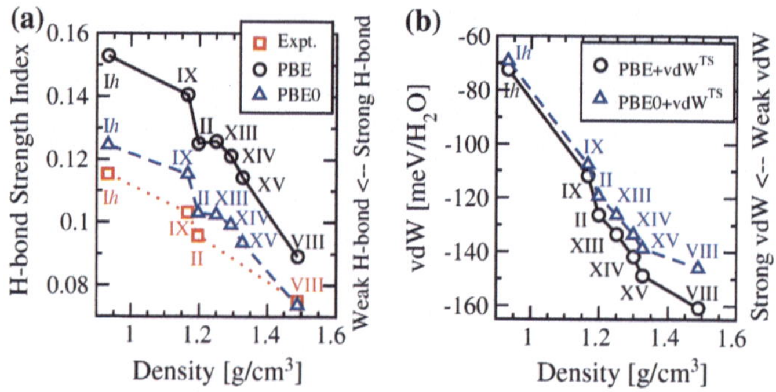


**Fig. 6.4** **a** H-bond strength index (relative shift of the H–O bond vibration frequency with respect to that of the monomer: [$\omega_H$ (water)—$\omega_H$ (ice)]/$\omega_H$ (monomer)) and **b** O:H nonbond (vdW) energy contributions as a function of ice density at zero pressure and structure phases. Experimental values are taken from [31–33]. The general trends remain irrespective algorithms used (Reprinted with permission from [30].)

interaction increases and that from the intramolecular H–O bond decreases when pressure is increased up to 2 GPa, see Fig. 6.4. Observations further evidence that compression shortens and stiffens the O:H nonbond but does the H–O bond oppositely.

### *6.4.2 Phonon Cooperative Relaxation*

Generally, compression stiffens all phonons of 'normal' substance such as carbon allotropes [34], ZnO [35], group IV [36], group III–V [37], and group II–VI [38] compounds without exception; however, for ice and water, compression stiffens the softer O:H phonons ($\omega_L < 300\ \text{cm}^{-1}$) but softens the stiffer H–O stretching phonons ($\omega_H > 3000\ \text{cm}^{-1}$) cooperatively [8, 25, 39–41]. This relaxation happens generally except for the No Man's Land regime that is subject to further quantitative verification.

Compression also lowers the $\omega_H$ of liquid water, as shown in Fig. 6.5 [41]. This observation follows the prediction that compression softens the $\omega_H$, which is acompanied with $\omega_L$ stiffening. Figure 6.1b features the Raman peak shift measured at 80 K cross the $I_C$, IX, II and XV phases [8]. The spectra revealed that compression below 5.0 GPa relaxes the O:H–O bond irregularly, which is opposite to the general trend measured at P > 5.0 GPa.

Figure 6.6a features the MD—derived phonon relaxation of ice-VIII phase as a function of pressure, which agree with trends probed using Raman and IR spectroscopy from ice-VIII at 80 K [25, 39, 40]. Compression softens the $\omega_H$ from 3520 to 3320 cm$^{-1}$ and stiffens the $\omega_L$ from 120 to 336 cm$^{-1}$, disregarding the possible phase change and other supplementary peaks nearby. Figure 6.6b

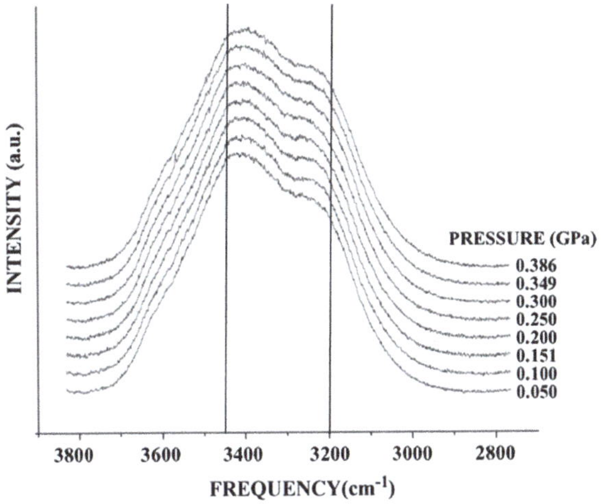


**Fig. 6.5** Compression softens the $\omega_H$ for water (3450 $cm^{-1}$ for the skin and 3200 $cm^{-1}$ for the bulk) at 296 K up to 0.4 GPa (Reprinted with permission from [41].)

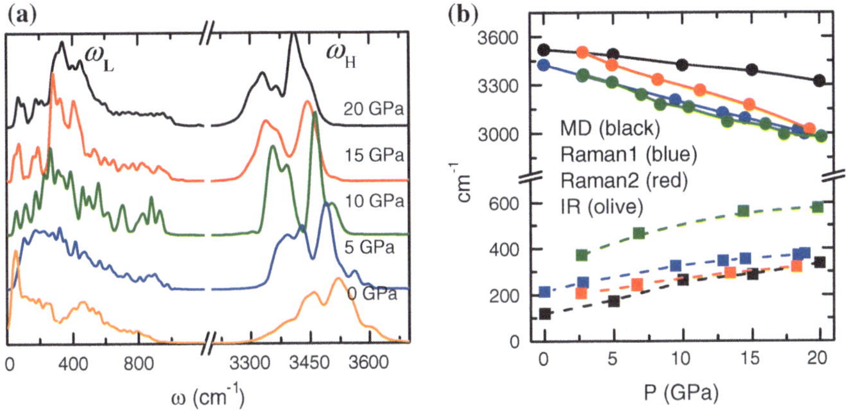


**Fig. 6.6 a** MD—derived $\omega_x$ relaxation and **b** trend agreement between Raman and IR measurements and calculations for the ice-VIII phase at 80 K [25, 39, 40] (Reprinted with permission from [11].)

compares the calculated and a collection of the measured $\Delta\omega_x$ for ice, despite of an offset of the calculation results. Consistency in the pressure-derived $\omega_x$ cooperativity in Fig. 6.6b for both water [41] and ice [39, 40, 42] confirms that compression shortens and stiffens the O:H nonbond and relaxes the H–O bond reversely in all forms of the liquid and solid water.

### 6.4.3 Liquid/Quasisolid Phase Boundary Dispersion

The $T_C(P_C)$ boundary for the Liquid/Quasisolid meets the following criterion,

$$\begin{cases} \frac{\Delta T_C(P)}{T_C(P_0)} & = -\sum \frac{s_x \int_{P_0}^{P} p\frac{dd_x}{dp}dp}{E_{x0}} < 0 \cdot \\ dd_x/dp & > 0 \end{cases} \tag{6.2}$$

With the known pressure dependence of the $d_x(P)$ in (6.1) [11], one can find that only the $d_H$ satisfies this criterion-compression elongation. Reproduction of the measured $T_m(P_C)$ curve and the facts of −22 °C depression at 210 MPa and +6.5 °C elevation at −95 MPa [5, 43] yields an $E_H$ value of 3.97 eV by taking the H atomic diameter of 0.106 nm as the H–O bond diameter [44], which confirms that the H–O energy relaxation dictates the $T_m$ for liquid/quasisolid transition.

### 6.4.4 O:H–O Bond Extraordinary Recoverability

Generally, external stimuli, such as stressing and heating modulate the length *d* (*T*, *P*) and energy *E*(*T*, *P*) of the representative bond along a path function *f* (*T*, *P*) [45]. For instance, compression stores energy into a substance by shortening and stiffening all bonds with possible plastic deformation when the compression is relieved. Tension does the opposite [46].

However, for the O:H–O bond, situation is different. As given in Table 6.2, compression raises the $E_L$ from 0.046 to 0.250 eV up to 40 GPa and then decrease to 0.16 eV at 60 GPa; the $E_H$ decreases monotonically from 3.97 to 1.16 eV at

**Table 6.2** Pressure-dependence of the O:H–O segmental cohesive energy $E_x$ and the net gain at each quasi-equilibrium state under compression

| *P* (GPa) | $E_L$ (eV) | $E_H$ (eV) | $E_{H+L}(P) - E_{H+L}(0)$ |
|---|---|---|---|
| 0 | 0.046 | 3.97 | 0 |
| 5 | 0.098 | 3.64 | −0.278 |
| 10 | 0.141 | 3.39 | −0.485 |
| 15 | 0.173 | 3.19 | −0.653 |
| 20 | 0.190 | 3.04 | −0.786 |
| 30 | 0.247 | 2.63 | −1.139 |
| 40 | 0.250 | 2.13 | −1.636 |
| 50 | 0.216 | 1.65 | −2.15 |
| 60 | 0.160 | 1.16 | −2.696 |

Unlike 'normal' substance that gains energy with possible plastic deformation under compression, O:H–O bond always losses energy and tends to recover from its higher energy state to lower initial state without any plastic deformation

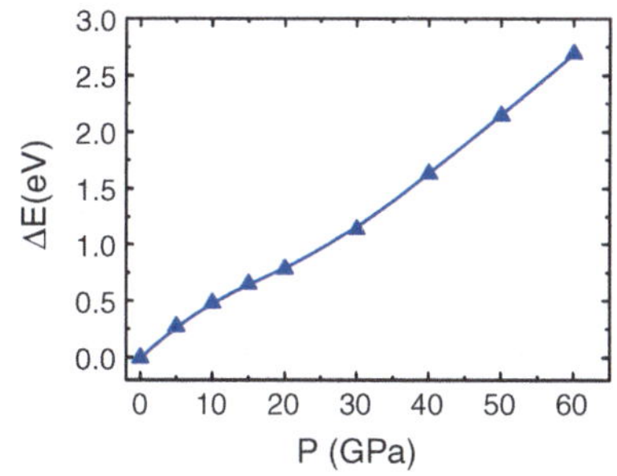


**Fig. 6.7** Compression raises the O:H–O bond total energy consistently, which drives the recoverability of O:H–O bond and the ensures ice regelation (Reprinted with permission from [9].)

60 GPa. The different $E_L$ and $E_H$ values at O:H and H–O length symmetrization under 60 GPa compression indicates compression does not dehybridize the sp orbits of oxygen, despite segmental length and containing angle relaxation.

Different from the 'normal' situation of bond relaxation, compression raises the total energy of the O:H–O bond rather than lowers it, as shown in Fig. 6.7, which entitles the O:H–O bond to fully recover its initial states once the compression is relieved without any plastic deformation. Therefore, ice melts under compression and freezes again when the pressure is relieved.

## *6.4.5 Mechanisms for Regelation*

### 6.4.5.1 Quasisolid Phase Boundary Dispersivity

The following clarifies the remaining questions regarding ice Regelation:

(1) Why does the $E_H$ instead of the $E_L$ dictates the $T_m$?
(2) Why does ice melt at the lowest limit of −22 °C?
(3) Why does the molten ice freeze again when the pressure is relieved?

In order to understand these mysteries, we need to clarify the following.

First, the $E_H$ and $\omega_H$ determine the $T_m$ that closes the upper boundary of the quasisolid phase. As the vibration frequency $\omega_x$ determines the Debye temperature $\Theta_{Dx}$ and the cohesive energy $E_x$ determines the integral of the respective segmental specific heat curve $\eta_x$ [10]. The supperposition of these two $\eta_x$ curves defines two intersecting points that correspond to the phase boundaries of the quasisolid phase or the extreme density temperatures. The high-temperature boundary is sensitive to the $\eta_H$ rather than the $\eta_L$ that has already saturated in magnitude, and therefore, the $E_H$ and the $\omega_H$ determine the $T_m$.

Second, quasisolid phase boundary is reversibly dispersive. Compression/tension ($\Delta P > 0$)/($\Delta P < 0$) disperses these boundaries simultaneously and reversely by modulating the $\Theta_{Dx} \propto \omega_{Dx}$, on the base of an extension of Einstein's relationship [9]: $\Theta_{DL}/\Theta_{DH} \approx \omega_L/\omega_H$. As shown in Fig. 6.5, compression stiffens the $\omega_L$ and meanwhile softens the $\omega_H$, which closes up the intersecting temperatures and lowers the $T_m$. Negative pressure does the opposite to raise the $T_m$. The entire process is reversible according to total energy change, as shown Fig. 6.7.

However, the $E_L$ determines the critical temperature not only for homogeneous freezing $T_N$ but also for vaporizing $T_V$. According to the superposition of the specific heats, the shift of the upper boundary of the quasisolid phase is associated with an opposite shift of the lower boundary of this quasisolid phase. On the other hand, if remove one $H_2O$ molecule from the bulk, one has to break four O:H nonbonds with energy of 0.38 eV per molecule at the ambient pressure [47]. As elaborated above, the $T_m$ depends on the $E_H$ though the $T_m$ is lower than the $T_V$ at the ambient pressure.

It is understood why Regelation happens under the conditions of $-95$ MPa $\leq P \leq 210$ MPa and 279.5 K $\geq T \geq 252$ K. If the pressure is higher than the critical value of 210 MPa, the quasisolid phase disappears and the liquid transits directly into phases III, VI and V. At temperature far below 252 K, the present rule of compression induced O:H–O bond relaxation may become invalid, as observed using Raman spectroscopy [8].

#### 6.4.5.2 O:H–O Bond Extraordinary Recoverability

Furthermore, molecular undercoordination shortens the H–O bond and lengthens the O:H nonbond. When two pieces of ice fuse, the skin $H_2O$ molecules will form hydrogen bonds with those of the other when they are brought in contact, which tends to recover their coordination numbers because of energetic favorability [48].

Once the O:H nonbond deforms or dissociates by perturbation, oxygen atoms will find new partners to retain the $sp^3$-orbital hybridization, which is the same to diamond oxidation and metal diffusive corrosion [12]. Therefore, O:H–O bond has the strong recoverability from deformation and dissociation without any plastic deformation. Compression stores energy into water ice by raising the total bond energy, $E_L + E_H$, through inter electron pair repulsion [49], once the pressure is relieved or the coordination number is recovered, the O:H–O bond will relax to its initially optimal energy states.

### 6.4.6 Compression Freezing, Melting, Dewing, and Boiling

The H–O bond relaxation determines the melting $T_m$ and the O:H relaxation decides the $T_H$ freezing. The O:H relaxation also determines the dewing and boiling $T_V$. Figure 6.8 shows that both air pressure and water boiling point drops as one goes higher. The lowered pressure lengthens and softens the O:H nonbond and the H–O bond relaxes in an opposite direction. Therefore, the critical temperatures for

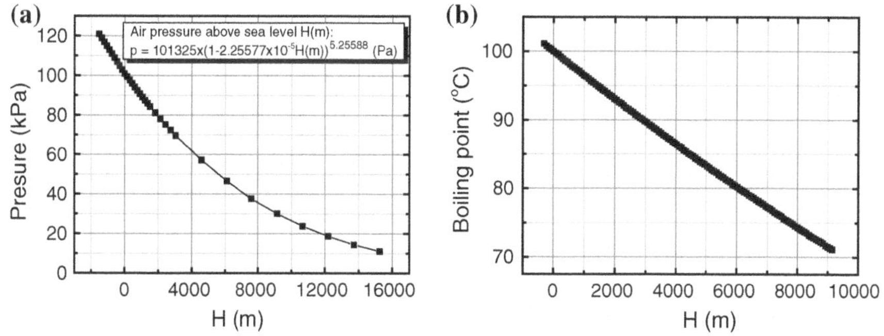


**Fig. 6.8** Altitude height dependence of the **a** pressure and **b** boiling point of water. The boiling and freezing pints will drop while the melting point will rise as one moves higher (Data sourced from [50, 51].)

boiling and freezing will drop but the freezing temperature will rise, as one goes higher.

It is true that the boiling point becomes lower as one move to high altitude. At 3000 m height, the air pressure is 70 % and the boiling point is 87 °C. At 20 km height (of a normal airplane flight) the pressure is almost zero and the boiling point will be 40 °C. The airplane must be pumped in air to maintain the cabinet pressure. The melting point, however, increases slightly only by 0.0072 °C at 20 km height. The freezing point will drop a little as goes to 2 km high.

## *6.4.7 Bandgap Expansion: Polarization*

### 6.4.7.1 Bandgap Enlargement

Generally, the band gap of a semiconductor is proportional to the cohesive energy of the representative bond with involvement of electron–phonon coupling [35, 52], according to the nearly—free electron approximation [53]. Mechanical compression shortens and stiffens the bond and hence enlarges the optical band gap [35].

Figure 6.9a and Table 6.2 feature the DFT—derived DOS evolution of ice-VIII with pressure varying from 1 to 60 GPa. The bottom edge of the valence band shifts down from −6.7 eV at 1 GPa to −9.2 eV at 60 GPa, but its upper edge at the Fermi level remains unchanged. The conduction band shifts up from 5.0 to 12.7 eV at 1 GPa to 7.4–15.0 eV at 60 GPa. The band gap expands further at higher pressures, from 4.5 to 6.6 eV, as shown in Fig. 6.9b, when the pressure is increased from 1 to 60 GPa.

Band gap expansion of ice follows the same pressure trend of "normal" substance but due to a different mechanism, because compression softens the H–O bond, instead and the contribution from the shortened O:H nonbond is insignificant [11].

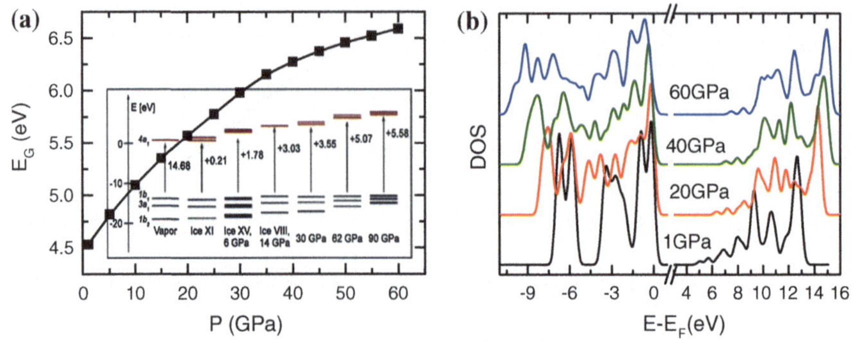


**Fig. 6.9** **a** Compression widens the band gap of ice-VIII through **b** entrapment of the bottom edge of the valence DOS and polarizes the conduction DOS. The upper edge of the valence band in at the $E_F$ remains unchanged. *Inset* **a** shows the occupied (*black lines*) and unoccupied (*red lines*) energy levels of water in vapor and molecular crystal structures, with a single-particle gap increase relative to vapor phase (Reprinted with permission from [11, 55].) (color online)

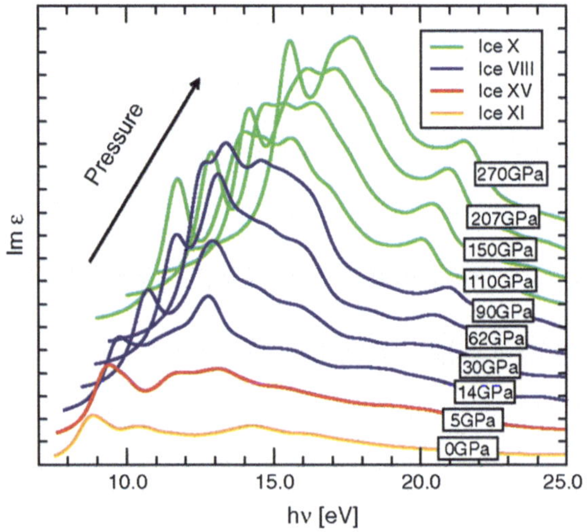


**Fig. 6.10** Compression enlarged optical absorption spectral edges of structured ice crystals. Spectra are offset vertically for clarity (Reprinted with permission from [55].)

The band gap expansion in compressed ice arises not from the $E_H$ reduction but is caused by strong polarization of the unoccupied states. The energy shift of the DOS above $E_F$ results from polarization of the lone pair by the entrapped core electrons. The energy shift of the valence DOS below $E_F$ arises from entrapment of the bonding states of oxygen [12, 54].

Figure 6.10 shows a collection of the UV absorption spectra by ice under different pressures [55]. The onset of UV absorption by ice, as an indication of band gap evolution, shifts positively with increasing pressure, making ice more transparent and blue (Fig. 6.11).

**Fig. 6.11** The bottom side (*left*) of an iceberg is strikingly *deep blue* in comparison with its upside that is white in Antarctica. The former is subject to band gap expansion reflecting *blue* light but the latter absorbs all wavelengths (*photo credit* Alex Cornell, 2015 [56]) (color online)

#### 6.4.7.2 Blue Iceberg

Usually icebergs are white because they are made of compressed snow, which reflects all frequencies of visible light so it is white. However, if high pressures squeeze the flakes together, or sea water freezes, air gaps between the snowflakes disappear, taking with them the equi-reflecting surfaces [56]. Long wavelengths of light from the sun (reds and yellows) are absorbed when passing through the ice, whereas blue light is scattered. Some of the scattered light is reflected to us, producing the blue color we associate with pure water—although microorganisms or chemicals can sometimes add a greenish tint.

The same thing happens when the water is frozen, provided all the air has been eliminated. This occurs a lot more often at the bottom of large, old blocks of ice than the top, and is usually hidden from view. Figure 6.11 shows the photo of an iceberg flipped over in Antarctica taken by Alex Cornell came across the aftermath of one such event, in comparison with the ordinary iceberg.

## 6.5 Relevant Issues

Ice Regelation is exceedingly interesting because of its relation to glacial action under nature circumstances [57], in its bearing upon molecular action [58], and damage recovery of living cells. Observations indicate that O:H–O bond has extraordinary ability of recovering its relaxation and damage [49].

### 6.5.1 Ice Cutting

One example of regelation can be seen by looping a fine wire over an ice cube and attaching a heavy weight to the wire (Fig. 6.1a). The pressure of the wire exerts on the ice will melt the local ice gradually until the wire passing through the entire block of ice. The water refreezes behind the path of the wire so one can pull the wire through the ice, while leaving the ice cube intact. While regelation is occurring, some of the melting may be caused by heating conduction of the wire under tension. A molecular-dynamics (MD) study of a nanowire cutting through ice suggests that the transition mode and the cutting rate depend on the wetting properties of the wire—hydrophobic and thicker wires cut ice faster [59]. A video clip shows that a copper wire cutting ice faster than a fishing wire because of thermal conductivity [60].

### 6.5.2 Glacier: Source of Rivers

Regelation occurs in glaciers, which forms the sources of Rivers. Glaciers of Himalayas Mountains (8881 km apex in the boundary of India and China) in Asia provide sources for the India River, Yaluzangbu River, and Heng River. The mass of a glacier allows it to exert a sufficient amount of pressure on its lower surface to lower the melting point of the ice at its base, melting the ice and allowing the glacier to slide over the liquid. Under the right conditions, liquid water can flow from the base of a glacier to lower altitudes when the temperature of the air is above the freezing point of water [1, 2] (left panel of Fig. 6.12).

### 6.5.3 Spiky Ice

The spikes of snow or ice are called penitentes, and some can be 4 m high. They are common on high-altitude glaciers, such as those in the Andes Mountains, where the air is dry, and the sun's rays can turn ice directly into water vapor without melting it first—a process called sublimation. An initially smooth snow surface first develops depressions as some regions randomly sublimate faster than others. The curved surfaces then concentrate sunlight and speed up sublimation in the depressions, leaving the higher points behind as forests of towering spikes. At the micro-scale, similar-looking spikes help solar cell surfaces maximize their sunlight absorption. Penitentes grow faster at temperature between −10 and −20 °C under flood lamp irradiation and RH 70 % coarsen [62].

According to the currently described premise, spiky ice formation requires suitable extrinsic conditions of temperature, pressure, humidity, and irradiation and intrinsic response of the O:H–O bond to these stimuli. Both the lower pressure at higher altitude and the undercoordination of molecules raises the $T_m$ of ice at apex so ice melts first at the dents, resulting in the spiky ice, as shown in Fig. 6.12 (right).

**Fig. 6.12** Glacier regelation and spiky ice. (*Left*) Meili Mountain in Yuannan, China, located at the junction of Nu River, Lanchang River, and Jinsha River (Photograph by Yi Sun, 2010). (*Right*) Spiked ice (called penitentes that can be 4 m high) formed naturally on high-altitude glaciers in the relative humidity of ~70 % and the −10 and −20 °C temperature range under irradiation (Reprinted with permission from [61, 62].)

### 6.5.4 *Summary*

Regelation arises from the O:H–O bond memory and the phase boundary dispersivity:

(1) Compression shortens and stiffens the O:H nonbond and lengthens and softens the H–O bond in all phases towards proton centralization, which lowers the compressibility of water ice and makes the O:H contribute positively, while the H–O contributes negatively to the lattice energy of compressed ice.
(2) Compression closes up the separation between boundaries of the quasisolid phase and depresses the $T_m$, resulting in ice regelation. Negative pressure does the opposite.
(3) Unlike the bond in a 'normal' substance that gains energy with plastic deformation, compression raises the O:H–O bond to higher energy states monotonically. When the pressure is relieved, the O:H–O bond recovers completely its initial states with memory.
(4) The O:H nonbond compression dictates the $T_C(P)$ phase boundaries with positive slopes like Liquid–Vapor transition-boiling; the H–O bond elongation dictates the $T_C(P)$ phase boundaries with negative slopes like melting and VII–VIII transition; O:H–O length and energy conserve at boundaries of zero or infinite slopes like (XII, XIII)—X and IC–VI transition.

## Appendix: Featured News

**Unlocking the mysteries of ice** Home/Chemistry World/News/2012/Marc
*By Erica Wise,* Editor for RSC Press
27 March 2012

The unusual properties of ice under compression are due to Coulomb repulsion between bonding and non-bonding electron pairs, say scientists from Singapore and China.

Frozen water behaves differently from other materials in response to pressure. It has abnormally low compressibility, and applying pressure decreases rather than increases the critical temperature for phase transitions. These anomalies have puzzled scientists for many years and satisfactorily modelling them has proven a great challenge.

Now, Chang Sun at Nanyang Technological University and his colleagues at Jilin and Xiangtan Universities have developed a new method to simulate these properties accurately. Their work has also helped to clarify the physical basis of the behaviour.

The key to their model is in considering O· · ·H–O as the basic structural unit of ice. The left hand oxygen forms a hydrogen bond using its lone pair of electrons to polarise electron density around the hydrogen. Meanwhile, the hydrogen shares its electron with the right hand oxygen to form a real bond.

Sun's model works better for the system than commonly used rigid non-polarisable models. Such models 'have a fixed molecular geometry so they cannot intrinsically account for changes in the molecular geometry,' according to Jose Abascal, an expert on the theoretical chemistry of water and ice, Universidad Complutense de Madrid, Spain. The rigid models 'approximate the $H_2O$ molecule as two point charges with a fixed bond length and bond angle,' explains Sun. However, 'what changes with the applied stimulus are the angle, length and energy of the hydrogen bond and the associated electron polarisation.'

Sun's results indicate that the repulsion between the lone pair and bonding pair causes the O···H hydrogen bond to shorten and the O-H real bond to lengthen. At sufficiently high pressure, the hydrogen bond and real bond become equivalent in length. The change in binding energy of the real bond dominates, causing the observed effects on physical quantities as it lengthens and weakens.

Sun now anticipates that there is further work to be done in unravelling the many other anomalies of ice, including why freezing water expands.

**Reference**

C.Q. Sun, X. Zhang, W Zheng, *Chem Sci.*, 2012, doi:10.1039/c2sc20066j

## References


1. J.D. Forbes, Illustrations of the viscous theory of glacier motion. part ii. an attempt to establish by observation the plasticity of glacier ice. Philos. Trans. R. Soc. London **136**, 157–175 (1846)
2. J.D. Forbes, Illustrations of the viscous theory of glacier motion. part i. containing experiments on the flow of plastic bodies, and observations on the phenomena of lava streams. Philos. Trans. R. Soc. London **136**, 143–155 (1846)
3. J. Thomson, Note on professor Faraday's recent experiments on regelation. Proc. R. Soc. London **10**, 151–160 (1859)
4. M. Faraday, Note on regelation. Proc. R. Soc. London **10**, 440–450 (1859)
5. J.L. Green, D.J. Durben, G.H. Wolf, C.A. Angell, Water and solutions at negative pressure: raman spectroscopic study to -80 megapascals. Science **249**(4969), 649–652 (1990)
6. M. Chaplin, *Water Structure and Science:* http://www.lsbu.ac.uk/water/
7. YouTube. *Does Pressure Melt Ice?,* https://www.youtube.com/watch?v=gM3zP72-rJE
8. Y. Yoshimura, S.T. Stewart, M. Somayazulu, H.K. Mao, R.J. Hemley, Convergent Raman features in high density amorphous ice, ice VII, and ice VIII under pressure. J. Phys. Chem. B **115**(14), 3756–3760 (2011)
9. Y. Huang, X. Zhang, Z. Ma, Y. Zhou, W. Zheng, J. Zhou, C.Q. Sun, Hydrogen-bond relaxation dynamics: resolving mysteries of water ice. Coord. Chem. Rev. **285**, 109–165 (2015)
10. C.Q. Sun, X. Zhang, X. Fu, W. Zheng, J.-L. Kuo, Y. Zhou, Z. Shen, J. Zhou, Density and phonon-stiffness anomalies of water and ice in the full temperature range. J. Phys. Chem. Lett. **4**, 3238–3244 (2013)
11. C.Q. Sun, X. Zhang, W.T. Zheng, Hidden force opposing ice compression. Chem. Sci. **3**, 1455–1460 (2012)
12. C.Q. Sun, Oxidation electronics: bond-band-barrier correlation and its applications. Prog. Mater. Sci. **48**(6), 521–685 (2003)
13. C.Q. Sun, Relaxation of the chemical bond, in *Springer Series in Chemical Physics 108*. vol. 108 (Springer, Heidelberg, 2014), 807 pp
14. T.B. James, Melting and regelation of ice. Nature (London) **5**, 185 (1872)

15. J.M. Thomas, *Michael Faraday and The Royal Institution: The Genius of Man and Place* (1991)
16. W. Holzapfel, On the symmetry of the hydrogen bonds in ice VII. J. Chem. Phys. **56**(2), 712 (1972)
17. A.F. Goncharov, V.V. Struzhkin, H.-K. Mao, R.J. Hemley, Raman spectroscopy of dense H2O and the transition to symmetric hydrogen bonds. Phys. Rev. Lett. **83**(10), 1998–2001 (1999)
18. M. Benoit, D. Marx, M. Parrinello, Tunnelling and zero-point motion in high-pressure ice. Nature **392**(6673), 258–261 (1998)
19. P. Loubeyre, R. LeToullec, E. Wolanin, M. Hanfland, D. Husermann, Modulated phases and proton centring in ice observed by X-ray diffraction up to 170 GPa. Nature **397**(6719), 503–506 (1999)
20. A.F. Goncharov, V.V. Struzhkin, M.S. Somayazulu, R.J. Hemley, H.K. Mao, Compression of ice to 210 gigapascals: infrared evidence for a symmetric hydrogen-bonded phase. Science **273**(5272), 218–220 (1996)
21. J. Teixeira, High-pressure physics - the double identity of ice X. Nature **392**(6673), 232–233 (1998)
22. I.A. Ryzhkin, "Symmetrical" phase and collective excitations in the proton system of ice. J. Exp. Theor. Phys. **88**(6), 1208–1211 (1999)
23. F.H. Stillinger, K.S. Schweizer, Ice under compression-transition to symmetrical hydrogen-bonds. J. Phys. Chem. **87**(21), 4281–4288 (1983)
24. L.N. Tian, A.I. Kolesnikov, J.C. Li, Ab initio simulation of hydrogen bonding in ices under ultra-high pressure. J. Chem. Phys. **137**(20), 204507 (2012)
25. Y. Yoshimura, S.T. Stewart, M. Somayazulu, H. Mao, R.J. Hemley, High-pressure x-ray diffraction and Raman spectroscopy of ice VIII. J. Chem. Phys. **124**(2), 024502 (2006)
26. K. Liu, J.D. Cruzan, R.J. Saykally, Water clusters. Science **271**(5251), 929–933 (1996)
27. R. Ludwig, Water: from clusters to the bulk. Angewandte Chemie-International Edition **40** (10), 1808–1827 (2001)
28. D. Kang, J. Dai, Y. Hou, J. Yuan, Structure and vibrational spectra of small water clusters from first principles simulations. J. Chem. Phys. **133**(1), 014302 (2010)
29. T. Yan, S. Li, K. Wang, X. Tan, Z. Jiang, K. Yang, B. Liu, G. Zou, B. Zou, Pressure-induced phase transition in N-H…O hydrogen-bonded molecular crystal oxamide. J. Phys. Chem. B **116**(32), 9796–9802 (2012)
30. B. Santra, J. Klimeš, D. Alfè, A. Tkatchenko, B. Slater, A. Michaelides, R. Car, M. Scheffler, Hydrogen bonds and van der Waals forces in ice at ambient and high pressures. Phys. Rev. Lett. **107**(18), 185701 (2011)
31. J.E. Bertie, E. Whalley, Infrared spectra of Ices II, III, and V in the Range 4000 to 350 cm—1. J. Chem. Phys. **40**(6), 1646–1659 (1964)
32. P.T.T. Wong, E. Whalley, Raman spectrum of ice VIII. J. Chem. Phys. **64**(6), 2359–2366 (1976)
33. J.E. Bertie, F.E. Bates, Mid-infrared spectra of deuterated ices at 10 °K and interpretation of the OD stretching bands of ices II and IX. J. Chem. Phys. **67**(4), 1511–1518 (1977)
34. W.T. Zheng, C.Q. Sun, Underneath the fascinations of carbon nanotubes and graphene nanoribbons. Energy Environ. Sci. **4**(3), 627–655 (2011)
35. J.W. Li, S.Z. Ma, X.J. Liu, Z.F. Zhou, C.Q. Sun, ZnO meso-mechano-thermo physical chemistry. Chem. Rev. **112**(5), 2833–2852 (2012)
36. M.X. Gu, Y.C. Zhou, L.K. Pan, Z. Sun, S.Z. Wang, C.Q. Sun, Temperature dependence of the elastic and vibronic behavior of Si, Ge, and diamond crystals. J. Appl. Phys. **102**(8), 083524 (2007)
37. M.X. Gu, L.K. Pan, T.C.A. Yeung, B.K. Tay, C.Q. Sun, Atomistic origin of the thermally driven softening of Raman optical phonons in group III nitrides. J. Phys. Chem. C **111**(36), 13606–13610 (2007)

38. C. Yang, Z.F. Zhou, J.W. Li, X.X. Yang, W. Qin, R. Jiang, N.G. Guo, Y. Wang, C.Q. Sun, Correlation between the band gap, elastic modulus, Raman shift and melting point of CdS, ZnS, and CdSe semiconductors and their size dependency. Nanoscale **4**, 1304–1307 (2012)
39. M. Song, H. Yamawaki, H. Fujihisa, M. Sakashita, K. Aoki, Infrared absorption study of Fermi resonance and hydrogen-bond symmetrization of ice up to 141 GPa. Phys. Rev. B **60** (18), 12644 (1999)
40. P. Pruzan, J.C. Chervin, E. Wolanin, B. Canny, M. Gauthier, M. Hanfland, Phase diagram of ice in the VII-VIII-X domain. Vibrational and structural data for strongly compressed ice VIII. J. Raman Spectrosc. **34**(7–8), 591–610 (2003)
41. T. Okada, K. Komatsu, T. Kawamoto, T. Yamanaka, H. Kagi, Pressure response of Raman spectra of water and its implication to the change in hydrogen bond interaction. Spectrochim. Acta A **61**(10), 2423–2427 (2005)
42. K. Aoki, H. Yamawaki, M. Sakashita, Observation of Fano interference in high-pressure ice VII. Phys. Rev. Lett. **76**(5), 784–786 (1996)
43. G. Malenkov, Liquid water and ices: understanding the structure and physical properties. J. Phys.-Condens. Matter **21**(28), 283101 (2009)
44. C.Q. Sun, H.L. Bai, B.K. Tay, S. Li, E.Y. Jiang, Dimension, strength, and chemical and thermal stability of a single C-C bond in carbon nanotubes. J. Phys. Chem. B **107**(31), 7544–7546 (2003)
45. C.Q. Sun, Thermo-mechanical behavior of low-dimensional systems: the local bond average approach. Prog. Mater Sci. **54**(2), 179–307 (2009)
46. X.J. Liu, M.L. Bo, X. Zhang, L.T. Li, Y.G. Nie, H. Tian, Y. Sun, S. Xu, Y. Wang, W. Zheng, C.Q. Sun, Coordination-resolved electron spectrometrics. Chem. Rev. **115**(14), 6746–6810 (2015)
47. M. Zhao, W.T. Zheng, J.C. Li, Z. Wen, M.X. Gu, C.Q. Sun, Atomistic origin, temperature dependence, and responsibilities of surface energetics: An extended broken-bond rule. Phys. Rev. B **75**(8), 085427 (2007)
48. C.Q. Sun, X. Zhang, J. Zhou, Y. Huang, Y. Zhou, W. Zheng, Density, elasticity, and stability anomalies of water molecules with fewer than four neighbors. J. Phys. Chem. Lett. **4**, 2565–2570 (2013)
49. X. Zhang, Y. Huang, P. Sun, X. Liu, Z. Ma, Y. Zhou, J. Zhou, W. Zheng, C.Q. Sun, Ice regelation: hydrogen-bond extraordinary recoverability and water quasisolid-phase-boundary dispersivity. Sci. Rep. **5**, 13655 (2015)
50. *Air Pressure and Altitude above Sea Level.* Engineering toolbox, http://www.engineeringtoolbox.com/air-altitude-pressure-d_462.html
51. *Boiling Points of Water at Various Elevations.* Engineering toolbox, http://www.engineeringtoolbox.com/boiling-points-water-altitude-d_1344.html
52. L.K. Pan, S.Q. Xu, W. Qin, X.J. Liu, Z. Sun, W.T. Zheng, C.Q. Sun, Skin dominance of the dielectric-electronic-phononic-photonic attribute of nanostructured silicon. Surf. Sci. Rep. **68** (3–4), 418–455 (2013)
53. C.Q. Sun, T.P. Chen, B.K. Tay, S. Li, H. Huang, Y.B. Zhang, L.K. Pan, S.P. Lau, X.W. Sun, An extended 'quantum confinement' theory: surface-coordination imperfection modifies the entire band structure of a nanosolid. J. Phys. D-Appl. Phys. **34**(24), 3470–3479 (2001)
54. C.Q. Sun, Dominance of broken bonds and nonbonding electrons at the nanoscale. Nanoscale **2**(10), 1930–1961 (2010)
55. A. Hermann, P. Schwerdtfeger, Blueshifting the onset of optical UV absorption for water under pressure. Phys. Rev. Lett. **106**(18), 187403 (2011)
56. S. Luntz, A. Cornell. *This Is What The Underside Of An Iceberg Looks Like.* 2015, http://www.iflscience.com/environment/underside-iceberg
57. J.D. Goddard, The viscous drag on solids moving through solids. AlChE J. **60**(4), 1488–1498 (2014)
58. D.T. Möhlmann, Are nanometric films of liquid undercooled interfacial water bio-relevant? Cryobiology **58**(3), 256–261 (2009)

59. T. Hynninen, V. Heinonen, C.L. Dias, M. Karttunen, A.S. Foster, T. Ala-Nissila, Cutting ice: nanowire regelation. Phys. Rev. Lett. **105**(8), 086102 (2010)
60. D. Petely, *Our strange desire to find a landslide trigger.* http://ihrrblog.org/2013/11/08/our-strange-desire-to-find-a-landslide-trigger/, (2013)
61. J. Corripio, *Spiked Ice* (Edinburgh, Scotland, 2001)
62. V. Bergeron, C. Berger, M. Betterton, Controlled irradiative formation of penitentes. Phys. Rev. Lett. **96**(9), 098502 (2006)

# Chapter 7
# Thermal Excitation

- *A superposition of the O:H and H–O specific heats defines two intersecting temperatures that divide the full temperature range into four regimes of different specific-heat ratios, which facilitates density oscillation in the phase of liquid, quasisolid, ice $I_{h+c}$, and ice XI.*
- *The segment of relatively lower specific heat serves as the "master" to follow the regular rule of thermal expansion, which drives the other "slave" segment to relax oppositely when they are subject to heating or cooling.*
- *Cooling shortens the O:H nonbond more than H–O expands in the liquid and the $I_{h+c}$ phase, which results in the seemingly "normal" thermal expansion of water and ice. The H–O bond and the O:H exchange roles in the quasisolid phase resulting in the negative thermal expansion of the quasisolid, which makes ice floats.*
- *Heating fluctuates the H–O radicals with increased abundance and redshift; the skin is thermally more stable than the bulk that undergoes H-O bond heating blueshift.*

**Abstract** Characterized by the Debye temperature that is proportional to the characteristic phonon frequency and by the thermal integration that is proportional to the segmental cohesive energy, the specific-heat for the O:H nonbond differs from that of the H–O bond. The thermodynamic disparity of the O:H and the H–O defines the cooperative angle-length-stiffness relaxation of the O:H–O bond in four different ways, which is responsible for the density and phonon-stiffness oscillation of water and ice in the full temperature range. Generally, cooling shortens and stiffens the part of relatively lower specific-heat, and meanwhile, lengthens and softens the other part of the O:H–O bond via O–O repulsion. Length contraction/elongation of a specific part always stiffens/softens its corresponding phonon. The O–O distance is longer in ice than it is in water, resulting in a lower density, so that ice floats. Phonon spectrometrics enabled the molecular-site-resolved O:H–O bond relaxation dynamics in terms of segmental length and stiffness, order of molecular fluctuation, and the abundance of phonons.

## 7.1 Challenge: Why Does Ice Float?

Water and ice respond abnormally to thermal excitation with the following puzzles:

(1) Ice floats [1]. Unlike argon and oxygen whose mass density increases by 12 % upon freezing, water drops its density by 9 % when turns into ice so an ice cube floats (Fig. 7.1a).
(2) Density oscillation [2]. The density ρ(T) oscillates over the full temperature range in four regions at different rates.



C.Q. Sun and Y. Sun, *The Attribute of Water*, Springer Series in Chemical Physics 113, DOI 10.1007/978-981-10-0180-2_7

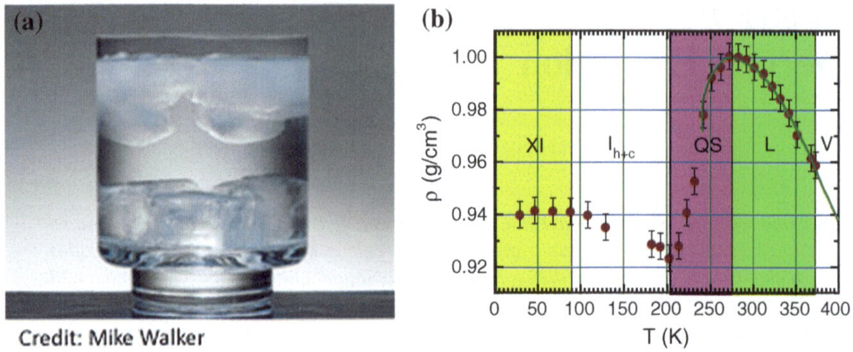


**Fig. 7.1** **a** Low density ice cubes float in a cup of water [1]. **b** Density ρ(T) profile for bulk water (T ≥ 273 K) and 1.4 nm water droplet (T < 273 K) [2] oscillates over the full temperature range and the transition temperatures disperse with applied stimulus [3]

(3) Bulk water ice reaches its extreme densities at 277 and 258 K and the Ic transits into XI phase at ~100 K [3]; these critical temperatures are subject to variation with stimulus such as droplet size and pressure.
(4) Debating continues since 1611 with yet unconsented mechanism for ice floats and the density oscillation mechanism remains unaware up to date.

## 7.2 Clarification: O:H–O Bond Specific-Heat Disparity

From the perspective of O:H–O bond thermodynamic disparity and its relaxation in its containing angle and segmental lengths, Fig. 7.2 clarifies the density thermal anomalies of water and ice [4]:

(1) A supposition of the specific heat $\eta_x(T, \Theta_{Dx})$ curves for the O:H nonbond (x = L) and the H–O bond (x = H) defines two intersecting temperatures that divide the full temperature range into four regimes with different $\eta_L/\eta_H$ ratios, see Fig. 7.2a.
(2) Cooling stretches the ∠O:H–O containing angle θ in different slopes (Fig. 7.2b), which lengthens the O–O distance and lowers the density to certain extents in the quasisolid and solid phases. The θ remains constant in liquid phase because of the strong fluctuation of molecular dynamics.
(3) Cooling shortens the O:H segment more than H–O bond elongates in the liquid and in the $I_{h+c}$ phase because $\eta_L/\eta_H < 1$, resulting cooling densification of water and ice at different rates.
(4) Cooling reverses the process of O:H–O relaxation because $\eta_L/\eta_H > 1$, toward density loss; neither segment changes its length in phase XI because $\eta_L \approx \eta_H \approx 0$, but cooling stretched θ expands volume slightly.

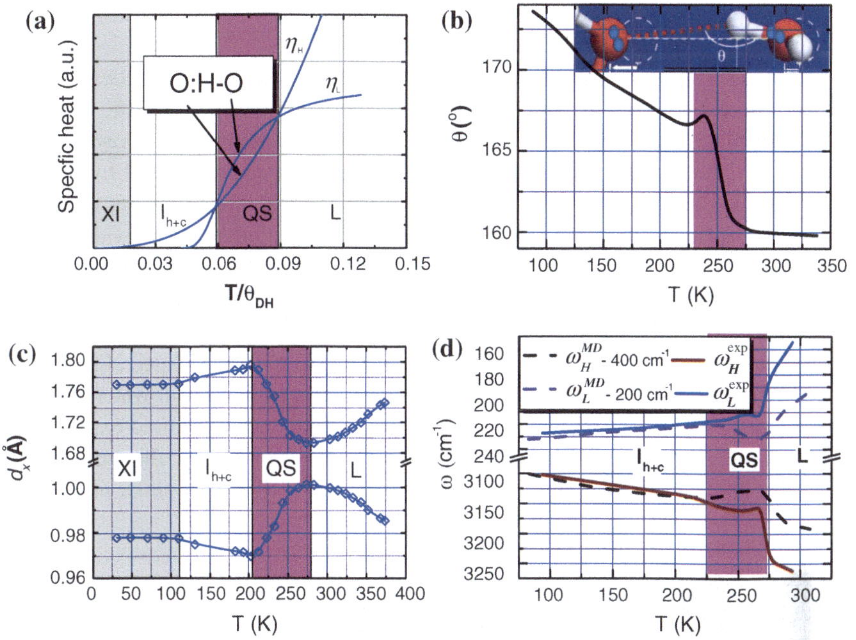


**Fig. 7.2** **a** Specific heat superposition defines the intersecting temperatures and different $\eta_L/\eta_H$ ratios in the Liquid, Quasisolid (QS), $I_{c+h}$ and XI phases. The segment of lower $\eta_x$ serves as the "master" to follow the general rule of thermal expansion and drives the other "slave" to relax oppositely by different amount. **b** Cooling stretches the angle θ in different regimes at different slopes. **c** The O:H and the H–O relaxation dynamics derived from measurements with θ involvement [5]. **d** Comparison of the measured (*solid lines*) and the calculated (*broken lines*) phonon relaxation dynamics. Indicated 200/400 $cm^{-1}$ offsets of the calculated $\omega_x$ to match observations. Inset in (**b**) illustrates the segmental cooling relaxation in the quasisolid phase and the containing angle θ (Reprinted with permission from [4].)

## 7.3 History Background

### *7.3.1 First Debate*

In 1611 [6], Galilei Galileo and Ludovico delle Colombe started a fierce, three-day debate on the topic "Why does ice float on water, when ice is itself water?" in front of dozens of wealthy spectators gathered in Florence. The focusing points of argument were bouncy, density, shape, and surface tension.

Ludovico's basic premise was that ice was the solid form of water, therefore it was denser than water. He argued that buoyancy was "a matter of shape only." "It had nothing to do with density." Ludovico presented a sphere of ebony to the audience. The sphere was placed on the surface of the water, and it began to sink. Then Ludovico took a thin wafer of ebony and placed it on the surface of the water, where it floated. Because the density of both the wafer and the sphere of ebony were the same, Ludovico announced that density had nothing to do with buoyancy and that an object's shape was all that mattered.

Galileo's primary argument for floating ice was based on Archimedes' density theory, wherein an object in water experiences a buoyant force equal to the weight of water it displaces. Because ice is less dense than liquid water, it will always float on liquid water. But Galileo then went too far. Aiming at the main thrust of Ludovico's argument, Galileo said that the shape of an object did not affect whether the object would sink or float. The reason ice floats on water has everything to do with density. Ice is a rare example of a solid that is less dense than its corresponding liquid.

Galileo had ignored the surface tension, however. Surface tension forces can help objects located on a liquid surface resist sinking on the basis of how much of that object is in contact with the liquid's surface. Consider a paper clip: If it is placed flat on the surface of water it can float, but if it is placed on water standing straight up, it sinks. The difference is the higher surface tension force experienced by the paper clip lying flat on the water's surface. So in a way, the shape of an object (in contact with the surface) does contribute to whether it sinks or floats.

Galilei Galileo (Pisa, Italy. 1564–1642) was a physicist, mathematician, engineer, astronomer, and philosopher. (Painting portrait of Galilei Galileo by Justus Susterman in 1636, free Wikimedia)

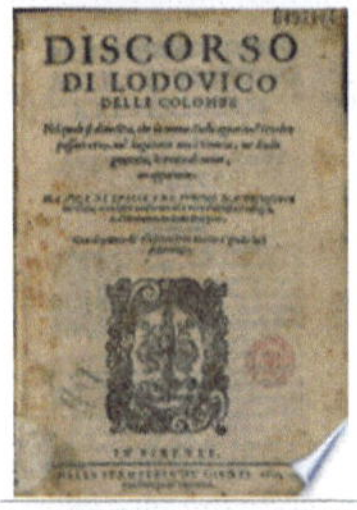
DISCORSO
DI LODOVICO

Ludovico delle Colombe (Florence, Italy, 1565–1616), a philosopher and a poet, is known above all for his opposition to Galileo, at first in the field of astronomy in his siding against the Copernican system (Earth revolved around the Sun), and then in the field of physics, on an issue concerning hydrostatics (buoyancy force)

Archimedes (Syracuse, Italy. 287–212 B.C.) was a Greek mathematician, physicist, engineer, inventor, and astronomer. (Painting portrait of Archimedes by Domenico Fetti in 1620, free Wikimedia)

The dispute became noisy and inconclusive, and the meeting was brought to a close. The patrons of both Ludovico and Galileo encouraged the two men to write up descriptions of the debate and their arguments, which led to publications of An Essay on Objects that Float in Water or that Move in It (Florence 1612); A Defence of Galileo's Essay (1612) and Considerations concerning Galileo's Essay (1613 by Ludovico). Both tracts attacked Galileo's theories on the basis of Aristotelian precepts. In 1615, Galileo published a book Response to the Disagreements of Ser Lodovico delle Colombe and Ser Vincenzo di Grazia against Signor Galileo's Treatise concerning Objects that Lie on Water.

To commemorate the 400th anniversary of this debate, two dozen brilliant minds met in Florence, Italy, for a week in July 2013 to discuss current unanswered questions in water research at a conference playfully dubbed Aqua Incognita (which can be translated as Water in Disguise or Unknowable Water). This discussion has led to a book edited by Ninham and Nostro [7].

The two deliberations about water, some 400 years apart, had similarities: both were multiday events featuring occasional raucous disagreement about experimental details or theoretical constructs. However, with the hindsight of four centuries, the earlier water debate provides a cautionary tale to water researchers—and in fact all scientists—about the double-edged sword of scientific arrogance.

Four hundred years after the debate, there are still many unresolved questions about water. The fundamental origin in terms of structure and dynamics of its many anomalous properties is still under debate. No model is currently able to reproduce these properties throughout the phase diagram. A four week symposium was held in Nordita, Stockholm, during October 13 and November 07 2014. This program, organized by Lars Pettersson, Anders Nilsson, and Richard Henchman, brought together hundreds experimentalists and theoreticians in strong synergy to explore interpretations and to provide a strong basis for further experimental and theoretical advances towards a unified picture of water. The primary objective of the program is to identify critical aspects of water's anomalous behavior that need to be included in new water models in order to give an overall encompassing agreement with experiments. It also aimed to stimulate further developments of models that can also include perturbations due to ion solvation, hydrophobic interactions as well as describe water at interfaces.

### 7.3.2 Available Mechanisms

Currently available mechanisms for density anomalies are mainly focused on the density change in the quasisolid regime. The mechanism behind the 'regular' process of cooling densification in the liquid and solid phase has attracted little attention. The following mechanisms address the freezing expansion in terms of supercooled liquid to mimics of ice:

(1) The mixed-phase scheme [2, 8–14] suggests that a competition between the randomly distributed, 'ice-like' nanoscale fragments, or the ring- or chain-like low-density liquid (LDL), and the tetrahedrally structured high-density liquid (HDL) fragments dictates the volume expansion in the supercooled liquid [10, 15]. Cooling increases the fraction of the LDL phase, and then ice floats. The many-body electronic structure and the non-local vdW interactions were suggested as possible forces driving volume expansion [16].
(2) The monophase notion [17–21] explains that water contains a homogeneous, three-dimensional, tetrahedrally coordinated structured phase with thermal fluctuation that is not quite random [21, 22]. The monophase model attributes freezing expansion to the O:H–O bond relaxation in length and angle in a fixed yet unclear manner.
(3) The linear correlation model [23] rationalizes that the local density changes homogeneously with the length and angle. Matsumoto used computer simulation to look at ways changing the O:H–O bond network volume: extension of the bonds, change in the containing angle between the bonds, and change in the network topology. He found that the O:H–O bond elongation is responsible for thermal expansion, while the angular distortion causes thermal contraction. The network topology does not contribute to volume change. Therefore, the competition between the O:H–O bond angle and its length relaxation determines the density anomalies of water ice.
(4) Two-kind of O:H bond [24, 25] model suggests that one kind of stronger and another kind of weaker O:H bond coexist randomly in the ratio of about 2:1 (they could be the weker O:H in the skin and the regiular O:H in the bulk regimes according to the present notion). By introducing these two types of O:H bond, Tu and Fang [25] reproduced a number of the anomalies, particularly the thermodynamic properties in the supercooled state. They found that the exchange between the strong and the weak O:H bonds enhance the competition between the open and the collapsed structures of liquid water.

### *7.3.3 Remaining Issues*

A number of issues on water density anomalies remain yet unattended [26–29]. Determination of the following attributes is beyond the scope of available models focusing on the phase composition. Solving the existing and the emerging challenges from the perspective of O:H–O bond relaxation could lead to a consistent resolution of the following issues:

(1) The 'regular' cooling densification in the liquid and in the solid phases
(2) The irregular volume expansion in the quasisolid and the XI phases
(3) O:H–O bond segmental length and stiffness cooperativity
(4) O:H–O bond segmental cooperative response to thermal excitation
(5) Correlation between the mass density $\rho$, segmental lengths, and phonon frequencies $\omega_x$ over the full temperature range

It is equally clear that the problem cannot be resolved in terms of the usual statistical mechanics, and two- and three-body molecular interaction potentials derived from quantum mechanical perturbation theory. Otherwise the problem would have been solved long since [30].

## 7.4 Quantitative Resolution

### 7.4.1 Buoyancy Versus Density

Ice floating follows Archimedes' principle. The upward buoyant force (B) that is exerted on a body immersed in a fluid, whether fully or partially submerged ($V + \Delta V$), is equal to the weight $Vg\rho$ of the liquid that the body has displaced . The $\Delta V$ is the unsubmerged volume. The following formulates the net force *f* floating the body:

$$\begin{aligned} f = B - Mg &= \left[V\rho_{liquid} - (V+\Delta V)\rho_{body}\right]g \\ &= V\rho_{liquid}g\left[1 - \frac{(V+\Delta V)\rho_{body}}{V\rho_{liquid}}\right] \geq 0 \end{aligned}$$

which requires,

$$\frac{\rho_{liquid} - \rho_{body}}{\rho_{body}} = \frac{\Delta\rho}{\rho_{body}} = \frac{\Delta V}{V} \tag{7.1}$$

That is, the density of the body is smaller than that of the liquid, $\Delta\rho = \rho_{liquid} - \rho_{body} > 0$. Therefore, ice is less dense than water, allowing it to float, as everybody knows.

One interesting question is that if a 2 kg wood block can float on 1 kg water. Neglecting the surface tension, one can assume the maximal flat contact area between the water and the block of wood, the buoyant force meets this relation where x is the weight ratio between wood and water,

$$B = \rho_w h_w = x\rho_l h_l = \rho_w\left(h_l + h_{up}\right) \tag{7.2}$$

where $h_w = h_l + h_{up}$ is the block height and $h_l$ is the depth of liquid. $L_{up}$ is the unmerged height of the block. Combining (7.1) and (7.2) yields,

$$x = \frac{\rho_w}{\rho_l}\left(1 + \frac{h_{up}}{h_L}\right) = \frac{\rho_w}{\rho_l}\left(1 + \frac{\rho_l - \rho_w}{\rho_w}\right) \equiv 1$$

Therefore, 1 kg water can only float the wood of the same weight disregarding the contacting area or the density of the wood block.

**Fig. 7.3** Because of cold temperatures (−25 °C), the ocean waves in Nantucket, an island 30 miles (50 km) south of Cape Cod, in the American state of Massachusetts, are rolling in as slush–quasisolid (nearly frozen) phase (*Photo credit* Jonathan Nimerfroh—Instagram: @jdnphotography), Feb 26, 2015 [31]

## *7.4.2 Quasisolid Phase*

Quasisolid phase exists between the liquid and the solid $I_h$ phases for highly pure water in the temperature range from 258 to 277 K [3]. The quasisolid phase is gel-like and viscous. Figure 7.3 shows the quasisolid phase-frozen waves, featured by CBS Boston News in February 2015. The record-setting winter of 2015 has left us with all kinds of remarkable images, most of them of snow and ice. But a photographer on Nantucket found something most of us have never seen-nearly frozen waves. Jonathan Nimerfroh was walking along a beach on the island when he saw these waves rolling in like slush. The waves were semi-frozen because they are much like jelly despite much ice inside them. The temperature was about −25°C that could not freeze sea water homogeneously.

## *7.4.3 Molecular Fluctuation Dynamics*

### 7.4.3.1 Bond Length Versus Mass Density

Figure 7.2 features the O:H–O bond segmental lengths $d_x$, containing angles, and the associated phonon relaxation dynamics derived from MD calculations [4] and

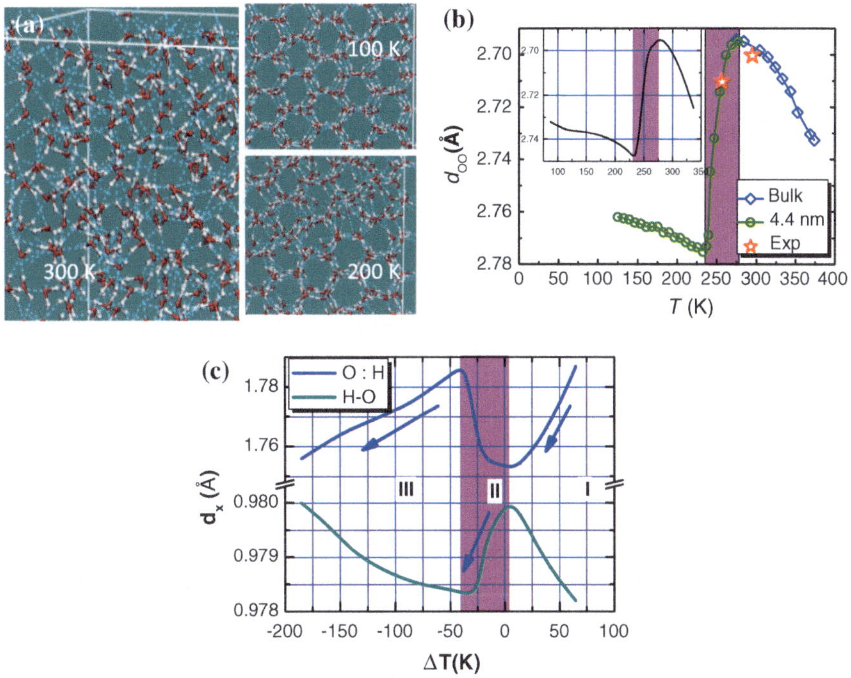


**Fig. 7.4** **a** Snapshots of MD trajectory show that the structural order decreases with increasing temperature from 100 to 300 K while the V-shaped $H_2O$ motifs remain intact at 300 K because of the stronger H–O bond (3.97 eV/bond) [3]. **b** O–O distance oscillation profiles derived from measurements [35] and computation (*inset*) agrees with the measured density trends of water ice except for transition temperatures at 202–258 K [2]. The $d_{OO}(T)$ profile also matches to the measured $d_{OO}$ at 25 and −16.8 °C [36]. **c** MD derived $d_x$ cooperative relaxation in different temperature regimes (Reprinted with permission from [4].)

experimental observations [2]. The MD derivatives show the purely $d_x$ relaxation whose accuracy is subject to numerical algorithm employed. The experimental derivative is a resultant relaxation of the $d_x$ and the θ angle.

The MD trajectory snapshots in Fig. 7.4a and the MD video in [4] show that the V-shaped $H_2O$ motifs remain intact at 300 K over the entire duration of recording. This configuration is accompanied by large fluctuations of the θ and the $d_L$ flashing in this regime but retain the mean of the tetrahedrally-coordinated structure of water molecules [32], even for a single molecule at 5 K temperature [33]. The MD-video in [4] shows that, in the liquid phase, the H and the O attract each other in the O:H interaction, but the O–O repulsion prevents this occurrence. The intact H–O–H motifs move ceaselessly like a complex pendulum because of the high fluctuation and frequent switching the O:H interaction on and off.

Figure 7.4b is the O–O distance as a function of temperature, which agrees qualitatively to measurements in Fig. 7.1b. Agreement between MD and experimental observations asserted that the shortening of the master segments (the part of relatively lower specific heat as denoted with arrows) is always coupled with a lengthening of the slaves during cooling. In the liquid and the solid $I_h$ and $I_c$ regimes, the O:H nonbond having a lower $\eta_L$ contracts more than the H–O bond elongates, resulting in a net loss of the O–O length. Thus, cooling-driven densification of $H_2O$ takes place in both the liquid and the solid phases. This mechanism differs completely from that conventionally adopted for the standard cooling densification of other 'normal' materials in which only one kind of chemical bond is involved [34].

In contrast, in the quasisolid phase, the master and the slave exchange roles. The H–O bond having a lower $\eta_H$ contracts less than the O:H bond expands, producing a net gain in the O–O length and resulting in density loss. Calculations reveal no feature of $d_x$ constant below 80 K as observed, due to the limitation of the algorithm.

The widening of the angle θ in Fig. 7.2b contributes consistently to volume expansion. In the liquid phase, the mean θ value of 160° remains almost constant, which has little to do with density change but the O:H cooling contraction and H–O elongation become dominant. In the quasisolid phase, cooling widens θ from 160° to 167°, which contributes a maximum of +1.75 % to the O:H–O bond elongation and about 5 % to the volume expansion. The volume expansion due to angle stretching is compatible to O:H–O cooling elongation, resulting in a 9 % less density.

In the $I_h$ and $I_c$ phases, θ increases from 167° to 174° and this trend results in a maximal value of −2.76 % to the volume contraction in ice. The cooling angle stretching compensates for the O:H–O contraction of bond, which explains why the density gains at a lower rate in the solid phase than it is in the liquid phase. An extrapolation of the θ widening in Fig. 7.2b results in the slight O–O lengthening in phase XI where the $d_x$ and its cohesive energy $E_x$ remain constant as $\eta_x \approx 0$, which explains the slight drop in density [37, 38] and the steady $\omega_L$($d_L$ and $E_L$) observed at temperatures below 80 K [39, 40]. Therefore, the angle stretching contributes only positively to the density loss in the quasisolid phase but negatively to density gain in phase $I_{h+c}$ without apparent influence on other physical properties such as the critical temperature for phase transition $T_C$, O 1 s energy shift $E_{1s}$, and phonon frequencies $\omega_x$, etc.

The O–O distance evolution shown in Fig. 7.4b agrees well in trend with the measured density evolution in the full temperate range, Fig. 7.1b [2]. In ice, the O–O distance is always longer than in water-hence ice floats, without necessary involvement of the mixed-phase configuration. Therefore, the entire process of density oscillation arises from O:H–O bond segmental length relaxation subject to the specific-heat disparity and bond angle cooling stretching.

The O–O distance dominates the mass density of water ice in the manner of $\rho \propto (d_{OO})^{-3} \propto (d_H + d_L)_{//}^{-3}$. The $d_x$ is the projection along the O–O line without contribution from the θ contribution, which remains >160° in all phases [4]. The angle difference between 160° and 180° deviates by only 3 % or less to the length scale [4].

When the structures are different, there are other possible volume changes. For example, the higher pressure ice VII phase has a smaller volume and longer intermolecular distance than ice $I_c$ because the former has double the network of the latter. Ice VII and VIII have similar network connectivities but different crystal symmetries [41]. Volume change by such structure variation contributes insignificantly to the O:H–O bond relaxation that dictates the physical anomalies of water ice at higher pressures.

#### 7.4.3.2 Phonon Spectrometrics of Phase Transition

Phonons of 'normal' materials undergo thermal softening because of the lengthening and softening of all bonds involved [42–49]. Figure 7.5 shows, however, heating stiffens the stiffer $\omega_H$ phonons of water [50–56] and ice [27, 28, 53, 57, 58]. The $\omega_H$ cooling blueshift starts at −1.5 °C [53] and 0 °C [4], shown in Fig. 7.5a, b, confirms the H–O cooling contraction in the quasisolid phase. The absent of $\omega_H$

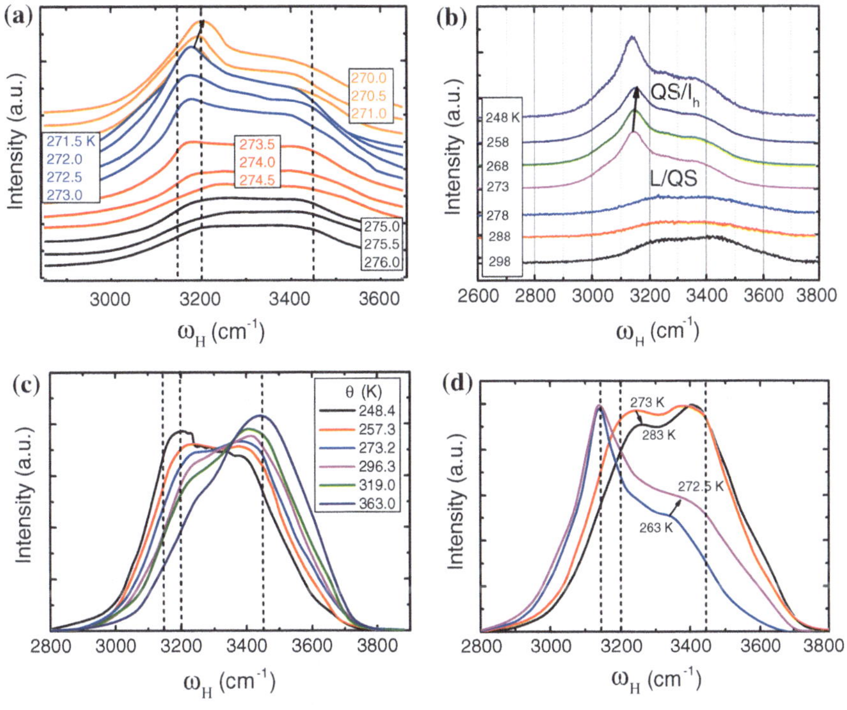


**Fig. 7.5** Thermal $\omega_H$ relaxation for bulk water. Cooling $\omega_H$ blueshift starting at **a** 271.5 K and **b** 273 K indicates liquid-quasisolid phase transition. The absence of $\omega_H$ cooling blueshift for **c** bulk water and **d** supercooled droplet suggests the essentiality of shorter temperature steps in measurements (Reprinted with permission from [4, 53, 59, 61].)

cooling blueshift for the supercooled water in the temperature range from −10 to 10 °C for bulk water, and from −34.6 to 90.0 °C for supercooled droplet, see Fig. 7.5c, d, suggests the essentiality of smaller probing steps. The $\omega_H$ heating blueshift proceeds consistently throughout the liquid phase [59]. The thermal $\omega_H$ stiffening and $\omega_L$ softening happen simultaneously, not only in liquid $H_2O$ but also in the liquid $D_2O$, despite an offset in the characteristic peak [60] because of the O:D-O dominance.

#### 7.4.3.3 Liquid Thermal Expansion

Fig. 7.6 shows the full-frequency range raman spectra as a function of heating. The inset shows the components of the H-O phonon. The components correspond to the coordination-resolved vibration frequency od the H-O bond in the bulk, skin and the H-O radicals.

#### 7.4.3.4 Quasisolid Cooling Expansion

In addition to Fig. 7.5a, b showing the $\omega_H$ cooling blueshift in the quasisolid phase, the structure factor S(q) for micrometer-sized water droplets, shown in Fig. 7.7, further evidence the H–O cooling contraction. In addition to the possible two phase structures, high- and low-density liquid in the supercooled quasisolid phase [63], the major S(q) peaks evolve oppositely with temperature. As the S(q) is proportional to the inverse of the character lengths, S1 and S2. The shifts of the S1 and S2 indicate that as temperature drops, the longer length becomes even longer and the shorter one becomes even shorter, which coincides to the O:H nonbond and the

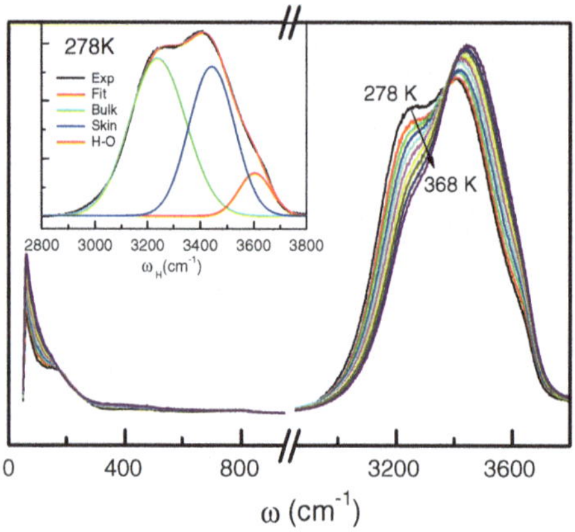


**Fig. 7.6** Full-frequency Raman spectroscopy of deionized water heated from 278 to 368 K. The *inset* shows the molecular-coordination-resolved bulk, skin, and H–O radical components (Reprinted with permission from [62].)

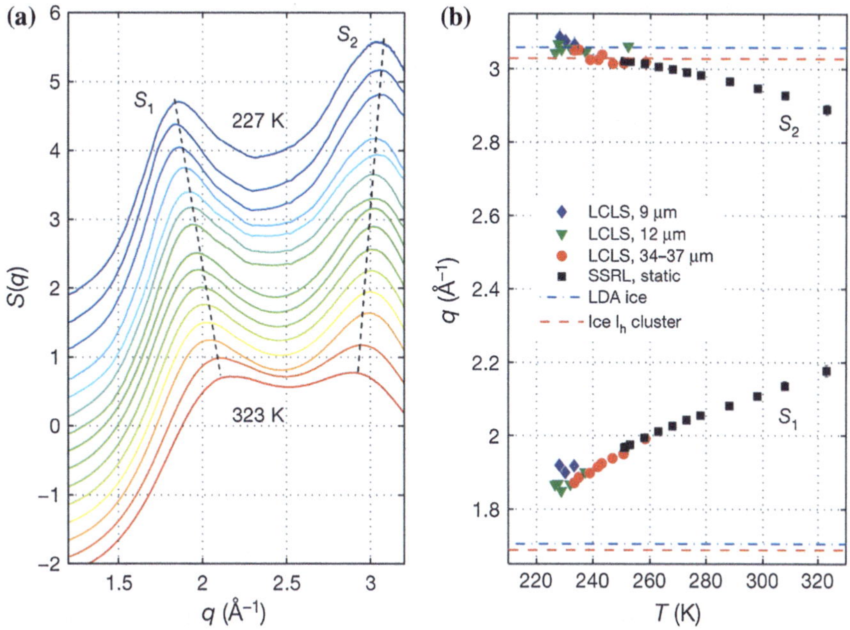


**Fig. 7.7** Structure factor S(q) for micrometer-sized water droplet obtained using coherent X-ray scattering with a single-shot selection scheme. **a** The S(q) changes from bottom to top in the order of 323, 298, 273, 268, 263, 258, 253, 251, 251, 247, 243, 239, 232, 229, 227 K. The principal S(q) maximum splits into two well-separated peaks, S1 and S2 (dashed lines). **b** Temperature dependence of the S1 and S2 peak positions from different sources show the same cooling trend of relaxation-longer one becomes even longer and shorter one turns to be even shorter (Reprinted with permission from [63].)

H–O bond cooling relaxation in the quasisolid phase. Cooling shortens the shorter H–O bond and lengthens the longer O:H nonbond [4].

#### 7.4.3.5 $I_{h+C}$ Phase Cooling Contraction

Figure 7.8 shows the T-dependent power spectra of $H_2O$ derived from MD calculations [4]. Despite the accuracy of temperatures, calculations reveled the transition from water and ice and the presence of the quasisolid phase. The splitting of the high-frequency peaks at 260 K indicates that water transits in the range of 200–260 K into ice. As expected, $\omega_L$ stiffening (softening) always couples with $\omega_H$ softening (stiffening) in the liquid and solid regions, which evidence the cooling densification of water and ice. The shaded areas in Fig. 7.8b covers the quasisolid phase that demonstrates cooling $\omega_H$ blueshift accompanied with $\omega_L$ redshift.

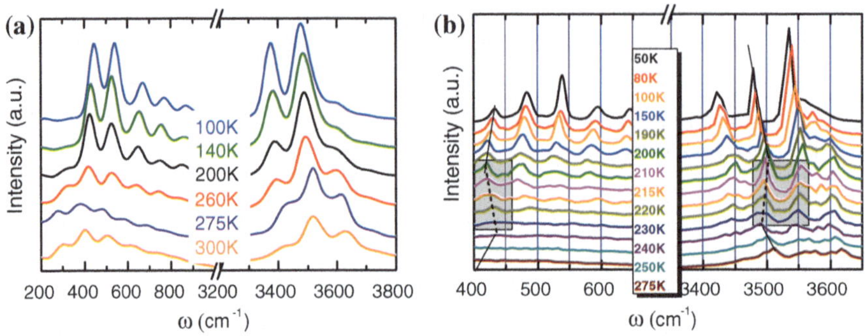


**Fig. 7.8** Temperature-dependent power spectra of $H_2O$ calculated using two algorithms. **a** Splitting of the high-frequency peaks at 260 K indicates the transition from water to ice at 200–260 K. **b** Phonon oscillation (indicated with *hatched lines*) holds the same trend as that of Raman measurements in the quasisolid phase. Both calculations revealed consistently O:H contraction and H–O expansion in the $I_{h+c}$ phase (Reprinted with permission from [4].)

#### 7.4.3.6 O:H–O Frozen in Phase XI

Figure 7.9 shows the temperature dependence of the unit cell volume in phase XI in the temperature of 100 K and below. Only slight cooling expansion occurs, dρ/dT > 0 (see b) [38]. However, the $\omega_x$ phonon frequencies remain almost constant, as Fig. 7.7 showed in Sect. 7.4. The observed $\omega_x$ frozen and volume expansion evidence that the O:H–O bond respond insensitively to thermal excitation because of the zero specific heat. The containing angle cooling stretching claims the responsibility for the slight volume expansion.

This observation also applies to volume cooling expansion of phase XVI. The (empty hydrate) cage structure of phase XVI has a density of 0.81 g $cm^{-3}$, which expands slightly when cooling in temperatures below 55 K due to undercoordination. This cage structure is mechanically more stable and has larger lattice

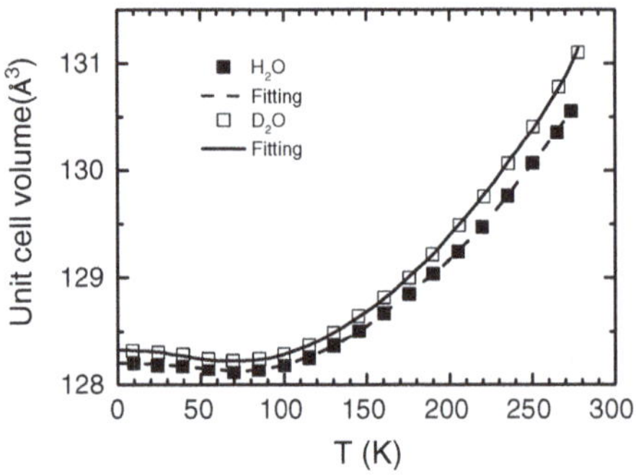


**Fig. 7.9** Slight cooling expansion of $H_2O$ and $D_2O$ in phase XI (T ≤ 100 K) but cooling contraction in phase $I_{c+h}$ [38] (Reprinted with permission from [35, 38].)

constants than the filled hydrate, because of the molecular undercoordination [64]. Therefore, phase XVI has dual characteristics of the supersolid phase due to undercoordination and phase XI of O:H–O bond frozen.

Consistency in volume cooling expansion for the XI and XVI phase evidence that neither the H–O nor the O:H undergoes thermal relaxation in length and stiffness but the ∠O:H–O angle is subject to cooling stretching in the very low temperature regime where $\eta_L \approx \eta_H \approx 0$ [5].

### 7.4.4 Site-Resolved Liquid O:H–O Bond Relaxation

The full-frequency Raman spectra in Fig. 7.6 and the componential DPS in Fig. 7.10 show consistently that heating stiffens the $\omega_H$ and softens the $\omega_L$ simultaneously, which confirms that heating shortens the H–O bond and stiffens its phonon but lengthens the O:H nonbond and softens the $\omega_L$ phonon because the O–O weakening. The differential phonon spectrometrics (DPS) is profoundly powerful and yet extremely simple in distilling the specifically concerned vibration features from their mixtures [65]. Subtraction of the referential spectrum from those of the conditioned specimen upon both spectra being background corrected and $\omega_x$ peak area normalized. Features above the x-axis are due to conditioning and the spectral valleys below the lateral axis shows clearly the intensities of these features evolve with temperature. Heating stiffens the $\omega_H$ from 3180 to 3550 $cm^{-1}$ and meanwhile softens the $\omega_L$ from 190 to 75 $cm^{-1}$. The phonon frequency varies only on with the O:H–O bond segmental length and energy without discriminating thermal excitation stimulus from molecular undercoordination.

Decomposition of the $\omega_H$ into three Gaussian components of bulk shifting from 3239 to 3287 $cm^{-1}$, skin shifting from 3443 to 3456 $cm^{-1}$, and the H–O radicals shifting from 3604 to 3588 $cm^{-1}$ when heated from 278 to 368 K as shown in

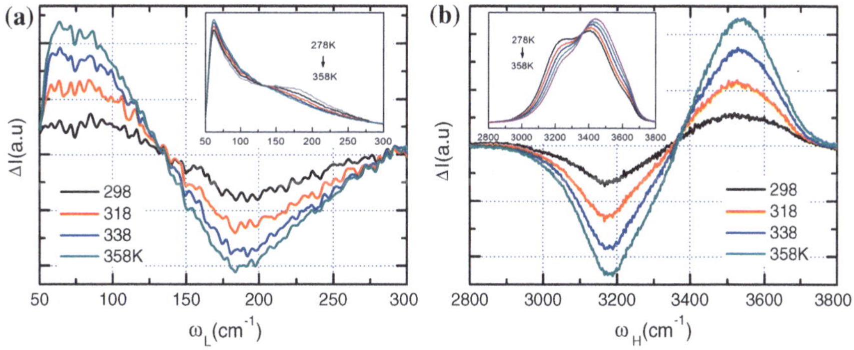


**Fig. 7.10** Temperature dependence of the $\omega_x$ DPS spectra for deionized water show the cooperative **a** $\omega_L$ heating softening and **b** $\omega_H$ heating stiffening in the liquid phase of water

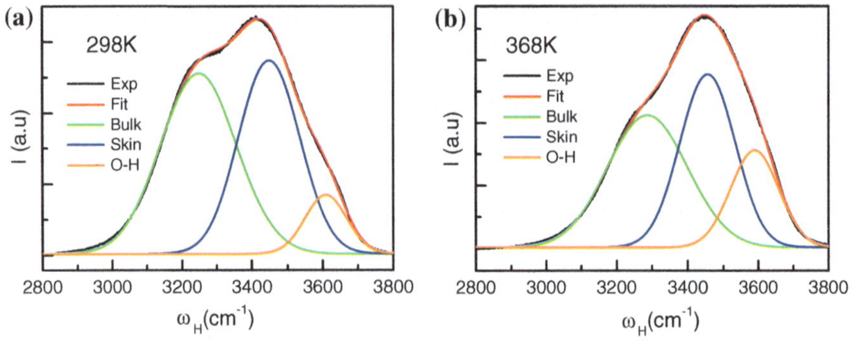


**Fig. 7.11** Raman $\omega_H$ spectra for **a** 298 K and **b** 368 K water decomposed using Gaussian components for the bulk, skin, and the H–O dangling bond (Fitting reliability: $R^2 \geq 0.9998$)

**Table 7.1** Gaussian decomposition of the $\omega_H$ for neat water at different temperatures (Reprinted with permission from [62].)

| | T (K) | Bulk | Skin | H–O radical | Notes |
|---|---|---|---|---|---|
| $\omega_H$ (cm$^{-1}$) (Bond stiffness) | 278 | 3238.85 | 3442.78 | 3603.56 | •Heating and molecular undercoordination stiffen the $\omega_H$ and shorten the H–O bond by different rates [4, 64] •Heating shifts the ωH of H–O radicals up and then downward—thermal fluctuation •H–O phonon lifetime is proportional to the $\omega_H$ [66] |
| | 298 | 3248.31 | 3448.89 | 3608.40 | |
| | 318 | 3264.31 | 3454.76 | 3602.85 | |
| | 338 | 3278.54 | 3459.32 | 3600.03 | |
| | 358 | 3284.31 | 3457.01 | 3592.34 | |
| | 368 | 3286.87 | 3456.02 | 3588.90 | |
| FWHM (cm$^{-1}$) (Fluctuation order) | 278 | 215.32 | 169.17 | 122.52 | •Undercoordination raises but heating depresses the order of molecular fluctuation in the bulk but not the supersolid skin |
| | 298 | 217.24 | 170.99 | 119.35 | |
| | 318 | 224.63 | 163.26 | 124.81 | |
| | 338 | 231.14 | 160.64 | 127.08 | |
| | 358 | 226.78 | 156.04 | 135.57 | |
| | 368 | 225.82 | 154.95 | 138.45 | |
| Peak integral (a.u.) (Abundance) | 278 | 0.53 | 0.39 | 0.08 | •Molecular undercoordination and heating depresses the abundance in the bulk and skin at different rates •Heating raises the $\omega_H$ abundance of the H–O radicals—thermal fluctuation effect |
| | 298 | 0.49 | 0.42 | 0.09 | |
| | 318 | 0.49 | 0.39 | 0.11 | |
| | 338 | 0.48 | 0.38 | 0.13 | |
| | 358 | 0.44 | 0.38 | 0.17 | |
| | 368 | 0.43 | 0.38 | 0.19 | |

Fig. 7.11. Decomposition of the components at other temperatures reveals details of the site resolved O:H–O bond relaxation, as summarized in Table 7.1.

Further refinement of the O:H–O bond relaxation is realized by subtracting the reference component collected at 278 K (inset of Fig. 7.6) from the respective

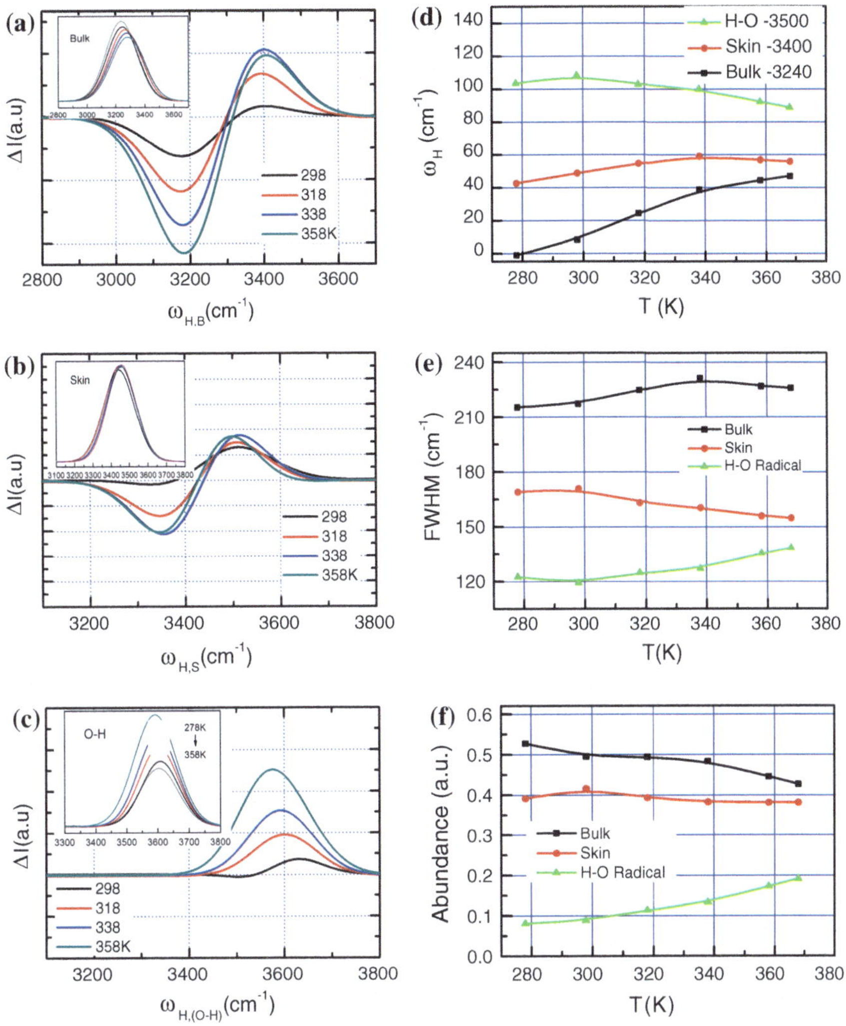


**Fig. 7.12** Molecular-site-resolved Raman $\omega_H$ DPS for the **a** bulk, **b** skin, and **c** H–O dangling bond and their respective **d** bond stiffness ($\omega_H$), **e** FWHM (fluctuation order), and **f** abundance (component integral) thermal relaxation. The sum of the spectral area losses is identical to the sum of spectral gain in each temperature as Fig. 7.10 showed. The *inset* (**d**) shows the thermal relaxation of the O:H and the H–O length in bulk water [2, 3] (Reprinted with permission from [62].)

component obtained at higher temperatures. The refined DPS shown in Fig. 7.12 revealed that all components undergo heating blueshift but the H–O radical component reverts slight with substantial intensity enhancement as the temperature is increased. The skin $\omega_H$ grows slower than the bulk with temperature. The individual DPS area may be asymmetric but their sum over all components reserves because of the normalization of the convoluted spectral peaks. The net abundance gain of the

H–O component results from the skin and the bulk due to the heating effect that shortens the H–O bond, but the redshift of the radical phonon frequency indicates the thermal fluctuation.

The heating stiffened $\omega_H$ in the skin and in the bulk is within the expectation that heating shortens the H–O bond and stiffens its phonon, which is associated with O: H nonbond elongation and $\omega_L$ redshift. The unexpected $\omega_H$ reverting of the H–O radicals indicates that heating lengthens and softens the H–O radicals slightly. Thermal fluctuation increases the probability of H–O exposing to the open surfaces with locally reduced Coulomb coupling, which lengthens and softens slightly the H–O radicals. Therefore, the H–O component undergoes a reverting redshift with higher peak intensity. The insignificant $\omega_H$ shift for the H–O radicals is in line with the dielectric spectroscopy measurements and MD calculations suggesting that heating does not change the skin H–O ordering but the length and strength. The skin component is relatively more thermally stable than others because of its supersolidity nature.

## 7.4.5 $\Delta E_{1s}$ and $\Delta\omega_H$ Correlation

The O 1 s binding energy shift from that of an isolated O atom measured using XPS is proportional to the H–O bond energy (4.0 eV) [67]. The O:H nonbond energy (0.10 eV) contributes negligibly. Information probed by the NEXAFS and XES, as shown in Fig. 7.13, is much more complicated as both the O 1 s (bottom) and the

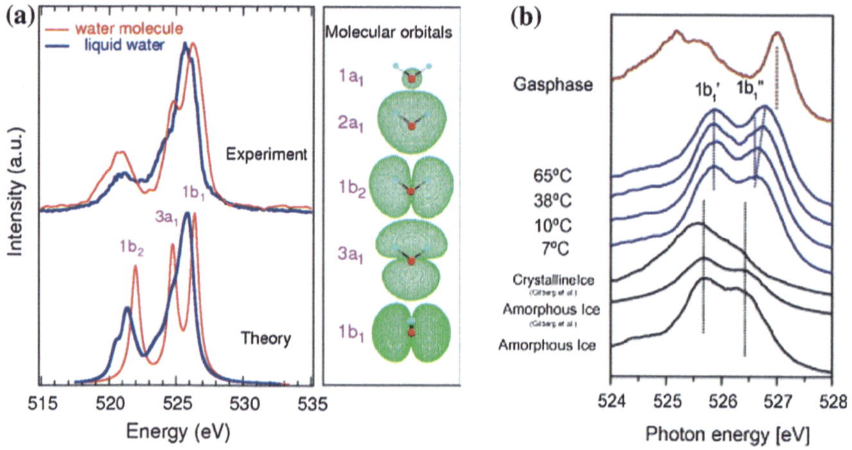


**Fig. 7.13** **a** The O 1 s orbital (*side panel*) energies of molecules and liquid water and **b** the O 1 s XES spectra of vapor, liquid water, and amorphous and crystalline ice at different temperatures, with an energy scale displaying the $1b_1$ orbitals. The $1b_1$ peak splits into a doublet of $1b_1'$(~525.5 eV) and $1b_1''$ (~526.5 eV) corresponding respectively to the $\omega_H$ for the bulk (3200 $cm^{-1}$) and skin (3450 $cm^{-1}$), which undergo thermal entrapment/stiffening consistently for crystals but thermal softening for amorphous ice transiting to crystals (Reprinted with permission from [68, 69].)

upper occupied and unoccupied levels are subject to shift in different amounts and manners [67]. Heating deepens the O 1 s energy in different phases towards that of gaseous molecules unless transition from amorphous to crystal [68, 69]. This trend is opposite to other usual materials.

Mechanisms for the O 1 s thermal entrapment are debated as consequence of the mixed-phase configuration, that is, ordered tetrahedral and distorted O:H–O bonded networks, with provision of the mixed-structure phase [15, 70].

In fact, the following correlates the O 1 s energy shift and the H–O phonon frequency shift $\Delta\omega_H$ [3]:

$$(d_H \Delta\omega_H)^2 \cong \Delta E_{1s}$$

This relation indicates that both the $\Delta E_{1s}$ and the $\Delta\omega_H$ always shift in the same direction, at different rates, when the specimen is excited. Therefore, the $1b_1''$ peak corresponds to the skin $\omega_H$ at 3450 $cm^{-1}$, and the $1b_1'$ to the bulk $\omega_H$ at 3200 $cm^{-1}$ for water (see Fig. 7.13). The O 1 s goes deeper in the crystal, and the $\omega_H$ shifts consistently higher at heating, because heating shortens and stiffens the H–O bond. When ice transits from amorphous to crystal, the trend is opposite, (see Fig. 7.14) [71]. The $\Delta E_{1s}$ will undergo thermal oscillation but its measurement in ultra-high vacuum is very difficult.

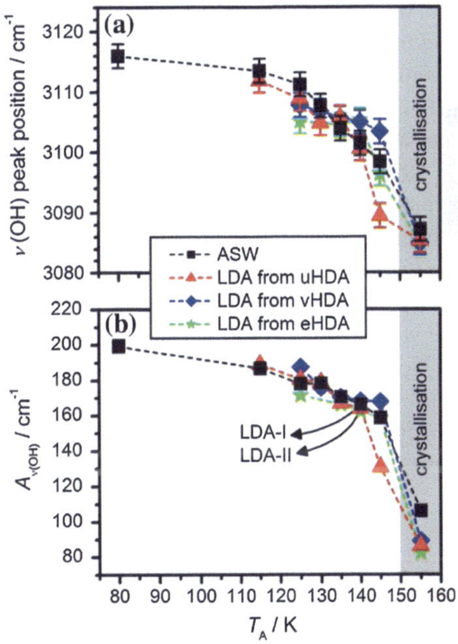


**Fig. 714** Annealing temperature dependence of **a** the $\omega_H$ peak position and **b** the integrated peak areas. LDA and HDA represent the low/high-density amorphous ice and amorphous solid water (ASW) (Reprinted with permission from [71].)

### 7.4.6 Thermal Relaxation of Amorphous Ice

Thermal annealing of low-density amorphous ice from 80 to 155 K softens $\omega_H$ from 3120 to 3080 $cm^{-1}$ instead [71], which is counter to the trend of $\omega_H$ heating stiffening in crystal ice. Thermal relaxation increases the structural order of the amorphous state on more extended length scales as the average O–O distance becomes shorter with narrower distribution.

Clearly, heating softens the $\omega_H$ of amorphous ice rather than stiffening it, as it occurs in crystalline ice. The $\omega_H$ redshift indicates H–O bond elongation. Molecular undercoordination shortens the H–O bond nearby defects [64], which distributes randomly in the amorphous phase. Annealing removes the defect and recovers the shortened H–O bond resulting in the $\omega_H$ redshift towards the value in ideal crystal. Therefore, amorphous ice annealing softens the $\omega_H$, which is within expectation of the BOLS and O:H–O bond cooperativity notions [3].

## 7.5 Insight Extension

### 7.5.1 Life Under and Above Ice

Most liquids expand when melting but not when freezing so the icy-cold water is less dense, so tends to move upwards. Most liquids exposed to cold air would cool, and the cold liquid would sink, forcing more liquid to rise and be cooled by the air. Eventually all the liquid would lose heat to the air and freeze, from the bottom up, till completely frozen. But with water, the cold regions, being less dense, stay on top, allowing the warmer regions to stay below and avoid losing heat to the air.

One can imagine what will happen if water contracts at freezing, instead of expansion. The consequences of this behavior go beyond practical issues in some winter sports to reach planetary significance: sinking ice on the ocean floor would, for example, disrupt the current flows that bring warmth to the European continent making most of it inhabitable. Although the surface is frozen, fish can still live in the water below. But if water were like other substances, large bodies of water, such as North America's Great Lakes, would be frozen solid, with dire effects on life on earth as a whole. Figure 7.15 shows two typical creatures surviving in winters. One can imagine what will happen to them without floating ice. The cold ice protects fishes underneath from being frozen and allows them move freely.

### 7.5.2 Rock Erosion: Freezing-Melting Cycle

As noted by the Chinese sage Lao Tzu in his ancient text: "There is nothing softer and flexible than water, and yet there is nothing better for attacking hard and strong

**Fig. 7.15** Floating ice enables surviving and reproducing of creatures in winter such whale (Credit to John Eggers, Bemidji, USA) and bears. Polar bears like this one are excellent swimmers but use floating sea ice as pathways to coastal areas and as platforms from which to hunt seals (Credited to Thomas Nilsen, The New York Times 2006)

**Fig. 7.16** Photographs of eroded rocks (National geographic park, Zhangyie, Gansu, China)

stuff." Erosion of rocks is the nature phenomenon as shown in Fig. 7.16 morphologies of the eroded rock blocks. Rainfall water penetrated into the rock through pores in the Autumn become ice at freezing in the Winter [72]. Volume expansion of ice enlarges the pore size, which exerts force nearby cracking the rocks. Melting of ice in the Spring and evaporation of the molten ice in the Summer leaves the damage behind. Repeated occurrence erodes the rocks.

Erosion is the act in which earth is worn away. A similar process, weathering, breaks down or dissolves rock, weakening it or turning it into tiny fragments. No rock is hard enough to resist the forces of erosion. Together, they shaped the sharp peaks of the Himalaya Mountains in Asia and sculpted the spectacular forest of rock towers of Bryce Canyon, in the U.S. state of Utah, as well as Zhang Ye in Gansu, China. The process of erosion moves bits of rock or soil from one place to another. Most erosion is performed by water, wind, or ice. These forces carry the rocks and soil from the places where they were weathered. When wind or water slows down, or ice melts, sediment is deposited in a new location. As the sediment builds up,

**Fig. 7.17** **a** Ice melting due to global warming (*left*) raises the sea levels worldwide at a rate of 3.5 mm per year since the early 1990s. The trend puts thousands of coastal cities, like Venice (*right*), Italy, (seen here during a historic flood in 2008), and even whole Greenland islands at the risk of being claimed by the ocean [75]

it creates fertile land. River deltas are made almost entirely of sediment. Delta sediment is eroded from the banks and bed of the river.

### *7.5.3 Watering Soil in Winter—Freezing Expansion*

Watering soil in winter has many advantages for keeping nutrition and fertilizing the soil [73]. Water molecules of snows or watering penetrate into the earth and freeze in cold weather. Water freezing expands its volume and the surrounding soil. The soil becomes loosen and soften with reservation of nutrition and moisture when the ice melts and evaporates, which is beneficial for the plant growth in the next Spring.

### *7.5.4 Sea Level Rise—Global Warming*

Every degree of global warming due to carbon pollution, global average sea level will rise by about 1.3 m in the long run by ice melting [74]. When multiplied by the current rate of carbon emissions, and the best estimate of global temperature sensitivity to pollution, this translates to a long-term sea level rise commitment that is now growing at about 0.3 m/decade. Such rates, if sustained, would realize the highest levels of sea level rise contemplated here in hundreds, not thousands of years-fast enough to apply continual pressure, as well as threaten the heritage, and very existence, of coastal communities everywhere.

Over the past century, the burning of fossil fuels and other human and natural activities has released enormous amounts of heat-trapping gases into the atmosphere. These emissions have caused the Earth's surface temperature to rise, and the oceans absorb about 80 % of this additional heat. The rise in sea levels is linked to three primary factors, all induced by this ongoing global climate change:

(1) Thermal expansion: When water heats up, it expands. About half of the past century's rise in sea level is attributable to warmer oceans simply occupying more space.
(2) Melting of glaciers and polar ice caps: Large ice formations, like glaciers and the polar ice caps, naturally melt back a bit each summer. But in the winter, snows, made primarily from evaporated seawater, are generally sufficient to balance out the melting. However, persistently higher temperatures caused by global warming have led to greater-than-average summer melting as well as diminished snowfall due to later winters and earlier springs. This imbalance results in a significant net gain in runoff versus evaporation for the ocean, causing sea levels to rise.
(3) Ice loss from Greenland and West Antarctica: As with glaciers and the ice caps, increased heat is causing the massive ice sheets that cover Greenland and Antarctica to melt at an accelerated pace. Melt water from above and seawater from below is seeping beneath Greenland's and West Antarctica's ice sheets, effectively lubricating ice streams and causing them to move more quickly into the sea. Moreover, higher sea temperatures are causing the massive ice shelves that extend out from Antarctica to melt from below, weaken, and break off.

When sea levels rise rapidly, as they have been doing, even a small increase can have devastating effects on coastal habitats. As seawater reaches farther inland, it can cause destructive erosion, flooding of wetlands, contamination of aquifers and agricultural soils, and lost habitat for fish, birds, and plants.

When large storms hit land, higher sea levels mean bigger, more powerful storm surges that can strip away everything in their path. In addition, hundreds of millions of people live in areas that will become increasingly vulnerable to flooding. Higher sea levels would force them to abandon their homes and relocate. Low-lying islands could be submerged completely, see Fig. 7.17, for instance.

## 7.6 Summary

Consistency in the specific heat $\eta_x$, mass density $\rho$, segmental length $d_x$, and phonon frequency $\omega_x$ oscillation evidences that the coupled O:H–O bond oscillator pair describes adequately the true situation of water and ice performance when cooling or heating. Consistency of numerical and experimental observations verifies the following:

(1) Interoxygen repulsion and the segmental specific-heat disparity of the O:H–O bond govern the change in the angle, length, and stiffness of the segmented O:H–O bond, and the oscillation of the mass density and the phonon-frequency of water ice over the full temperature range.

(2) The segment with relatively lower specific heat contracts and drives the O:H–O bond cooling relaxation. Cooling stretching of the ∠O:H–O angle contributes positively to volume expansion in the quasi-solid phase but it contributes negatively to cooling densification in the solid phase. Angle relaxation has no direct influence on the physical properties, with the exception of mass density.

(3) In the liquid and solid phases, the O:H bond contracts more than the H–O bond elongates, resulting in the cooling densification of water and ice. This mechanism is completely different from the process experienced by other 'normal' materials where only one type of chemical bond is involved.

(4) In the quasisolid phase, H–O bond contracts less than the O:H bond lengthens, resulting in volume expansion during freezing. Stretching of the O:H–O bond angle lowers the density slightly at $T < 100$ K as the length and energy of the O:H–O conserve.

(5) The O–O distance is larger in ice than it is in water, and therefore ice floats.

(6) The site-resolved DPS revealed that the H–O bond undergo heating contraction in the skin and in the bulk but the dangling H–O radicals undergo heating expansion with higher probability of exposing to the open end of the skin because of the thermal fluctuation.

## References


1. H. Perlman, Ice is less dense than water (2014)
2. F. Mallamace, C. Branca, M. Broccio, C. Corsaro, C.Y. Mou, S.H. Chen, The anomalous behavior of the density of water in the range 30 K < T < 373 K. Proc. Natl. Acad. Sci. U.S.A. **104**(47), 18387–18391 (2007)
3. Y. Huang, X. Zhang, Z. Ma, Y. Zhou, W. Zheng, J. Zhou, C.Q. Sun, Hydrogen-bond relaxation dynamics: resolving mysteries of water ice. Coord. Chem. Rev. **285**, 109–165 (2015)
4. C.Q. Sun, X. Zhang, X. Fu, W. Zheng, J.-L. Kuo, Y. Zhou, Z. Shen, J. Zhou, Density and phonon-stiffness anomalies of water and ice in the full temperature range. J. Phys. Chem. Lett. **4**, 3238–3244 (2013)
5. Y. Huang, X. Zhang, Z. Ma, Y. Zhou, J. Zhou, W. Zheng, C.Q. Sun, Size, separation, structure order, and mass density of molecules packing in water and ice. Sci. Rep. **3**, 3005 (2013)
6. S. Everts, Galileo on ice. Chem. Eng. News **91**(34), 28–29 (2013)
7. P.L. Nostro, B.W. Ninham, *Aqua Incognita: Why Ice Floats on Water and Galileo 400 Years on* (Connor Court Publishing Pty Ltd, 2014)
8. A.J. Stone, Water from first principles. Science **315**(5816), 1228–1229 (2007)
9. K. Stokely, M.G. Mazza, H.E. Stanley, G. Franzese, Effect of hydrogen bond cooperativity on the behavior of water. Proc. Natl. Acad. Sci. U.S.A. **107**(4), 1301–1306 (2010)
10. C. Huang, K.T. Wikfeldt, T. Tokushima, D. Nordlund, Y. Harada, U. Bergmann, M. Niebuhr, T.M. Weiss, Y. Horikawa, M. Leetmaa, M.P. Ljungberg, O. Takahashi, A. Lenz, L. Ojamäe, A.P. Lyubartsev, S. Shin, L.G.M. Pettersson, A. Nilsson, The inhomogeneous structure of water at ambient conditions. Proc. Natl. Acad. Sci. **106**(36), 15214–15218 (2009)

11. F. Mallamace, M. Broccio, C. Corsaro, A. Faraone, D. Majolino, V. Venuti, L. Liu, C.Y. Mou, S.H. Chen, Evidence of the existence of the low-density liquid phase in supercooled, confined water. Proc. Natl. Acad. Sci. U.S.A. **104**(2), 424–428 (2007)
12. O. Mishima, H.E. Stanley, The relationship between liquid, supercooled and glassy water. Nature **396**(6709), 329–335 (1998)
13. E.B. Moore, V. Molinero, Structural transformation in supercooled water controls the crystallization rate of ice. Nature **479**(7374), 506–508 (2011)
14. V. Molinero, E.B. Moore, Water modeled as an intermediate element between carbon and silicon. J. Phys. Chem. B **113**(13), 4008–4016 (2009)
15. P. Wernet, D. Nordlund, U. Bergmann, M. Cavalleri, M. Odelius, H. Ogasawara, L.A. Naslund, T.K. Hirsch, L. Ojamae, P. Glatzel, L.G.M. Pettersson, A. Nilsson, The structure of the first coordination shell in liquid water. Science **304**(5673), 995–999 (2004)
16. A. Nilsson, C. Huang, L.G.M. Pettersson, Fluctuations in ambient water. J. Mol. Liq. **176**, 2–16 (2012)
17. G.N.I. Clark, C.D. Cappa, J.D. Smith, R.J. Saykally, T. Head-Gordon, The structure of ambient water. Mol. Phys. **108**(11), 1415–1433 (2010)
18. A.K. Soper, J. Teixeira, T. Head-Gordon, Is ambient water inhomogeneous on the nanometer-length scale? Proc. Natl. Acad. Sci. U.S.A. **107**(12), E44–E44 (2010)
19. T. Head-Gordon, M.E. Johnson, Tetrahedral structure or chains for liquid water. Proc. Natl. Acad. Sci. U.S.A. **103**(21), 7973–7977 (2006)
20. G.N. Clark, G.L. Hura, J. Teixeira, A.K. Soper, T. Head-Gordon, Small-angle scattering and the structure of ambient liquid water. Proc. Natl. Acad. Sci. U.S.A. **107**(32), 14003–14007 (2010)
21. V. Petkov, Y. Ren, M. Suchomel, Molecular arrangement in water: random but not quite. J. Phys. Condens. Matter **24**(15), 155102 (2012)
22. N.J. English, J.S. Tse, Density fluctuations in liquid water. Phys. Rev. Lett. **106**(3), 037801 (2011)
23. M. Matsumoto, Why does water expand when it cools? Phys. Rev. Lett. **103**(1), 017801 (2009)
24. J.C. Li, A.I. Kolesnikov, Neutron spectroscopic investigation of dynamics of water ice. J. Mol. Liq. **100**(1), 1–39 (2002)
25. Y.S. Tu, H.P. Fang, Anomalies of liquid water at low temperature due to two types of hydrogen bonds. Phys. Rev. E **79**(1), 016707 (2009)
26. Y. Yoshimura, S.T. Stewart, M. Somayazulu, H. Mao, R.J. Hemley, High-pressure x-ray diffraction and Raman spectroscopy of ice VIII. J. Chem. Phys. **124**(2), 024502 (2006)
27. Y. Yoshimura, S.T. Stewart, M. Somayazulu, H.K. Mao, R.J. Hemley, Convergent Raman features in high density amorphous ice, ice VII, and ice VIII under pressure. J. Phys. Chem. B **115**(14), 3756–3760 (2011)
28. Y. Yoshimura, S.T. Stewart, H.K. Mao, R.J. Hemley, In situ Raman spectroscopy of low-temperature/high-pressure transformations of $H_2O$. J. Chem. Phys. **126**(17), 174505 (2007)
29. M. Song, H. Yamawaki, H. Fujihisa, M. Sakashita, K. Aoki, Infrared investigation on ice VIII and the phase diagram of dense ices. Phys. Rev. B **68**(1), 014106 (2003)
30. S. Andersson, B.W. Ninham, Why ice floats on water. Solid State Sci. **5**(5), 683–693 (2003)
31. J. Nimerfroh, News: nearly frozen waves captured on camera by nantucket photographer. CBC Boston (2015), http://boston.cbslocal.com/2015/02/26/nearly-frozen-waves-captured-on-camera-by-nantucket-photographer/
32. T.D. Kuhne, R.Z. Khaliullin, Electronic signature of the instantaneous asymmetry in the first coordination shell of liquid water. Nat. Commun. **4**, 1450 (2013)
33. J. Guo, X. Meng, J. Chen, J. Peng, J. Sheng, X.Z. Li, L. Xu, J.R. Shi, E. Wang, Y. Jiang, Support: real-space imaging of interfacial water with submolecular resolution-supp. Nat. Mater. **13**, 184–189 (2014)
34. M.X. Gu, Y.C. Zhou, C.Q. Sun, Local bond average for the thermally induced lattice expansion. J. Phys. Chem. B **112**(27), 7992–7995 (2008)

35. M. Erko, D. Wallacher, A. Hoell, T. Hauss, I. Zizak, O. Paris, Density minimum of confined water at low temperatures: a combined study by small-angle scattering of X-rays and neutrons. PCCP **14**(11), 3852–3858 (2012)
36. U. Bergmann, A. Di Cicco, P. Wernet, E. Principi, P. Glatzel, A. Nilsson, Nearest-neighbor oxygen distances in liquid water and ice observed by x-ray Raman based extended x-ray absorption fine structure. J. Chem. Phys. **127**(17), 174504 (2007)
37. A. Falenty, T.C. Hansen, W.F. Kuhs, Formation and properties of ice XVI obtained by emptying a type sII clathrate hydrate. Nature **516**(7530), 231–233 (2014)
38. K. Rottger, A. Endriss, J. Ihringer, S. Doyle, W.F. Kuhs, Lattice-constants and thermal-expansion of $H_2O$ and $D_2O$ Ice ih between 10 and 265 K. Acta Crystallogr. B **50**, 644–648 (1994)
39. G.P. Johari, H.A.M. Chew, T.C. Sivakumar, Effect of temperature and pressure on translational lattice vibrations and permittivity of ice. J. Chem. Phys. **80**(10), 5163 (1984)
40. C. Medcraft, D. McNaughton, C.D. Thompson, D. Appadoo, S. Bauerecker, E.G. Robertson, Size and temperature dependence in the far-ir spectra of water ice particles. Astrophys. J. **758** (1), 17 (2012)
41. V.F. Petrenko, R.W. Whitworth, *Physics of Ice* (Clarendon Press, 1999)
42. I. Calizo, A.A. Balandin, W. Bao, F. Miao, C.N. Lau, Temperature dependence of the Raman spectra of graphene and graphene multilayers. Nano Lett. **7**(9), 2645–2649 (2007)
43. X.X. Yang, J.W. Li, Z.F. Zhou, Y. Wang, L.W. Yang, W.T. Zheng, C.Q. Sun, Raman spectroscopic determination of the length, strength, compressibility, Debye temperature, elasticity, and force constant of the C–C bond in graphene. Nanoscale **4**(2), 502–510 (2012)
44. H.Q. Zhou, C.Y. Qiu, H.C. Yang, F. Yu, M.J. Chen, L.J. Hu, Y.J. Guo, L.F. Sun, Raman spectra and temperature-dependent Raman scattering of carbon nanoscrolls. Chem. Phys. Lett. **501**(4–6), 475–479 (2011)
45. X.X. Yang, J.W. Li, Z.F. Zhou, Y. Wang, W.T. Zheng, C.Q. Sun, Frequency response of graphene phonons to heating and compression. Appl. Phys. Lett. **99**(13), 133108 (2011)
46. J.W. Li, L.W. Yang, Z.F. Zhou, X.J. Liu, G.F. Xie, Y. Pan, C.Q. Sun, Mechanically stiffened and thermally softened Raman modes of ZnO crystal. J. Phys. Chem. B **114**(4), 1648–1651 (2010)
47. M.X. Gu, Y.C. Zhou, L.K. Pan, Z. Sun, S.Z. Wang, C.Q. Sun, Temperature dependence of the elastic and vibronic behavior of Si, Ge, and diamond crystals. J. Appl. Phys. **102**(8), 083524 (2007)
48. M.X. Gu, L.K. Pan, T.C.A. Yeung, B.K. Tay, C.Q. Sun, Atomistic origin of the thermally driven softening of Raman optical phonons in group III nitrides. J. Phys. Chem. C **111**(36), 13606–13610 (2007)
49. M.X. Gu, L.K. Pan, B.K. Tay, C.Q. Sun, Atomistic origin and temperature dependence of Raman optical redshift in nanostructures: a broken bond rule. J. Raman Spectrosc. **38**(6), 780–788 (2007)
50. P.C. Cross, J. Burnham, P.A. Leighton, The Raman spectrum and the structure of water. J. Am. Chem. Soc. **59**, 1134–1147 (1937)
51. J.D. Smith, C.D. Cappa, K.R. Wilson, R.C. Cohen, P.L. Geissler, R.J. Saykally, Unified description of temperature-dependent hydrogen-bond rearrangements in liquid water. Proc. Natl. Acad. Sci. U.S.A. **102**(40), 14171–14174 (2005)
52. F. Paesani, Temperature-dependent infrared spectroscopy of water from a first-principles approach. J. Phys. Chem. A **115**(25), 6861–6871 (2011)
53. M. Paolantoni, N.F. Lago, M. Albertí, A. Laganà, Tetrahedral ordering in water: Raman profiles and their temperature dependence†. J. Phys. Chem. A **113**(52), 15100–15105 (2009)
54. M. Smyth, J. Kohanoff, Excess electron localization in solvated DNA bases. Phys. Rev. Lett. **106**(23), 238108 (2011)
55. Y. Marechal, Infrared-spectra of water. 1. Effect of temperature and of H/D isotopic dilusion. J. Chem. Phys. **95**(8), 5565–5573 (1991)
56. G.E. Walrafen, Raman spectral studies of the effects of temperature on water structure. J. Chem. Phys. **47**(1), 114–126 (1967)

57. I. Durickovic, R. Claverie, P. Bourson, M. Marchetti, J.M. Chassot, M.D. Fontana, Water-ice phase transition probed by Raman spectroscopy. J. Raman Spectrosc. **42**(6), 1408–1412 (2011)
58. K. Furic, V. Volovsek, Water ice at low temperatures and pressures: new Raman results. J. Mol. Struct. **976**(1–3), 174–180 (2010)
59. H. Suzuki, Y. Matsuzaki, A. Muraoka, M. Tachikawa, Raman spectroscopy of optically levitated supercooled water droplet. J. Chem. Phys. **136**(23), 234508 (2012)
60. Y. Marechal, The molecular structure of liquid water delivered by absorption spectroscopy in the whole IR region completed with thermodynamics data. J. Mol. Struct. **1004**(1–3), 146–155 (2011)
61. X. Xue, Z.-Z. He, J. Liu, Detection of water-ice phase transition based on Raman spectrum. J. Raman Spectrosc. **44**(7), 1045–1048 (2013)
62. Y. Zhou, Y. Huang, Y. Gong, C.Q. Sun, Molecular site resolved O:H–O bond thermal relaxation in liquid water. Communicated (2015)
63. J.A. Sellberg, C. Huang, T.A. McQueen, N.D. Loh, H. Laksmono, D. Schlesinger, R.G. Sierra, D. Nordlund, C.Y. Hampton, D. Starodub, D.P. DePonte, M. Beye, C. Chen, A.V. Martin, A. Barty, K.T. Wikfeldt, T.M. Weiss, C. Caronna, J. Feldkamp, L.B. Skinner, M.M. Seibert, M. Messerschmidt, G.J. Williams, S. Boutet, L.G. Pettersson, M.J. Bogan, A. Nilsson, Ultrafast X-ray probing of water structure below the homogeneous ice nucleation temperature. Nature **510**(7505), 381–384 (2014)
64. C.Q. Sun, X. Zhang, J. Zhou, Y. Huang, Y. Zhou, W. Zheng, Density, elasticity, and stability anomalies of water molecules with fewer than four neighbors. J. Phys. Chemis. Lett. **4**, 2565–2570 (2013)
65. C.Q. Sun, Atomic scale purification of electron spectroscopic information (Patent publication: US20130090865). 18 June 2010: USA
66. S.T. van der Post, C.S. Hsieh, M. Okuno, Y. Nagata, H.J. Bakker, M. Bonn, J. Hunger, Strong frequency dependence of vibrational relaxation in bulk and surface water reveals sub-picosecond structural heterogeneity. Nat. Commun. **6**, 8384 (2015)
67. X.J. Liu, M.L. Bo, X. Zhang, L.T. Li, Y.G. Nie, H. TIan, Y. Sun, S. Xu, Y. Wang, W. Zheng, C.Q. Sun, Coordination-resolved electron spectrometrics. Chem. Rev. **115**(14), 6746–6810 (2015)
68. T. Tokushima, Y. Harada, O. Takahashi, Y. Senba, H. Ohashi, L.G.M. Pettersson, A. Nilsson, S. Shin, High resolution X-ray emission spectroscopy of liquid water: the observation of two structural motifs. Chem. Phys. Lett. **460**(4–6), 387–400 (2008)
69. J.H. Guo, Y. Luo, A. Augustsson, J.E. Rubensson, C. Såthe, H. Ågren, H. Siegbahn, J. Nordgren, X-ray emission spectroscopy of hydrogen bonding and electronic structure of liquid water. Phys. Rev. Lett. **89**(13), 137402 (2002)
70. A. Nilsson, L.G.M. Pettersson, Perspective on the structure of liquid water. Chem. Phys. **389** (1–3), 1–34 (2011)
71. J.J. Shephard, J.S.O. Evans, C.G. Salzmann, Structural relaxation of low-density amorphous ice upon thermal annealing. J. Phys. Chem. Lett. 3672–3676 (2013)
72. C. Lister, On the penetration of water into hot rock. Geophys. J. Int. **39**(3), 465–509 (1974)
73. F. Van der Paauw, Effect of winter rainfall on the amount of nitrogen available to crops. Plant Soil **16**(3), 361–380 (1962)
74. R. Winkelmann, A. Levermann, M.A. Martin, K. Frieler, Increased future ice discharge from Antarctica owing to higher snowfall. Nature **492**(7428), 239–242 (2012)
75. A. Pattero, Sea level rise—national geographic (2008) http://ocean.nationalgeographic.com/ocean/critical-issues-sea-level-rise/

# Chapter 8
# Molecular Undercoordination: Supersolidity

- *Undercoordinated water molecules fascinate bubbles, clusters, hydration shells, hydrophobically confined droplets, and water skins even further.*
- *Molecular undercoordination shortens the H–O bond and stiffens its phonon but relaxes the O:H nonbond oppositely associated with nonbonding electron dual polarization.*
- *Phonon cooperative relaxation disperses boundaries of the quasisolid phase, which not only raises the $T_m$ but also lowers the $T_N$, resulting in the "supercooling" at freezing and "superheating" at melting.*
- *Most strikingly, molecular undercoordination also creates a supersolid phase that is hydrophobic, frictionless, less dense, ice like, and viscoelastic with long-lived H–O phonons.*

**Abstract** As an often overlooked degree of freedom, molecular undercoordination shortens the H–O polar-covalent bond and stiffens its phonon but lengthens and softens the O:H nonbond more significantly through the Coulomb repulsion between the electron pairs of adjacent oxygen. This process shrinks those $H_2O$ molecules having fewer-than-four neighbors such as molecular clusters, hydration shells, and the surface skin of water and ice. The shortening of the H–O bond raises the local density of bonding electrons, which in turn polarizes the lone pairs of electrons on oxygen forming anchored dipoles pointing outwardly. The stiffening of the shortened H–O bond increases the magnitude of the O 1s binding energy shift, causes the blueshift of the H–O phonon frequencies, and elevates the melting point of molecular clusters and ultrathin films of water, which gives rise to their elastic, hydrophobic, ice-like, frictionless, and low-density behavior irrespective of temperature.

## 8.1 Challenge: Why Are Skins and Clusters so Special?

Undercoordinated water molecules are referred to those with fewer than four nearest neighbors (CN < 4) as they occur in the bulk interior of water and ice [1–6]. Molecular undercoordination occurs in the terminated O:H–O bonded networks, in the skin of a large volume of water and ice, in hydration shells, molecular clusters, ultrathin films, snowflakes, clouds, fogs, nanodroplets, nanobubbles, and in the gaseous state, see Fig. 8.1 for example. As a new degree of freedom,



C.Q. Sun and Y. Sun, *The Attribute of Water*, Springer Series in Chemical Physics 113, DOI 10.1007/978-981-10-0180-2_8

Fig. 8.1 Molecular undercoordination brings excessive anomalies to water and ice. **a** A snow sculpture at the Harbin International Ice and Snow Sculpture Festival held on December, 2007—"Romantic Feelings" in Harbin, China. (*Credit* Timeea Vinerean 2011, Public domain.) **b** Clouds, fogs, and snow observed at Meili Mountains, Yunnan, China (photo taken by Yi Sun, 2010)

undercoordination makes water molecules perform even more fascinating than they do in large trunks typically in the following [4, 7–15]:

(1) Why are water skins and droplets less dense, elastic, stiffer, and thermally more stable?
(2) Ice is most slippery of ever known—is it covered by a sheet of liquid-water lubricant?
(3) Water skin is elastic, hydrophobic, and tough—is it covered by a layer of solid ice?
(4) Why do nanodroplets and nanobubbles exhibit "supercooling" at freezing and "superheating" at melting?

## 8.2 Clarification: O:H–O Bond Elongation and Polarization

As illustrated in Figs. 8.2, the following mechanisms drive undercoordinated molecules to perform differently from themselves in the bulk [6, 16, 17]:

(1) Molecular undercoordination shortens the H–O bond and raises its phonon frequency $\omega_H$, Debye temperature $\Theta_{DH}$, the O 1s energy shift $\Delta E_{1s}$, and the energy $E_H$ for H atomic dissociation.
(2) The associated O:H nonbond elongation lowers its phonon frequency $\omega_L$, Debye temperature $\Theta_{DL}$, and the $E_L$ for molecular dissociation.
(3) O:H–O elongation and dual polarization creates the supersolid phase that is hydrophobic, viscoelastic, repulsive, ¼ less dense, slowed molecular dynamics with long-lived H–O phonon.

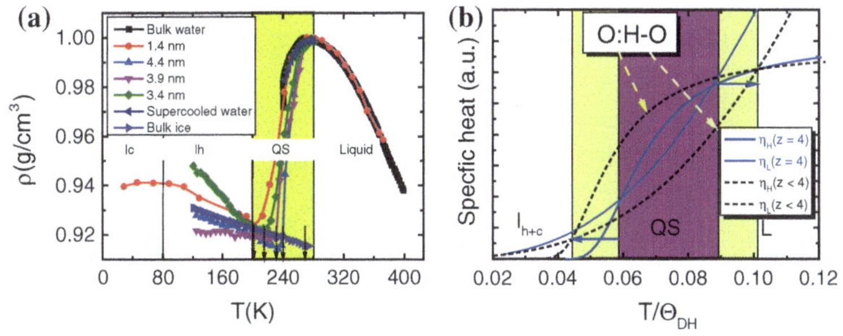


**Fig. 8.2** **a** Droplet size dependence of the $T_N$ [18–22] and **b** molecular undercoordination stretched quasisolid-phase boundary, which raises the $T_m$ and lowers the $T_N$ [23]

(4) Phonon cooperative relaxation disperses boundaries of the quasisolid phase, elevating the $T_m$ for quasisolid melting and depressing the $T_N$ for quasisolid transiting into homogeneous ice.

## 8.3 Wonders of Molecular Undercoordination

Molecular undercoordination has resulted in the following excessive anomalies of water and ice compared with the bulk water and ice, which will be addressed in subsequent sections:

(1) Structure geometry varies with the number of molecules gathered together.
(2) O–O distance becomes longer when the droplet size is reduced, see Fig. 8.3.
(3) H–O phonon blueshift and O:H phonon redshift—stiffer becomes even stiffer and softer even softer as the fraction of the undercoordinated molecules increases.
(4) O 1s energy level shifts positively to stronger binding energy when the droplet size is reduced.
(5) Molecular dipole moment increases with the drop of cluster size and the water droplet becomes non-additive to other substance—highly hydrophobic.
(6) Nanodroplet becomes thermally stable, “supercooling” and “superheating” take place at freezing and melting.
(7) Nanobubbles are long lived, mechanical stronger and thermally more stable.
(8) Skins of ice and water are hydrophobic and frictionless, making droplet moving faster in microchannels.
(9) Skins are viscoelastic, rigid, less dense, and molecules become less mobile.
(10) Undercoordinated molecules have longer lifetime extending from the bulk value of 200 fs to some 700 fs.

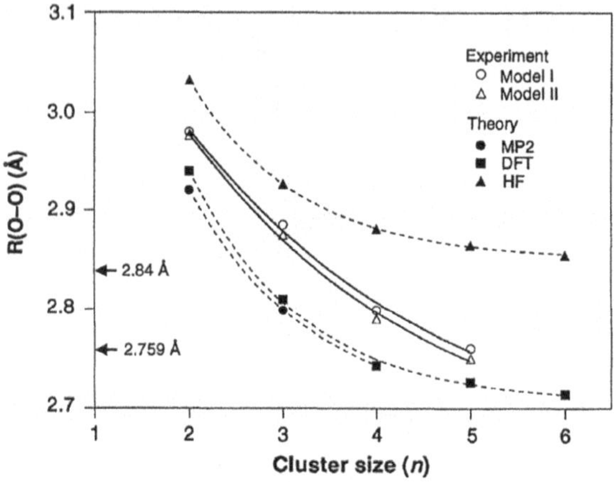


**Fig. 8.3** $(H_2O)_N$ cluster size dependence of O–O distance derived using various calculation methods [7]

However, as an important degree of freedom [6], the effect of molecular undercoordination on the structural and physical properties of the undercoordinated systems has long been overlooked [24]. Molecular undercoordination induced H–O bond contraction and O:H nonbond elongation and their electronic energetic consequences reconcile the observable properties associated with hydration shells, nanobubbles, nanodroplets, hydrophobic capillary confined droplets, ice and water skins, molecular clusters [17]. One of the most appealing observations is that the melting of ice in porous glass having different distribution of tempetatures. The confined water crystallizes only partially and that an interface layer, between the ice crystallites and the surface of the pore, remains liquid. Nuclear magnetic resonance and differential scanning calorimetry measurements revealed a 0.5 nm thick interface [25].

## 8.4 Quantitative Resolution

### *8.4.1 Extended BOLS-NEP Notion*

Bonds in the skins of metals, alloys, semiconductors, insulators, and nanostructures contract globally with respect to their bulk counterparts [24]. The first interlayer spacing of these systems contracts by (12 ± 2)% relative to the bulk. For nanostructures, the relaxation extends radially inward to deeper layers [26, 27]. For the one-dimensional atomic chains and edges of the two-dimensional atomic ribbons such as graphene, bonds contract by up to 30 % [24, 28]. Atomic undercoordination induced bond contraction and the associated quantum entrapment and nonbonding electron polarization claim the full responsibility for the unusual behavior of the undercoordinated adatoms, point defects, and nanostructures of different shapes, including the size-dependence of the known bulk properties and the size-induced emergence of properties that the bulk parent never shows [29]. Atomic undercoordination induced local bond contraction and the associated energy change follow the notion of BOLS correlation [24, 29].

The curvature $K^{-1}$ dependence of the effective atomic coordination ($z_i$), bond length ($d_i$), charge density ($n_i$), energy density (proportional to the elastic modulus, $B_i$), and the potential trap depth ($E_i$) in the ith atomic site follow the relationships [29].

$$\begin{cases} z_1 = 4(1-0.75K^{-1}); z_2 = z_1 + 2; z_3 = 12 & \text{(Effective CN)} \\ d_i = C_i d_b = 2d_b[1+\exp((12-z_i)/(8z_i))]^{-1} & \text{(Bond contraction coefficient)} \\ E_i = E_b C_i^{-m} & \text{(Bond energy; Potential trap depth)} \\ B_i = B_b C_i^{-(m+3)} & \text{(Energy density; Elastic modulus)} \\ n_i = n_b C_i^{-m} & \text{(Charge density)} \end{cases} \quad (8.1)$$

K being the dimensionless form of size is the number of atoms lined along the radius of a spherical dot. The bond nature index *m* varies approximately from one to four, which is a constant for a specific substance. The subscript *i* denotes the *i*th atomic layer counted from the outermost inward and the subscript *b* denotes the corresponding bulk values. Therefore, one can focus on the energetic and electronic behavior of the skin bonds of a certain weightage over the entire object while keeping in mind that the core interior ($i > 3$) of a nanostructure remains its bulk nature [26, 30].

The involvement of the lone-pair ':' interaction and the O–O repulsion prevents, however, the O:H and the H–O from following the BOLS notion simultaneously because the lone pairs screen an $H_2O$ molecule from strong interacting with its neighbours. The O:H–O bind disparity, see Fig. 8.2c, means that the stronger H–O bond serves as the "master" to contract by a different amount from what the BOLS notion predicts. The contraction of the H–O bond lengthens and softens the "slave" O:H nonbond through Coulomb repulsion, with a dual process of polarization.

## *8.4.2 Geometries: Rings, Cages, and Clusters*

Figure 8.4 and Table 8.1 feature the bond geometry, ∠O:H–O containing angle, O:H–O bond segmental length and energy for the optimal $(H_2O)_N$ structures derived from DFT computations [17]. The $(H_2O)_N$ structure varies from a dimer (N = 2), to rings (N = 3–5), cages (N = 6–10) and solid clusters (N = 12–20). N = 6 derives the book, prism, and cage structures of the same binding energy [13]. The O:H–O bridge holds for any geometry despite the ∠O:H–O containing angle varying from 160 to 177° and the segmental lengths being subject to cooperative relaxation. The effective CN of the $H_2O$ also varies from situation to situation. For the same N value, the CN

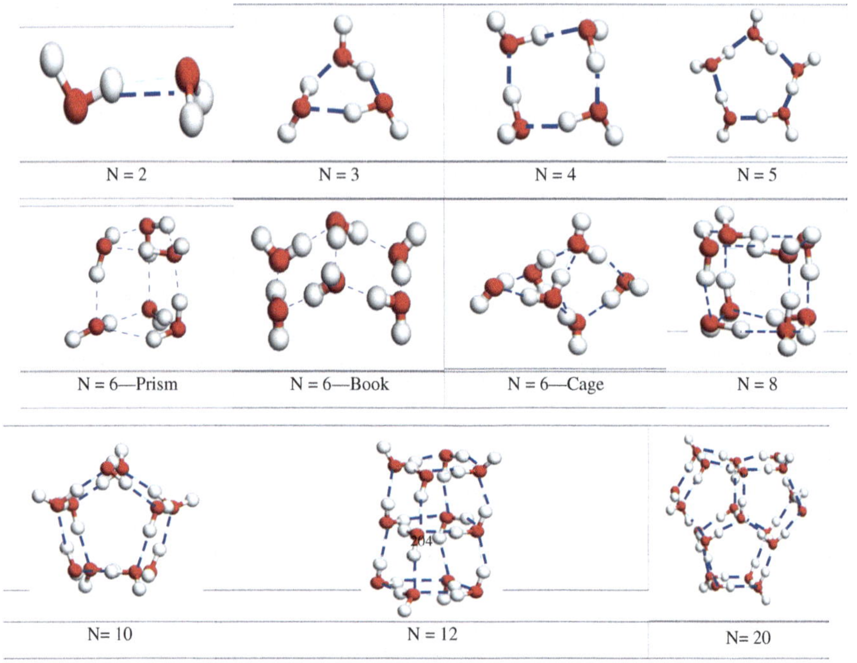


**Fig. 8.4** DFT optimized $(H_2O)_N$ crystal structures showing the chain like (N = 2), ring-like (N < 6), cage-like (N = 6–10), and solid clusters (N = 12–20). N = 6 creates three structures of nearly identical binding energy [13, 17]

**Table 8.1** DFT-derived segmental length $d_x$, ∠O:H–O containing angle θ, and phonon frequency $\omega_x$ for $(H_2O)_N$ clusters [17][a]

| | Monomer | Dimer | Trimer | Tetramer | Pentamer | Hexamer | Bulk [18] |
|---|---|---|---|---|---|---|---|
| N | 1 | 2 | 3 | 4 | 5 | 6 | Ih |
| $d_H$ (Å) | 0.969 | 0.973 | 0.981 | 0.986 | 0.987 | 0.993 | 1.010 |
| $d_L$ (Å) | – | 1.917 | 1.817 | 1.697 | 1.668 | 1.659 | 1.742 |
| θ (°) | – | 163.6 | 153.4 | 169.3 | 177.3 | 168.6 | 170.0 |
| $\omega_L$ ($cm^{-1}$) | – | 184 | 190 | 200 | 210 | 218 | 220 |
| $\omega_H$ ($cm^{-1}$)[50–52,55] | 3650 | 3575 | 3525 | 3380 | 3350 | 3225 | 3150 |
| $\Theta_{DL}$ (K) | – | 167 | 171 | 180 | 189 | 196 | 198 [31] |
| $\Theta_{DH}$ (K) | 3650 | 3575 | 3525 | 3380 | 3350 | 3225 | 3150 |
| $E_L$ (meV) | – | 34.60 | 40.54 | 66.13 | 69.39 | 90.70 | 95 |
| $E_H$ (eV) | 5.10 | 4.68 | 4.62 | 4.23 | 4.20 | 3.97 | 3.97 |
| $T_m$ (K) | – | 322 | 318 | 291 | 289 | 273 | 273 |
| $T_N$ (K) | – | 94 | 110 | 180 | 188 | 246 | 258 |

[a]$T_m$ = 325 K (monolayer) [32]; 310 K (skin of bulk) [33]; $T_N$ = 242 K (4.4 nm droplet) [19]; 220 K (3.4 nm droplet) [19]; 205 K (1.4 nm droplet) [20]; 172 K (1.2 nm droplet) [21]; <120 K (1–18 molecules) [22]

varies with the dimensionality of chains, rings, cages, and solid clusters. Therefore, the O:H–O segmental lengths and the associated containing angle are the key identities in all possible geometrical configurations of water and ice.

### *8.4.3 O:H–O Length, Cohesive Energy, and Mass Density*

X-Ray absorption spectrosocpy revealed that the skin O–O distance for water expands by 5.9 % to 2.965 Å, compared to a 4.6 % contraction of the skin O–O distance for liquid methanol [34], which differentiates the surface tension (should be compression instead) of 72 mN/m for water from 22 mN/m for methanol. The O–O distance in the skin and between a dimer is about 3.00 Å; the O–O distance in the bulk varies from 2.70 [35] to 2.85 Å [36], depending on experimental conditions. The ideal O–O distance at 4 °C is 2.6950 Å [37].

The volume of water confined in 5.1 and 2.8 nm sized $TiO_2$ pores expands by 4.0 and 7.5 %, respectively, with respect to the bulk water [38]. MD calculations also reveal that the $d_H$ contracts from 0.9732 Å at the center to 0.9659 Å at the skin of a free-standing water droplet containing 1000 molecules [39]. X-ray scattering, neutron reflection, and SFG spectroscopy investigations suggested that the boundary layer in the vicinity of hydrophobic surface water consists of a ~0.5 nm depletion layer with a density of 0.4 g/cm$^3$ and a considerable amount (25–30 %) of water molecules with free OH groups [40]. The 0.4 g·cm$^{-3}$ density correpsonds to O–O distance of $d_{OO}$ = 3.66 Å, which is in the gaseous phase.

Figure 8.5a and Table 8.1 show O:H–O bond segmental relaxation as a function of N for the $(H_2O)_N$ clusters derived from calculations using the PW and OBS algorithms [17]. As N is reduced from 24 (an approximation to the bulk) to two (a dimer), the H–O bond contracts by 4 % from 0.101 to 0.097 nm, and the O:H bond expands by 17 % from 0.158 to 0.185 nm, according to the OBS derivatives. Figure 8.5b plots

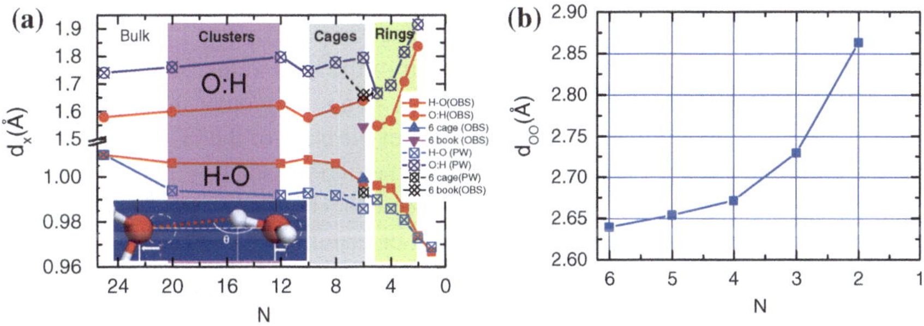


**Fig. 8.5** Cluster size dependence of **a** $d_x$ in the $(H_2O)_N$ clusters optimized using the PW [41] and the OBS [42] methods. The non-monotonic size trend of $d_x$ stems from the effective molecular CN that also changes with geometrical configuration. N dependence of **b** the $d_{OO}$ for N = 2–6 gives the mass density in the form of $\rho \propto (d_{OO})^{-3}$. (Reprinted with permission from [17])

the $N$-dependence of the O–O distance that expands by 8 %, when the $N$ is reduced from 20 to 3, which is compatible to the value of 5.9 % (subject to the reference used) measured in the water skin at 25 °C [34]. This O:H–O cooperative relaxation expands the O–O distance by 13 % and lowers the density by 30 % for the dimer. The O:H–O length change non-monotonically with N because of the altered dimensionality. We will focus on the monotonic relaxation profiles for the $d_x$ at $N \le 6$ in the subsequent discussions without rendering the generality of conclusion.

Consistency between BOLS predictions and experimental [3, 4] and numerical [17] observations confirms the following for undercoordinated $H_2O$ molecules:

(1) The H–O bond shortening (lengthening) is always coupled with the O:H lengthening (shortening), irrespective of the algorithm used, which evidence the expected O:H–O bond cooperativity—one segment contracts and the other must expand because of O–O Coulomb coupling.
(2) The non-monotonic change of $d_x$ results from the effective CN that varies not only with the number of molecules N but also with the geometrical configuration of the $(H_2O)_N$ cluster. The effective CN of a ring-like cluster is smaller than that of a cage for the same N value.
(3) Molecular undercoordination increases the $E_H$ and reduces the $E_L$ in magnitude, as the BOLS notion predicts and Sect. 8.5 demonstrates.
(4) Computation algorithms deviated outcomes suggest that one should focus more on the trend and the natural origin than on the accuracy of the derived values. Numerical derivatives and experimental observation serve as only means for concept verification.

### 8.4.4 H–O Bond Charge Densification and Entrapment

Following the same trend as "normal" materials, molecular undercoordination imparts to water local charge densification [11, 12, 43–46], binding energy entrapment [10, 43, 47, 48], and nonbonding electron polarization [45]. Figure 8.6a shows that the O 1s level shifts more deeply from the bulk value of 536.6 to 538.1 and 539.7 eV when move from bulk water to its skin to monomer in gaseous phase [49–51]. The O 1s binding energy shift is a direct measure of the H–O bond energy and the contribution from the O:H nonbond is negligibly small.

Atomic undercoordination lowers the atomic cohesive energy, a product of the bondenergy and the atomic CN(z), $zE_z$, that determines the thermal stability of 'normal' materials in general. The energy necessary for dissociating a $(H_2O)_N$ cluster into $(H_2O)_{N-1} + H_2O$ increases, conversely, when the cluster size is reduced to a trimer (Fig. 8.6b) [52], which conflicts with the traditional understanding of 'normal' material behavior.

Figure 8.7 compares the NEXFAS spectra of nanobubbles [53], vapor, liquid skin, and bulk water [54]. The spectra exhibit three majors at 535.0 to 536.8 and 540.9 eV corresponding, respectively, to the bulk, skin, and H–O dangling bond

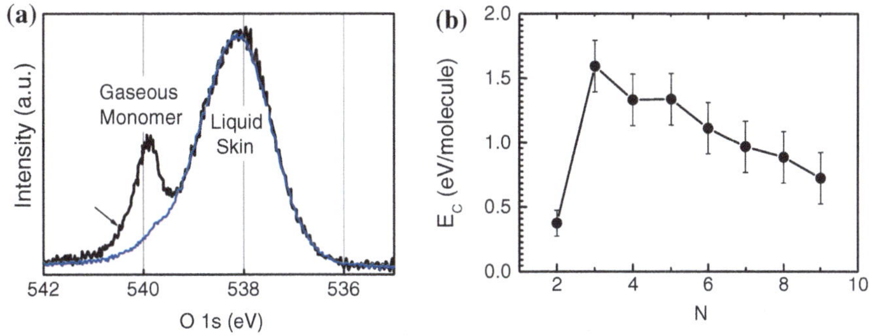


**Fig. 8.6** **a** XPS O 1s spectra of water containing emissions from the liquid skin at 538.1 eV and from the gaseous phase at 539.9 eV (reprinted with permission from [51]); **b** energy required for dissociating a $(H_2O)_N$ cluster into $(H_2O)_{N-1} + H_2O$ (1 kJ/mol = 0.02 eV/molecule). (Reprinted with permission from [52])

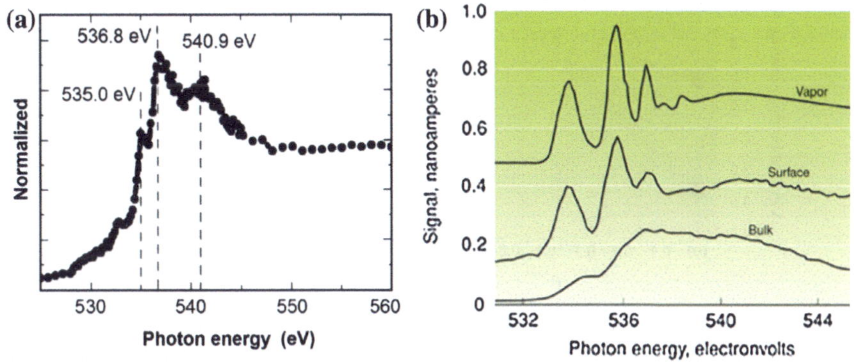


**Fig. 8.7** NEXFAS spectra of **a** nanobubbles [53] and **b** vapor, liquid skin, and bulk liquid [54]. The spectra resolve discrete peaks that correspond to the bulk, skin, and H–O dangling bond radicals

radicals. The energy conservation mechanism in the NEXFAS measurements is different from that of the XPS. The NEXFAS involves the shift of both the valence 2p and the O 1s core band but the XPS involves the O 1s only. Analysis of the electron spectroscopic adata is much more complicated.

### *8.4.5 Nonbonding Electron Dual Polarization*

Molecular undercoordination induced skin polarization can be detected using an ultrafast liquid-jet UPS [45]. Results shown in Fig. 8.8, resolved the vertical bound energies (being equivalent to work function) of 1.6 and 3.3 eV for the solvated electrons in the skin and in the bulk interior of water solution, respectively. The

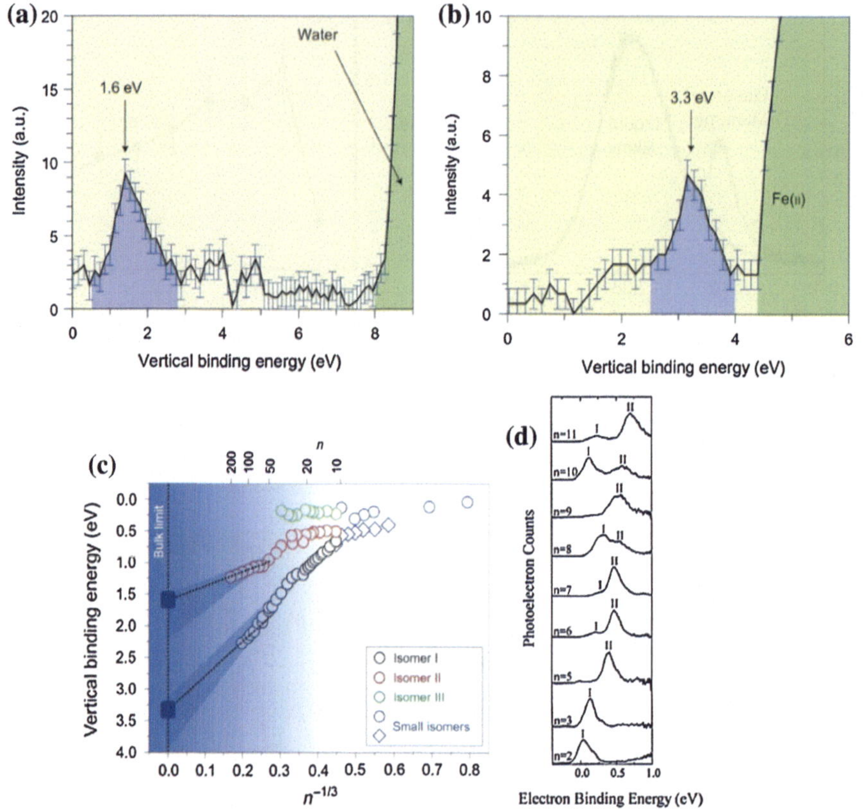


**Fig. 8.8** Molecular undercoordination polarizes nonbonding solvent electrons [6]. The vertical bound energy for solvated electrons drops to **a** 1.6 eV in the skin from **b** 3.3 eV in the bulk of the liquid water. **c** The bound energy of solvated electrons in the skin and in the bulk reduces further with the number $n$ of $(H_2O)_n$ clusters toward zero [55–57] **d** n = 2–11 [55]. (Reprinted with permission from [45])

bound energy decreases with the number $n$ of the $(H_2O)_n$ clusters toward zero [55–57]. The hydrated electrons live longer than 100 ps near the surface compared with those solvated inside liquid bulk interior. Observations evidence that molecular undercoordination substantially enhances nonbonding electron polarization [17], which increases the viscoelasticity and hence lowers the skin molecular mobility. The anchored skin dipoles allow nanodroplet interacting with other substance through electrostatic, repulsive, and hydrophobic interactions without exchanging electrons or bond formation, named nonadditivity [58].

The nonbonding electrons are subject to dual polarization when the molecular CN is reduced [6]. Firstly, H–O bond contraction deepens the H–O potential well and entraps and densifies electrons in the H–O bond and those in the core orbitals of oxygen. This locally and densely entrapped electrons polarize the lone pair of oxygen from the net charge of −0.616 e to −0.652 eV according to DFT calculation for ice skin [33]. The increased charge of O ions further enhances the O–O repulsion as the second round of polarization. This dual polarization raises the valence band energy up, as shown in Fig. 8.8d.

Further reduction of cluster size, or the molecular CN, enhances this dual polarization, resulting observations in Fig. 8.8c—cluster trend of the solvate electron polarization. Therefore, electron dipoles formed on the flat and the curved skins enhances such polarization, which creates the repulsive force, making liquid water hydrophobic and ice slippery. Most strikingly, engineered water nanostructures (25 nm) can inactivate and kill foodborne microorganisms effectively [59]. There is only two choices for charge energy, either being trapped or polarized in energy domains, which is responsible for the unusual behaviour of substnce at the nanometer scale.

### 8.4.6 Phonon Cooperative Relaxation

Normally, the loss of neighboring atoms softens the phonons of 'normal' materials such as diamond and silicon except for the G mode (1550 $cm^{-1}$) for graphene [60] and the $A_1$ mode (141 $cm^{-1}$) for $TiO_2$ [61] as these two modes arise from dimer vibration. However, water molecular undercoordination stiffens its stiffer $\omega_H$ phonon significantly [62, 63]. The $\omega_H$ has a peak centered at 3200 $cm^{-1}$ for bulk water, and at 3450 $cm^{-1}$ for the skins of water ice [64]. The $\omega_H$ for gaseous molecules is around 3650 $cm^{-1}$ [65–68]. The $\omega_H$ shifts from 3200 to 3650 $cm^{-1}$ when the N of the $(H_2O)_N$ cluster drops from 6 to 1 (Fig. 8.9a) [65, 69, 70]. Encapsulation by Kr

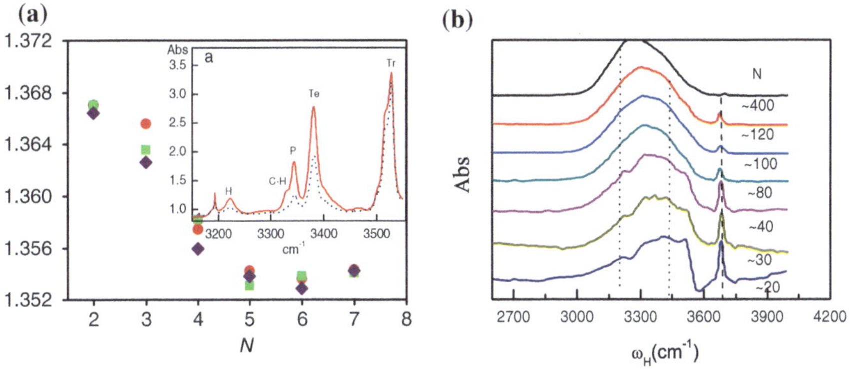


**Fig. 8.9** Size-dependent $\omega_H$ of **a** $(H_2O)_N$ clusters (in the frequency ratio of $\omega_H/\omega_D$) and **b** large clusters. *Line* (N = 2) corresponds to a dimer [73], (3, Tr) to a trimer [74], (4, Te) to a tetramer, (5, P) to a pentamer, (6, c–H) corresponds to a cyclic hexamer, (7, H) corresponds to a cage hexamer. *Red circles* correspond to He matrix, *green squares* correspond to Ar matrix, and *blue diamond* corresponds to p–$H_2$ measured at 2.8 K. *Inset* **a** denotes the sharp $\omega_H$ peaks for the small clusters. Size-reduction stiffens the H–O bonds with little disturbance to the dangling H–O bonds at 3700 $cm^{-1}$ in (**b**). (Data adopted from [65, 67])

and Ar matrices softens the $\omega_H$ slightly by 5–10 $cm^{-1}$ due to the involvement of interface interaction [70]. Size-reduction caused $\omega_H$ stiffening also occurs in large molecular clusters [67] (see Fig. 8.9b). When N drops from 475 to 85, $\omega_H$ transits from the dominance of the 3200 $cm^{-1}$ component (bulk attribute) to the dominance of the 3450 $cm^{-1}$ component (skin attribute) [71]. The high frequency at approximately 3700 $cm^{-1}$ corresponds to the vibration of the free H–O radicals, with possible charge transportation in the skin of water and ice [14, 72].

Figure 8.10 features the cluster-size dependence of the calculated vibration spectra of $(H_2O)_N$ with respect to the ice-$I_h$ phase. As expected, N reduction stiffens the $\omega_H$ from 3100 to 3650 $cm^{-1}$ and meanwhile softens the $\omega_L$ from 250 to 170 $cm^{-1}$ as the bulk water turns into dimers. The ∠O:H–O bending mode $\omega_{B1}$ (400–1000 $cm^{-1}$) shifts to a slightly lower value, but the ∠H–O–H libration mode $\omega_{B2}$ (≈1600 $cm^{-1}$) remains unchanged [75].

The N-reduction-stiffened $\omega_H$ is consistent with spectroscopic measurements (Fig. 8.9a). For instance, reduction of the $(H_2O)_N$ cluster from N = 6 to 1 stiffens the $\omega_H$ from 3200 to 3650 $cm^{-1}$ [69]. The skin $\omega_H$ of 3450 $cm^{-1}$ corresponds to an effective cluster size of N = 2–3. Indeed, molecular undercoordination shortens and stiffens the H–O bond, and lengthens and softens the O:H nonbond consistently.

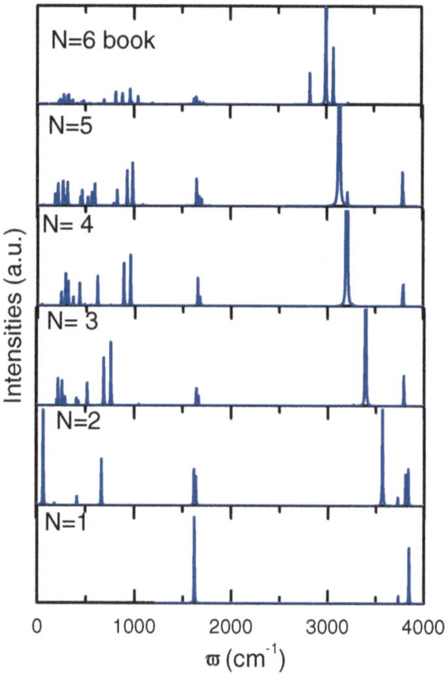


**Fig. 8.10** Cluster size (N) dependence of the phonon $\omega_x$ relaxation in $(H_2O)_N$ clusters. Features centered at 200 $cm^{-1}$ corresponds to O:H stretching vibration, centered at 500 $cm^{-1}$ correspond to ∠O:H–O bending, centered at 600 $cm^{-1}$ correspond to ∠O–H–O libration bending, and features >3000 $cm^{-1}$ correspond to H–O stretching vibration. (Reprinted with permission from [17])

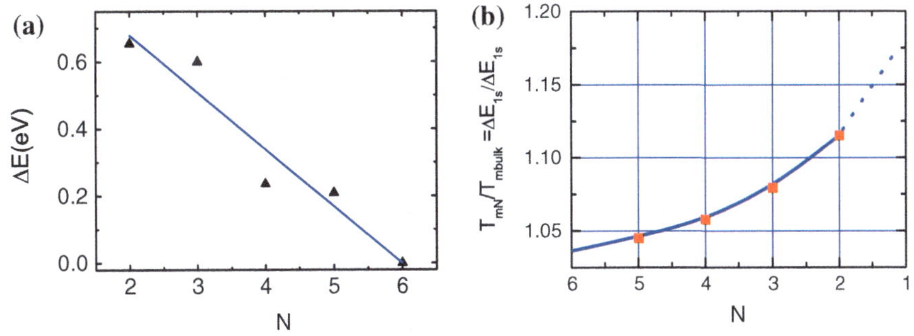


**Fig. 8.11** **a** The total energy shift, $\Delta E_L + \Delta E_H > 0$, of the O:H–O bond as a function of $(H_2O)_N$ size, ensures the strong recoverability of the O:H–O bond between undercoordination molecules [16]. **b** *N*-dependence of the melting point, $T_{mN}$, (to $N = 2$ for dimers) and the O 1s core-level shift (to $N = 1$ for gas monomers) of $(H_2O)_N$ clusters based on DFT derived $d_{HN}/d_{HB}$ values and the expression of $T_{mN}/T_{mB} = \Delta E_{1sN}/\Delta E_{1sB} = E_{HN}/E_{HB} = (d_{HN}/d_{HB})^{-m}$, (m = 4 for covalent bond)

### *8.4.7 O:H–O Bond Potentials*

Lagrangian solution to the O:H–O bond oscillating dynamics [76] transforms the known segmental length and phonon frequency ($d_x$, $\omega_x$) for the H–O bond (x = H) and the O:H nonbond (x = L) [17] into their force constants and bond energies ($k_x$, $E_x$) turned out the potential paths for the O:H–O bond relaxing with $(H_2O)_N$ cluster size, as described in Chap. 55.

The combination of Coulomb repulsion and molecular undercoordination not only reduces the molecular size ($d_H$) with enhanced intra-molecular interaction but also enlarges the molecular separation ($d_L$) with attenuated inter-molecular nonbond strength. The relaxation increases the H–O cohesive energy, as measured, from the bulk value of 3.97 eV [33], to the skin of 4.66 eV [6], and to the gaseous monomers 5.098 of $H_2O$ [77], 5.10 eV of $H_2O$ [78] and $D_2O$ [79] as well. The $E_L$ for N = 6 is around the bulk value of 0.095 eV [31]. The total O:H–O bond energy gain Fig. 8.11a with a linear dependence on the CN, which plays a role in ice regelation [16]. The O:H–O bond recovers its bulk length and energy when the skin is subject to contact with recovery of molecular CN.

## 8.5 Nanodroplet and Nanobubble Thermodynamics

### *8.5.1 Supercooling or Superheating?*

#### 8.5.1.1 $T_m$ Elevation and $T_N$ Depression

Water nanodroplets and nanobubbles undergo not only "superheating" at melting but also "supercooling" at freezing. The latter creates the second critical point corresponding to the $T_N$ for homogeneous ice nucleation. Water droplets encapsulated in hydrophobic capillaries [80, 81] or ultrathin water films deposited on graphite, silica, and certain metals [9, 82–89] behave like ice at room temperature. Molecules at the air/water interface and those at the hydrophobic contacting interface performed the same as both are subjected to undercoordination.

Observations shown in Fig. 8.2c revealed at least three $\rho(T)$ transition temperatures [6] that vary with water droplet size [19, 20] or the fraction of undercoordinated molecules [90]:

(1) The maximal density ($\rho_M$ at temperatures close to the $T_m$) transits from 277 K for the bulk to 315 K for the monolayer skin [18]. The transition temperature for liquid formation shifts from the bulk value of 273 K [18] up to 310 K for the skin of bulk water [33]. A monolayer of ice melts at 325 K according to MD calculations [32].
(2) The minimal density ($\rho_m$ at temperature nearby freezing) transits from the bulk value of 258 K [18] to 242 K for 4.4 nm, 220 K for 3.4 nm [19], 205 K for 1.4 nm [20] and 172 K for 1.2 nm sized droplets [21]. Freezing transition for clusters containing ≤18 molecules cannot be observed at temperature even down to 120 K [22]
(3) The second maximal density (XI/Ic phase boundary) transits from the bulk value of 100–55 K for the XVI phase [90].
(4) The transition temperatures vary with the droplet size of water, which is often regarded as the droplet size induced "supercooling in freezing" and "superheating at melting" [91].

Figure 8.11b shows the expected cluster size dependence of the $T_m$ and the O 1s energy shift ($\Delta E_{1s}$) as a function of cluster size. As both $T_m$ and $\Delta E_{1s}$ are proportional to the H–O bond energy in the form of:

$$T_{mN}/T_{mB} = \Delta E_{1sN}/\Delta E_{1sB} = E_{HN}/E_{HB} = (d_{HN}/d_{HB})^{-4}.$$

Subscript $B$ denotes the bulk. One can derive from the plots that when the $N$ is reduced from a value of infinitely large to two, the $T_m$ will increase by 12 % from 273 to 305 K, which explains why the ultrathin water films [9, 82, 84–88] or water droplets encapsulated in hydrophobic nanopores [80, 81] behave like ice at room temperature. The expected O 1s energy shift ($C_z^{-4}$–1) of water clusters also agrees with the size trend of measurements. For instance, the O 1s core level shifts from 538.2 to 538.6 eV and to 539.8 eV, when the water cluster size is reduced from $N$ = 200 to 40 and to free water molecules [49, 92].

Melting point elevation is more apparent for the even undercoordinated molecules at the curved skin. SFG spectroscopy reveals that outermost two molecular layers are highly ordered at the hydrophobic contacts compared with those at a flat water–air interface [93]. A tiny difference in the molecular CN makes a big impact on the performance of the even undercoordinated molecules.

#### 8.5.1.2 Supercooling and Superheating

The droplet size effect on $T_m$ and $T_N$ is often regarded as "superheating" at melting and "supercooling" at freezing. The $T_m$ and $T_N$ dispersion is an intrinsic phenomenon determined by the O:H–O bond relaxation in length and energy. Freezing point depression occurs when a solution is cooled below the freezing point of the corresponding pure liquid due to the presence of the aqueous solute or due to droplet size reduction. The freezing point depression occurs when sodium chloride is added to pure water. The local electric fields of ions will cluster, strecth, and polarize water molecules, translating into O:H–O bond eleongation—H–O bond contraction and O:H nonbond elongation—the former raises the $T_m$ and the latter depresses the $T_N$ by dispersing the quasisolid phase boundaries. In fact, $T_m$ elevation is different from "superheating" and $T_N$ depression is different from "supercooling". The former is intrinsic and the latter is process dependent.

Supercooling, also known as undercooling [91], is the process of lowering the temperature of a liquid or a gas below its freezing point without it becoming a solid. Once the supercooled water is disturbed, it soonly become ice, see Fig. 8.12. Supercooled water occurs in the form of small droplets in clouds and plays a key role in the processing of solar and terrestrial radiative energy fluxes. Supercooled water is also important for life at subfreezing conditions for the commercial preservation of proteins and cells, and for the prevention of hydrate formation in nature gas pipelines. Superheating is the opposite. Once the superheated water is disturbed by adding sugar, the superheated water will explode.

### 8.5.2 *Quasisolid Phase Boundary Dispersion*

Generally, melting a specific atom inside a normal substance requires heat that is a fraction of its cohesive energy, $E_C = zE_z$, i.e., the sum of bond energy $E_z$ over its coordination neighbors (z or CN). The $T_m$ of a solid changes with the solid size because of the skin atomic undercoordination and the varied fraction of

**Fig. 8.12** Ice making from supercooled water [94] and superheated water exploding by adding sugar. The gel-like quasisolid water stays on a hydrophobic substrate. (public domain)

undercoordinated skin atoms [24]. However, the presence of the critical temperatures at 273 K ($T_m$) and 258 K ($T_N$), see Fig. 8.11, indicates that bulk liquid transits into the quasisolid and then into solid [18]. Traditionally, the quasisolid phase is absent from existing phase diagram and has received little attention.

Why does droplet size effects on the $T_m$ and the $T_N$?

One can imagine what will happen to the quasisolid phase boundaries by raising the $\Theta_{DH}$ and meanwhile lowering the $\Theta_{DL}$. The $\eta_L$ will saturate quicker and the $\eta_H$ slower than they were in the bulk standard. This process will raise the $T_m$ and lower the $T_N$, as illustrated in Fig. 8.2d. According to the $\Theta_{Dx} \propto \omega_x$ relationship ($\Theta_{DL} = 198$ K, $\Theta_{DH} = 3200$ K) and data in Table 8.1, One can estimate $\Theta_{DL} = 198$ (for bulk)/260(calculated bulk) × 195(calculated cluster) = 149 K and $\Theta_{DH} = 3200/3200 \times 3550 = 3550$ K, for N = 2 [18].

With the known bulk values of $\Theta_{DL} = 198$ K, $T_m = 273$ K, $T_N = 258$ K, and the respective $\omega_x$ and $E_x$ one can estimate the cluster size dependence of the $\Theta_{Dx}$, $T_m$, and $T_N$ using the following relationships [24]:

$$\begin{cases} \Theta_{Dx} \propto \omega_x \\ T_{N,m} \propto E_{L,H} \end{cases}$$

Numerical reproduction of the $T_m(P)$ profiles indicates that the $T_m$ is proportional to $E_H$ and that the $T_N$ is proportional to $E_L$. In order to minimize calculation artifacts, a modification of the $\omega_L(N)$ curve in Fig. 8.2b is made with respect to the measured value of 220 $cm^{-1}$ for bulk water and to that the calculated $\omega_x$ matches the measured value at N = 2. This modification improves the precision of estimating cluster size dependence of the $\Theta_{DL}$.

## 8.5.3 *Nanodroplet and Nanofilm Supersolidity*

### 8.5.3.1 Mobility Depression of Water Molecules

Kim et al. [95] measured the mechanical properties of the nanometric water column by using noncontact AFM, see Fig. 8.13a. They observed coincidently that the relaxation time τ is associated with the meniscus increases with its elongation and ruptures at the same value of τ, independent of the meniscus volume. The elongation of τ between formation and rupture of the meniscus is indicative of the increased solid-like response, similar to that observed in nanoconfined water layers. The longer and thinner water column is more solid-like, which suggest that:

(i) water skin is more solid-like than that inside the meniscus and
(ii) the lowered mobility of the skin water molecules is responsible for the structural stability of the water meniscus.

Khan and co-workers [96] and also measured the same trend of mechanical relaxation time versus number of confined layers in water squeezed under AFM

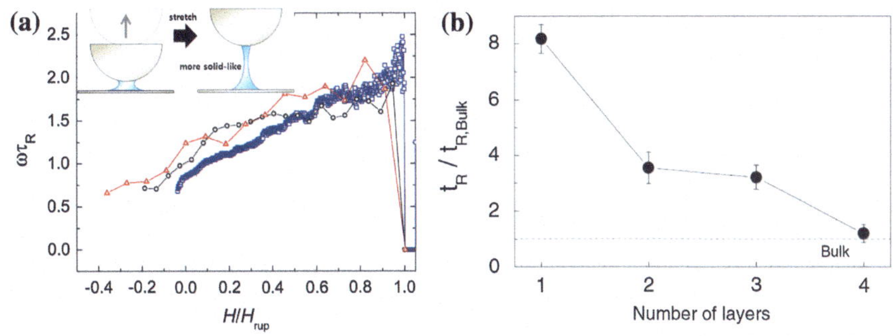


**Fig. 8.13** Elongation thinning induced supersolidity of water meniscus. Molecular mobility depression raises the relaxation time $\tau_R$ under constant oscillation frequency $\omega$ of the AFM tip. $H/H_{rup}$ is the relative height of the column to the maxim at rupture. (Reprinted with permission from [95])

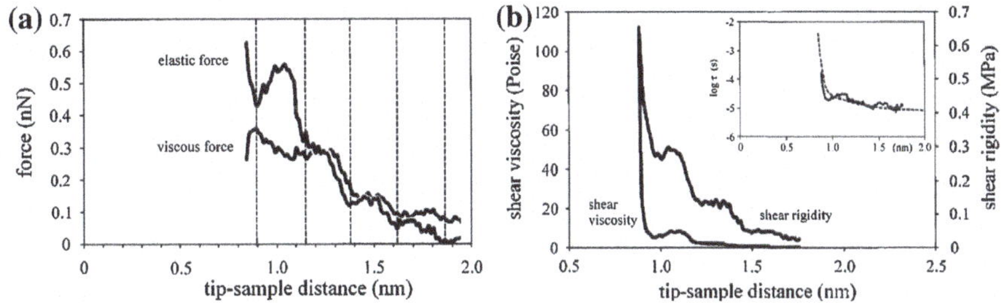


**Fig. 8.14** **a** Viscous and elastic forces and **b** shear rigidity and viscosity of water film. The *inset* shows the mechanical relaxation time. (reprinted with permission from [99])

tip. At a certain thickness, the film becomes solid like with transition from viscosity to elasticity [97]. Using Ultrafast Infrared Vibrational Echo Spectroscopy, Tan et al. [98] examined the dynamics of water, confined on a nanometer length scale (1.7–4.0 nm) in comparison to bulk water and salt solution. Results show that the confined water dynamics is substantially slower than bulk water and the mobility is droplet size dependent. The fastest dynamics ($\sim$50 fs) approaches to bulk water, while the slowest time scale dynamics is much slower than water.

#### 8.5.3.2 Elasticity, Viscosity, Rigidity of Nanofilms

Figure 8.14 shows the thickness dependence of the viscous and elastic forces, shear viscosity and shear rigidity, and the relaxation time of water films probed using an AFM by Antognozzi et al. [99], which is consistent with aforementioned experimental results and the presently expected supersolidity. These observations further confirm the supersolidity of undercoordinated water molecules.

## 8.5.4 *Nanobubbles and Nanodroplets*

### 8.5.4.1 Nanobubble Formation

Nanobubbles (<200 nm in diameter) have unique properties such as long lifetime in liquid owing to its negatively charged surface, and its high gas solubility into the liquid owing to its high internal pressure. They are used in a variety of fields including diagnostic aids, drug delivery, water treatment, biomedical engineering, degradation of toxic compounds, water disinfection, and cleaning/defouling of solid surfaces including membrane [100–102]. For example, Oh and co-workers [103] fabricated hydrogen nanobubbles in gasoline by a generator which consists of gas–liquid dispersion system. The mean diameter is 159 ± 32 nm and the concentration is $11.25 \pm 2.77 \times 10^8$ particles/ml, which changes insignificantly after 121 days of aging. Hydrogen nanobubbles in gasoline fuel can improve engine efficiency and emission performance [103].

Figure 8.15 features the key differences among bubbles at different size scales [104]. Microbubbles tend to gradually decrease in size and subsequently collapse due to long stagnation and dissolution of interior gases into the surrounding water, whereas nanobubbles remain as they were for months and do not burst out at once. It has been revealed that the interface of nanobubbles consists of hard H–O bonds similar to those found in ice and gas hydrates due to the undercoordination effect. This in turn leads to their reduced diffusivity that helps to maintain adequate kinetic balance of the bubbles against high internal pressure. Nanobubbles can form freely and remain stable for long periods of time under the right conditions (Fig. 8.16).

On the other hand, gas-filled nanoscale bubbles can form on a variety of solid surfaces [102, 105, 106], see Fig. 8.17, being attributed to the heterogeneous

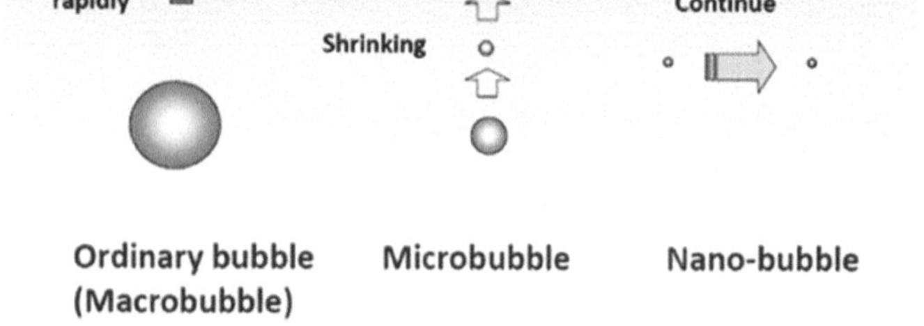


**Fig. 8.15** Schematic illustration of the life of macro-, micro- and nanobubbles. (Reprinted with permission from [104])

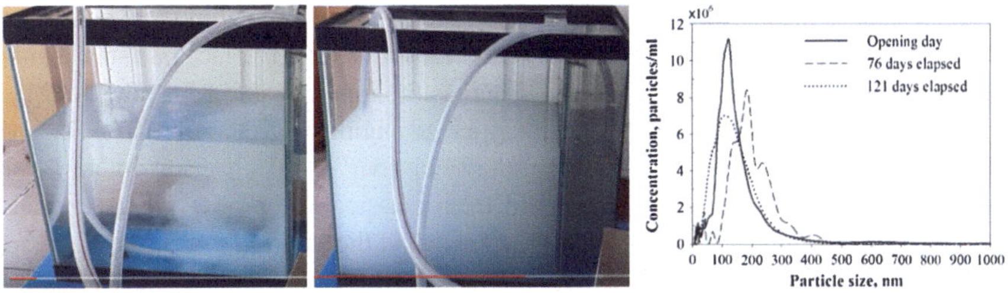


**Fig. 8.16** Nanobubble formation by dissolving gases and the bubble size dependence of life time at 298 K. (*Credit* https://www.youtube.com/watch?v=qt_a_TvguaM; Reprinted with permission from [103])

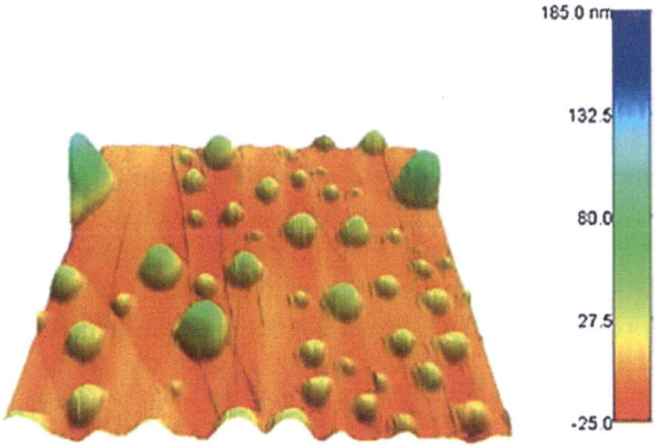


**Fig. 8.17** AFM image of nanobubbles on a hydrophobic surface. (*Credit* Xuehua Zhang, University of Melbourne) [107]

nucleation of dissolved gas. Nanobubbles can form easily by dissolving gases like argon, hydrogen, nitrogen, oxygen, and methane in bulk water and the size and density distribution of the bubbles made by dissolving hydrogen at 298 K [103].

These nanobubbles appear to be stable for very long times, of the order of days to months. AFM investigation [107] revealed that surface nanobubbles are kinetically stable and the liquid/gas interface is gas impermeable, which suggest that surface nanobubbles be stabilized by a layer which has a great diffusive resistance. Theoretical investigation [108] suggests that the limited gas diffusion through the water in the far field, the cooperative effect of nanobubble clusters, and the pinned contact line of the nanobubbles lead to the slow dissolution rate stabilize the nanobubbles that are stable for many hours or even up to days rather than the expected microseconds. Nanobubbles have a much larger contact angle and longer lifetime than predicted by classical theory.

Chan et al. [106] reported the dynamics of a three phase contact line moving over surface nanobubbles, polymeric droplets, and hydrophobic particles. The dynamics is distinct: across polymeric droplets the contact line quickly jumps and hydrophobic particles pin the contact line, while surface nanobubbles rapidly shrink once merging with the contact line [105].

According to the classical view of the air–water interface, such bubbles should not exist at all, since their small radius of curvature implies a high Laplace pressure inside the bubble that should drive gas diffusion across the interface and cause the bubbles to dissolve almost instantly [109]. The unexpected stability was thought an awkward but conspicuous instance of "surface misbehaviour". A consensus theory for nanobubble stability could not be derived from classical thermodynamics schemes [110]. The intrinsic behavior of the O:H–O bond is the key controlling the performance of nanobubbles. Therefore, the skin supersolidity is responsible for the unusual thermal and mechanical stability of the interface and surface nanobubbles [23].

### 8.5.4.2 Skin Supersolidity of Bubbles

A nanobubble is the inverse of a droplet of the same size. Molecular undercoordination not only produces the supersolidity but also disperses the quasisolid phase boundary, which is responsible for the thermodynamic behavior of water droplets and gas bubbles, particularly at the nanometer scale [23]. These systems of undercoordinated molecular dominance have far-reaching physical, chemical, and biological effects [111] because molecular undercoordination induced unusual bond–electron–phonon behavior, as discussed. They are hardly destroyed and thermally much more stable than bubbles at the millimeter scale [112] in addition to long lifetime [108]. Water nanodroplets and nanobubbles do follow the trend of $T_m$ elevation and $T_N$ depression because of the dominant fraction of undercoordinated skin molecules. Droplet size reduction raises the $\Theta_{DH}(\omega_H)$ and stretches the $\eta_H(T)$ curve and meanwhile, lowers the $\Theta_{DL}(\omega_L)$ and compresses the $\eta_L(T)$ curve, which disperses the extreme-density temperatures.

A bubble is just the inversion of a droplet; a hollow sphere like a soap bubble contains two skins—the inner and the outer. Both skins are in the supersolid phase and the volume fraction of such supersolid phase over the entire liquid-shell volume is much greater than simply a droplet. Therefore, bubbles demonstrate more significantly the supersolidity nature—elastic, hydrophobic, and less dense, which makes bubbles mechanically stronger and thermally more stable [33]. It is the strong polarization that prevents gas molecular diffusion across the skin of the bubble. The gas molecules are also strongly polarized. The supersolidity is a notion that appeals to old ideas about hydrophobic particles creating a highly ordered, ice-like hydration shell [113], or high-charge density ions acting as "structure-makers" in aqueous solution [114].

Bulk water prefers uniquely the mono-phase of fluctuating tetrahedral structure with the addition of the low-density supersolid skin that is a negligible fraction of the entire body. The skin is 0.04–0.12 nm thick, according to small-angle X-ray scattering and TIP4P/2005 force-field calculations at 7, 25 and 66 °C and atmospheric pressure [115]. However, the skin contains at least two molecular layers that are subject to undercoordination effect. The geometrical configuration of the skin molecules remains the bulk tetrahedral attribute, but the length scale changes with the CN loss—smaller size larger separation. H–O free radicals or dangling bonds are present at the surface; these are even shorter and stiffer than the skin H–O bond, with characteristic phonon frequency of 3650 $cm^{-1}$ or around. However, for a sufficiently small water droplet or a bubble, the volume of the skin and the core is compatible. Therefore, a small droplet holds a bi-phase structure of low-density skin and high-density core. It is now clear that water droplet at larger scale is mono-phase dominance, and at smaller bi-phase dominance in the core-shell configuration.

### 8.5.5 Quasisolid Versus Supersolid: Second Critical Point

Droplet size reduction has two efffrects. It not only increases the fraction of undercoordinated molecules but also decreases the effective CN of the skin molecules. The former determines the quantity and the latter the nature of the size effect. Size reduction raises the $\omega_H(\Theta_{DH})$ and lowers the $\omega_L(\Theta_{DL})$. The droplet-size-induced $\Theta_{Dx}(\omega_x)$ relaxation mediates the specific heat and hence disperses the extreme-density temperatures or boundaries of the quasisolid phase, which depresses the freezing point and elevates the melting temperature. Thus it is not surprising that nanodroplets and nanobubbles demonstrate the "superheating" and "supercooling" phenomena with their extent being droplet size dependent.

One needs to note that molecular undercoordination not only expands the temperature range of the quasisolid phase via an $\omega_x$ relaxation mechanism but also creates the hydrophobic, frictionless, less dense, viscoelastic, and supersolid phase. The quasisolid due to specific disparity is different from the supersolid due to molecular undercoordination. The temperature range of quasisolid phase also expands under tension and electrification. The supersolidity is more sensitive to the size and curvature of the droplet and presents throughout the full temperature range.

Quasisolid is subject to negative thermal expansion when transits from liquid into solid. Supersolid is polarized, much less dense. Both phases are gel-like and viscoelastic. Quasisolid cooling, liquid and solid heating further shorten the shortened H–O bond, which should enhance the supersolidity, but liquid heating depolarizes the O:H–O bond. Dual polarization is the key to the supersolid but cooling expansion is the key to quasisolid.

The superposition of the quasisolidity and supersolidity should be responsible for the anomalous second critical point $T_H$ and the Widom line as well as the simultaneous $\omega_H$ and $\omega_L$ compression softening, though the actual mechanism is subject to further verification.

These observations explain the thermodynamic behavior in the "no man's land" temperature regime of a water droplet that is subject to certainty of homogeneous monophase structure or the inhomogeneous low- and high-density structure [116–118]. The size dependence of the quasisolid phase boundary dispersion results in the drop of the critical temperature for the homogeneous crystallization and the rise of the temperature for melting.

## 8.6 Summary

Molecular undercoordination induced O:H–O bond cooperative relaxation and the associated dual polarization clarify the origin of the observed length scale, binding energy, phonon frequency, and the thermodynamics of water molecules with fewer-than-four nearest neighbors such as molecular clusters, hydration shells, snowflakes, and surface skins of liquid water. This notion also reconciled the anomalies of O–O expansion, O 1s electron densification and entrapment, surface electron polarization, high-frequency phonon stiffening, and the ice like and hydrophobic nature of such undercoordinated water molecules. Agreement between numerical calculations and experimental observations has verified our hypothesis and predictions:

(1) Molecular undercoordination not only disperses the quasisolid phase but also generates a hydrophobic, less dense, frictionless supersolid phase. Superposition of both abnormal phases makes water ice even stranger.
(2) Undercoordination—induced contraction of the H–O bond and inter-electron-pair repulsion driven O:H elongation dictate the unusual behaviour of water molecules in the nanoscale O:H–O networks and in the skin of water.
(3) The shortening of the H–O bond raises the density of the core and bonding electrons in the under-coordinated molecules, which in turn polarizes the nonbonding electron lone pairs on oxygen.
(4) The stiffening of the H–O bond increases the O 1s core-level shift, causes the blue-shift of the H–O phonon frequency, and elevates the melting point of water molecular clusters, surface skins, and ultrathin films of water.
(5) Undercoordinated water molecules could form an ice-like, low-density phase that is hydrophobic, stiffer, longer phonon lifetime, and thermally more stable than the bulk water.
(6) Undercoordination—induced O:H–O relaxation results in the supersolid phase that is elastic, hydrophobic, thermally more stable, and less dense, which dictates the unusual behaviour of water molecules at the boundary of the O:H–O networks or in the nanoscale droplet.
(7) H–O bond contraction densifies and entraps the core and bonding electrons; H–O bond stiffening shifts positively the O 1s energy, the $\omega_H$ and the $T_m$ of molecular clusters, surface skins, and ultrathin films of water.
(8) The dual polarization makes the skins hydrophobic, viscoelastic, and frictionless.
(9) H–O bond contraction elevates the melting point and O:H nonbond elongation depresses the freezing temperature of water droplets and bubbles of which the undercoordinated molecules become dominant.
(10) The H–O relaxation time is longer for nanodroplets and nanobubbles than it is in the bulk.

# References


1. T.D. Kuhne, R.Z. Khaliullin, Electronic signature of the instantaneous asymmetry in the first coordination shell of liquid water. Nat. commun. **4**, 1450 (2013)
2. V. Petkov, Y. Ren, M. Suchomel, Molecular arrangement in water: random but not quite. J. Phys.: Condens. Matter **24**(15), 155102 (2012)
3. F.N. Keutsch, R.J. Saykally, Water clusters: untangling the mysteries of the liquid, one molecule at a time. Proc. Natl. Acad. Sci. U.S.A. **98**(19), 10533–10540 (2001)
4. J.K. Gregory, D.C. Clary, K. Liu, M.G. Brown, R.J. Saykally, The water dipole moment in water clusters. Science **275**(5301), 814–817 (1997)
5. F.Y. Li, Y. Liu, L. Wang, J.J. Zhao, Z.F. Chen, Improved stability of water clusters (H2O) (30-48): a Monte Carlo search coupled with DFT computations. Theor. Chem. Acc. **131**(3), 1163 (2012)
6. Y. Huang, X. Zhang, Z. Ma, Y. Zhou, W. Zheng, J. Zhou, C.Q. Sun, Hydrogen-bond relaxation dynamics: resolving mysteries of water ice. Coord. Chem. Rev. **285**, 109–165 (2015)
7. K. Liu, J.D. Cruzan, R.J. Saykally, Water clusters. Science **271**(5251), 929–933 (1996)
8. R. Ludwig, Water: from clusters to the bulk. Angew. Chem. Int. Ed. **40**(10), 1808–1827 (2001)
9. A. Michaelides, K. Morgenstern, Ice nanoclusters at hydrophobic metal surfaces. Nat. Mater. **6**(8), 597–601 (2007)
10. L. Turi, W.S. Sheu, P.J. Rossky, Characterization of excess electrons in water-cluster anions by quantum simulations. Science **309**(5736), 914–917 (2005)
11. J.R.R. Verlet, A.E. Bragg, A. Kammrath, O. Cheshnovsky, D.M. Neumark, Observation of large water-cluster anions with surface-bound excess electrons. Science **307**(5706), 93–96 (2005)
12. N.I. Hammer, J.W. Shin, J.M. Headrick, E.G. Diken, J.R. Roscioli, G.H. Weddle, M.A. Johnson, How do small water clusters bind an excess electron? Science **306**(5696), 675–679 (2004)
13. C. Perez, M.T. Muckle, D.P. Zaleski, N.A. Seifert, B. Temelso, G.C. Shields, Z. Kisiel, B.H. Pate, Structures of cage, prism, and book isomers of water hexamer from broadband rotational spectroscopy. Science **336**(6083), 897–901 (2012)
14. T. Ishiyama, H. Takahashi, A. Morita, Origin of vibrational spectroscopic response at ice surface. J. Phys. Chem. Lett. **3**, 3001–3006 (2012)
15. F.Y. Li, L. Wang, J.J. Zhao, J.R.H. Xie, K.E. Riley, Z.F. Chen, What is the best density functional to describe water clusters: evaluation of widely used density functionals with various basis sets for ($H_2O$) (n) (n = 1-10). Theor. Chem. Acc. **130**(2–3), 341–352 (2011)
16. Y. Huang, X. Zhang, Z. Ma, G. Zhou, Y. Gong, C.Q. Sun, Potential paths for the hydrogen-bond relaxing with ($H_2O$)N cluster size. J. Phys. Chem. C **119**(29), 16962–16971 (2015)
17. C.Q. Sun, X. Zhang, J. Zhou, Y. Huang, Y. Zhou, W. Zheng, Density, elasticity, and stability anomalies of water molecules with fewer than four neighbors. J. Phys. Chem. Lett. **4**, 2565–2570 (2013)
18. C.Q. Sun, X. Zhang, X. Fu, W. Zheng, J.-L. Kuo, Y. Zhou, Z. Shen, J. Zhou, Density and phonon-stiffness anomalies of water and ice in the full temperature range. J. Phys. Chem. Lett. **4**, 3238–3244 (2013)
19. M. Erko, D. Wallacher, A. Hoell, T. Hauss, I. Zizak, O. Paris, Density minimum of confined water at low temperatures: a combined study by small-angle scattering of X-rays and neutrons. PCCP **14**(11), 3852–3858 (2012)
20. F. Mallamace, C. Branca, M. Broccio, C. Corsaro, C.Y. Mou, S.H. Chen, The anomalous behavior of the density of water in the range 30 K < T < 373 K. Proc. Natl. Acad. Sci. U.S.A. **104**(47), 18387–18391 (2007)

21. F.G. Alabarse, J. Haines, O. Cambon, C. Levelut, D. Bourgogne, A. Haidoux, D. Granier, B. Coasne, Freezing of water confined at the nanoscale. Phys. Rev. Lett. **109**(3), 035701 (2012)
22. R. Moro, R. Rabinovitch, C. Xia, V.V. Kresin, Electric dipole moments of water clusters from a beam deflection measurement. Phys. Rev. Lett. **97**(12), 123401 (2006)
23. X. Zhang, P. Sun, Y. Huang, Z. Ma, X. Liu, J. Zhou, W. Zheng, C.Q. Sun, Water nanodroplet thermodynamics: quasi-solid phase-boundary dispersivity. J. Phys. Chem. B **119** (16), 5265–5269 (2015)
24. C.Q. Sun, *Relaxation of the Chemical Bond*. Springer Series in Chemical Physics. 108, vol. 108 (Springer, Heidelberg, 2014), 807 pp
25. J. Rault, R. Neffati, P. Judeinstein, Melting of ice in porous glass: why water and solvents confined in small pores do not crystallize? Eur. Phys. J. B **36**(4), 627–637 (2003)
26. W.J. Huang, R. Sun, J. Tao, L.D. Menard, R.G. Nuzzo, J.M. Zuo, Coordination-dependent surface atomic contraction in nanocrystals revealed by coherent diffraction. Nat. Mater. **7**(4), 308–313 (2008)
27. D.D.D. Ma, C.S. Lee, F.C.K. Au, S.Y. Tong, S.T. Lee, Small-diameter silicon nanowire surfaces. Science **299**(5614), 1874–1877 (2003)
28. X.J. Liu, M.L. Bo, X. Zhang, L.T. Li, Y.G. Nie, H. TIan, Y. Sun, S. Xu, Y. Wang, W. Zheng, C.Q. Sun, Coordination-resolved electron spectrometrics. Chem. Rev. **115**(14), 6746–6810 (2015)
29. C.Q. Sun, Size dependence of nanostructures: impact of bond order deficiency. Prog. Solid State Chem. **35**(1), 1–159 (2007)
30. F. Matsui, T. Matsushita, Y. Kato, M. Hashimoto, K. Inaji, F.Z. Guo, H. Daimon, Atomic-layer resolved magnetic and electronic structure analysis of Ni thin film on a Cu(001) surface by diffraction spectroscopy. Phys. Rev. Lett. **100**(20), 207201 (2008)
31. M. Zhao, W.T. Zheng, J.C. Li, Z. Wen, M.X. Gu, C.Q. Sun, Atomistic origin, temperature dependence, and responsibilities of surface energetics: an extended broken-bond rule. Phys. Rev. B **75**(8), 085427 (2007)
32. H. Qiu, W. Guo, Electromelting of confined monolayer ice. Phys. Rev. Lett. **110**(19), 195701 (2013)
33. X. Zhang, Y. Huang, Z. Ma, Y. Zhou, W. Zheng, J. Zhou, C.Q. Sun, A common supersolid skin covering both water and ice. PCCP **16**(42), 22987–22994 (2014)
34. K.R. Wilson, R.D. Schaller, D.T. Co, R.J. Saykally, B.S. Rude, T. Catalano, J.D. Bozek, Surface relaxation in liquid water and methanol studied by x-ray absorption spectroscopy. J. Chem. Phys. **117**(16), 7738–7744 (2002)
35. U. Bergmann, A. Di Cicco, P. Wernet, E. Principi, P. Glatzel, A. Nilsson, Nearest-neighbor oxygen distances in liquid water and ice observed by x-ray Raman based extended x-ray absorption fine structure. J. Chem. Phys. **127**(17), 174504 (2007)
36. K.R. Wilson, B.S. Rude, T. Catalano, R.D. Schaller, J.G. Tobin, D.T. Co, R.J. Saykally, X-ray spectroscopy of liquid water microjets. J. Phys. Chem. B **105**(17), 3346–3349 (2001)
37. Y. Huang, X. Zhang, Z. Ma, Y. Zhou, J. Zhou, W. Zheng, C.Q. Sun, Size, separation, structure order, and mass density of molecules packing in water and ice. Sci. Rep. **3**, 3005 (2013)
38. E.G. Solveyra, E. de la Llave, V. Molinero, G. Soler-Illia, D.A. Scherlis, Structure, dynamics, and phase behavior of water in $TiO_2$ Nanopores. J. Phys. Chem. C **117**(7), 3330–3342 (2013)
39. R.M. Townsend, S.A. Rice, Molecular dynamics studies of the liquid–vapor interface of water. J. Chem. Phys. **94**(3), 2207–2218 (1991)
40. Y.I. Tarasevich, State and structure of water in vicinity of hydrophobic surfaces. Colloid J. **73**(2), 257–266 (2011)
41. J.P. Perdew, Y. Wang, Accurate and simple analytic representation of the electron-gas correlation-energy. Phys. Rev. B **45**(23), 13244–13249 (1992)
42. F. Ortmann, F. Bechstedt, W.G. Schmidt, Semiempirical van der Waals correction to the density functional description of solids and molecular structures. Phys. Rev. B **73**(20), 205101 (2006)

43. O. Marsalek, F. Uhlig, T. Frigato, B. Schmidt, P. Jungwirth, Dynamics of electron localization in warm versus cold water clusters. Phys. Rev. Lett. **105**(4), 043002 (2010)
44. S. Liu, J. Luo, G. Xie, D. Guo, Effect of surface charge on water film nanoconfined between hydrophilic solid surfaces. J. Appl. Phys. **105**(12), 124301–124304 (2009)
45. K.R. Siefermann, Y. Liu, E. Lugovoy, O. Link, M. Faubel, U. Buck, B. Winter, B. Abel, Binding energies, lifetimes and implications of bulk and interface solvated electrons in water. Nat. Chem. **2**, 274–279 (2010)
46. D.H. Paik, I.R. Lee, D.S. Yang, J.S. Baskin, A.H. Zewail, Electrons in finite-sized water cavities: hydration dynamics observed in real time. Science **306**(5696), 672–675 (2004)
47. R. Vacha, O. Marsalek, A.P. Willard, D.J. Bonthuis, R.R. Netz, P. Jungwirth, Charge transfer between water molecules as the possible origin of the observed charging at the surface of pure water. J. Phys. Chem. Lett. **3**(1), 107–111 (2012)
48. F. Baletto, C. Cavazzoni, S. Scandolo, Surface trapped excess electrons on ice. Phys. Rev. Lett. **95**(17), 176801 (2005)
49. M. Abu-Samha, K.J. Borve, M. Winkler, J. Harnes, L.J. Saethre, A. Lindblad, H. Bergersen, G. Ohrwall, O. Bjorneholm, S. Svensson, The local structure of small water clusters: imprints on the core-level photoelectron spectrum. J. Phys. B **42**(5), 055201 (2009)
50. K. Nishizawa, N. Kurahashi, K. Sekiguchi, T. Mizuno, Y. Ogi, T. Horio, M. Oura, N. Kosugi, T. Suzuki, High-resolution soft X-ray photoelectron spectroscopy of liquid water. PCCP **13**, 413–417 (2011)
51. B. Winter, E.F. Aziz, U. Hergenhahn, M. Faubel, I.V. Hertel, Hydrogen bonds in liquid water studied by photoelectron spectroscopy. J. Chem. Phys. **126**(12), 124504 (2007)
52. L. Belau, K.R. Wilson, S.R. Leone, M. Ahmed, Vacuum ultraviolet (VUV) photoionization of small water clusters. J. Phys. Chem. A **111**(40), 10075–10083 (2007)
53. L.-J. Zhang, J. Wang, Y. Luo, H.-P. Fang, J. Hu, A novel water layer structure inside nanobubbles at room temperature. Nucl. Sci. Tech. **25**, 060503 (2014)
54. K.R. Wilson, B.S. Rude, T. Catalano, R.D. Schaller, J.G. Tobin, D.T. Co, R. Saykally, X-ray spectroscopy of liquid water microjets. J. Phys. Chem. B **105**(17), 3346–3349 (2001)
55. J. Kim, I. Becker, O. Cheshnovsky, M.A. Johnson, Photoelectron spectroscopy of the `missing' hydrated electron clusters ($H_2O$) − n, n = 3, 5, 8 and 9: Isomers and continuity with the dominant clusters n = 6, 7 and ⩾11. Chem. Phys. Lett. **297**(1–2), 90–96 (1998)
56. J.V. Coe, S.M. Williams, K.H. Bowen, Photoelectron spectra of hydrated electron clusters vs. cluster size: connecting to bulk. Int. Rev. Phys. Chem. **27**(1), 27–51 (2008)
57. A. Kammrath, G. Griffin, D. Neumark, J.R.R. Verlet, Photoelectron spectroscopy of large (water)[sub n][sup −] (n = 50–200) clusters at 4.7 eV. J. Chem. Phys. **125**(7), 076101 (2006)
58. C.A. Silvera Batista, R.G. Larson, N.A. Kotov, Nonadditivity of nanoparticle interactions. Science **350**(6257) (2015)
59. G. Pyrgiotakis, A. Vasanthakumar, Y. Gao, M. Eleftheriadou, E. Toledo, A. DeAraujo, J. McDevitt, T. Han, G. Mainelis, R. Mitchell, P. Demokritou, Inactivation of foodborne microorganisms using engineered water nanostructures (EWNS). Environ. Sci. Technol. **49** (6), 3737–3745 (2015)
60. X.X. Yang, J.W. Li, Z.F. Zhou, Y. Wang, L.W. Yang, W.T. Zheng, C.Q. Sun, Raman spectroscopic determination of the length, strength, compressibility, Debye temperature, elasticity, and force constant of the C–C bond in graphene. Nanoscale **4**(2), 502–510 (2012)
61. X.J. Liu, L.K. Pan, Z. Sun, Y.M. Chen, X.X. Yang, L.W. Yang, Z.F. Zhou, C.Q. Sun, Strain engineering of the elasticity and the Raman shift of nanostructured TiO(2). J. Appl. Phys. **110**(4), 044322 (2011)
62. U. Buck, F. Huisken, Infrared spectroscopy of size-selected water and methanol clusters. Chem. Rev. **100**(11), 3863–3890 (2000)
63. K.E. Otto, Z. Xue, P. Zielke, M.A. Suhm, The Raman spectrum of isolated water clusters. Phys. Chem. Chem. Phys. (2014). doi:10.1039/c3cp54272f
64. T.F. Kahan, J.P. Reid, D.J. Donaldson, Spectroscopic probes of the quasi-liquid layer on ice. J. Phys. Chem. A **111**(43), 11006–11012 (2007)

65. J. Ceponkus, P. Uvdal, B. Nelander, Water tetramer, pentamer, and hexamer in inert matrices. J. Phys. Chem. A **116**(20), 4842–4850 (2012)
66. Y.R. Shen, V. Ostroverkhov, Sum-frequency vibrational spectroscopy on water interfaces: polar orientation of water molecules at interfaces. Chem. Rev. **106**(4), 1140–1154 (2006)
67. V. Buch, S. Bauerecker, J.P. Devlin, U. Buck, J.K. Kazimirski, Solid water clusters in the size range of tens-thousands of $H_2O$: a combined computational/spectroscopic outlook. Int. Rev. Phys. Chem. **23**(3), 375–433 (2004)
68. P.C. Cross, J. Burnham, P.A. Leighton, The Raman spectrum and the structure of water. J. Am. Chem. Soc. **59**, 1134–1147 (1937)
69. Q. Sun, The Raman OH stretching bands of liquid water. Vib. Spectrosc. **51**(2), 213–217 (2009)
70. S. Hirabayashi, K.M.T. Yamada, Infrared spectra and structure of water clusters trapped in argon and krypton matrices. J. Mol. Struct. **795**(1–3), 78–83 (2006)
71. C.C. Pradzynski, R.M. Forck, T. Zeuch, P. Slavicek, U. Buck, A fully size-resolved perspective on the crystallization of water clusters. Science **337**(6101), 1529–1532 (2012)
72. X. Wei, P. Miranda, Y. Shen, Surface vibrational spectroscopic study of surface melting of ice. Phys. Rev. Lett. **86**(8), 1554–1557 (2001)
73. J. Ceponkus, P. Uvdal, B. Nelander, Intermolecular vibrations of different isotopologs of the water dimer: experiments and density functional theory calculations. J. Chem. Phys. **129**(19), 194306 (2008)
74. J. Ceponkus, P. Uvdal, B. Nelander, On the structure of the matrix isolated water trimer. J. Chem. Phys. **134**(6), 064309 (2011)
75. S.A. Deshmukh, S.K. Sankaranarayanan, D.C. Mancini, Vibrational spectra of proximal water in a thermo-sensitive polymer undergoing conformational transition across the lower critical solution temperature. J. Phys. Chem. B **116**(18), 5501–5515 (2012)
76. Y. Huang, X. Zhang, Z. Ma, Y. Zhou, G. Zhou, C.Q. Sun, Hydrogen-bond asymmetric local potentials in compressed ice. J. Phys. Chem. B **117**(43), 13639–13645 (2013)
77. P. Maksyutenko, T.R. Rizzo, O.V. Boyarkin, A direct measurement of the dissociation energy of water. J. Chem. Phys. **125**(18), 181101 (2006)
78. S.A. Harich, D.W.H. Hwang, X. Yang, J.J. Lin, X. Yang, R.N. Dixon, Photodissociation of $H_2O$ at 121.6 nm: A state-to-state dynamical picture. J. Chem. Phys. **113**(22), 10073–10090 (2000)
79. S.A. Harich, X. Yang, D.W. Hwang, J.J. Lin, X. Yang, R.N. Dixon, Photodissociation of D2O at 121.6 nm: a state-to-state dynamical picture. J. Chem. Phys. **114**(18), 7830–7837 (2001)
80. M.L. Lakhanpal, B.R. Puri, Boiling point of capilary-condensed water. Nature **172**(4385), 917–917 (1953)
81. L. Li, Y. Kazoe, K. Mawatari, Y. Sugii, T. Kitamori, Viscosity and wetting property of water confined in extended nanospace simultaneously measured from highly-pressurized meniscus motion. J. Phys. Chem. Lett. 2447–2452 (2012)
82. K. Xu, P.G. Cao, J.R. Heath, Graphene visualizes the first water Adlayers on mica at ambient conditions. Science **329**(5996), 1188–1191 (2010)
83. D. Xu, K.M. Liechti, K. Ravi-Chandar, Mechanical probing of icelike water monolayers. Langmuir **25**(22), 12870–12873 (2009)
84. P.B. Miranda, L. Xu, Y.R. Shen, M. Salmeron, Icelike water monolayer adsorbed on mica at room temperature. Phys. Rev. Lett. **81**(26), 5876–5879 (1998)
85. F. McBride, G.R. Darling, K. Pussi, A. Hodgson, Tailoring the structure of water at a metal surface: a structural analysis of the water bilayer formed on an alloy template. Phys. Rev. Lett. **106**(22), 226101 (2011)
86. A. Hodgson, S. Haq, Water adsorption and the wetting of metal surfaces. Surf. Sci. Rep. **64** (9), 381–451 (2009)
87. S. Meng, E.G. Wang, S.W. Gao, Water adsorption on metal surfaces: a general picture from density functional theory studies. Phys. Rev. B **69**(19), 195404 (2004)

88. C. Wang, H. Lu, Z. Wang, P. Xiu, B. Zhou, G. Zuo, R. Wan, J. Hu, H. Fang, Stable liquid water droplet on a water monolayer formed at room temperature on ionic model substrates. Phys. Rev. Lett. **103**(13), 137801–137804 (2009)
89. J.C. Johnston, N. Kastelowitz, V. Molinero, Liquid to quasicrystal transition in bilayer water. J. Chem. Phys. **133**(15) (2010)
90. A. Falenty, T.C. Hansen, W.F. Kuhs, Formation and properties of ice XVI obtained by emptying a type sII clathrate hydrate. Nature **516**(7530), 231–233 (2014)
91. P.G. Debenedetti, H.E. Stanley, Supercooled and glassy water. Phys. Today **56**(6), 40–46 (2003)
92. O. Bjorneholm, F. Federmann, S. Kakar, T. Moller, Between vapor and ice: free water clusters studied by core level spectroscopy. J. Chem. Phys. **111**(2), 546–550 (1999)
93. S. Strazdaite, J. Versluis, E.H. Backus, H.J. Bakker, Enhanced ordering of water at hydrophobic surfaces. J. Chem. Phys. **140**(5), 054711 (2014)
94. E. Bittman. Supercooling water. 2014
95. J. Kim, D. Won, B. Sung, W. Jhe, Observation of Universal solidification in the elongated water nanomeniscus. J. Phys. Chem. Lett. 737–742 (2014)
96. S.H. Khan, G. Matei, S. Patil, P.M. Hoffmann, Dynamic solidification in nanoconfined water films. Phys. Rev. Lett. **105**(10), 106101 (2010)
97. G. Zhao, Q. Tan, L. Xiang, D. Cai, H. Zeng, H. Yi, Z. Ni, Y. Chen, Structure and properties of water film adsorbed on mica surfaces. J. Chem. Phys. **143**(10), 104705 (2015)
98. H.-S. Tan, I.R. Piletic, R.E. Riter, N.E. Levinger, M.D. Fayer, Dynamics of water confined on a nanometer length scale in reverse micelles: ultrafast infrared vibrational echo spectroscopy. Phys. Rev. Lett. **94**(5), 057405 (2005)
99. M. Antognozzi, A.D.L. Humphris, M.J. Miles, Observation of molecular layering in a confined water film and study of the layers viscoelastic properties. Appl. Phys. Lett. **78**(3), 300 (2001)
100. A. Agarwal, W.J. Ng, Y. Liu, Principle and applications of microbubble and nanobubble technology for water treatment. Chemosphere **84**(9), 1175–1180 (2011)
101. K. Ebina, K. Shi, M. Hirao, J. Hashimoto, Y. Kawato, S. Kaneshiro, T. Morimoto, K. Koizumi, H. Yoshikawa, Oxygen and air nanobubble water solution promote the growth of plants, fishes, and mice. Plos One **8**(6) (2013)
102. X. Zhang, D. Lohse, Perspectives on surface nanobubbles. Biomicrofluidics **8**(4), 041301 (2014)
103. S.H. Oh, J.G. Han, J.-M. Kim, Long-term stability of hydrogen nanobubble fuel. Fuel **158**, 399–404 (2015)
104. M. Takahashi, Base and technological application of micro-bubble and nano-bubble. Mater. Integr. **22**, 2–19 (2009)
105. Y. Liu, X. Zhang, A unified mechanism for the stability of surface nanobubbles: contact line pinning and supersaturation. J. Chem. Phys. **141**(13), 134702 (2014)
106. C.U. Chan, L.Q. Chen, M. Arora, C.D. Ohl, Collapse of surface nanobubbles. Phys. Rev. Lett. **114**(11), 114505 (2015)
107. S. Wang, M. Liu, Y. Dong, Understanding the stability of surface nanobubbles. J. Phys.: Condens. Matter **25**(18), 184007 (2013)
108. J.H. Weijs, D. Lohse, Why surface nanobubbles live for hours. Phys. Rev. Lett. **110**(5), 054501 (2013)
109. V.S.J. Craig, Very small bubbles at surfaces-the nanobubble puzzle. Soft Matter **7**(1), 40–48 (2011)
110. P. Ball, Nanobubbles are not a superficial matter. Chem. Phys. Chem. **13**(8), 2173–2177 (2012)
111. M. Chaplin. Water structure and science: http://www.lsbu.ac.uk/water/
112. J.W. Tyrrell, P. Attard, Images of nanobubbles on hydrophobic surfaces and their interactions. Phys. Rev. Lett. **87**(17), 176104 (2001)

113. H.S. Frank, M.W. Evans, Free volume and entropy in condensed systems III. Entropy in binary liquid mixtures; partial molal entropy in dilute solutions; structure and thermodynamics in aqueous electrolytes. J. Chem. Phys. **13**(11), 507–532 (1945)
114. K.D. Collins, M.W. Washabaugh, The Hofmeister effect and the behaviour of water at interfaces. Q. Rev. Biophys. **18**(04), 323–422 (1985)
115. C. Huang, K.T. Wikfeldt, D. Nordlund, U. Bergmann, T. McQueen, J. Sellberg, L.G.M. Pettersson, A. Nilsson, Wide-angle X-ray diffraction and molecular dynamics study of medium-range order in ambient and hot water. PCCP **13**(44), 19997–20007 (2011)
116. A.K. Soper, Supercooled water continuous trends. Nat. Mater. **13**(7), 671–673 (2014)
117. J.A. Sellberg, C. Huang, T.A. McQueen, N.D. Loh, H. Laksmono, D. Schlesinger, R.G. Sierra, D. Nordlund, C.Y. Hampton, D. Starodub, D.P. DePonte, M. Beye, C. Chen, A.V. Martin, A. Barty, K.T. Wikfeldt, T.M. Weiss, C. Caronna, J. Feldkamp, L.B. Skinner, M.M. Seibert, M. Messerschmidt, G.J. Williams, S. Boutet, L.G. Pettersson, M.J. Bogan, A. Nilsson, Ultrafast X-ray probing of water structure below the homogeneous ice nucleation temperature. Nature **510**(7505), 381–384 (2014)
118. D. Paschek, R. Ludwig, Advancing into water's "no man's land": two liquid states? Angew. Chem. Int. Ed. Engl. **53**(44), 11699–11701 (2014)

# Chapter 9
# Superlubricity of Ice

- *Electrostatic repulsivity (locally pinned dipoles) and phononic elasticity (high-magnitude and low-frequency vibration) ensure the superlubricity at contacts with soft phonons and dipoles involvement.*
- *Molecular undercoordination associated elastic, ice like, less dense and polarized supersolid skin lubricates ice.*
- *O:H phonon resonance and interface Regelation enhance ice on ice friction.*
- *Pressure melting and friction heating are insufficient to create a lubricant skin—supersolidity instead of quasiliquidity dominance.*

**Abstract** Superlubricity means non-sticky and frictionless when two bodies are set contacting motion. Although this occurrence has been extensively investigated since 1859 when Faraday firstly proposed a quasiliquid skin on ice, the mechanism behind the superlubricity remains debating. This chapter features a consistent understanding of the superlubricity pertaining to the slipperiness of ice, self-lubrication of dry solids, and aqueous lubricancy from the perspective of skin bond-electron-phonon adaptive relaxation. The presence of nonbonding electron polarization, atomic or molecular undercoordination, and solute ionic electrification of the hydrogen bond as an addition, ensures the superlubricity. Nonbond vibration creates soft phonons of high magnitude and low frequency with extraordinary adaptivity and recoverability of deformation. Molecular undercoordination shortens the covalent bond with local charge densification, which in turn polarizes the nonbonding electrons making them localized dipoles. The locally pinned dipoles provide force opposing contact, mimicking maglev and hovercraft.

## 9.1 Challenge: Why Is Ice so Slippery?

Ice is most slippery of ever known at temperatures even below its melting limit at −22 °C under 2000 atmospheric pressure (200 MPa) pressure. All sorts of surfaces can get slick and slippery if ice and snow abound in winter weather. Slipperiness of snow and ice forms the platform of Winter Olympic Games and many kinds of outdoor entertainments in winter like the jealous skating on ice, see Fig. 9.1. However, slipperiness of snow and ice has two sides effect. If you are a driver, this



C.Q. Sun and Y. Sun, *The Attribute of Water*, Springer Series in Chemical Physics 113, DOI 10.1007/978-981-10-0180-2_9

**Fig. 9.1** Is ice covered with a quasiliquid sheet or a supersolid skin? **a** An early 1820's print for the ice-skating scene (*Credit* W. Belen, Free Wikipedia). **b** The friction coefficient of steel-pin on ice-disc under $10^{-10}$ Pa vacuum condition shows linear temperature dependence in the regime of bulk ice [7]. The *inset* shows friction coefficients in the quasisolid phase regime (258–273 K) [8] under different conditions [9] (Reprinted with permission from [7].)

is quite troublesome. Ice and snow can make driving treacherous. Slipperiness of ice is one of the unanswered puzzles since 1859 when Faraday [1] proposed that a quasiliquid kin serves as the lubricant. Debating is still going on with the following possible mechanisms:

(1) Pressure melting creates the quasiliquid lubricant [2, 3].
(2) Friction heating melts ice [4].
(3) Quasiliquid skin forms due to molecular undercoordination [5].
(4) Low-frequency and high-magnitude vibrations adapt sliding motion [6].

## 9.2 Clarification: Supersolid Lubricant Skin

Instead of a quasiliquid layer, friction heating, or pressure melting, ice is covered with a supersolid skin that is elastic, polarized, less dense, and thermally more stable [10–12], as illustrated in Fig. 9.2:

(1) Molecular undercoordination shortens and stiffens the H–O bond, and meanwhile, lengthens and softens the O:H nonbond with dual polarization of electron lone pairs on oxygen ions (H–O contraction polarizes the lone pair electrons in the first round and then enhances O–O repulsion in the second).
(2) H–O bond stiffening raises the melting point from 273 to 310 K and the H–O phonon frequency from 3200 to 3450 $cm^{-1}$; O–O elongation lowers the local mass density from 1.0 to 0.75 $g/cm^3$.

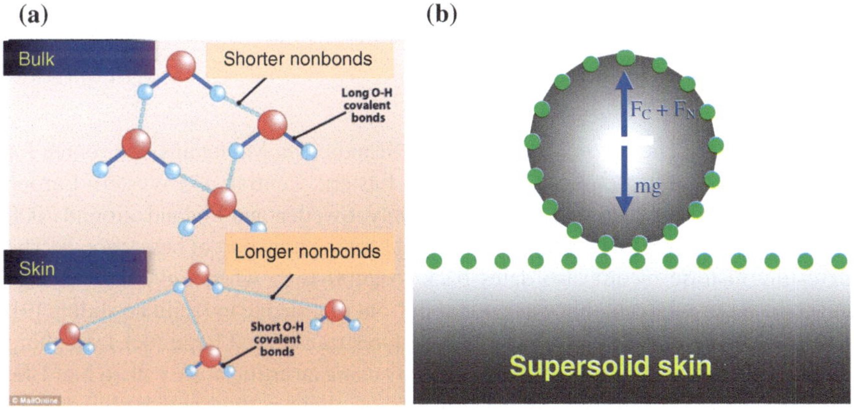


**Fig. 9.2** Elastic Coulomb repulsion makes ice a supersolid skin capable of superlubricity. **a** Undercoordination of $H_2O$ molecules reduces their sizes but enlarges their separations, which softens the O:H nonbond by lowering the frequency and enhancing the amplitude of O:H vibration [11]. **b** The softer O:H springs attached with dipoles not only levitate the object on it but also recover readily from deformation, which make the supersolid skin elastic and slippery. *Arrows* in (**b**) denote the force acting on the load: $F_N + F_C - mg = 0$, with $F_N$, $F_C$, and mg being the normal force, the Coulomb levitation force, and the weight of the object, respectively. *Green dots* represent dipoles associated with O:H soft springs and surrounding the rim of the load. The skin of any subject is subject to global BOLS quantum entrapment and subjective polarization [12], particular at the nanometer scale, which makes the subject nonadaptivity [13]

(3) The O:H nonbond softening and the O–O dual polarization enhance the viscoelasticity and hydrophobicity of the skin.
(4) Interface Coulomb repulsion between the locally pinned dipoles and the skin O:H phonon elasticity lower the friction at contacts, making ice slippery, which is the same in principle to maglev and hovercraft, see Fig. 9.3.

**Fig. 9.3** Maglev train and Hovercraft move frictionlessly because of the interface noncontact repulsion of magnetic force and the air ejection (Public domain)

# 9.3 History Background

## *9.3.1 Wonders of Ice Friction*

The first report of sliding on ice comes from Scandinavia Mountains, the source for repeated glaciation that covered much of the Eastern, Central, and Western Europe with a particular emphasis on Denmark, Norway, Sweden and Finland, around 7000 B.C. Rock carvings illustrate the use of a sledge for the transport of heavy goods. The interesting historic record also dates back to 2400 B.C. when Egyptian carvings employed water lubricant that was poured in front of a sledge to facilitate sliding [9].

In the 15th century [14], Chinese architectures transported large rocks weighing hundreds of tons to the site from 70 km away by using an artificial ice path to build the Forbidden City, an imperial palace, consists of about a thousand buildings, see Fig. 9.4, for a typical stone sculpture. The artificial path was made by pouring water from wells dig aside the path, or along the artificially Great Transportation River, in winter. This kind of ice path overcame limitations of other transport means available at that point of time. For instance, using wooden rollers would require creating a smooth surface on tricky, winding roads. Wheeled carriages would not have been able to transport such heavy blocks, even with the technology of the late 1500s.

Understanding the mechanism of friction on ice is particularly important in a broad field of applications, such as motorized vehicle traffic in winter road conditions, glacial movements, and cargo transportation through northern sea ways, design of offshore structures and ice breakers, and ice sports. High friction on ice is desired for motorized vehicle traffic in winter road conditions and the grip of shoe soles on ice to avoid accidents. However, in the field of cargo transportation through northern sea ways and the design of offshore structures low friction materials are desired to limit maintenance and operation costs, e.g., 70 % of the power of an ice breaker ship is consumed to overcome ice friction.

Furthermore, friction and its consequences are of great concern from both a sustainability and quality-of-life point of view, and the economic impact is massive. Indeed, by one estimate, improved attention to tribology (the study of friction, lubrication, and wear) would save developed countries up to 1.6 % of their gross national product, or close to $225 billion annually in the USA alone [15]. Therefore, understanding of ice friction would help in regulating the friction coefficient is solid dry friction and aqueous solute lubrication.

**Fig. 9.4** The Large Stone Carving is the heaviest stone in the Forbidden City in Beijing. It weighed more than 300 tons when it was first transported to the site between 1407 and 1420 (*Credit* W. Buss and De Agostini)

#### 9.3.1.1 Factors Dominating Friction

In 1785, Charles Augustin Coulomb examined five main factors for frictional resistance that involves the nature of materials in contact and their surface coatings, the extent of the contacting surface area, the normal pressure, the length of time that surfaces stay in contact, and the frictional behavior under vacuum as well as under varying ambient conditions namely temperature and humidity [9]. Besides, surface roughness, surface structure, wettability, sliding velocity, and thermal conductivity affect the friction behavior of ice.

| | |
|---|---|
|  | Charles-Augustin de Coulomb (14 June 1736–23 August 1806), a French physicist, was best known for developing Coulomb's law, the definition of the electrostatic force of attraction and repulsion, and on friction. He examined multiple factors affecting the friction coefficient (Free Wikipedia) |
|  | Captain Robert Falcon Scott (6 June 1868–29 March 1912) was a British Royal Navy officer and explorer who led two expeditions to the Antarctic regions. He noted skiing easily at −30 °C though the snow surface is sand-like at −46 °C (Free Wikipedia) |
|  | Frank Philip Bowden FRS (2 May 1903–3 September 1968) was an Australian physicist and proposed friction heating mechanism for slipperiness of ice. In 1956, Bowden was made a CBE and awarded the Rumford Medal of the Royal Society "In recognition of his distinguished work on the nature of friction" |

#### 9.3.1.2 Two-Regime Friction Coefficient

Figure 9.2b shows that the friction coefficient of steel-pin on ice-disc in $10^{-10}$ Pa vacuum depends linearly on temperature in the regime of solid bulk phase [7] but the coefficient (inset) exhibits insignificant temperature dependence in the bulk quasisolid phase regime [8] under different conditions [9]. However, the kinetic friction coefficient of sea ice on sea ice varies from 0.05 (at −20 °C) to 0.5 (at −2 °C) [16]. These temperature trends indicate the intrinsic behavior of ice at different temperature regimes or its structure phases, which is closely related to the manner of O:H–O bond relaxation in different regimes.

#### 9.3.1.3 Ice on Ice: Pressure, Temperature, and Velocity

One may expect that the friction coefficient of ice sliding on ice be lower, but observed the opposite. The friction coefficient is sensitive to many factors such as pressure, temperature, and the velocity of sliding. Sukhorukov and Loset [16] examined the effects of sliding velocity (6–105 mm/s), air temperatures (−2 to −20 °C), normal load (300–2000 N), presence of sea water in the interface, and ice grain orientation with respect to the sliding direction on the friction coefficient of sea ice on itself. The kinetic friction coefficient of sea ice on sea ice varies from 0.05 (at −20 °C) to 0.5 (at −2 °C), regardless of the presence of sea water in the sliding interface. The friction coefficient is independent of the velocity when sliding occurs between natural ice surfaces. As the contacting surfaces became smoother, the kinetic friction coefficient started to depend on the velocity, as predicted by existing ice friction models [9].

Kennedy and coworkers [17] reported that the friction coefficient μ of ice on ice varies with sliding velocity, 0.03 at $5 \times 10^{-2}$ m/s and 0.58 at $5 \times 10^{-7}$ m/s within the temperature range of −3 and −40 °C under normal pressure of 0.007–1.0 MPa. Generally, the μ decreases with increasing velocity and temperature, but it is relatively insensitive to both pressure and grain size. The friction coefficients for freshwater and saltwater ice were almost indistinguishable at higher temperatures (−3 and −10 °C), but saline ice lowers the friction coefficient at lower temperatures with unknown reasons.

Schulson and Fortt [18] measured the friction coefficient of freshwater polycrystalline ice sliding slowly ($5 \times 10^{-8}$ to $1 \times 10^{-3}$ m/s) upon itself at temperatures from −175 to −10 °C under low normal stresses (≤98 kPa). The coefficient of kinetic friction of smooth surfaces varies from $\mu_k = 0.15$ to 0.76 and, at elevated temperatures (≥−50 °C), which exhibits both lower velocity strengthening (<$10^{-5}$ to $10^{-4}$ m/s) and higher velocity weakening of the friction. At intermediate temperatures of −100 and −140 °C, the kinetic coefficient appears to not exhibit significant dependence upon velocity. However, at the low temperature of −175 °C the coefficient of kinetic friction exhibits moderate velocity strengthening at both the lowest and the highest velocities but velocity independence over the range of intermediate velocities.

### 9.3.2 *Quasiliquid Skin Notion*

Scientists have heavily debated the seemingly simple question of why ice is slippery since 1850 when Faraday [1] firstly proposed that a liquid or a quasiliquid layer serves as the lubricant making ice slippery after his experiment: he pressed two cubes of ice against each other submerged in 0 °C water, and they fused together. Faraday argued that the liquid layers remain on a surface but they froze when they were at the interface. He also used this mechanism to explain the observation of ice regelation—ice melts under compression and freezes again when the pressure is relieved [19].

Intuition indicates that liquids are mobile whereas solid surfaces are relatively rigid. Asking why ice is slippery is thus roughly equivalent to asking how a liquid or quasiliquid layer can occur on the ice surface in the first place. The presence of liquid reduces friction between solids, which is why water spilled on a kitchen floor or rainwater on asphalt or concrete can create the same kinds of hazards for walkers and drivers that ice can. Therefore, in order to make that solid slippery a liquid must form on it that allows skates to slip. Therefore, Fraday's proposal of quasiliquid skin was deemed true up to date [10].

How is that thin layer of liquid water going to appear if ice's temperature is well under its melting point? Rosenberg [20], an emeritus professor of chemistry at Lawrence University in Appleton, Wisconsin, featured in *Physics Today* 2005 on the history and progress on "why ice is slippery" in terms of pressure melting [2], frictional heating [4], and intrinsic quasiliquid forming or premelting [20].

#### 9.3.2.1 Pressure Melting

The conventional explanation, pressure melting, was suggested by James Thomson [2] in 1850 and lately experimentally approved by his brother, William Thomson, Later Lord Kevin [3], in 1850 as a consequence of the higher density of liquid water relative to ice. James Thomason calculated that a pressure of 46.6 MPa would lower the melting point by −3.5 °C. Kelvin verified that result experimentally. However, James was not able to explain how hockey players and figure skaters were able to slide at temperatures below −35 °C at which temperature no pressure melting takes place. Skating is possible at very cold from around −30 °C, so how is it possible for skaters to skate at this very cold temperature? The player's own weight would not be able to pressure the ice enough to drop the melting temperature of ice and create a thin layer of liquid water. The pressure-melting explanation also fails to explain why someone wearing flat-bottom shoes, with a much greater surface area that exerts even less pressure on ice, can also slip on the ice.

**Fig. 9.5** Ice skating lowers the melting point by only 0.24 °C that is insufficient to provide lubricant liquid (Galina Barskaya/Dreamstime.com). With negligible friction or compression, a penguin can hardly stand still on a slightly sloped skin of ice (BBC online, Jun 24, 2013)

The optimum temperature for figure skating is −5.5 °C and for hockey, it is −9 °C; figure skaters prefer slower, softer ice for their landings, whereas hockey players exploit the harder, faster ice. Indeed, skating is possible in climates as cold as −30 °C and skiing waxes are commercially available for such low temperatures. In his 1910 account of his last expedition to the South Pole, Robert Falcon Scott [20] tells of skiing easily at −30 °C though the snow surface is sand-like at −46 °C. But surprisingly, even with little evidence in its favor, pressure melting was dominant for more than a century and still remains as the dominant explanation of the slipperiness of ice in many text books.

Ice skating is given as an example of regelation to create liquid lubricating ice; however the pressure required is much greater than the weight of a skater. Additionally, regelation does not explain how one can skate at temperatures below the limit of −22 °C. If the contacting area of the skate to ice is $150 \times 10^{-6}$ $m^2$ (1 mm wide and 150 mm long) and the skater weighs 500 N, the pressure applied will be 3.3 MPa. As the melting point of ice falls by 0.0072 °C for each additional atm (0.1 MPa) of pressure applied, the melting point will drop by 0.24 °C only. Therefore, skating provides insufficient pressure for melting ice (Fig. 9.5).

#### 9.3.2.2 Friction Heating

Bowden and Hughes [4] proposed in 1939 the frictional heating as an alternative mechanism. Friction is the force that generates heat whenever two objects slide against each other. If you rub your hands together, you can warm them up. When a skate moves on the surface of ice, the friction between the skate and the ice generates heat that melts the outermost layer of ice. Bowden and Hughes did an experiment at a research station in Switzerland to maintain temperatures below −3 °C using solid $CO_2$ and liquid air. Using surfaces of wood and metal, they measured the effects of static and kinetic friction on ice melting. They concluded that frictional heating was responsible for melting ice. Although frictional heating may answer why ice is slippery when moving, this theory does not explain why ice can be so slippery even for someone, such as a penguin, standing still on it, see Fig. 9.5.

### 9.3.3 *Quasiliquid Skin Formation*

#### 9.3.3.1 Surface Premelting

Faraday [1] suggested that a film of water on ice would remain liquid on the surface of a single piece but the water layer would freeze when placed between two pieces of ice. However, he was not able to reason why the liquid layer forms at the molecular level without pressure melting or friction heating being presented.

In 1949, Gurney [5] suggested that an intrinsic liquid film forms on ice. Gurney hypothesized that molecules, inherently unstable at the surface due to the lack of molecules above them, migrate into the solid until the surface becomes stable, which prompts the formation of a liquid phase. If appreciable atomic migration takes place, the surface of a crystalline solid melts, like surface melting point depression happened to most substance [21], and the solid is covered with a thin liquid film under a tension force greater than that of the corresponding supercooled liquid. This tension force is numerically equal to the free energy of the surface. If such a solid is subsequently cooled to a temperature at which atomic migration effectively ceases, it will have frozen its surface with a tension force corresponding to thermal equilibrium at some higher temperatures.

#### 9.3.3.2 Interface Phonons and Electrons

At the atomic scale, Krim [6, 15] proposed that interface atomic lattice vibration and the electronic charge play significant roles in friction. When atoms close one surface are set into motion, atoms in both surfaces create waves in terms of phonons. The amount of mechanical energy transformed into phonons depends on the sliding substances. Solids are much like musical instruments in the way that they can vibrate only at certain distinct frequencies, so the amount of mechanical energy consumed will depend on the frequencies actually excited. If the "plucking" action of the atoms in the opposite surface resonates with one of the frequencies of the other, then friction arises. If it is not resonant with any of the other surface's own frequencies, then sound waves are effectively generated.

On the other hand, the smaller the resulting amplitude of vibration is, the greater the friction will be from the "rubbing" action of the film sliding about on the substrate. For insulating surfaces, friction arises from the attraction of unlike charges attached to the surfaces, like a balloon being rubbed on hair and left to cling to a wall. In 1989, Krim and coworkers [6] found the friction coefficient of Krypton films on crystalline gold surfaces is lower when dry; adding a liquid film raises the coefficient by five times, instead. Applying electric field cross the contacting interface can also affect the coefficient of friction [22].

#### 9.3.3.3 Diffraction Examination of Premelting

Since 1960s, a variety of experimental approaches, performed under various conditions, has been brought to bear on the premelting problem to determine the temperature range and thickness of the postulated quasiliquid layer. In 1969, Orem and Adamson [23] found that impurity adsorption promotes surface melting. Physical adsorption of simple hydrocarbon vapors on ice creates a liquid-like layer on the surface of ice. The adsorption of *n*-hexane on the surface of ice can form liquid-like layer at temperatures above −35 °C. These researchers interpreted their results as indicating that the onset of ice's surface premelting is at −35 °C. In the 1990 s, chemistry Nobel laureate Molina and coworkers [24] attributed the adsorption of hydrochloric acid on polar stratospheric clouds to the existence of a liquid-like layer on ice, which plays a role in the destruction of ozone.

Nuclear magnetic resonance (NMR) spectroscopy measurements [25] suggested liquid layer formation on ice: below the melting point there is a narrow absorption line, not the broad line one would expect from a periodic solid. Molecules at the surface between 20 and 0 °C rotate at a frequency five orders of magnitude greater than those in bulk ice and about 1/25 as fast as those in liquid water. The self-diffusion coefficient of molecules is two orders of magnitude larger than that in bulk ice. Using proton backscattering, Golecki and Jaccard [26] found in 1977 that surface vibrations of the oxygen atoms are roughly 3.3 times the amplitude of their bulk value, and estimated an amorphous layer 10 times thicker than what NMR measurements had estimated. But, unlike NMR, the proton backscattering measurements were made under high vacuum, a condition markedly different from the finite vapor pressures at which surface melting typically occurs. Molecules perform differently under the ambient vapor pressure.

X-ray diffraction (XRD) study [27] conducted in 1987 suggested that the intermolecular distance on the ice surface is shorter than that of ice's bulk interior and slightly shorter than it is in liquid water. However, X-Ray absorption spectroscopy revealed that the skin O–O distance for water expands by 8.9 % to 2.965 Å from the bulk value of 2.70 Å [28]. In the mid-1990s, Dosch and coworkers [29] found a liquid-like layer on the different crystallographic ice surfaces between −13.5 and 0 °C. The surface layer exhibits rotational disorder with intact long-range positional order well below the surface melting temperature. At the surface-melting temperature, a completely disordered layer exists on the surface above the rotationally disordered layer.

Low-energy electron diffraction (LEED) experiments conducted in 1996 by Somorjai and coworker [30] also suggested the presence of quasiliquid layer when they probed the surface of thin layers of ice. LEED is a technique that uses electrons to determine the surface structure of a crystal in the same way as XRD reveals the crystal structure of a solid. By observing how electrons bounced off ice surface, they suggested that the rapidly vibrating oxygen ions actually make the surface of ice slippery. These "liquid-like" water molecules do not move from side to side—only up and down. If the atoms moved from side to side, the layer would actually become liquid, which is what happens when the temperature rises above 0 °C.

#### 9.3.3.4 AFM Sliding and Scratching Friction

In 1998, using an atomic force microscopy (AFM), Döppenschmidt and Butt [31], measured the thickness of the liquid-like layer on ice, in temperatures above −35 °C. As illustrated in Fig. 9.6, capillary contacting forces on the liquid surface prompted the cantilever tip of the AFM to jump into contact with the solid ice once it reached the much softer layer's level. The upper limit in thickness of the liquid-like layer varied from 70 nm at −0.7 °C, 32 nm at −1 °C, to 11 nm at −10 °C. Their results indicated that at about −33 °C surface melting starts. The temperature dependence of the thickness follows roughly a rule, $d \propto -\log \Delta T$, where $\Delta T$ is the difference between the melting temperature and the temperature of detection.

The addition of salt could increase the thickness of the liquid-like layer. However, dragging the AFM tip across the surface of ice derived high friction of

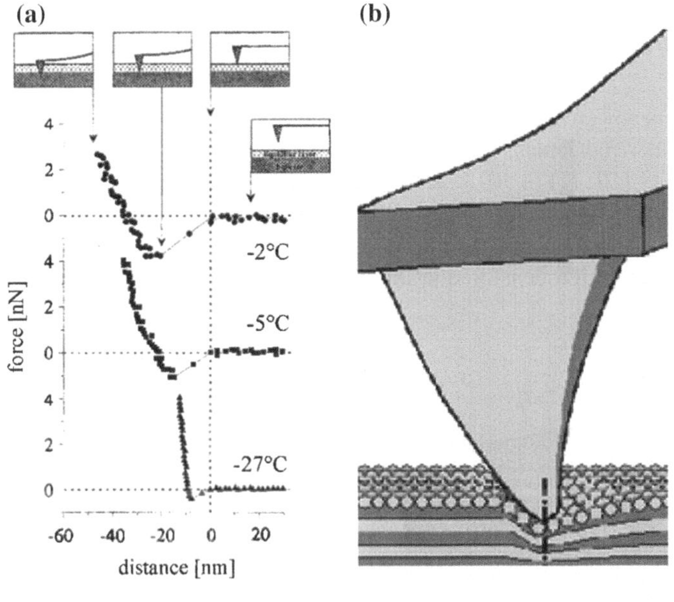


**Fig. 9.6** Temperature dependence of the liquid-layer thickness in the atomistic scratching (plastic dislocation) and sliding (elastic deformation) friction. **a** Force curves measured at different temperatures. "Zero distance" is the reference at the surface of the liquid-like layer. **b** The assumed position of the tip is indicated schematically for the force curve taken at −2 °C (Reprinted with permission from [31].)

ice, which indicates that while the top layer of ice may be liquid, it is too thin to contribute much to slipperiness except near the melting temperature.

#### 9.3.3.5 X-ray Reflection at Ice-$SiO_2$ Interface

Engemann and coworkers [32] examined in 2004 the X-ray reflectivity at the interface between ice and solid silicon dioxide and calculated the thickness and density of the liquid layer at temperatures between −25 and 0 °C, as illustrated in Fig. 9.7. They derived that the skin is a "high-density form of amorphous ice"—the density of the quasiliquid skin varied from that of liquid water at its melting point to 1.16 g/cm$^3$ at −17 °C. The thickness of the quasiliquid layer follows the relationship,

$$L(T) = (0.84 \pm 0.02) Ln \frac{17 \pm 3}{T_m - T} \text{ (nm)}$$

This experiment supports quasiliquid skin mechanism as the main cause of ice's slipperiness observed at −17 °C and above.

### *9.3.4 A Common Supersolid Skin Covers Both Water and Ice*

From the perspective of O:H–O bond cooperative relaxation between undercoordinated water molecules, Sun and coworkers [33] proposed and verified the skin supersolidity [10, 11] in 2013 using quantum calculations and electron and phonon spectrometrics. Molecular undercoordination not only disperses the quasisolid phase boundaries but also results in a temperature independent supersolid skin. Instead of the high-density quasiliquid skin, the elastic, less dense, polarized

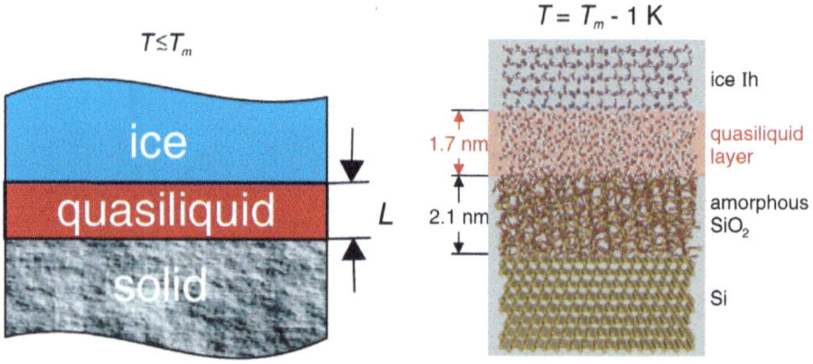


**Fig. 9.7** High-density quasiliquid skin forms between ice and amorphous $SiO_2$ at T ≥ $T_m$ −17 K and its thickness increases with temperature (Reprinted with permission from [32].)

supersolid phase presents due to molecular undercoordination at the skins of water and ice. An elastic Coulomb-levitation mechanism is responsible not only for ice slipperiness and water hydrophobicity but also for low friction of dry solid such as graphite, nitrides, oxides, and fluorides because the presence of nonbonding electrons [33]. Supersolid skin also presents at interface between hydrophobic substrate such as $SiO_2$ and water or ice [32], addition to a 0.5–1.0 nm air gap between them [34].

## 9.4 Quantitative Resolution

### *9.4.1 Skin O:H–O Bond Relaxation*

Generally, bond order loss shortens and stiffens the bond between undercoordinated atoms by up to 12 % for a flat skin of an fcc geometry, which enhances the bond energy by 45 % and depresses the atomic cohesive energy by 62 % for a metal such as gold and copper [35]. The enhanced bond energy raises the skin elasticity by 67 % and lowers the local meting temperature by 62 % [12]. However, for water and ice, molecular undercoordination shortens the H–O bond and stiffens its phonon. The O:H nonbond responds to undercoordination oppositely in length and phonon frequencies because of its weak interaction and the O–O repulsion. No electron exchange exists in the O:H nonbond as observed using AFM [36].

Figure 9.8 features the differential length spectra (DLS) for the MD-derived $d_x$ of ice [10]. Subtracting the length spectrum calculated using the 360-molecular unit cell without skin from that with a skin resulted in the DLS. Features above the lateral x-axis represent the length gain and features below the axis represent the

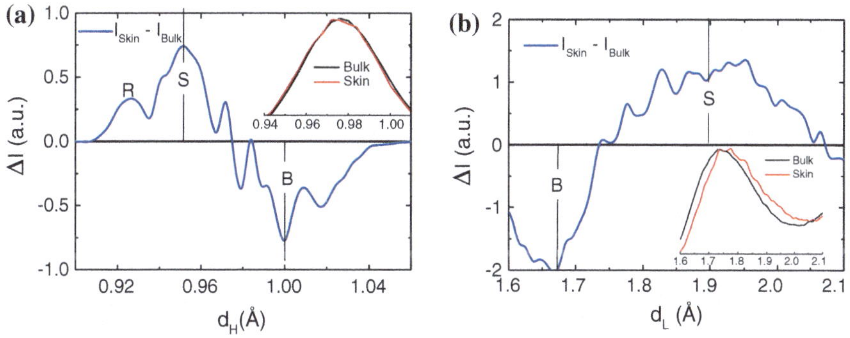


**Fig. 9.8** MD-derived DLS reveals that **a** $d_H$ contracts from the bulk value (B) 1.00 to 0.95 Å for the skin (S) and to 0.93 Å for the H–O free radicals (R), which is coupled with **b** $d_L$ elongation from the bulk value (B) of 1.68 to 1.90 Å, with high fluctuation. Insets show the raw length spectra of the unit cell with (denoted 'skin') and without skin presence (denoted 'bulk') (Reprinted with permission from [10].)

length loss due to the presence of the skin. The DLS shows that the $d_H$ contracts from the bulk value of about 1.00 to about 0.95 Å at the skin, while the $d_L$ elongates from the bulk value of 1.68 to 1.90 Å with high fluctuation as a broad peak. This cooperative relaxation lengthens the O–O by 6.8 % (=1 − (0.95 +1.90)/(1.0 + 1.68)) or lowers the mass density to 82 % from the bulk standard on the base of $\rho \propto d_{OO}^{-3}$. The $d_H$ = 0.93 Å peak even corresponds to the undercoordinated H–O radicals, whose vibration frequency is around 3650 $cm^{-1}$ [11]. The standard length is $d_H$ = 1.0004 and $d_L$ = 1.6946 Å at 4 °C [37].

According to the density-geometry-length correlation of molecules packed in water and ice [37], the measured $d_{OO}$ of 2.965 Å for liquid water skin [28] gives rise to $d_H$ = 0.8406 Å and $d_L$ = 2.1126 Å, which correspond to a 0.75 g/$cm^3$ skin mass density [10]. In comparison, the MD derived from Fig. 9.8 a density of 0.82 g/$cm^3$. These values, 0.75–0.82 g/$cm^3$, are much lower than 0.92 g/$cm^3$ for bulk ice or 1.0 g/$cm^3$ for bulk water at 4 °C, which is much lower than the 1.16 g $cm^{-3}$ for amorphous quasiliquid skin, derived from X-ray reflection [32].

### 9.4.2 Identical $\omega_H$ for Skins of Water and Ice

Figure 9.9 shows the differential phonon spectra (DPS) in comparison to the measured $\omega_H$ DPS for both water and ice given in Fig. 9.10 [38]. The valleys of the DPS represent the bulk feature, while peaks feature the skin attributes. A proper offset of the calculated DPS is necessary, as the MD code overestimates the intra- and intermolecular interactions [8]. As expected, the $\omega_L$ undergoes a redshift, while the $\omega_H$ undergoes a blueshift with multiple components. The $\omega_H$ blueshift results from the stiffening of the skin H–O bonds (S) and the free H–O radicals (R). The $\omega_L$ redshift arises from O–O repulsion and polarization. The polarization in turn

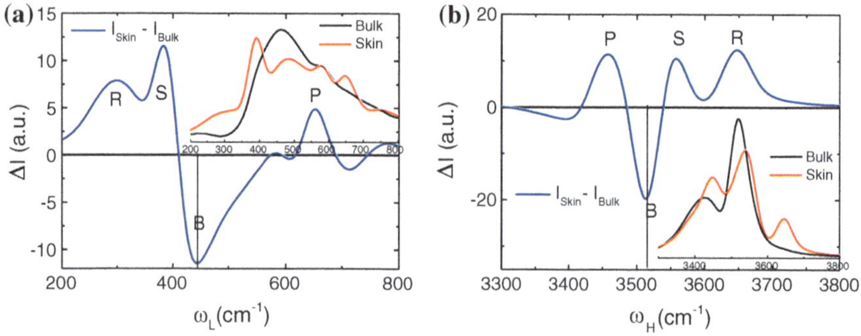


**Fig. 9.9** MD-derived DPS for the **a** $\omega_L$ and **b** $\omega_H$ of 200 K ice [10]. *Insets* show the raw spectra of calculations. Features S corresponds to the skin H–O bond; R corresponds to the free H–O radicals; the P component arises from the screening and splitting of the crystal potential by the polarized nonbonding electrons

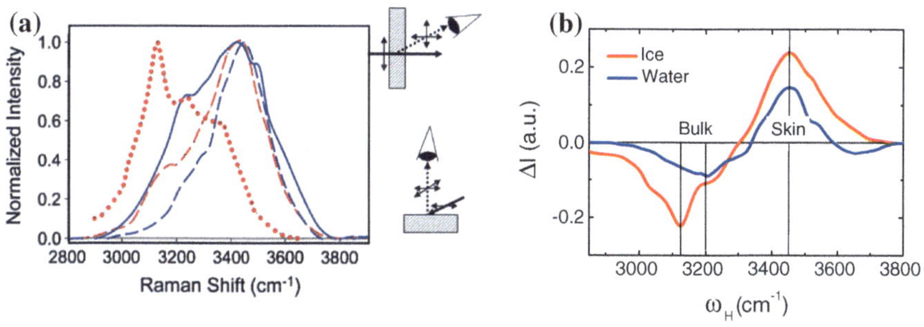


**Fig. 9.10** **a** Raman $\omega_{\mathrm{H}}$ spectroscopy of water (in *blue*, at 25 °C) and ice (*red*, at −20 and −15 °C) [38] collected at 87° (peaks toward higher frequency) and 0° with respect to the surface normal and water (*side views*). **b** The experimental DPS of water and ice distills the skin peak from the bulk as valley contribution to the spectra [10] (color online)

screens and splits the intramolecular potential, which adds another $\omega_{\mathrm{H}}$ peak (denoted P as polarization) with a frequency being lower than that of the bulk valley (B), which was ever regarded as a second type of the O:H nonbond.

Most strikingly, the measured DPS in Fig. 9.10 shows that the skins of 25 °C water and -(15-20) °C ice share the same $\omega_H$ value of 3450 $cm^{-1}$, which indicates that the H–O bond in both skins is identical in length and energy, according to the relationship $\omega_{\mathrm{H}} \propto (E_{\mathrm{H}}/d_{\mathrm{H}}^2)^{1/2}$. The peak intensity changes with the scattering from ice and water. The skin $\omega_L$ of ice may deviate from that of liquid water because of the extent of polarization, which is subject to experimental verification. Nevertheless, the skin $\omega_{\mathrm{H}}$ stiffening agrees with the DFT-MD derivatives that the $\omega_{\mathrm{H}}$ shifts from $\approx$ 3250 $cm^{-1}$ at 7 Å depth to $\approx$ 3500 $cm^{-1}$ of the 2 Å skin of liquid water [39]. Therefore, it is neither the case that an ice skin forms on water nor the case that a liquid skin covers ice. Rather, an identical supersolid skin covers both. In the supersolid skin, molecules shrink their size and enlarge separations, the O:H vibration frequency becomes lower and the amplitude is expected greater, which promotes the slipperiness of ice against other objects.

### 9.4.3 Skin H–O Bond Contraction and Electron Entrapment

Table 9.1 features the DFT-derived Milliken charge accumulation at the skin and in the bulk of water. O increases its net charge from the bulk value from −0.616 to −0.652 e when located at the skin. The net charge of a water molecule increases from 0.022 to −0.024 e correspondingly, which confirms the first round polarization of the electron lone pair by the entrapped O 1s core electrons due to H–O bond contraction [10].

**Table 9.1** DFT-derived charge localization at the skin and in the bulk of ice and derivatives (in bold) based on the referenced data using (9.1)

| | Skin | Bulk | $(H_2O)_1$ | O atom |
|---|---|---|---|---|
| $q_O$ | −0.652 | −0.616 | – | – |
| $q_H$ | 0.314 | 0.319 | – | – |
| $q_{H2O}$ | −0.024 | 0.022 | – | – |
| $E_{1s}$ (eV) [40–42] | 538.1 | 536.6 | 539.7 | 525.71 |
| $E_H$ (eV) | 4.52/4.66 | 3.97 [37] | 5.10 [43] | – |
| $T_m$ (K) | 311/320 | 273 | – | – |

Negative sign represents net electron gain

The following formulates the skin H–O bond energy $E_H$(Skin) and the atomic O 1s energy $E_{1s}(0)$. Table 9.1 lists the derivatives [21]:

$$\frac{\Delta E_{1s}(N)}{\Delta E_{1s}(\infty)} = \frac{E_{1s}(N) - E_{1s}(0)}{E_{1s}(\infty) - E_{1s}(0)} = \frac{E_H(N)}{E_H(\infty)} = \frac{T_C(N)}{T_C(\infty)} = \left(\frac{d_H}{d_{H0}}\right)^{-m}. \quad (9.1)$$

The $E_H(\text{Skin}) = 3.97 \times (538.1 - 525.71)/(536.6 - 525.71) = 4.518$ eV is compatible with the value of 4.66 eV for breaking the H–O bond of $H_2O$ molecules deposited on a $TiO_2$ surface in less than a monolayer coverage using laser excitation [43]. A laser beam of 267 nm can break the H–O bond in liquid [44] and the exclusion zone showed 270 nm UV light absorption peak [45]. The 267–270 nm wavelengths correspond to 4.64–4.59 eV energy for dissociating the H–O bond in the supersolid skin and the bulk. The deviation $\Delta E_H$(Skin) = 0.14 eV (about 3 %) arises mainly from molecular undercoordination in these two situations—one is the water skin and the other is the even less coordinated water molecules deposited on $TiO_2$ surface, which indicates that interaction between water molecules and the hydrophobic $TiO_2$ surface is very weak because the presence of an 5–10 Å thick air gap in the hydrophobic contacts [34].

With the known values of $(d_H, E_H)_{Skin}$ = (0.84 Å, 4.52 eV) and $(d_H, E_H)_{Bulk}$ = (1.0 Å, 3.97 eV) and the $E_H(1)$ = 5.10 eV, the bond nature index is estimated as $m = 0.744$ and the $d_H(1) = 0.714$ Å of a monomer. It seems that the m value for the H–O bond remains no constant but only for estimation herewith. The densely and locally entrapped core electrons of the undercoordinated water molecules polarize in a dual-process the nonbonding electrons [44].

### *9.4.4 Skin Thermal Stability*

Generally, atomic undercoordination depresses the critical temperature for phase transition of many substances because of the undercoordination reduced atomic cohesive energy, $T_C \propto zE_z$, where z is the atomic coordination number and $E_z$ is the cohesive energy per bond. The phase transition includes liquid–solid, liquid–vapor, ferromagnetic, ferroelectric, and superconductive transitions [12]. The skin melting temperature $T_{m,s}$ drops or rises depending the nature of the chemical bond, $T_{m,s}/T_{m,b} = z_s/z_b C_z^{-m}$, where m is the bond nature index and $C_z = 2\{1 + \exp[(z-12)/(8z)]\}^{-1}$ is the contraction coefficient of bond between undercoordinated atoms. According to the BOLS notion, the skin $T_{m,s}$ is 40 and 62 % of the bulk metal (m = 1) and silicon (m = 4.88) as the effective atomic CN of the top layer is 4 and the bulk is 12 for an fcc structure standard [21].

However, for water molecules, the $T_C$ is dominated by, and proportional to either $E_H$ or the $E_L$ only, depending on the nature of phase change, because of the 'isolation' of the $H_2O$ molecule by its surrounding lone pairs. For instance, $E_L$ determines the $T_C$ for evaporation $T_V$, as this process dissociates the O:H nonbond. The $E_L$ also determines the freezing temperature as defined by the specific heat disparity [46]. The $E_H$ dictates $T_m$(Skin) that is estimated from the correlation between the $T_C$(N) and the $\Delta E_{1s}$(N) from (9.1):

$$\frac{T_C(Skin)}{T_C(\infty)} = \frac{T_m(Skin)}{273} = \frac{E_H(Skin)}{E_H(Bulk)} = \frac{4.59 \pm 0.07}{3.97},$$

which yields the skin melting temperatures in the range of 315 ± 5 K. It is therefore not surprising that water skin performs like ice or gel at room temperature but evaporate at 333 K [47].

### 9.4.5 Skin Viscoelasticity

The polarization of molecules enhances the skin repulsivity and viscoelasticity. The high viscoelasticity and the high-density skin dipoles are essential to the superhydrophobicity and superlubricity at contacts [48]. According to the BOLS–NEP notion [12], the local energy densification stiffens the skin and the densely and deeply entrapped bonding charges polarize nonbonding electrons in dual process to form anchored skin dipoles [33]. The negative charge gain and the nonbonding electron polarization provide electrostatic repulsive forces lubricating ice.

Table 9.2 features the MD-derived thickness-dependent $\gamma$, $\eta_s$, and $\eta_v$ of ice films. Reducing the number of molecular layers increases them all. The O:H–O cooperative relaxation and associated electron entrapment and polarization enhances the surface stress to reach the value of 73.6 mN/m for five layers, which approaches the measured value of 72 mN/m for water skin at 25 °C. Generally, the viscosity of water reaches its maximum at temperatures around $T_m$ [49].

**Table 9.2** Thickness-dependent surface tension $\gamma$ and viscosity $\eta$

| Number of layers | 15 | 8 | 5 |
|---|---|---|---|
| $\gamma$ (mN/m) | 31.5 | 55.2 | 73.6 |
| $\eta_s$ ($10^{-2}$ mN s/m$^2$) | 0.007 | 0.012 | 0.019 |
| $\eta_v$($10^{-2}$ mN s/m$^2$) | 0.027 | 0.029 | 0.032 |

### 9.4.6 Skin Hydrophobicity and Electrostatic Repulsivity

Measurements, shown in Fig. 9.11, verified the presence of the repulsive forces between a hydrated mica substrate and the tungsten tip contacts at 24 °C in AFM measurements [50]. Such repulsive interactions appear at 20–45 % relative humidity (RH). The repulsion corresponds to an elastic modulus of 6.7 GPa. Monolayer ice also forms on a graphite surface at 25 % RH and 25 °C [51]. These observations and the present numerical derivatives evidence the presence of the supersolidity with repulsive forces because of bonding charge densification, surface polarization, and $T_m$ elevation due to undercoordination induced O:H–O bond cooperative relaxation.

### 9.4.7 Phononic Elasticity and Atomistic Friction

It is convenient to adapt the concept of supersolidity from the superfluidity of solid $^4$He at mK temperatures. The skins of $^4$He fragments are highly elastic and frictionless with repulsion between them when set in motion [33]. It is clarified in 2012 that the "supersolidity" arises from the shear elasticity of the $^4$He fragment [52–54]. But the interface repulsivity between fragments is essential to ensure the frictionless motion of the fragment. The skins of water and ice form an extraordinary "supersolid" phase [10] that is elastic [38], hydrophobic [47, 55], polarized [44, 56], less dense [28], and thermally stable [57], because of the densely entrapped bonding electrons [40–42, 58] and the dual polarization. The fewer the molecular neighbors there are, the smaller the water molecule size and the greater the molecular separation is, and therefore, the greater extent the repulsivity and supersolidity will be.

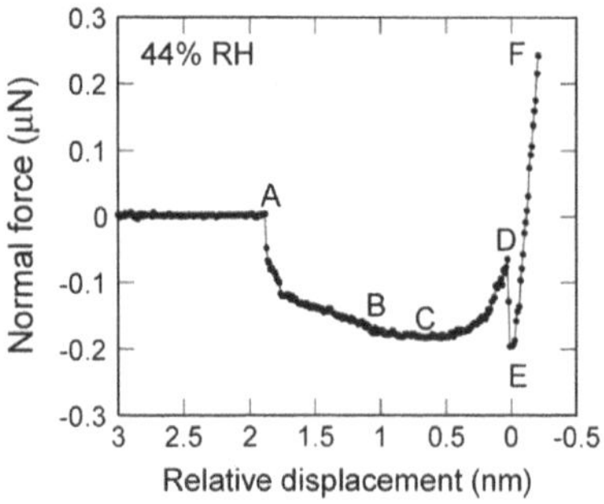


**Fig. 9.11** Normal force profiles between mica and tungsten tip at 44 % RH. Point *A* is the initiation of water nucleation and condensation; *B* and *C* are the formation of a complete water bridge cross the tip and substrate; *D* is the maximum attractive force before the tip–substrate contact; *E* denotes the sudden drop of force; and *F* indicates the tip–substrate contact repulsive force (Reprinted with permission from [50].)

#### 9.4.7.1 O:H Phonon Frequency and Vibration Amplitude

According to the BOLS–NEP notion [59], molecular undercoordination shortens and stiffens the intramolecular H–O bond and meanwhile, lengthens and softens the intermolecular O:H–O bond because of the Coulomb repulsion between electron pairs on adjacent oxygen ions. The H–O will vibrate faster and the $(H_2O):(H_2O)$ oscillate slower at the skin. The dual polarization increases the local charge of O ions at the skin.

MD (Figs. 9.8 and 9.9) and DFT (Table 9.1) calculations confirmed so. The O:H nonbond contracts from the bulk value of 1.0 to 0.95 Å for the skin and 0.93 for H–O radical and the H–O expands from 1.65 to 1.90 Å. The O:H phonon frequency shifts from the bulk value of 450 to 400 for the skin and to 300 $cm^{-1}$ for those close to free H–O radicals. The H–O phonon shifts from 3500 to 3550 and 3650 $cm^{-1}$ for the skin and H–O radicals, disregarding the artifact of the potential splitting and polarization effect.

The curvature of an interatomic potential generally approximately conserves [12], which correlates the amplitude and frequency of an oscillator at vibration,

$$\mu(\omega x)^2 = \left( \frac{\partial^2 u(r)}{\partial r^2} \bigg|_{r=d} \right) x^2 = const.,$$

The O:H–O bond segmental vibration amplitudes at the skin or associated with a radical vary from that of the bulk as,

$$\frac{x_{skin}}{x_{bulk}} = \frac{\omega_{bulk}}{\omega_{skin}} \cong \begin{cases} 200/100 = 2 & \text{(skin O:H)} \\ 3200/3450 = 64/69 & \text{(skin H–O)} \end{cases}$$

The estimated $x_{skin}$ agrees with that probed using proton backscattering, the vibration amplitude of surface oxygen atoms are roughly 3.3 times of their bulk [26]. Therefore, in addition to the stronger dual polarization, the greater amplitudes and the lower frequencies of the skin O:H oscillators are responsible for the slipperiness of ice, as illustrated in Fig. 9.2b. The soft springs deform and recover easily when they are compressed. If the compression force is too large, the O:H nonbond breaks, the friction coefficient increases sharply-scratching other than sliding.

#### 9.4.7.2 High Friction Coefficient of Ice on Ice

As shown in Fig. 9.1b, the kinetic friction coefficient of steel on ice ranges from 0.01 to 0.1. The friction coefficient of ice on ice varies unexpectedly from 0.03 at 0.05 m/s and 0.58 at $5 \times 10^{-7}$ m/s sliding velocity, within the temperature range of −3 and −40 °C under normal pressure of 0.007–1.0 MPa [17]. The following rules govern the unexpected high friction coefficient for ice sliding on ice:

(1) Regelation takes place when two pieces of ice contact at temperatures above −22 °C. As observed by Faraday [1], 0 °C water can fuse two pieces of ice under a slight compression.
(2) Water molecules tend to recover their unoccupied neighbors, reserving energetically favorable $sp^3$ bonding configurations of oxygen [60, 61].
(3) O:H phonon resonant coupling occurs when two pieces of ice are brought contact, as noted by Krim [6]. Resonance of higher vibration amplitude and identically lower frequency O:H oscillators hinders their sliding motion.
(4) However, the friction coefficient for saline ice on normal ice is expected lower because the solute ionic electrification enhances the polarization and shifts the O:H phonon frequency even lower, which decouples the interface phonon resonant between identical ice—pure ice motion against sea ice could improve the friction.

#### 9.4.7.3 Two-Regime Friction: Phonon Criterion

Figure 9.12 illustrates the O:H–O bond relaxation dynamics in the solid and in the quasisolid phases of bulk water, which clarifies why the friction coefficient shows two temperature regimes. In the quasisolid phase of 258–273 K range, the relative specific heat $\eta_H/\eta_L < 1$, the H–O bond contracts at cooling, which lengthens the O:H

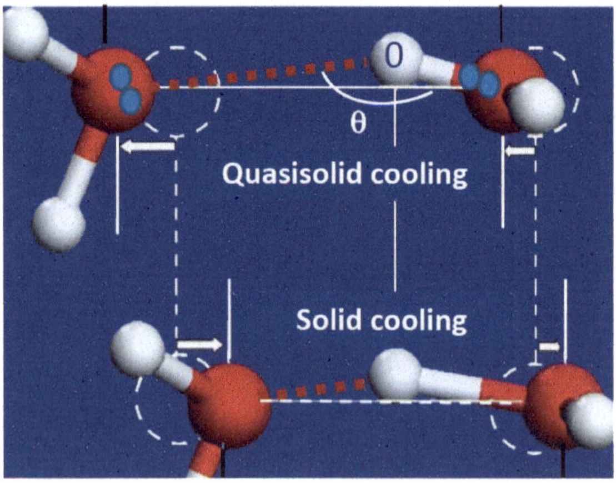


**Fig. 9.12** O:H–O bond relaxation in the solid (T < 258 K) and in the quasisolid (258 ≤ T ≤ 277 K) phases because of the segmental specific disparity [8]. O:H nonbond elongation and its vibration amplitude elevation lowers the friction coefficient in the quasisolid phase at cooling; O:H cooling contraction and its vibration amplitude reduction raise the friction coefficient when ice cools (Reprinted with permission from [59].)

nonbond and soften its phonons with increasing vibrating amplitude, which enhances the slipperiness of ice as it cools though measurement varies from situation to situation.

In the solid phase below 258 K, $\eta_H/\eta_L > 1$, the O:H nonbond contracts at cooling, which results in its higher frequency and smaller magnitude of vibration, as it cools. The O:H nonbond contraction and its vibration amplitude reduction increase the friction coefficient of ice. This trend carries on as it cools so the friction coefficient increases when temperature is lowered in the crystalline ice to an expected temperature of 100 K at which the O:H-O bond freezes, as shown in Fig. 9.1b.

#### 9.4.7.4 AFM Friction: Scratching or Sliding?

Generally, one talks about friction of an object sliding on ice, which gives a lower kinetic friction coefficient because of the elastic and adaptive atomic deformation. However, as shown in Fig. 9.6, an AFM in contacting mode derived high kinetic friction coefficient of 0.6 in the temperature range of –20 and –40 °C, which is compatible to the static coefficient measured in macroscopic experiments [62]. The AFM tip scratching into the skin of several nanometers thick breaks the skin O:H nonbond with resistance of the high viscosity during scratching. The tip does not entertain the superlubricancy for sliding but experienced the creep and viscosity resistance because of the plastic dislocation. Therefore, care needs to be taken when one measures the atomistic friction coefficient using an AFM.

## 9.5 Solid Dry Friction: Elasticity and Repulsivity

### *9.5.1 $^4$He Supersolidity: Elasticity and Repulsivity*

Helium is the noblest amongst all elements: the interaction between even its own atoms is so weak that it solidifies only under intense pressure and extremely low temperatures. If the pressure is reduced to 2.5 MPa at the absolute zero temperature, quantum-mechanical fluctuations of the atoms' positions become so large that the solid melts, becoming a 'quantum liquid'. No crystalline solid is perfect—there are always some vacancies in the crystal lattice where atoms are missing—and in 1969 Andreev and Lifshitz [63] proposed that helium's large quantum fluctuations might, at zero temperature, stabilize a dilute gas of vacancies within the solid. Atoms of the prevalent isotope $^4$He are Bosons (they have zero spin), and so vacancies in solid $^4$He can also be thought of as Bosons. The vacancies can thus condense to form an exotic phase known as a Bose–Einstein condensate that suffuses the solid. This 'supersolid' phase, lately referred to shear elasticity, would share some commensurateness with a superfluid—namely, frictionless flow—but at the same time have a non-zero shear modulus, a defining characteristic of a solid.

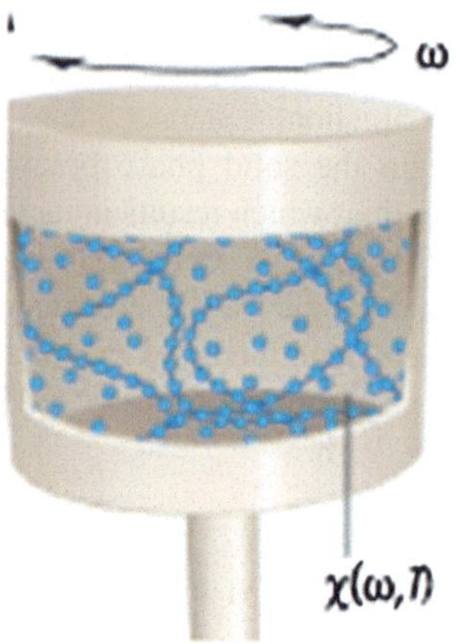


**Fig. 9.13** Supersolidity of $^4$He at 2 K temperatures or below [64]. The torsional oscillator is a disk filled with solid $^4$He of multiple fragments, as denoted with *color lines* (color online)

Figure 9.13 illustrates the supersolid state of solid $^4$He. Only 1 % undercoordinated $^4$He atoms presenting at boundaries of the fragments makes the fragments frictionless at torsional motion.

Supersolidity describes the coexistence of solid and superfluid properties in a quantum crystal. The phenomenon was discovered in 2004 by Kim and Chan [52, 65] when they measured the resonance period of a small cylindrical box oscillating around a torsion rod. The box contained solid $^4$He at temperature below 100 mK, the oscillation frequency increases as if 1 % of the helium mass had ceased moving with the box. To run the $^4$He supersolidity experiment, they hang the disk from a stiff rod and oscillate the disk back and forth. By measuring the frequency of oscillation, they detected whether the solid $^4$He behaves like a supersolid—high shear elasticity and repulsivity in the fragment contacting normal. An oscillating disk of normal matter, for example, behaves as expected: because the atoms are rigidly linked, they rotate together. In an oscillating disk of supersolid matter, many of the atoms rotate, but some do not. Instead, those atoms slip through the lattice like a superfluid, without friction whatsoever, and sit motionless. That reduces the mass of the disk, which allows it to oscillate faster.

The same method had been widely used for the detection of superfluidity in a liquid in the absence of viscosity, the liquid in the box remains at rest while the box walls move. At temperature below 200 mK, $^4$He crystal is readily decoupled into fragments in a torsional oscillator to exhibit superfluidic nature-frictionless motion without viscosity [52–54]; meanwhile, the $^4$He crystal fragments are stiffer than expected and hence react elastically to a shear stress applied [66]. The individual $^4$He segment would be thus both superelastic and superfluidic in motion—the supersolidity meant.

The 'supersolid' form of $^4$He is stiffer, more elastic and frictionless than the normal solid [67]. The superfluidity of $^4$He solid is usually described in terms of Bose–Einstein condensation or quantum statistics in energy domain. All particles occupy the lowest energy states simultaneously. A scenario in real space is infancy though the unercoordinated crystal defects have been recognized as the key to the supersolidity [68]. The superfluidity of $^4$He solid is related to the quantum defects such as atomic vacancies of 1 nm size or around [69] and the supersolidity is related to structural disorder [70] such as dislocations, grain boundaries, or ill-crystallized regions where the undercoordinated atoms become dominant. According to Pollet et al. [69], inside a dislocation or a grain boundary, the local stress is anisotropic, which is sufficient to bring the vacancy energy to zero, so that the defect is invaded by vacancies that are mobile and superfluidic. Solid $^4$He could contain a network of defects and if these defects are connected to each other, mass could flow from one side of the crystal to the other frictionlessly. On the other hand, the disorder—induced stiffening could be the result of dislocations becoming pinned by isotropic impurities (i.e., $^3$He atoms even at very small concentrations).

Later ultrasound and torsional oscillator studies [71, 72], however, evidence that shear modulus stiffening is responsible for at least a fraction of the period drop found in bulk solid helium samples. The experimental configuration of Kim and Chan makes it unavoidable to have a small amount of bulk solid inside the torsion cell containing the Vycor disk. The results of a new helium in Vycor experiment with a design that is completely free from any bulk solid shear modulus stiffening effect [73].

According to Anderson [68], "*Crystal defects enhance the local density of vacancies*". Observations are conjectured to be describable in terms of a rarified Gross-Pitaevskii superfluid of vacancies, with a transition temperature of about 50 mK, whose density is locally enhanced by crystal imperfections. The observations can be affected by this density enhancement. Therefore, disorder and defects that could enhance the local mass density appear to play an important yet uncertain role in the supersolidity of $^4$He crystals [74].

The interatomic "bond" breaks easily for $^4$He crystals, which requires energy at the critical point of 4.2 K for liquid–vapor transition in the order of 1/3000 eV, much smaller than a typical van der Waals bond of 0.1 eV or around. The extremely weak interatomic interaction without charge sharing makes the $^4$He atoms or grains are stickingless—more like hard spheres with closed electronic shell packing together. The stickingless interaction between grains will lower the friction coefficient.

The understanding of slipperiness of ice provides a feasible mechanism for the superfluidity and supersolidity of $^4$He crystal. Repulsion between the "electric monopoles pinned in the elastic skins" of the small grains could resolve this puzzle. Broken-bond-induced local strain and quantum entrapment leads to a densification of charge and energy in the skin of a few atomic layers thick. The densification of energy corresponds to the enhancement of the elasticity, which stiffens the solid skin allowing the $^4$He segment to react elastically to a shear stress. The repulsion between the densely entrapped electrons makes the motion frictionless. $^4$He crystals lack the nonbonding electrons because of the close atomic shells. Therefore, the broken bonds that serve as not only centers that initiate structure failure but also provide sites for pinning dislocations by charge and energy entrapment, which could be responsible for the superfluidity and supersolidity as observed. Its 'supersolid' behavior results just from atomic CN imperfection that changes the bulk properties of the crystal [75]—tomic undercoordination induces local quantum entrapment and polarization. Lattice contraction of the supersolid $^4$He segments is expected to happen, though this contraction is measured tiny [53].

### 9.5.2 Superlubricity in Dry Sliding: Atomistic Friction

The ultralow-friction linear bearing of carbon nanotubes (CNTs) and the superlubricity at dry nanocontacts sliding in high vacuum [76, 77] are fascinating. As shown in Fig. 9.14a, the velocity of the liquid water moving in the CNTs is inversely proportional to the diameter under constant pressure applied to the CNT ends [78], which is beyond theory expectations. Transition electron microscopy revealed that the inner walls of a multi-walled CNT can slide back and forth with respect to the outer walls of the CNT, being free from wear for all cycles of motion (see Fig. 9.14b) [79]. Surface energy calculations suggested that the force retracting the core nanotubes back into the outer tubes was 9 nN, being much lower than the van der Waals forces pulling the core nanotubes back into their sheath. The removal of the outer walls of the MWCNT corresponded to the highly localized dissipation at defect scattering sites, located primarily at the ends of the tube. The demonstration of ultralow friction between multi-walled CNT layers confirms that they will be useful mechanical components in molecular nanotechnology such as molecular bearing.

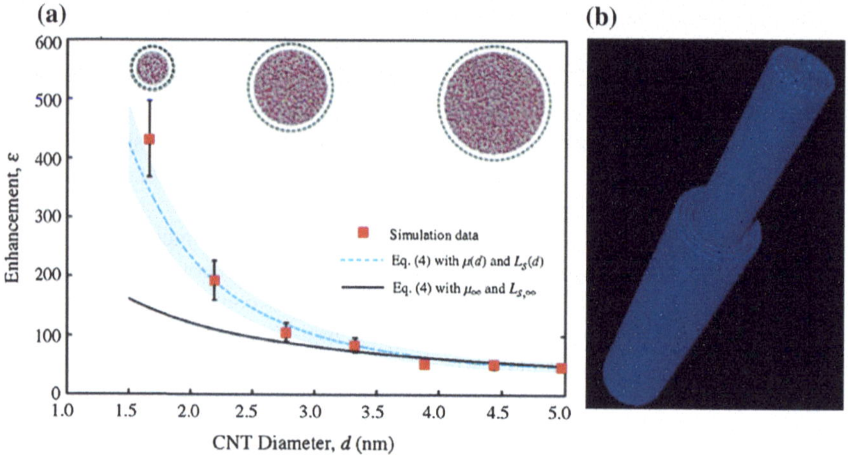


**Fig. 9.14** **a** Superfluidity of water droplet in CNTs of different diameters [78] and **b** ultralow-friction nanoscale linear bearing made of multi-walled CNT [79]

### 9.5.3 Quantum Friction: Charging and Isotopic Effect

The occurrence of quantum friction is a kinetic process of energy dissipation ($E = f_r \cdot s$ with $f_r$ being the friction force and s the sliding distance) due to the phonon (heat) and electron excitation (electron–hole pair production) during sliding [80]. A state of ultralow friction is reached when a sharp tip slides over a flat surface and the applied pressure is below a certain threshold, whose value is dependent on the surface potential sensed by the tip and the stiffness of the contacting materials [81–83].

A comparative study of hydrogen- and deuterium-terminated single-crystal diamond and silicon surfaces revealed that the hydrogenated surface (terminated with $H^+$) exhibited higher friction than the surface passivated with $^2H^+$, as illustrated in Fig. 9.15. The additional neutron in the $^2H^+$ plays a certain yet unclear role of significance because of the possible mass difference between the H and $^2H$ adsorbates [80]. In fact, adsorption of the isotope lowers the vibration frequency by $2^{-1/2}$ of the adsorbate on substrate by folding the reduced mass of the oscillator, which reduces the vibration phonon frequency and therefore the friction coefficient [59].

However, if changes the tip to ice sliding on ice with added Deuterium, situation may reverse—the friction coefficient, or the shear strength of the O:D nonbond should be lower than ice on ice because of the uncoupled resonance of the identical phonons of the tip and substrate (see Sect. 4.7).

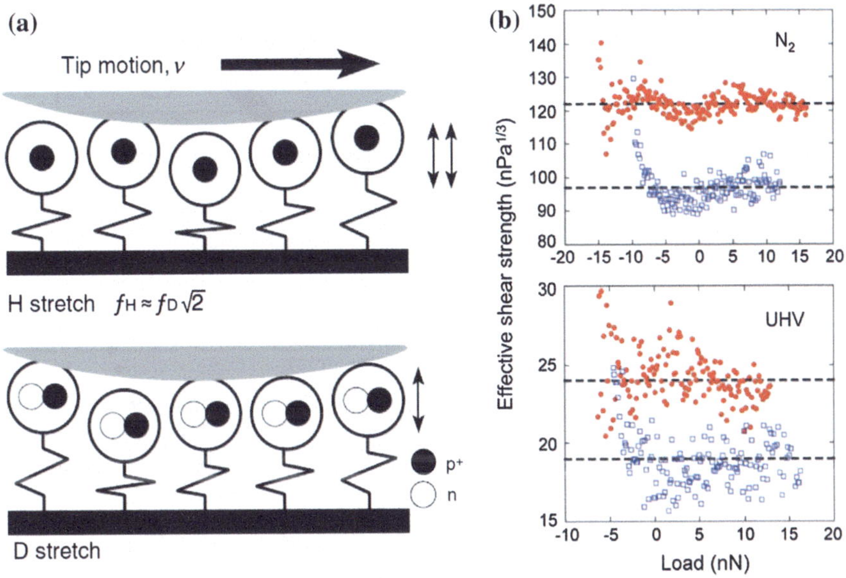


**Fig. 9.15** **a** A schematic of the frictional interface. Vibrating adsorbates collide with and dissipate kinetic energy from the moving tip at a rate that depends on the adsorbates frequency and thus its mass. **b** The shear strength of the H–C bond (*red symbols*) is higher than that of the D–C bond (*blue symbols*) measured in the $N_2$ and vacuum conditions (Reprinted with permission from [80].) (color online)

Park et al. [84] found a remarkable type and concentration dependence of the friction force on doped silicon. The friction force between the AFM sliding conductive TiN tip and the doped Si is substantially different, as given in Fig. 9.16. Charge depletion or accumulation on a Si substrate with patterned *p* and *n* stripes contributes differently to the friction force under a bias. A positive bias applied to the *p*-region causes a substantial increase of the friction force. If the n-region is biased positively, an accumulation of holes (+ charged) in the p-region. No variation of friction force was resolvable between *n* and *p* regions under negative bias.

Both observations [80, 84] indicate clearly that the positively charged ($H^+$) tip or substrate (electronic holes +) would induce high friction force [22].

The superlubricity phenomenon was explained using the classical Prandtl–Tomlinson (PT) model [85, 86] and its extensions, including thermal activation, temporal and spatial variations of the surface corrugation, and multiple-contact effects [81]. Observation suggests that the friction force depends linearly on the number of atoms that interact chemically across the contact [87]. According to the one-dimensional PT model, the slider atoms feel the periodic potential of the substrate surface atoms as they slide over them, experiencing a net force that is the sum of individual instantaneous friction force on each atom resulting from the gradient of the periodic potential.

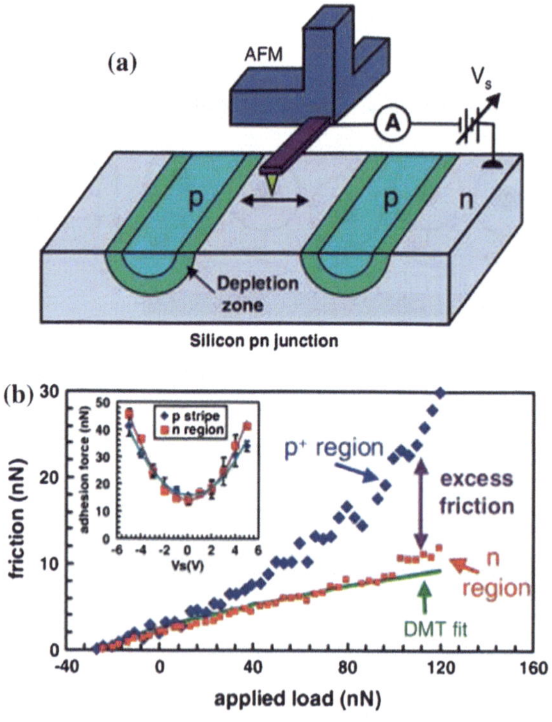


**Fig. 9.16** **a** Schematic of AFM measurements on a silicon p-n junction device. **b** Plot of friction force as a function of applied load at +4 V sample bias. The *inset* shows the pull-off force as a function of sample bias (Reprinted with permission from [84].)

### 9.5.4 Solid Nonbonding Electron Self-lubricancy

The mechanism of elastic interface Coulomb repulsion also applies to the frictionless CNT linear bearing and the superlubricity of micro channels. In fact, bond contraction happens to the CNT of limited number of walls. Bonds near the open ends contract even further [88]. Densification of the σ-bond electrons and polarization of the π-electrons take place to all the CNT walls; the repulsion between the densely packed and locally polarized like charges will reduce the friction force substantially, while the electrostatic forces of the additionally densely charged CNT ends may provide force for retracting motion and oscillation. The saturated potential barrier due to the skin charge entrapment of nanocontacts also provides a repulsion force between the contacts.

Skins of nitrides, oxides, and fluorides share the similarity of water and ice. Figure 9.17 illustrates the bonding rules for superlubricity of nitrides, oxides, and fluorides. The difference between these compounds is the number of lone pairs associated with each electronegative atom and their group symmetry and geometrical orientations. The key gradient of ice slipperiness is the presence of electron lone pairs and undercoordinated molecules. The O:H nonbond softening is associated with vibration amplitude enlargement and charge density elevation due to the dual polarization. N reacts with a solid B skin preferring $C_{3v}$ symmetry, such as fcc(111) and hcp(0001) planes [89]. The N atom is located in a place between the top two layers and the lone pair is directed into the substrate. The surface is hence networked with the smaller $B^+$ and the saturate bonded $N^{3-}$cores with densely

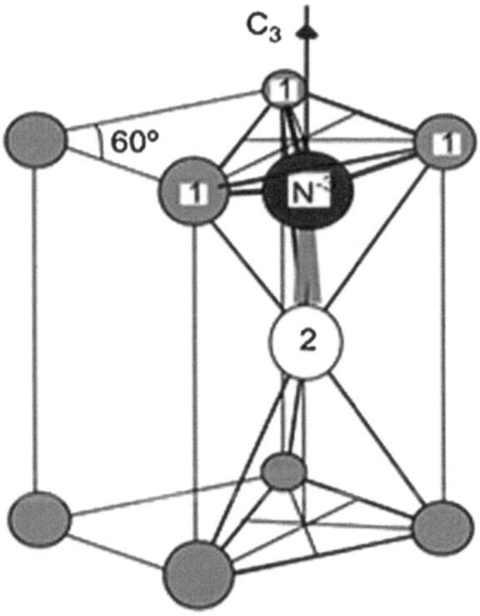


**Fig. 9.17** **a** $NB_4$ nitride quasi-tetrahedron structure and **b** the associated valence density-of-states [90]. Smaller ions (*labeled 1*) donate electrons to the central N acceptor of which the sp orbits hybridize with production of a nonbonding lone pair (*labeled 2*). N induced valence DOS with four features representing the states of bonding electron pairs, nonbonding lone pairs, antibonding dipoles, and electronic holes. The interlayer soft ":" phonon makes the top layer adaptive and elastic and the intralayer bonds make the top layer harder and stronger. (see Fig. 3.3 in Chap. 3)

packed electrons. Hence, the top skin layer is chemically inert as mechanical stronger and harder. Electrons in the saturated bond should be more stable compared with the otherwise in the neutral host atoms. The interlayer soft phonon makes it adaptive and elastic.

When react with other electropositive atoms, the $sp^3$-orbital hybridization occurs with creation of the lone pairs that polarize the neighboring atoms becoming dipoles. There are four additional features in the valence band. The nonbonding lone pairs are responsible for the phononic elasticity—low vibration frequency and high amplitude. The localized antibonding dipoles stem the surface repulsivity.

The high intra-surface stress due to the ionic network could be responsible for the hardness of the top layer of a nitride. On the other hand, the $N^{3-}$–$A^+$ network at the surface is connected to the substrate mainly through the nonbonding lone pairs. The nonbonding interaction is rather weak ($\sim$0.1 eV per bond) compared with the original metallic bonds ($\sim$1.0 eV per bond) or the intra-surface ionic bond (2–3 eV per bond). The weak lone-pair interaction is highly elastic within a critical load at which the weak interaction will break. Therefore, the enhanced intra-layer strength makes a nitride usually harder ($\sim$20 GPa), and the weakened inter-layer bonding makes the nitride highly elastic and self-lubricate. This mechanism also applies to graphite because of the weak $\pi$-nonbonding interaction along the [0001] direction.

Nanoindentation profiles from TiCrN surface and sliding friction measurements from CN and TiN surfaces have confirmed the predicted high elasticity and high hardness at lower pressing load and the existence of the critical scratching load [91]. As compared in Fig. 9.18a, under 0.7 mN load of indentation, the elastic recoverability and hardness for a GaAlN film are higher than that of an amorphous carbon film [91]. The GaAlN surface is also much harder than the amorphous-C

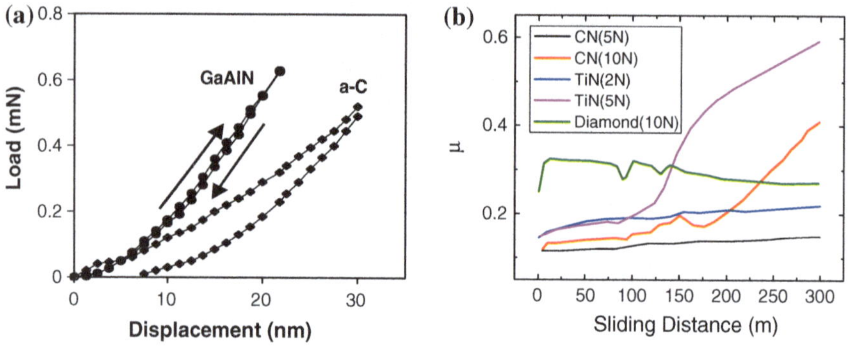


**Fig. 9.18** Electron lone pairs serve as solid lubricants. **a** GaAlN/$Al_2O_3$ exhibits higher hardness and full elastic recoverability in comparison to amorphous carbon films under the same friction load. **b** The pin-on-disc measurements of the sliding friction coefficients of nitrides and diamond under different loads. Lowering the operating temperature from the ambient (**b**) may reduce nitrides' friction coefficient to be compatible to that of ice. The abrupt increase of the coefficient indicates the presence of the critical load at which the lone pair nonbond breaks (Reprinted with permission from [91] and references therein.)

film under the lower indentation load. Figure 9.18b shows the profiles of pin-on-disk sliding friction test. The abrupt increase of the friction coefficient of nitride films represents the critical load. For polycrystalline diamond thin films, no such abruption in friction coefficient is observed though the friction coefficient is generally higher than nitride films. The absence of lone pairs in a–C film makes the film less elastic than a nitride film under the same pressing load. The abrupt change in the friction coefficient evidences the existence of critical load that breaks the nitride interlayer bonding-lone pair interaction. Therefore, the non-bonding interaction enhances the elasticity of nitride surfaces. Such high elasticity and high hardness by nature furnishes the nitride surfaces with self-lubricate for tribological applications.

The mechanism of slipperiness of ice is analogous to the self-lubrication of metal nitride [91, 92] and oxide [93] skins with electron lone pairs becoming dominant. TiCrN, GaAlN and $\alpha$-$Al_2O_3$ skins exhibit a 100 % elastic recovery at nanoindentation load under the critical friction load (e.g., <5 N for carbon nitride) at which the lone pair breaks.

Figure 9.19 shows the zone-selective electron spectrometrics (ZPS) profiles for Ti–O and Ti–N. The spectral difference between the chemisorbed surface and the clean reference show both oxide and nitride share the same valence DOS features as expected in Fig. 9.17. Raman spectra in Fig. 9.20 further evidence the presence of lone pairs in oxides and nitrides with character frequencies ranged below 1000 $cm^{-1}$. However, carbon and carbide manifest no such features. The lone-pair features of oxides are stronger than those of nitrides, which result from the number of lone pairs of an oxygen and nitrogen atom.

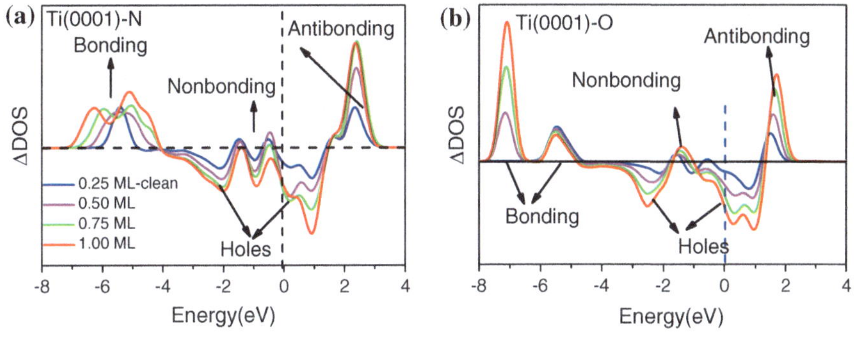


**Fig. 9.19** Comparison **a** Ti(0001)–N with **b** Ti(0001)–O calculated ZPS profiles of n(Ti + X)−n (Ti), at 0.25, 0.50, 0.75, 1.00 ML with respect to that of clean Ti(0001) surface. Where n(Ti + X) is the DOS of the adsorbate system Ti(0001)–X (X = N, O), and n(Ti) the clean Ti(0001). Both reveal four DOS features corresponding to antibonding, nonbonding, bonding states and holes

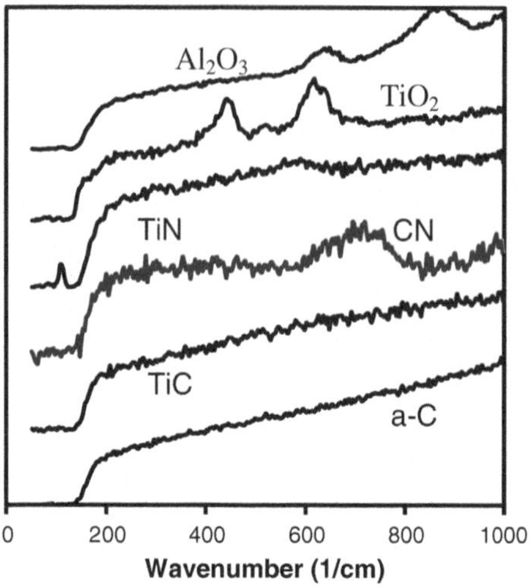


**Fig. 9.20** Low-frequency Raman shifts indicate that nonbonding lone pair interaction exists in oxides and nitrides but carbides. Peak intensities of oxides are stronger than that of nitrides because of the number of the lone pairs that follow the 4-n rule (Reprinted with permission from [91].)

## 9.6 Wet Lubricant Superlubricity

In order to lower the friction coefficient at moving contacts, one often appeals to solute grease lubricant or detergent like graphite and sulfide powers. Little attention has been paid to the acid (H+↔H+ antibond quantum breaker with point repulsion) and alcohol (O:↔:O quantum point compressor dominance in base-like solutions) solutions but a group of researchers at Tsinghua University has been focused on the mechanism of such solute lubricants. The following shows two excellent lubricants containing acids and alcohols that have ensured the superlubricity with extremely low friction coefficients. Electrification of the O:H–O bond by the excessive $H^+$ in the contacting interface and the molecular undercoordination induced skin supersolidity play significant roles in promoting the lubrication, according to the present knowledge. Readers may be referred to [94] for a comprehensive review on the fluid friction dynamics at the nanometer scale from the perspectives of diffusion, molecular cooperativity, phase transition, during motion at the hydrophobic and hydrophilic interfaces.

### 9.6.1 *Acid Solutions*

Phosphoric acid solutions exhibit superlubricity effect as a lubricant [95] with a friction coefficient around 0.004 after a short running-in period. During sliding test, $H^+$ ions bond to the friction surfaces through tribochemical reaction, and meantime, the phosphoric acid and water molecules can form a stable hydrogen bonded network and then superlubricity appears [96]. The superlubricity of the aqueous acid arises from the polarization of the nonbonding electron lone pairs by the Phosphoric anion and fragilated by the $H^+$↔$H^+$ repulsion between the hydromnium $H_3O$

having one lone pair short and one of its four neighbours. Ionic electrification will align, stretch, and polarize the O:H–O bond, which enhances the skin supersolidity. The ionic electrification and molecular undercoordination promote superlubricity of the acid solution. This is also true for saline ice [17] (Fig. 9.21).

The friction force between a silica particle and silica wafer changes with the lubricant of pure water and electrolyte solutions of LiCl, NaCl, and CsCl salts [98]. Figure 9.22 shows that smaller and more hydrated cations have higher lubrication capacities than the larger and less hydrated cations. Furthermore, the friction force also drops with the increases of solute concentration.

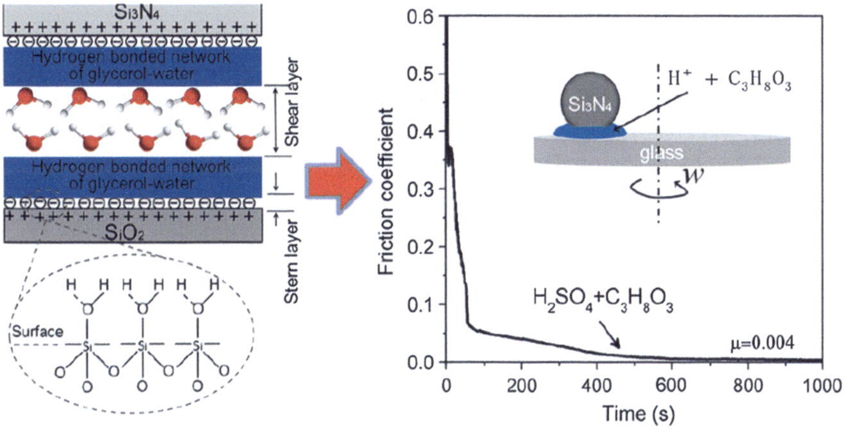


**Fig. 9.21** Schematic of the experimental set up and the friction behavior of solution lubricants (acids or alcohols) that lowers the friction coefficient to 0.004 (Reprinted with permission from [97].)

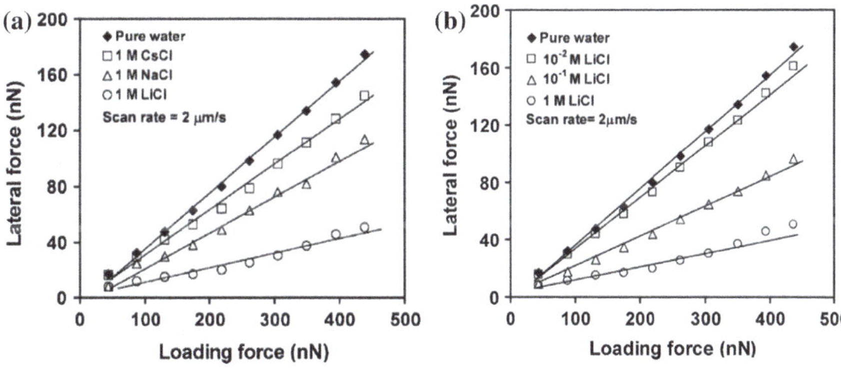


**Fig. 9.22** Lateral force of a 6.8-μm silica particle interacting with a silica wafer using **a** $H_2O$, and CsCl, NaCl, and LiCl solutions of 1 M at a fixed scan rate of 2 μm/s, and **b** its variation with the LiCl solution at $10^{-2}$, $10^{-1}$, and 1.0 M concentrations (Reprinted with permission from [98].)

Observations agree with Raman phonon spectrometric results showing that the ionic electrification stiffens the $\omega_H$ phonon in the order $Na^+ > K^+ > Rb^+ > Cs^+$ and the bulk $\omega_H$ shifts more significantly than those in the skin [99]. The $\omega_H$ blue shift is associated with the O:H $\omega_L$ softening. The phonon cooperative relaxation indicates the electrification shortens the H–O bond and meanwhile lengthens the O:H bond through O–O Coulomb repulsion and polarization. This bond and phonon relaxation takes place throughout the solution so the bulk $\omega_H$ shift more than it is in the skin. O:H–O bond electrification by the short fields of ions raises the viscoelasticity of the lubricant. The softened $\omega_L$ phonon could lower the friction coefficient.

### *9.6.2 Glycerol and Alcohols*

Glycerol is another efficient medium that promotes hydrogen-bonded network lubricancy. Ma et al. [100] found that a mixture of glycerol and boric acid results in the superlubricity behavior. The adsorbed diglycerin borate and the hydration layer polarize water molecules acting as the lubricant in the contact region. Strikingly, addition of glycerol can promote the superlubricity of numerous acid solutions [97]. The ultra-low friction coefficient is closely related to the pH value of acid and the concentration of glycerol.

Furthermore, a replacement of glycerol with polyhydroxy, the same family of glycerol, can also raise the superlubricity of the lubricant [101]. Therefore, the hydrated water layer between the hydrogen-bonded networks of polyhydroxy alcohol and water molecules on the positively charged surfaces forms a promising kind of lubricant for wet friction applications—polarization of the electron lone pairs in the bond network seems very effective.

In order to examine the O:H nonbond phonon elasticity and the electrostatic repulsivity in the wet lubricants, we can examine the O:H–O bond phonon frequency relaxation as a function of glycerol volume fraction in $H_2O$ solution. Figure 9.23 compares the full-frequency IR absorption and Raman spectra for $H_2O$, Glycerol, and their mixture [102]. Glycerol IR spectra show features at 3250 and 3450 $cm^{-1}$ overlapping that of water with additional features in frequency range lower than 3000 $cm^{-1}$. An addition of the glycerol depresses the intensity of the H–O bond vibration feature. Chapter 12 will examines phonon relaxation dynamics of glycerol, ethanol and methanol soltions in comparison with acids, bases, and salts.

(1) The O:H phonon frequency $\omega_L$ shifts from the skin component centered at 60 $cm^{-1}$ and the bulk component at 175 to the 325 $cm^{-1}$ centered peak, which indicates that the O:H nonbond undergoes length contraction and polarization because of the O:↔:O quantum point compression that will be addressed shortly in the Chapt dealing with aqueoussolutions.
(2) The H–O phonon $\omega_H$ shifts cooperatively from the H–O radical component centered 3620 to 3330 $cm^{-1}$, which corresponds to H–O bond elongation driven by O–O Coulomb coupling. The components centered at 3200 and 3450 $cm^{-1}$ corresponds, respectively, to the bulk and skin component for pure water.

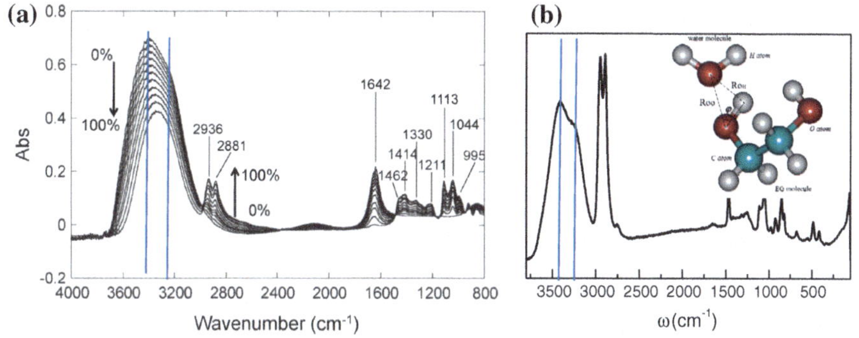


**Fig. 9.23** **a** IR spectra of glycerol/water solutions from 0 to 100 volume % glycerol [102] and **b** the Raman spectrum of pure glycerol [103]. The *inset* (**b**) illustrates the molecular structure of the ethylene glycol–$H_2O$ hydrogen bond [104] (Reprinted with permission from [102, 103].)

(3) The narrowing of the $\omega_H$ peak indicates the viscoelasticity enhancement of the solution, which also lowers the molecular mobility.
(4) The $\omega_L$ blue shift and the $\omega_H$ redshift disperse the quasisolid phase boundary inwardly, which raises the temperature for quasisolid/ice (freezing) transition and lowers the temperature for liquid/quasisolid (melting) transition.
(5) The trend of glycerol addition is opposite to salting [105] that stiffens the $\omega_H$ and softens the $\omega_L$ associated with $\omega_L$ peak narrowing and contact angle increment.

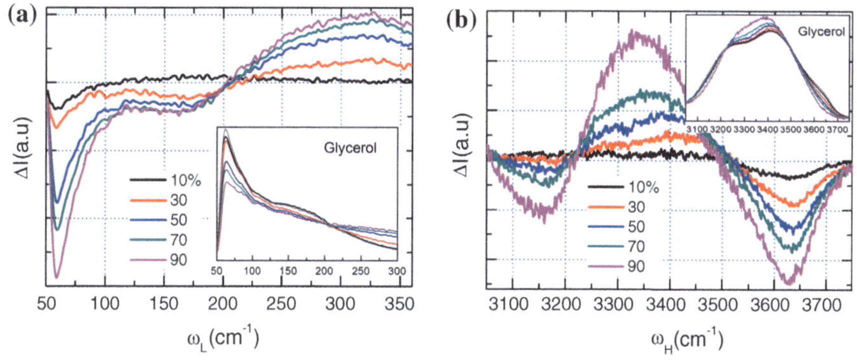


## 9.7 Summary

Molecular undercoordination-induced O:H–O bond relaxation and the enhanced nonbonding electron polarization and skin elasticity clarify the skin supersolidity of ice. Consistence between theory and observations clarifies the following bonding rules for the superlubricancy of ice, aqueous solutions, and sold self-lubrication:

(1) Undercoordination-induced O:H–O relaxation results in the supersolid phase that is elastic, hydrophobic, thermally more stable, and less dense, which dictates the unusual frictionless behaviour of ice skin.
(2) The dual polarization makes ice skin hydrophobic, viscoelastic, and frictionless. Interface Coulomb repulsion and elasticity is essential to lower the friction force.
(3) The supersolid skin causes slipperiness of ice through the elastic Coulomb-levitation mechanism. The elastic, soft O:H nonbond springs of low frequency and high amplitude of vibration attached with pinned dipoles have high recoverability of reformation.
(4) These understanding extend to the superfluidity of $^4$He and the lubricity of water droplet flow in carbon nanotubes as well as nitrides and oxides.
(5) Lon pair interactions and the skin polarization play the key role in determining the dry and wet lubricity in lowering the friction coefficient.

The presence of phononic elasticity and electronic repulsivity is essential for superlubricity. Nonbond vibration creates soft phonons of low frequency and high magnitude with extraordinary recoverability of deformation. Localized polarization by the electron lone pairs and the densely entrapped core and bonding electrons provide the repulsivity at contacts. O:H–O bond electrification by charged ions would be an promising means for lowering the friction coefficient to realize superlubricity.

## Appendix: Featured News

**Why is ice slippery?—New Scientist** 2 September 2015 by Gilead Amit

*Most think it's down to a liquid layer, but can't agree on how it forms. One theory insists it's a "supersolid skin" capable of electrostatic repulsion.*

*(Image: Josh Haner/eyevine)*

FOR physicists no less than figure skaters, ice is remarkably hard to get a grip on. The overwhelming consensus is that ice has low friction because of a thin film of liquid water coating its surface. Hence skaters balanced on thin metal blades can glide smoothly across the ice rink, but grind to a halt on the wooden floor beyond. The tricky part is how this liquid layer forms. More than a century of research has brought us little closer to a definitive answer.

It all started in June 1850, when Michael Faraday told an audience at London's Royal Institution of how pressing two ice cubes together led to them forming a single block. He attributed this to the appearance of an intervening film of water that quickly refreezes. For many years, the appearance of this layer of water was put down to pressure. In fact, even a person of above-average weight on a single skate generates far too little pressure to account for the observed melting, says Anne-Marie Kietzig of McGill University in Montreal, Canada. "The mathematics doesn't work out."

Instead, Kietzig argues that the main player is frictional heating. The movement of a blade across the ice, for instance, easily generates enough heat to melt some of it.

You might think that would be the end of it. But Changqing Sun of Nanyang Technical University in Singapore has other ideas. He argues that since ice is slippery even when you're standing still, friction cannot be the whole story. "Mechanisms such as friction heating and pressure melting have been ruled out," he says.

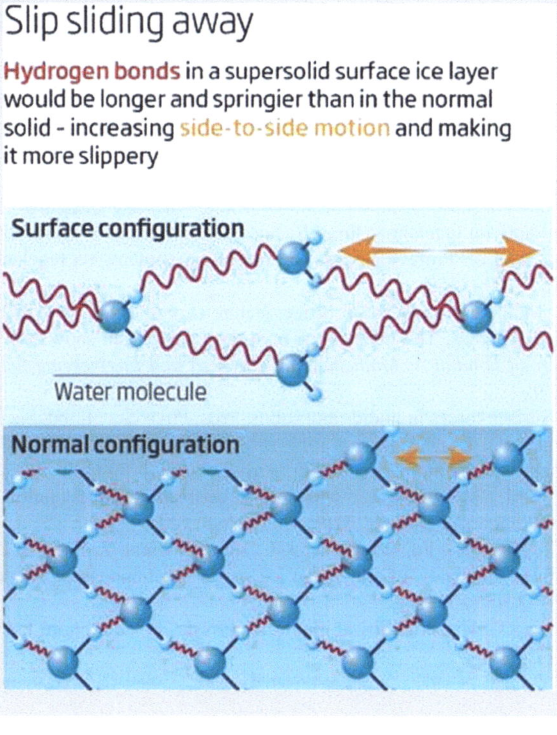

https://www.newscientist.com/article/mg22730371-000-why-is-ice-slippery/

According to Sun, the assumption that the slippery layer coating ice is a liquid is also fundamentally flawed. He says this layer should properly be called a "supersolid skin" because the weak bonds between $H_2O$ molecules at the surface are stretched, but unlike in liquid water none of them are broken. He also argues that this elongation of bonds ultimately produces a repulsive electrostatic force between the surface layer and anything it comes into contact with (see diagram).

He compares the effect to the electromagnetic force that levitates Maglev trains, or the air pressure a hovercraft generates beneath its hull. If he's right, his model helps to explain many of the layer's properties, including its remarkably low friction. "I believe the problem has been completely resolved," says Sun.

Most in the ice field are not convinced. Gen Sazaki at Hokkaido University in Sapporo, Japan, who made the first direct observations of this layer in 2013, prefers to call it a quasi-liquid. He thinks it represents a transitional stage between solid and liquid as the temperature goes up.

For Sazaki, understanding how this mysterious sheet of $H_2O$ forms is still some way off. Even when it comes to something as familiar as slipping on ice, he says, "reality is much more complicated than we expected".

(Reprinted with permission from [106].)

**Notes to this piece of news.**

The perspective of "quasiliquid skin" returns this issue to the very beginning of Faraday' suggestion with challenging physical origin.

## References


1. M. Faraday, Note on regelation. Proc. R. Soc. London **10**, 440–450 (1859)
2. J. Thomson, Note on Professor Faraday's recent experiments on regelation. Proc. R. Soc. Lond. **10**, 151–160 (1859)
3. T.B. James, Melting and regelation of ice. Nature (London) **5**, 185 (1872)
4. F. Bowden, T. Hughes, The mechanism of sliding on ice and snow, in *Proceedings of the Royal Society of London A: Mathematical, Physical and Engineering Sciences* (The Royal Society, 1939)
5. C. Gurney, Surface forces in liquids and solids. Proc. Phys. Soc. Lond. Sect. A **62**(358), 639–655 (1949)
6. J. Krim, Friction at the atomic scale. Sci. Am. **275**(4), 74–80 (1996)
7. H. Liang, J.M. Martin, T.L. Mogne, Experimental investigation of friction on low-temperature ice. Acta Mater. **51**(9), 2639–2646 (2003)
8. C.Q. Sun, X. Zhang, X. Fu, W. Zheng, J.-L. Kuo, Y. Zhou, Z. Shen, J. Zhou, Density and phonon-stiffness anomalies of water and ice in the full temperature range. J. Phys. Chem. Lett. **4**, 3238–3244 (2013)
9. A.-M. Kietzig, S.G. Hatzikiriakos, P. Englezos, Physics of ice friction. J. Appl. Phys. **107**(8), 081101-15 (2010)

10. X. Zhang, Y. Huang, Z. Ma, Y. Zhou, W. Zheng, J. Zhou, C.Q. Sun, A common supersolid skin covering both water and ice. PCCP **16**(42), 22987–22994 (2014)
11. C.Q. Sun, X. Zhang, J. Zhou, Y. Huang, Y. Zhou, W. Zheng, Density, elasticity, and stability anomalies of water molecules with fewer than four neighbors. J. Phys. Chem. Lett. **4**, 2565–2570 (2013)
12. C.Q. Sun, *Relaxation of the Chemical Bond*. Springer Series in Chemical Physics, vol. 108 (Springer, Heidelberg, 2014), 807 pp
13. C.A. Silvera Batista, R.G. Larson, N.A. Kotov, Nonadditivity of nanoparticle interactions. Science, **350**(6257) (2015)
14. J. Li, H. Chen, H.A. Stone, Ice lubrication for moving heavy stones to the Forbidden City in 15th- and 16th-century China. Proc. Natl. Acad. Sci. U.S.A. **100**(50), 20023–20027 (2013)
15. J. Krim, Friction and energy dissipation mechanisms in adsorbed molecules and molecularly thin films. Adv. Phys. **61**(3), 155–323 (2012)
16. S. Sukhorukov, S. Loset, Friction of sea ice on sea ice. Cold Reg. Sci. Technol. **94**, 1–12 (2013)
17. F. Kennedy, E. Schulson, D. Jones, The friction of ice on ice at low sliding velocities. Philos. Mag. A **80**(5), 1093–1110 (2000)
18. E.M. Schulson, A.L. Fortt, Friction of ice on ice. J. Geophys. Res.-Solid Earth **117**, B12204 (2012)
19. X. Zhang, Y. Huang, P. Sun, X. Liu, Z. Ma, Y. Zhou, J. Zhou, W. Zheng, C.Q. Sun, Ice regelation: hydrogen-bond extraordinary recoverability and water quasisolid-phase-boundary dispersivity. Sci. Rep. **5**, 13655 (2015)
20. R. Rosenberg, Why is ice slippery? Phys. Today **58**(12), 50 (2005)
21. C.Q. Sun, Size dependence of nanostructures: impact of bond order deficiency. Prog. Solid State Chem. **35**(1), 1–159 (2007)
22. E. Strelcov, R. Kumar, V. Bocharova, B.G. Sumpter, A. Tselev, S.V. Kalinin, Nanoscale lubrication of ionic surfaces controlled via a strong electric field. Sci. Rep. **5**, 8049 (2015)
23. M.W. Orem, A.W. Adamson, Physical adsorption of vapor on ice: II. n-alkanes. J. Colloid Interface Sci. **31**(2), 278–286 (1969)
24. M.J. Molina, Heterogeneous chemistry on polar stratospheric clouds. Atmos. Environ. Part A. General Topics **25**(11), 2535–2537 (1991)
25. V.I. Kvlividze, V.F. Kiselev, A.B. Kurzaev, L.A. Ushakova, The mobile water phase on ice surfaces. Surf. Sci. **44**(1), 60–68 (1974)
26. I. Golecki, C. Jaccard, The surface of ice near 0 C studied by 100 keV proton channeling. Phys. Lett. A **63**(3), 374–376 (1977)
27. Y. Furukawa, M. Yamamoto, T. Kuroda, Ellipsometric study of the transition layer on the surface of an ice crystal. J. Cryst. Growth **82**(4), 665–677 (1987)
28. K.R. Wilson, R.D. Schaller, D.T. Co, R.J. Saykally, B.S. Rude, T. Catalano, J.D. Bozek, Surface relaxation in liquid water and methanol studied by x-ray absorption spectroscopy. J. Chem. Phys. **117**(16), 7738–7744 (2002)
29. H. Dosch, A. Lied, J.H. Bilgram, Glancing-angle X-ray scattering studies of the premelting of ice surfaces. Surf. Sci. **327**(1–2), 145–164 (1995)
30. Y. Li, G.A. Somorjai, Surface premelting of ice. J. Phys. Chem. C **111**(27), 9631–9637 (2007)
31. A. Döppenschmidt, H.-J. Butt, Measuring the thickness of the liquid-like layer on ice surfaces with atomic force microscopy. Langmuir **16**(16), 6709–6714 (2000)
32. S. Engemann, H. Reichert, H. Dosch, J. Bilgram, V. Honkimaki, A. Snigirev, Interfacial melting of ice in contact with $SiO_2$. Phys. Rev. Lett. **92**(20), 205701 (2004)
33. C.Q. Sun, Y. Sun, Y.G. Ni, X. Zhang, J.S. Pan, X.H. Wang, J. Zhou, L.T. Li, W.T. Zheng, S. S. Yu, L.K. Pan, Z. Sun, Coulomb repulsion at the nanometer-sized contact: a force driving superhydrophobicity, superfluidity, superlubricity, and supersolidity. J. Phys. Chem. C **113** (46), 20009–20019 (2009)

34. A. Uysal, M. Chu, B. Stripe, A. Timalsina, S. Chattopadhyay, C.M. Schlepütz, T.J. Marks, P. Dutta, What X rays can tell us about the interfacial profile of water near hydrophobic surfaces. Phys. Rev. B **88**(3), 035431 (2013)
35. X.J. Liu, M.L. Bo, X. Zhang, L.T. Li, Y.G. Nie, H. TIan, Y. Sun, S. Xu, Y. Wang, W. Zheng, C.Q. Sun, Coordination-resolved electron spectrometrics. Chem. Rev. **115**(14), 6746–6810 (2015)
36. J. Zhang, P. Chen, B. Yuan, W. Ji, Z. Cheng, X. Qiu, Real-space identification of intermolecular bonding with atomic force microscopy. Science **342**(6158), 611–614 (2013)
37. Y. Huang, X. Zhang, Z. Ma, Y. Zhou, J. Zhou, W. Zheng, C.Q. Sun, Size, separation, structure order, and mass density of molecules packing in water and ice. Sci. Rep. **3**, 3005 (2013)
38. T.F. Kahan, J.P. Reid, D.J. Donaldson, Spectroscopic probes of the quasi-liquid layer on ice. J. Phys. Chem. A **111**(43), 11006–11012 (2007)
39. M. Sulpizi, M. Salanne, M. Sprik, M.-P. Gaigeot, Vibrational sum frequency generation spectroscopy of the water liquid-vapor interface from density functional theory-based molecular dynamics simulations. J. Phys. Chem. Lett. **4**(1), 83–87 (2013)
40. M. Abu-Samha, K.J. Borve, M. Winkler, J. Harnes, L.J. Saethre, A. Lindblad, H. Bergersen, G. Ohrwall, O. Bjorneholm, S. Svensson, The local structure of small water clusters: imprints on the core-level photoelectron spectrum. J. Phys. B **42**(5), 055201 (2009)
41. K. Nishizawa, N. Kurahashi, K. Sekiguchi, T. Mizuno, Y. Ogi, T. Horio, M. Oura, N. Kosugi, T. Suzuki, High-resolution soft X-ray photoelectron spectroscopy of liquid water. PCCP **13**, 413–417 (2011)
42. B. Winter, E.F. Aziz, U. Hergenhahn, M. Faubel, I.V. Hertel, Hydrogen bonds in liquid water studied by photoelectron spectroscopy. J. Chem. Phys. **126**(12), 124504 (2007)
43. S.A. Harich, D.W.H. Hwang, X. Yang, J.J. Lin, X. Yang, R.N. Dixon, Photodissociation of H2O at 121.6 nm: a state-to-state dynamical picture. J. Chem. Phys. **113**(22), 10073–10090 (2000)
44. K.R. Siefermann, Y. Liu, E. Lugovoy, O. Link, M. Faubel, U. Buck, B. Winter, B. Abel, Binding energies, lifetimes and implications of bulk and interface solvated electrons in water. Nat. Chem. **2**, 274–279 (2010)
45. G.H. Pollack, *The Fourth Phase of Water: Beyond Solid, Liquid, and Vapor* (Ebner & Sons Seattle, USA, 2013)
46. X. Zhang, P. Sun, Y. Huang, Z. Ma, X. Liu, J. Zhou, W. Zheng, C.Q. Sun, Water nanodroplet thermodynamics: quasi-solid phase-boundary dispersivity. J. Phys. Chem. B **119** (16), 5265–5269 (2015)
47. M. James, T.A. Darwish, S. Ciampi, S.O. Sylvester, Z.M. Zhang, A. Ng, J.J. Gooding, T.L. Hanley, Nanoscale condensation of water on self-assembled monolayers. Soft Matter **7**(11), 5309–5318 (2011)
48. J. Li, Y.X. Li, X. Yu, W.J. Ye, C.Q. Sun, Local bond average for the thermally driven elastic softening of solid specimens. J. Phys. D-Appl. Phys. **42**(4), 045406 (2009)
49. M.J. Holmes, N.G. Parker, M.J.W. Povey, Temperature dependence of bulk viscosity in water using acoustic spectroscopy. J. Phys: Conf. Ser. **269**, 012011 (2011)
50. D. Xu, K.M. Liechti, K. Ravi-Chandar, Mechanical probing of icelike water monolayers. Langmuir **25**(22), 12870–12873 (2009)
51. K.B. Jinesh, J.W.M. Frenken, Experimental evidence for ice formation at room temperature. Phys. Rev. Lett. **101**(3), 036101 (2008)
52. E. Kim, M.H.W. Chan, Probable observation of a supersolid helium phase. Nature **427** (6971), 225–227 (2004)
53. S. Balibar, F. Caupin, Supersolidity and disorder. J. Phys.: Condens. Matter **20**(17), 173201 (2008)

54. B. Hunt, E. Pratt, V. Gadagkar, M. Yamashita, A.V. Balatsky, J.C. Davis, Evidence for a superglass state in solid $^4$He. Science **324**, 632–636 (2009)
55. C. Wang, H. Lu, Z. Wang, P. Xiu, B. Zhou, G. Zuo, R. Wan, J. Hu, H. Fang, Stable liquid water droplet on a water monolayer formed at room temperature on ionic model substrates. Phys. Rev. Lett. **103**(13), 137801–137804 (2009)
56. T. Ishiyama, H. Takahashi, A. Morita, Origin of vibrational spectroscopic response at ice surface. J. Phys. Chem. Lett. **3**, 3001–3006 (2012)
57. H. Qiu, W. Guo, Electromelting of confined monolayer ice. Phys. Rev. Lett. **110**(19), 195701 (2013)
58. R. Vacha, O. Marsalek, A.P. Willard, D.J. Bonthuis, R.R. Netz, P. Jungwirth, Charge transfer between water molecules as the possible origin of the observed charging at the surface of pure water. J. Phys. Chem. Lett. **3**(1), 107–111 (2012)
59. Y. Huang, X. Zhang, Z. Ma, Y. Zhou, W. Zheng, J. Zhou, C.Q. Sun, Hydrogen-bond relaxation dynamics: resolving mysteries of water ice. Coord. Chem. Rev. **285**, 109–165 (2015)
60. C.Q. Sun, Oxidation electronics: bond-band-barrier correlation and its applications. Prog. Mater. Sci. **48**(6), 521–685 (2003)
61. J. Guo, X. Meng, J. Chen, J. Peng, J. Sheng, X.-Z. Li, L. Xu, J.-R. Shi, E. Wang, Y. Jiang, Real-space imaging of interfacial water with submolecular resolution. Nat. Mater. **13**, 184–189 (2014)
62. H. Bluhm, T. Inoue, M. Salmeron, Friction of ice measured using lateral force microscopy. Phys. Rev. B **61**(11), 7760 (2000)
63. A. Andreev, I. Lifshits, Quantum theory of defects in crystals. ZHUR EKSPER TEORET FIZIKI **56**(6), 2057–2068 (1969)
64. T.L. Schindler, *A Possible New Form of 'Supersolid' Matter,* National Science Foundation News (2005), http://www.nsf.gov/news/news_videos.jsp?org=NSF&cntn_id=100323&preview=false&media_id=51151
65. E. Kim, M.H. Chan, Observation of superflow in solid helium. Science **305**(5692), 1941–1944 (2004)
66. J. Day, J. Beamish, Low-temperature shear modulus changes in solid He-4 and connection to supersolidity. Nature **450**(7171), 853–856 (2007)
67. S. Balibar, Supersolid helium: stiffer but flowing. Nat. Phys. **5**(8), 534–535 (2009)
68. P.W. Anderson, A Gross-Pitaevskii treatment for supersolid helium. Science **324**, 631–632 (2009)
69. L. Pollet, M. Boninsegni, A.B. Kuklov, N.V. Prokof'ev, B.V. Svistunov, M. Troyer, Local stress and superfluid properties of solid He-4. Phys. Rev. Lett. **101**(9), 097202 (2008)
70. S. Sasaki, R. Ishiguro, F. Caupin, H.J. Maris, S. Balibar, Superfluidity of grain boundaries and supersolid behavior. Science **313**(5790), 1098–1100 (2006)
71. H.J. Maris, Effect of elasticity on torsional oscillator experiments probing the possible supersolidity of helium. Phys. Rev. B **86**(2), 020502 (2012)
72. D.Y. Kim, M.H.W. Chan, Absence of supersolidity in solid helium in porous vycor glass. Phys. Rev. Lett. **109**(15), 155301 (2012)
73. D.Y. Kim, M.H.W. Chan, Upper limit of supersolidity in solid helium. Phys. Rev. B **90**(6), 064503 (2014)
74. J. Saunders, A glassy state of supersolid helium. Science **324**, 601–602 (2009)
75. A.T. Dorsey, D.A. Huse, Condensed-matter physics: shear madness. Nature **450**(7171), 800–801 (2007)
76. A. Socoliuc, E. Gnecco, S. Maier, O. Pfeiffer, A. Baratoff, R. Bennewitz, E. Meyer, Atomic-scale control of friction by actuation of nanometer-sized contacts. Science **313** (5784), 207–210 (2006)
77. E. Gnecco, S. Maier, E. Meyer, Superlubricity of dry nanocontacts. J. Phys.-Condens. Matter **20**(35), 354004 (2008)
78. J.A. Thomas, A.J.H. McGaughey, Reassessing fast water transport through carbon nanotubes. Nano Lett. **8**(9), 2788–2793 (2008)

79. J. Cumings, A. Zettl, Low-friction nanoscale linear bearing realized from multiwall carbon nanotubes. Science **289**(5479), 602–604 (2000)
80. R.J. Cannara, M.J. Brukman, K. Cimatu, A.V. Sumant, S. Baldelli, R.W. Carpick, Nanoscale friction varied by isotopic shifting of surface vibrational frequencies. Science **318**(5851), 780–783 (2007)
81. A. Socoliuc, R. Bennewitz, E. Gnecco, E. Meyer, Transition from stick-slip to continuous sliding in atomic friction: entering a new regime of ultralow friction. Phys. Rev. Lett. **92**(13), 134301 (2004)
82. A. Erdemir, J.M. Martin (eds.), *Superlubricity* (Elsevier, 2007)
83. Q.Z. Yuan, Y.P. Zhao, Hydroelectric voltage generation based on water-filled single-walled carbon nanotubes. J. Amer. Chem. Soc. **131**(18), 6374 (2009)
84. J.Y. Park, D.F. Ogletree, P.A. Thiel, M. Salmeron, Electronic control of friction in silicon pn junctions. Science **313**(5784), 186–186 (2006)
85. G. Tomlinson, CVI. A molecular theory of friction. Lond. Edinburgh Dublin Philos. Mag. J. Sci. **7**(46), 905–939 (1929)
86. L. Prandtl, Mind model of the kinetic theory of solid bodies. Zeitschrift Fur Angewandte Mathematik Und Mechanik **8**, 85–106 (1928)
87. Y.F. Mo, K.T. Turner, I. Szlufarska, Friction laws at the nanoscale. Nature **457**(7233), 1116–1119 (2009)
88. C.Q. Sun, H.L. Bai, B.K. Tay, S. Li, E.Y. Jiang, Dimension, strength, and chemical and thermal stability of a single C-C bond in carbon nanotubes. J. Phys. Chem. B **107**(31), 7544–7546 (2003)
89. W.T. Zheng, C.Q. Sun, Electronic process of nitriding: mechanism and applications. Prog. Solid State Chem. **34**(1), 1–20 (2006)
90. C.Q. Sun, A model of bonding and band-forming for oxides and nitrides. Appl. Phys. Lett. **72**(14), 1706–1708 (1998)
91. C.Q. Sun, B.K. Tay, S.P. Lau, X.W. Sun, X.T. Zeng, S. Li, H.L. Bai, H. Liu, Z.H. Liu, E.Y. Jiang, Bond contraction and lone pair interaction at nitride surfaces. J. Appl. Phys. **90**(5), 2615–2617 (2001)
92. C.Q. Sun, Thermo-mechanical behavior of low-dimensional systems: the local bond average approach. Prog. Mater Sci. **54**(2), 179–307 (2009)
93. C. Lu, Y.W. Mai, P.L. Tam, Y.G. Shen, Nanoindentation-induced elastic-plastic transition and size effect in alpha-Al2O3(0001). Philos. Mag. Lett. **87**(6), 409–415 (2007)
94. Y.P. Zhao, *Physical Mechanics of Surface and Interface (in Chinese)* (Scientific Press, Beijing, 2012)
95. J. Li, C. Zhang, J. Luo, Superlubricity behavior with phosphoric acid-water network induced by rubbing. Langmuir **27**(15), 9413–9417 (2011)
96. J. Li, C. Zhang, L. Sun, X. Lu, J. Luo, Tribochemistry and superlubricity induced by hydrogen ions. Langmuir **28**(45), 15816–15823 (2012)
97. J. Li, C. Zhang, L. Ma, Y. Liu, J. Luo, Superlubricity achieved with mixtures of acids and glycerol. Langmuir **29**(1), 271–275 (2012)
98. B.C. Donose, I.U. Vakarelski, K. Higashitani, Silica surfaces lubrication by hydrated cations adsorption from electrolyte solutions. Langmuir **21**(5), 1834–1839 (2005)
99. Y. Sun, C.Q. Sun, *The Soul of Water: Single Notion, Multiple Myths*. Springer Series of Chemical Physics (Springer, 2016)
100. M. Zhi-Zuo, Z. Chen-Hui, L. Jian-Bin, L. Xin-Chun, W. Shi-Zhu, Superlubricity of a mixed aqueous solution. Chin. Phys. Lett. **28**(5), 056201 (2011)
101. J. Li, C. Zhang, J. Luo, Superlubricity achieved with mixtures of polyhydroxy alcohols and acids. Langmuir **29**(17), 5239–5245 (2013)

102. Y. Kataoka, N. Kitadai, O. Hisatomi, S. Nakashima, Nature of hydrogen bonding of water molecules in aqueous solutions of glycerol by attenuated total reflection (ATR) infrared spectroscopy. Appl. Spectrosc. **65**(4), 436–441 (2011)
103. Y. Zhou, Y. Huang, Y. Gong, C.Q. Sun, Molecular site resolved O:H-O bond thermal relaxation in liquid water. Communicated (2015)
104. L. Weng, C. Chen, J. Zuo, W. Li, Molecular dynamics study of effects of temperature and concentration on hydrogen-bond abilities of ethylene glycol and glycerol: implications for cryopreservation. J. Phys. Chem. A **115**(18), 4729–4737 (2011)
105. X. Zhang, T. Yan, Y. Huang, Z. Ma, X. Liu, B. Zou, C.Q. Sun, Mediating relaxation and polarization of hydrogen-bonds in water by NaCl salting and heating. PCCP **16**(45), 24666–24671 (2014)
106. G. Amit, *Why is ice slippery?*, in *New Scientist* (London, 2015)

# Chapter 10
# Water Supersolid Skin

- *Skin molecular undercoordination not only disperses the quasisolid phase boundary but also derives supersolidity that is hydrophobic, less dense, viscoelastic, repulsive, and thermally stable.*
- *The supersolidity and quasisolidity defines the anomalies of water skin when interact with other objects.*
- *The skin supersolidity increases with its curvature but drops when heated due to depolarization.*
- *Electrostatic repulsivity and elasticity claims the superhydrophobicity, superfluidity, superlubricity, and supersolidity at the contacting interface.*

**Abstract** Consistency in experimental observations, numerical calculations, and theoretical predictions revealed that skins of 25 °C water and −(15-20) °C ice share the same attribute of supersolidity characterized by the identical H–O vibration frequency of 3450 $cm^{-1}$. Molecular undercoordination and inter-electron-pair repulsion shortens the H–O bond and lengthen the O:H nonbond, leading to a dual process of nonbonding electron polarization. This relaxation-polarization process enhances the dipole moment, elasticity, viscosity, thermal stability of these skins with 25 % density loss, which is responsible for the hydrophobicity and toughness of water skin and the superfluidity in a microchannel.

## 10.1 Challenge: Why Is Water Skin Unusual?

Water on water, water on certain kinds of object, or the inverse exhibits numerous anomalies as epitomized in Figs. 10.1 and 10.2:

(1) Why is the stress of water skin abnormally high and why the stress drops when heated?
(2) Why is the curved skin more elastic, hydrophobic, less dense, tough, and thermally more stable?
(3) Why are some contacting interfaces superhydrophobic and some others superhydrophilic?
(4) Does a solid skin form on water or a liquid sheet cover ice?


C.Q. Sun and Y. Sun, *The Attribute of Water*, Springer Series in Chemical Physics 113, DOI 10.1007/978-981-10-0180-2_10

**Fig. 10.1** **a** A strider can stand still and slide effortlessly on water and **b** an ant is totally unable to escape from a tiny drop of water when trapped (*Credit* Adam Gormley, Queensland, Australia, 2011) [1]

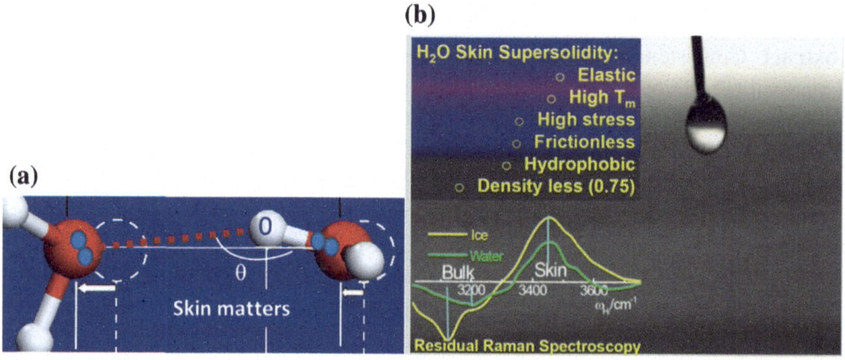


**Fig. 10.2** **a** Molecular undercoordination shortens the H–O bond and stiffens its phonon, meanwhile, lengthens and softens the O:H nonbond associated with dual polarization, which result in the supersolidity of water and ice skins [2] and dispersion of the quasisolid phase boundaries [6]. **b** A video clip [7] shows that a water droplet bounces continuously on a flat water surface, which evidences the elasticity and hydrophobicity of both skins. The Residual Raman spectra of *inset* (**b**) revealed identical $\omega_H = 3450\ cm^{-1}$ for the skins water and ice, and the $3200\ cm^{-1}$ for 25 °C bulk water and $3150\ cm^{-1}$ for –15 °C bulk ice, see Fig. 10.6 in Sect. 10.4 [3]

## 10.2 Clarification: Undercoordinated Molecular Supersolidity

Molecular undercoordination and interoxygen repulsion result in O:H–O bond cooperative relaxation, which claims the full responsibility for water skin anomalies (see Fig. 10.2) [2]:

(1) Undercoordinated water molecules shrink their sizes ($d_H$) and expand their separations ($d_L$) through interoxygen repulsion and polarization; H–O contraction and O:H expansion reduce the skin mass density by up to 25 % [3].

(2) H–O bond stiffening increases the frequency and lifetime of the H–O phonon $\omega_H$, melting point $T_m$, O 1s energy shift $E_{1s}$, and the $E_H$ for H–O atomic dissociation; O:H softening lowers the O:H phonon frequency $\omega_L$, critical temperatures for boiling $T_V$ and freezing $T_N$, and the $E_L$ for molecular dissociation [4].
(3) Polarization raises the hydrophobicity, viscoelasticity, repulsivity, and molecular stability with compressive instead of tensile stress, which make water and ice skins to be supersolid characterized by an identical $\omega_H$ of 3450 $cm^{-1}$ [2].
(4) Undercoordinated skin molecules are subject to "intermolecular repulsion (compressive stress)" rather than "stress (tensile stress)" because of the O–O repulsion and polarization [5].

## 10.3 History Background

### *10.3.1 Wonders of Water's Tough Skin*

#### 10.3.1.1 Skin Stress

Water skin stress is amazingly high [8]. Small insects such as a strider can stand, walk and glide on water effortlessly. A water strider statically standing on water can bear a load up to ten times its body weight with its middle and hind legs, which tread deep puddles without piercing the water skin [9]. If carefully placed on the skin, a small needle, or a coin, floats on water even though its density is times higher than that of water because: (i) it weighs insufficiently to penetrate the skin and (ii) the interface between its paddle and the skin of water is hydrophobic. If the surface is agitated to break up the stress, then the needle will sink quickly. The extraordinary hydrophobicity and toughness of water skin are attributed to the presence of a layer of molecules in the solid state [10, 11].

Skin stress helps seeds bury themselves by causing awns to coil and uncoil. It enables a floating fern to maintain an air layer, even when submerged. It also makes a beetle fly in two dimensions, not three. Surface stress also allows human and agricultural pathogens to travel long distances in tiny, buoyant droplets. The hardly noticed skin stress does play an important role in life at large [12].

#### 10.3.1.2 Wetting and Dewetting

The interaction between a liquid and a solid involves three interfaces: the solid–liquid interface, the liquid–vapor interface and the solid–vapor interface. The $\gamma_{SG}$, $\gamma_{SL}$ and $\gamma_{LG}$ are used to describe stresses of the solid/gas, solid/liquid, and liquid/gas interfaces, respectively. The free surface of the liquid droplet meets a solid to form the so-called three-phase-contact line. The contact line can move along the solid surface, leading to “wetting” or “de-wetting”.

Wetting is the ability of a liquid to maintain contact with a solid surface, resulting from intermolecular interactions when the two are brought together. The degree of wetting (wettability) is determined by a force to balance the interface adhesive and the intradroplet cohesive forces. Wetting deals with the three phases of materials: gas, liquid, and solid. Regardless of the extent of wetting, the shape of a liquid drop on a rigid surface is roughly a truncated sphere. Controlling the wettability of solid materials is a classical and key issue in surface engineering. Typical examples for wetting-dependent processes in daily life, biology and industry include adhesion, cleaning, lubricating, painting, printing, and many more.

Figure 10.3 illustrates the contact of water droplet on patterned substrate. The contact angle between the droplet and the substrate depends on the pattern of the substrate. One may fabricates the substrate to adjust the wettability of the interface.

#### 10.3.1.3 Young’s Equation

In 1804, Thomas Young firstly developed the theory of capillary phenomena on the principle of surface stress. He also observed the constancy of the angle of contact of a liquid surface with a solid surface, and showed how from these two principles to deduce the phenomena of capillary action. Young formulated the surface wetting from the perspective of force equilibrium, as illustrated in Fig. 10.4 [15]. Each of these interfaces has an associated surface stress, $\gamma$, which represents the energy required to create a unit area of that particular interface. A different approach is to regard $\gamma$ as force acting on the water drop. At equilibrium, force equilibrium along

**Fig. 10.3** Substrate patterns and the superhydrophobicity of water droplets [13]. Atomic undercoordination at the sharp proxy undergoes bonding electron quantum entrapment and densification, which polarizes the nonbonding electrons if exist, which results in the high elasticity and strongly polarized skin, being responsible for the superhydrophobicity of the substrate, If nonbonding electrons are absent, the skin will be hydrophilic. Therefore, roughened skin enhances the initial hydrophilicity and hydrophilicity of the parent substance of the substrate [14]

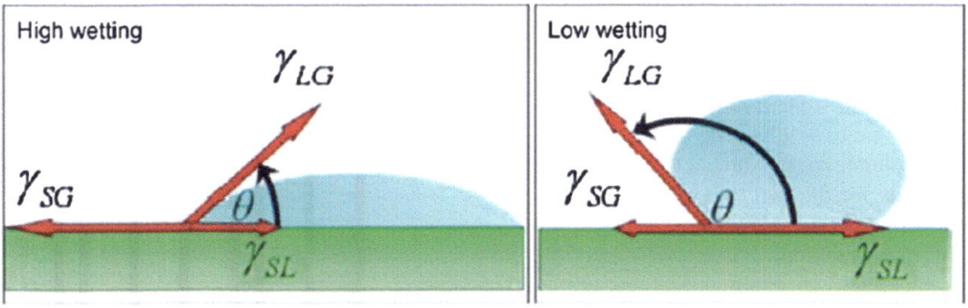


**Fig. 10.4** Illustration of Young equation for the surface energy. On the *left*, there is much wetting and the contact angle is small. On the *right*, little wetting and the contact angle is large. The drawing derives Young force equations, $\cos \theta = (\gamma_{SG} - \gamma_{SL})/\gamma_{LG}$ [17]

the X axis correlates the contact angle θ, and the surface stresses of the three interfaces [16].

Young's equation describes the contact angle of a liquid drop on a plane solid surface as a function of the surface free energy, the interfacial free energy, and the surface stress of the liquid. The magnitude of Young's contact angle is the result of energy minimization. If the liquid–gas surface stress is smaller than the solid–gas surface stress ($\gamma_{LG} < \gamma_{SG}$), the liquid–solid interface will increase to minimize energy. As the drop wets the surface, the contact angle approaches zero, leading to complete wetting. Other ratios of $\gamma_{LG}$ and $\gamma_{SG}$ will lead to the formation of drops of different shapes. A hydrophilic surface is defined as a surface where $0^o < \theta < 90^o$, and hydrophobic surface is a surface where $\theta \geq 90^o$. The surface stresses of different substances in contact with the gas phase vary over a wide range.

#### 10.3.1.4 Young–Laplace–Gauss Equation

In 1805, Pierre-Simon Laplace discovered the significance of meniscus radii with respect to capillary action. He introduced a nonlinear partial differential equation to describe the capillary pressure difference sustained across the interface between two static fluids, such as water and air, due to the surface stress. The Young–Laplace equation relates the pressure difference to the shape of the surface or wall and it is fundamentally important in the study of static capillary surfaces. The equation is named after Young, who developed the qualitative theory of surface stress in 1805, and Laplace who mathematied the description in the following year. It is sometimes also called the Young–Laplace–Gauss equation, as Gauss unified the work of Young and Laplace in 1830, deriving both the differential equation and boundary conditions. Figure 10.5 illustrates the capillary rise described by Young–Laplace–Gauss equation.

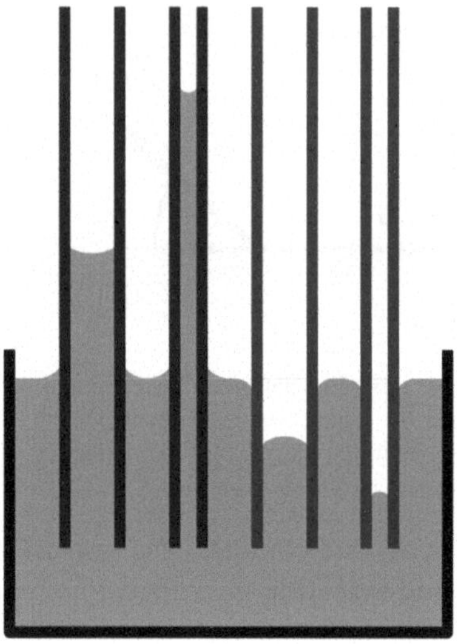

**Fig. 10.5** Schematic of capillary rise in the Young–Laplace–Gauss equation. *Red* hydrophilic (θ < 90°); *blue* hydrophobic (θ > 90°). The liquid column height varies oppositely for the two situations with the cylinder radius (color online)

#### 10.3.1.5 The Young–Dupré Equation

Young's equation was further developed in 1890s by Lewis Dupré to account for thermodynamic effects, and this is known as the Young–Dupré equation. The Young–Dupré equation indicates that neither $\gamma_{SG}$ nor $\gamma_{SL}$ can be larger than the sum of the other two surface energies. The consequence of this restriction is the prediction of complete wetting when $\gamma_{SG} > \gamma_{SL} + \gamma_{LG}$ and zero wetting when $\gamma_{SL} > \gamma_{SG} + \gamma_{LG}$. The lack of a solution to the Young–Dupré equation is an indicator that there is no equilibrium configuration with a contact angle between 0 and 180° for those situations. A parameter for gauging wetting is the spreading parameter $S = \gamma_{SG} - (\gamma_{SL} + \gamma_{LG})$. When $S > 0$, the liquid wets the surface completely. When $S < 0$, partial wetting occurs.

The following summarizes the three equations for the surface stress, capillary effect, and wetting in terms of surface or interface energies.

$$\begin{cases} \cos\theta = \dfrac{\gamma_{SG} - \gamma_{SL}}{\gamma_{LG}} & \text{(Young, 1804)} \\ \Delta p = \rho g h - \gamma(1/R_1 + 1/R_2) & \text{(Young–Laplace–Gauss, 1830)} \\ S = \gamma_{LG}(\cos\theta - 1) & \text{(Young–Dupre, 1890)} \end{cases} \quad (10.1)$$

In the Young–Laplace equation, $\Delta p$ is the pressure difference across the fluid interface, $\gamma$ is the surface stress and $R_1$ and $R_2$ are the principal radii of curvatures.

| | |
|---|---|
|  | Thomas Young (13 June 1773–10 May 1829) was an English polymath and physician. Young made notable scientific contributions to the fields of vision, light, solid mechanics, energy, physiology, language, musical harmony, and Egyptology. In 1804, he developed the qualitative theory of capillary phenomena on the principle of surface stress |
|  | Pierre-Simon, marquis de Laplace (23 March 1749–5 March 1827) was an influential French scholar who made important contribution to mathematics, statistics, physics, and astronomy. Laplace built upon the work of Young the theory of capillary action and the Young–Laplace equation |
|  | Johann Carl Friedrich Gauss (30 April 1777–23 February 1855) was a German mathematician who contributed significantly to many fields, including number theory, algebra, statistics, analysis, differential geometry, geodesy, geophysics, mechanics, electrostatics, astronomy, matrix theory, and optics |

### *10.3.2 Known Mechanisms for Wettability*

One may note that the hydrophobicity, superfluidity, superlubricity, and supersolidity (for $^4$He solid) (called 4S) share considerable similarities. They are non-wetting and frictionless at motion and involved water skin except for $^4$He solid. The following theories describe the 4S in terms of surface stress:

(1) Young–Laplace–Gauss–Dupré theories [16] in terms of surface stress and interface energies.

(2) Wenzel–Cassie–Baxters' law [18, 19] in terms of substrate surface roughness for enhancement of superhydrophobicity and superhydrophilicity without the nature of the liquid water being involved.
(3) Electrical double layer (EDL) scheme [20] for the superfluidity.
(4) Prandtl–Tomlinson (PT) theory [21, 22] of the superposition of the slope of atomic potential and multiple-contact effects [23] for atomic scale quantum friction.
(5) BOLS-NEP notion [2, 24] from the perspective of skin phononic elasticity and electronic repulsivity at the 4S contacting interface. Higher curvature enhances the quantum entrapment and/or polarization of both the water droplet and the fine structures of the solid substrate.

Figure 10.6 illustrates the Wenzel–Cassie–Baxter's law. Wenzel suggests that water will penetrate into the grooves of the rough surface to enhance whatever wetting trend the flat surface exhibits. The nanoscaled roughening makes a hydrophobic surface even more hydrophobic and a hydrophilic surface more hydrophilic. According to Cassie–Baxter's notion, the water drop doesn't penetrate into the surface grooves instead lies on top of them, so that air bubbles are trapped inside. The trapped air increases the contact angle and the surface becomes superhydrophobic.

Many superhydrophobic materials found in nature appeared to fulfill Wenzel–Cassie–Baxters' laws [19] stating that the surface contact angle can be modulated by simply roughing up the solid substrate surface. For instance, fluids can slip freely past pockets of air between textured surfaces with micrometer-scale grooves or posts of tiny distances [25]. By doing so, the slip length for water is much longer than previously achieved, indicating that engineered surfaces can significantly reduce drag in fluid systems. On the base of Cassie–Baxters' law and thermodynamics considerations, one can design tunable superhydrophobic surfaces to control the directional motion of water droplets by varying the pillar width and spacing [26, 27]. Varying the gradient of the stiffness of a micro-beam could also drive directional movement of liquid droplets on a microbeam [28].

A water strider statically standing on water can bear a load up to ten times its body weight with its middle and hind legs, which tread deep puddles without piercing the

**Fig. 10.6** Skin roughness dependence of the contact angles and models for smooth skin, grooved skins with water penetration (Wenzel) and gas bubble entrapment (Cassie–Baxter) [19]. The BOLS-NEP notion [2, 24] elaborates the Wenzel–Cassie–Baxters' law as resulting from local densification, quantum entrapment, and polarization due to atomic/molecular undercoordination

water surface [9]. This fact illustrates the superhydrophobicity of the water strider legs due to "bio-wax" coatings that are composed partly of electron lone pairs. Another comparative experiment [29] using the real water strider legs and artificial legs made of wax-coated steel wires revealed that the adaptive-deformation capacity of the real leg through its three joints makes a more important contribution to the superior load-bearing ability than the superhydrophobicity.

The existing equations or laws correlate the contact angle to the roughness and interface energies without involvement of the nature and relaxation of the chemical bonds and the energetics of electrons at the contacting interface. Furthermore, the classical concepts of surface stress or interface energies often refer to the cost in creating a unit area of surface and interface [30]. In fact, the local energy density and the atomic cohesive energy in the surface and interface region and the associated dynamics, densification, localization, entrapment, and polarization of electrons dictates the performance of a surface and an interface [31]. Most strikingly, the hydrophobicity–hydrophilicity transition by energetic beam radiation is beyond the scope of all the above laws or equations.

### *10.3.3 Selection Rules for Wetting Interface*

The adhesive force between the solid and the liquid causes the drop to spread and wet the surface, and the cohesive force within the liquid drop causes it to ball up and avoid contact with the surface. For example, let us consider the contact angle between water and three types of surfaces—polymers, compounds, and metals.

- Some polymers, such as PVC and Teflon, form mainly van der Waals bonds with a water drop placed on them. These bonds are weak relative to the hydrogen bonds within the drop, so that the water prefers to bond with itself and not with the surface. The result is a bead shaped drop that almost doesn't wet the surface. These are hydrophobic surfaces.
- Compounds containing N, O, and F elements have electron lone pairs that polarize their neighboring atoms. Interface repulsion dominates so the interface is hydrophobic except the skin dipole is artificially removed.
- Metals such as Pt and Co should be hydrophilic because of the dominance of skin charge quantum entrapment. However, Rh, Au, W, and Ag are polarization dominance so they are expected to be hydrophobic [31]. Point defects or rough surfaces will enhance such effects.

### *10.3.4 Contact Angle Manipulation*

Water's surface stress and the contact angle can be controlled through different methods:

(1) Thermal excitation: generally, surface stress drops as temperature increases. The dependence of surface stress on temperature is usually approximately linear reaching zero at the critical temperature, as in Fig. 10.2b.
(2) Surface roughening: water may penetrate into or may form air gap on the grooved surfaces. As afore discussed, the fine structures of a rough surface enhace the undercoordination effect on the quantum entrapment and/or polarization to enhace the hydrophobic/hydrophilic property of the substrate.
(3) Electro-wetting: when a potential difference is applied across, the electrostatic force may pull the droplet down into the surface, lowering the contact angle.
(4) Chemical modification: chemical modification of the solid surface can lead to a change in its surface stress. For example, an addition of polar, hydrophilic groups to the surface will lower the contact angle. However, salt addition into liquid can raise the contact angle by enhancing polarization [32].

In order to understand why the contact angle changes with the geometry and substance, one needs to consider interaction not only between the liquid and the solid but also between the molecules/atoms themselves in the skin of each component. Let us examine the water skin first and then the solid skin next.

## 10.4 Quantitative Resolution

### *10.4.1 Segmental Length—Phonon Frequency—Binding Energy*

Experimental and numerical observations [3, 33] revealed consistantly that the skin O–O distance of water expands by 5.9–6.4 %, corresponding to a density loss by 16–17 % with respect to the density of 4 °C water. However, the skin O–O distance of liquid methanol contracts by 4.6 %. The O–O length relaxation differentiates the surface stress of 72 mN/m for water from the values of 22 mN/m for the liquid methanol. According to the literature record, the density of water skin can be lowered to 0.4 g/cm$^3$ (corresponding to the $d_{OO}$ = 3.66 Å) [34, 35].

Following the same trend of 'normal' materials, molecular undercoordination imparts water local charge densification [36–41], binding energy entrapment [37, 42–44], and nonbonding electron polarization [39]. For instance, the O 1s level shifts more deeply from the bulk value of 536.6–538.1 eV and to 539.7 eV when the bulk water is transformed into skin or into the gaseous monomers [45, 46]. The H–O bond energy is 3.97 eV for bulk water [47], and it is 4.52 and 5.10 eV for the skin and for the gaseous monomers [48], respectively.

DFT calculations estimated the Milliken charge accumulation at the skin and in the bulk of water [3]. O increases its net charge from the bulk value of −0.616 to −0.652 e for the skin. The net charge of a skin water molecule increases from the bulk value of 0.022 to −0.024 e. The densification and entrapment of the bonding electrons polarize the nonbonding charge. An ultra-fast liquid jet vacuum UV photoelectron spectroscopy probed that the bound energy for an nonbonding electron in solution changes from a bulk value of 3.3 to 1.6 eV at the water skin. The bound energy of nonbonding electrons, as a proxy of work function and surface polarization, decreases further with molecular cluster size [39].

Water molecular undercoordination stiffens the stiffer $\omega_H$ significantly [49, 50]. The $\omega_H$ has a peak centered at 3200 $cm^{-1}$ for bulk water, and at 3450 $cm^{-1}$ for the skins of both water and ice (see Fig. 10.1 inset) [51]. The $\omega_H$ for the gaseous molecules is around 3650 $cm^{-1}$ [52–55]. DFT–MD calculations derived that the $\omega_H$ shifts from ~3250 $cm^{-1}$ at a 7 Å depth to ~3500 $cm^{-1}$ of the 2 Å thick skin of liquid water [56].

With the known O–H and H:O bond length relaxation and the tetrahedrally–coordinated structure [4], we obtained the size $d_H$, separation $d_{OO}$, and mass density ρ of molecules packing in water ice in the following relationships with the $d_{H0}$ and the $d_{L0}$ being the references at 4 °C [47],

$$\begin{cases} d_{OO} = 2.6950\rho^{-1/3} & (Molecular\ separation) \\ \frac{d_L}{d_{L0}} = \frac{2}{1+exp[(d_H-d_{H0})/0.2428]}; & (d_{H0} = 1.0004\ and\ d_{L0} = 1.6946\ at\ 4\ °C) \end{cases} \quad (10.2)$$

With the measured $d_{OO}$ of 2.965 Å [33] as an input, this relation yields the segmental lengths of $d_H$ = 0.8406 Å and $d_L$ = 2.1126 Å, which turns out a 0.75 g $cm^{-3}$ skin mass density. This value is much lower than the least bulk value of 0.92 g $cm^{-3}$ for ice. Indeed, the mass density of both skins suffers loss due to undercoordination induced O–O elongation. Table 10.1 summarizes experimental information of the bond length $d_x$, phonon frequency $\omega_x$, and bond energy $E_x$ of water and ice under different coordination environments in comparison to those of ice at 80 K and water dimers.

### 10.4.2 *Hydrophobicity, Repulsivity, and Viscoelasticity*

The polarization of molecules enhances the skin repulsivity and viscoelasticity. The high viscoelasticity and the high density of skin dipoles are essential to the hydrophobicity and lubricity at contacting interfaces [60]. According to the

**Table 10.1** Experimentally-derived skin supersolidity ($\omega_x$, $d_x$, ρ) of water and ice[a]

| | Water (298 K) | Water (298 K)/Ice (253 K) | Ice (253 K) | Ice (80 K) | Vapor |
|---|---|---|---|---|---|
| | Bulk | Skin | Bulk | Bulk | Dimer |
| $\omega_H$ ($cm^{-1}$) | 3200 [51] | 3450 [51] | 3125 [51] | 3090 [57] | 3650 [53] |
| $\omega_L$ ($cm^{-1}$) [57] | 220 | ~180 [4] | 210 | 235 | 0 |
| $d_{OO}$(Å) [47] | 2.700 [58] | 2.965 [33] | 2.771 | 2.751 | 2.980 [33] |
| $d_H$ (Å) [47] | 0.9981 | 0.8406 | 0.9676 | 0.9771 | 0.8030 |
| $d_L$(Å) [47] | 1.6969 | 2.1126 | 1.8034 | 1.7739 | ≥2.177 |
| ρ ($g \cdot cm^{-3}$) [47] | 0.9945 | 0.7509 | 0.92 [59] | 0.94 [59] | ≤0.7396 |

[a]The referenced data are used as input for deriving the rest using (10.2)

BOLS-NEP notion, the local energy densification stiffens the skin and the densely and tightly entrapped bonding charges polarize nonbonding electrons to form anchored skin dipoles [14]. If the substrate is quantum entrapment dominance, the positive skin will attratct the lone pairs of the liquid, making the skin hydrophilic; if the solid skin is polarization dominance, the negative skin will repell the lone pairs of the liquid. The liquid skin always has excessive negative charge due to the undercoordination effect.

Reducing the number of molecular layers of the skin increases local surface stress $\gamma$, and viscosity $\eta_s$ and $\eta_v$ [3]. The O:H–O cooperative relaxation and associated electron entrapment and polarization enhances the surface stress from 31.5 for 15 layers to 73.6 mN/m for five layers, which approaches the measured value of 72 mN/m for water skin at 25 °C. The skin viscosity increases from 0.007 to 0.019 $10^{-2}$ mN s/m$^2$. The bulk $\eta_v$ changes insignificantly from 0.027 to 0.032 for five layer thick skin. Generally, the viscosity of water reaches its maximum at a temperature around the $T_m$ [61].

The negative charge gain and the nonbonding electron polarization provide electrostatic repulsive force that not only lubricates ice but also the hydrophobicity of water skin. Measurements of an elastic modulus of 6.7 GPa have verified the presence of the repulsive forces between a hydrated mica substrate and the tungsten contacts at 24 °C under 20–45 % relative humidity (RH) [62]. Monolayer ice also forms on a graphite surface at 25 % RH and 25 °C [63]. These observations and the present numerical derivatives evidence the presence of the supersolidity with repulsive forces because of bonding charge densification, surface polarization and $T_m$ elevation.

As justified above, the skins of water and ice form an extraordinary supersolid phase that is elastic [51], hydrophobic [64, 65], polarized [39, 66] and thermally stable [67], with densely entrapped bonding electrons [42, 45, 46, 68] and ultra-low-density [33]. The fewer the molecular neighbors there are, the smaller the water molecular size is, the greater the molecular separation is, and therefore the greater the supersolidity will be. The supersolid skin is responsible not only for the slipperiness of ice but also for the hydrophobicity and toughness of water skin.

### 10.4.3 Skin Curvature Dispersed $T_m$ and $T_N$

Water droplets encapsulated in hydrophobic nanopores [69] and point defects [70, 71] are thermally even more stable than the bulk water even because of the undercoordinated molecules in the curved skin. SFG spectroscopy revealed that the skin of two adjacent molecular layers are highly ordered at the hydrophobic contacts compared with those at the flat water–air interface [72]. MD simulations suggested that freezing preferentially starts in the subsurface of water instead of the outermost layer that remains ordered during freezing [70]. The subsurface accommodates better than the bulk the increase of volume connected with freezing.

Furthermore, bulk melting is mediated by the formation of topological defects which preserve the coordination of the tetrahedral network. Such defected clusters

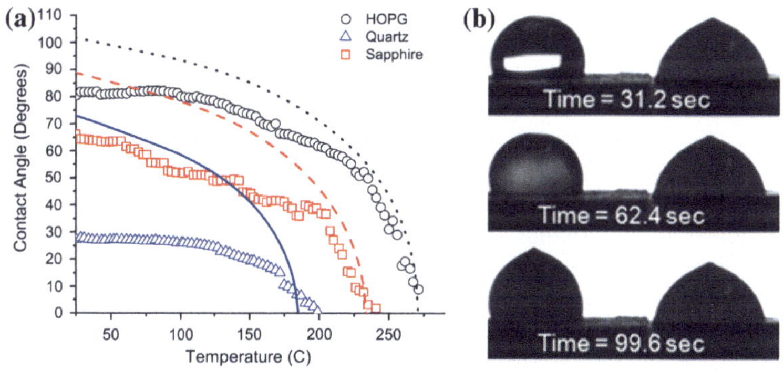


**Fig. 10.7** **a** Evolution of water contact angle on quartz, sapphire and graphite as a function of temperature (°C). **b** Water droplet on rough (*left*) Ag skin freezes 68.4 s later than that on the smooth Ag. The proxy tip of the droplet indicates frozen (Reprinted with permission from [73, 74].)

form a defective region involving about 50 molecules with a surprisingly long lifetime [71]. These observations verify the BOLS-NEP expectations that the undercoordinated water molecules are indeed thermally stable. Therefore, a liquid layer never forms on ice [70] or surrounding defects [71].

Figure 10.7a shows the substrate dependence of transition of the initial contact angles of a droplet to zero [73]. Droplet of initially highly curved skin spreads over the substrates requires higher temperature because curvature raised skin $T_m$ of the droplet.

The higher $T_m$ is always associated with a lower $T_N$ for freezing because the O:H–O relaxation dispersed boundary of the quasisolid phase. As expected and shown Fig. 10.7b, a water droplet on a roughened more hydrophobic Ag skin (with nanocolumnar structures) having a greater contact angle and higher curvature, freezes 68.4 s later than on a smooth Ag surface at –4 °C [74]. The formation of the proxy tip due to volume expansion at the top of the droplet indicates frozen that proceeds from the bottom of the droplet. The delayed freezing indicates the $T_N$ depression as well. Molecules at the curved skin are thermally more stable than those at the flat skin and need longer time to freeze.

Chen et al. [75] examined ice nucleation temperatures of 10 mL water droplets on a series of sol–gel coatings with different roughness and surface energies. They found that the wetting mode at low temperatures is strongly correlated with the icing behavior of the droplets on the surfaces. Ice-phobic coatings (large contact angle) can lower the icing temperature of the droplet on the surface by up to 6.9 °C compared with non-ice phobic ones. The dominant nucleation sites are along the substrate–water–vapor three-phase contact line.

## 10.4.4 *Skin Supersolidity: Elasticity and Hydrophobicity*

### 10.4.4.1 Air Gap at Contact

SFG spectral measurements and MD calculations suggested that the outermost two layers of water molecules have an 'ice-like' order at room temperature [76]. At the ambient temperature, ultrathin films of water perform like ice with a hydrophobic nature [11, 64]. Water at temperatures of 7, 25, and 66 °C and atmospheric pressure has an ordered skin 0.04–0.12 nm thick [77].

Besides, an air gap presents at hydrophobic contacts. Using specular x-ray reflectivity analysis, Uysal et al. [78] suggested that an air gap of 0.5–1.0 nm thick exists between the water and the hydrophobic substrate. The air gap increases with the contact angle, the droplet curvature or with the lowering of the effective CN of molecules at the skin.

Wetting studies on mesoporous silica nanochannels [79] (5) indicate that water is separated from the hydrophobic walls by a vapor gap of 0.3–0.4 nm thick and the presence of ~0.6 nm layer of low-density fluid (vapor) separating water from the hydrophobic solid. MD simulations predict density oscillations extending up to 1.0 nm into the adjacent water accompanied by a molecular orientation order affecting a water layer of 0.7 nm. The best fit of X-ray reflection from water/$SiO_2$ interface derived that the density of the 3.8 Å thick water skin is 0.71 that of water and the hydrophobic gap between water and $SiO_2$ is 0.6 nm [80].

A video clip [7] (see Fig. 10.2b) shows that a water droplet bounces continuously and repeatedly on water, which evidences straightforwardly the elasticity and hydrophobicity of water skin regardless of curvature. Consistency between theoretical calculations and measurements further confirmed that a monolayer film of water manifests 'quasi-solid' behavior at room temperature, and a hydrophobic nature that prevents it from being wetted itself by a water droplet [64, 65], see Fig. 10.8.

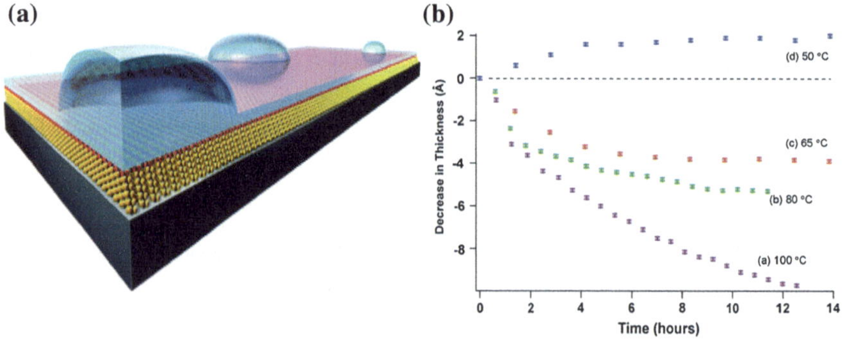


**Fig. 10.8** **a** Nanoscaled water droplet in contact with a continuous water layer deposited on a hydrophilic COOH-terminated monolayer is "ice-like" at room temperature. **b** Time and temperature dependent thickness of water films shows the hydrophobicity and thermal stability of ultrathin water layer at room temperature [64, 65]. The $T_m$ is between 50 and 65 °C and evaporation occurs at temperature around 65 °C, which means the $T_m$ (constant thickness) increase is associated with a depression of the evaporation temperature (thickness reduction). Thickness increase means continuation of ice formation (Reprinted with permission from [65].)

#### 10.4.4.2 $T_m$ Elevation and $T_V$ Depression

The time and temperature dependent thickness of water films deposited on a hydrophilic COOH-terminated monolayer is "ice-like" at room temperature. The hydrophobicity and thermal stability of the ultrathin water layer retains at a critical temperature between 50 and 65 °C [64, 65]. The H–O bond energy dictates melting temperature yet the O:H nonbond energy determines temperature of evaporation. The thickness reduction at 65 °C means that the evaporation temperature is lowered from 100 to 65 °C for ultrathin films though the $T_m$ reaches 50 °C and above.

Water is almost universally present on apparently dry self-assembled monolayers, even on those considered almost hydrophobic by conventional methods such as water contact goniometry, as observed by James and co-workers [65] using AFM, X-ray and neutron reflectometry. Condensation of water on hydrophilic surfaces under ambient conditions formed a dense sub-nanometer surface layer; the thickness of which increased with exponentially limiting kinetics. Tapping mode AFM measurements show the presence of nanometer-sized droplets that covered about 2 % of the total surface area, and which became fewer in number and larger in size with time. While high vacuum ($\sim 10^{-3}$ Pa) at room temperature could hardly remove the adsorbed water droplet from these monolayers; heating to temperatures above 65 °C under atmospheric conditions results in evaporation from the surface.

One may expect that water wet the hydrophilic surface like that of platinum metal. It is expected that the Pt skin is hydrophilic and the Rh skin hydrophobic as electron quantum entrapment is dominant the former but polarization dominates the latter [31]. However, water coats a hydrophobic layer on the hydrophilic surface by hydration and turns the hydrophilic surface hydrophobic [64, 81]. Hydration and drying transitions are highly sensitive to local geometry (for example, concave or convex surfaces), which are "felt" by the three-dimensional hydrogen-bonded network.

It has been shown that the H–O bond energy dictates temperature of melting while the O:H nonbond energy determines evaporation. Findings of James and co-workers [65] confirmed the predicted hydrophobicity and thermal stability of ultrathin water film at room-temperature [64]. The $T_m$ is higher and the $T_V$ of the ultrathin water film is lower compared with the bulk water.

#### 10.4.4.3 Droplet Dancing

Water droplets also dance on solid surfaces, regardless of substrate temperatures and materials (−79 °C $CO_2$; 22 °C superhydrophobic surface, and 300 °C Al plate) [82]. These observations were attributed to the Leidenfrost effect first reported in 1756 [83], and the superhydrophobicity [84] known from the late 1950s for room temperature and substrate sublimitation effects. In the Leidenfrost condition on a hot substrate, the impacting liquid drop rapidly forms a vapor layer at the liquid-substrate interface. This vapor layer (with a thickness typically in the range of 10–100 μm) acts both as cushion and as thermal insulator, causing a freely floating and gradually evaporating drop. Under very low pressure, water droplet on hydrophobic substrate can jump up because the vaporization underneath the suspended droplet [85].

On a superhydrophobic surface, the contact between the water drop and the solid substrate is only partial, due to the presence of a composite air–liquid–solid interface, where air pockets prevent full surface wetting. The wetting area is typically less than 20 % of the total solid surface area. As a result, water drops move easily on the surface, due to low adhesion capillary forces between liquid drop and the featured solid substrate. The lower contact areas and air pockets at the interface result from the substrate atomic undercoordination effect. Atomic undercoordination shortens the bond causing local densification and entrapment of bonding and core electrons, which in turn polarize the nonbonding electrons pertaining to undercoordinated edge atoms, providing repulsive force and making substrate superhydrophobic. Drop rebounds from the substrate at –79 °C temperatures, which indicates sublimation of the solid surface occurs by frost formation, preventing the surface from being contacted to the solid $CO_2$ skin [82] (Fig. 10.9).

**Fig. 10.9** Water droplets impacting on **a** Al sample at 300 °C (subject to Leidenfrost effect—vapor formed below the droplet), **b** superhydrophobic surface (subject to superhydrophobicity—atomic undercoordination induced substrate quantum entrapment and polarization [86]), and **c** solid $CO_2$ at −79 °C (subject to sublimating—frost formation between contacts) (Reprinted with permission from [82].)

### 10.4.5 T-Dependent Surface Stress: $\Theta_{DL}$ and $E_L$

Instead of the energy loss upon surface formation—conventionally called surface energy, the energy gain of a unit volume or the cohesive energy remnant of a discrete atom in the skin of certain thickness governs the performance of a surface [30]. For instance, elasticity and yield strength is proportional to the local energy density according to their dimensionality [Pa = Force/Area = energy/volume], $Y_z \propto E_z/d_z^3$ and the thermal stability characterized by the critical temperature of phase transition, varies with the atomic cohesive energy $T_{Cz} \propto zE_z$. The surface stress and surface energy are proportional to the skin energy density [30].

Water has the highest surface stress of known, which drops at heating. Numerical reproduction of the temperature dependent stress turns out the Debye temperature and the molecular cohesive energy—the sum of binding energy of one molecule with all its neighbors. The following equation formulates the temperature-dependence of the elastic modulus Y and the surface stress $\gamma$, both of which are proportional to the local energy density [30],

$$\gamma(T) \propto \frac{E_S(T)}{d^3(T)} = \frac{E_S(0) - \int_0^T \eta(t)dt}{d^3\left(1 + \int_0^T \alpha(t)dt\right)^3},$$

where $\eta$ is the specific heat in Debye approximation; $\alpha$ is the coefficient of thermal expansion. The following formulates the temperature-dependent $\gamma_s$ [30]:

$$\frac{\gamma_S(T)}{\gamma_S(0)} \cong \left(1 + \int_0^T \alpha_S(t)dt\right)^{-3} \times \begin{cases} 1 - \frac{\int_0^T \eta_S(t/\theta_D)dt}{E_S(0)}, & (T \le \theta_{DL}) \\ 1 - \frac{\eta_S T}{E_S(0)}, & (T > \theta_{DL}) \end{cases}. \quad (10.3)$$

Reproduction of the measured temperature-dependence of surface energy [99], see Fig. 10.10b, results in the molecular cohesive energy $E_S(0) = 4E_L = 0.38$ eV/molecule and the Debye temperature $\Theta_{DL} = 198$ K [30] with the known $\alpha = 0.162$ mJ/m$^2$K for bulk water as input. This $\Theta_{DL}$ value is compatible with $185 \pm 10$ K as derived from helium scattering from ice in the temperature range of 150–191 K [87].

### 10.4.6 Frequency-Resolved H–O Bond Relaxation Time

The time-resolved IF and SFG spectroscopies provide the lifetime of the H–O bond in water under external stimulus, such as salting and electrification. Figure 10.11 shows the excitation frequency dependence of the H–O vibration decay curves and the SFG profiles of the bulk $H_2O$ [88]. The spectra were obtained by monitoring the vibrational dynamics after excitation with a narrowband infrared pump pulse in

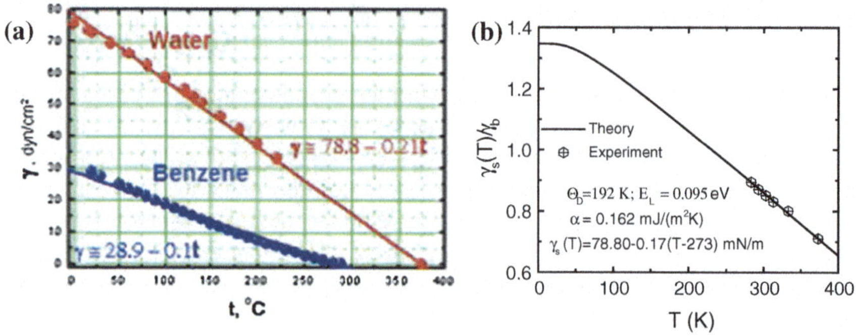


**Fig. 10.10** **a** Temperature dependence of the surface stress $\gamma_s$ for water and benzene. **b** Reproduction of the $\gamma_s(T)$ [7] for water results in an estimation of the Debye temperature $\Theta_{DL} = 198$ K and skin molecular cohesive energy $E_S(0) = 0.38$ eV, which gives rise to $E_L = E_b(0)/4 = 0.095$ eV/nonbond for the skin O:H nonbond dissociation (Reprinted with permission from [30].)

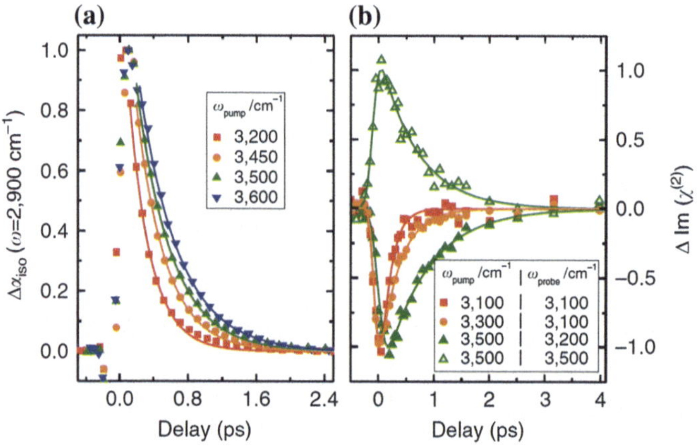


**Fig. 10.11** Frequency dependence of the H–O stretching vibrational decay curves. **a** Normalized infrared pump/probe data for H–O bonds in bulk $H_2O$. **b** Dynamics of the interfacial water molecules obtained using an infrared pump/HD-SFG probe. The SFG probe frequency is set to spectral ranges where the contribution of thermalization to the signal is negligible (Reprinted with permission from [88].)

bulk using infrared pump-probe spectroscopy and at the air/$H_2O$ interface using interfacial IF pump-SFG probe experiments by probing the vibrational response.

Figure 10.12 discriminates the excitation-frequency dependence of the H–O phonon relaxation time in the bulk and in the supersolid skin. As expected, the lifetime of the H–O phonon at the skin is longer than that of the same frequency in the bulk, which evidences the skin supersolidity- smaller, less mobile molecules having longer life time because of the strong polarization.

These experimental observations revealed a remarkably strong dependence of the vibrational relaxation time on the frequency of the H–O stretching vibration of

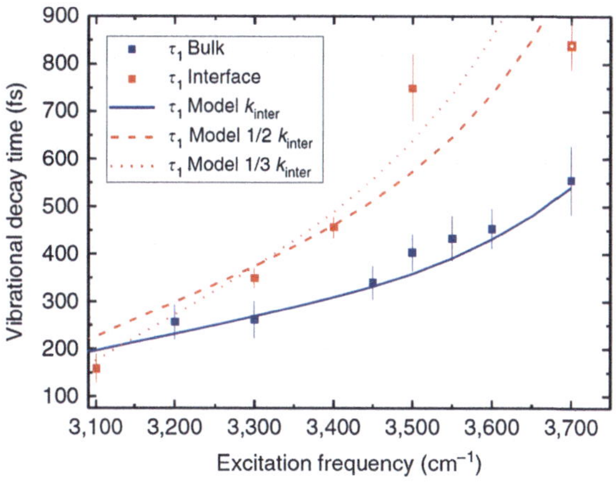


**Fig. 10.12** Excitation frequency dependence of the vibrational relaxation times τ of bulk and skin $H_2O$. The *open-red symbol* corresponds to the τ of the free H–O radicals. The *curves* represent the calculated τ using different models (Reprinted with permission from [88].)

liquid water in the bulk and at the air/water interface (skin). For bulk water, the vibrational relaxation time increases continuously from 250 to 550 fs when the frequency is increased from 3,100 to 3,700 cm$^{-1}$. For $H_2O$ at the supersolid skin, the frequency dependence is even stronger.

According to the present O:H–O cooperativity premise, the relaxation time is a reflection of the molecular dynamics. It is true that the higher the H–O vibration frequency is, the longer the relaxation time will be. The long-life H–O phonon corresponds to lower degree of fluctuation or the higher degree of viscosity. The peak widths result from fluctuation perturbation, and therefore, one cannot decompose these components into infinitely number of discrete spectral lines representing the heterogeneity of the O:H–O bonds. Therefore, the liquid water possesses limited structural heterogeneity in the bulk and in the skin. This spectral temporal feature is general to water molecules disregarding the type of stimulus. Molecules in the supersolid skin are less mobile because of the undercoordination induced supersolidity.

### *10.4.7 Supersolid Skin Rigidity*

Zhao and co-workers [89] developed a surface force apparatus to detect the skin rigidity of the monolayer skin of the adsorbed water film on a mica surface under conditions with different degrees of relative humidity. They confirmed that the first layer of the adsorbed water film is ice-like, including a lattice constant similar with ice crystal, a high bearing capacity that can support normal pressure as high as 4 MPa, a creep behavior under the action of even a small normal load, and a character of hydrogen bond. Adjacent to the first layer of the adsorbed water film, the water molecules in the outer layer are liquid-like that can flow freely under the action of external loads.

Figure 10.13 shows the experimental set up and the force-distance profile. The interaction force changes with the separation between the two mica surfaces upon exposing to RH 80 % water vapor for over 10 h. The bottom mica surface is driven towards the upper one at a constant velocity about 10 nm/s. At large separation denoted as I, the interaction force is negligible due to the weak surface interaction between these mica surfaces. The two mica surfaces are then observed to jump from D = 170 Å into contact with D = 18 Å due to the van der Waals attraction. A repulsive force is measured to increases in stage II with the film thickness decreasing from D = 16–12 Å. At D = 12 Å, another slight jump-in occurs and the film thickness decreases to D = 9 Å. The force-distance profiles in these two stages imply that a monolayer of water (about 3–4 Å) is squeezed out in each step [90].

With the confined film thickness further decreasing from D = 12–8.5, and 7.3 Å, the normalized repulsion F/R drastically increases from ~440–900 mN/m (~1.6 MPa), and 2400 mN/m (~4.0 MPa), respectively. R is the radius of the probing discs. The high bearing capacity of the adsorbed water film at D < 9 Å further confirms that the water molecules in this part are similar to solid that cannot flow easily. The 7.3 Å height corresponds to two layer spacings of the water molecules packed in the tetrahedron configuration, which corresponds to $d_{OO}$ = 3.165, $d_L$ = 2.3692, $d_H$ = 0.7958 Å; and density of $\rho$ = 0.6174 g/cm$^3$ (see Sect. 10.2).

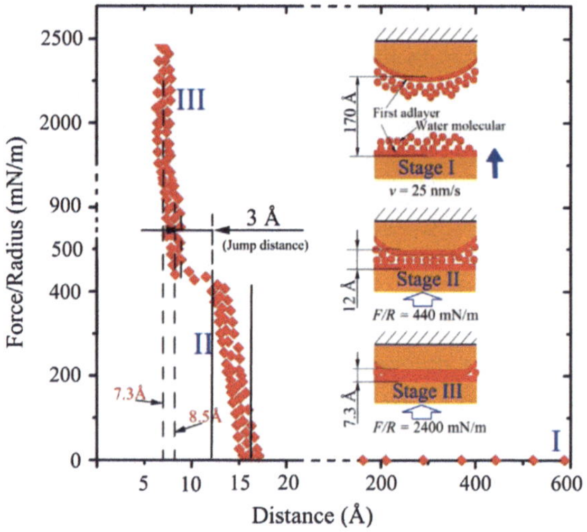


**Fig. 10.13** The force–distance relation for two mica sheets brought together at 298 K. Thin water films formed on each mica substrate by adsorption under 80 % relative humidity for 10 h. The repulsive force increases in two steps (*inset*) and it reaches a maximum of 2400 mN/m (4.0 MPa) at a separation of 7.3 Å (Reprinted with permission from [89].)

Therefore, the experimental results confirmed that the high rigidity of the adsorbed water film exhibiting the supersolid nature—solid-like properties including low flow ability, water insolubility, high carrying capacity in normal direction (more than 4 MPa), and the low mass density, smaller size and larger separation of molecules.

## 10.5 Superhydrophobicity, Superlubricity, Superfluidity, and Supersolidity

### *10.5.1 Common Attributes*

The phenomena of superhydrophobicity, superfluidity, superlubricity and supersolidity (4S) at the nanometer-sized contacts of liquid–solid or solid–solid share the common characteristics of chemically non-stick, mechanically elastic, and kinetically frictionless in motion. Although the 4S occurrences have been extensively investigated, mechanism behind the common characteristics remains in its infancy. The BOLS-NEP notion provides an energetic and electronic mechanism indicating that Coulomb repulsion between "dipoles pinned in the elastic skins or the supersolid covering sheets of liquid droplets" dictates the 4S in common [14].

The transport of fluid in and around nanometer-sized objects with at least one characteristic dimension below 100 nm enables the superfluidic occurrence that is impossible on bigger length scales [91]. Nanofluids have significantly greater thermal and mass conductivity in nanochannels compared with their base fluids [92]. The difference between the nanofluid and the base fluid is the high value of surface-to-volume ratio that increases with the miniaturization of the dimensions of both the fluid and the channel cavity in which the fluid is flowing. This high ratio in nanochannels results in surface-charge-governed transport, which allows ion separation and is described by an electrokinetic theory of electrical double layer (EDL) scheme [20]. The EDL channel can be operated as field-effect transistors to detect chemical and biological species label-free, and transport through nanochannels leads to analyte separation and new phenomena when the EDL thickness becomes comparable to the smallest channel opening.

On the other hand, the rate of the pressure-driven water flow through CNTs is orders higher than predictions by conventional fluid-flow theory [93]. The thinner the channel cavity is, the faster the fluid-flow rate will be under the same pressure [94]. This high fluid velocity results from an almost frictionless interface between the CNT wall and the fluid droplets [95, 96]. A MD calculation [97] suggested that water flow in CNT could generate a constant voltage difference of several mV between the two ends of a CNT, due to interactions between the water dipole chains and charge carriers in the CNT, which might also contribute to the abnormal frictionless fluid flow in the CNT.

Although the crystal defects have been recognized as the key to the supersolidity of $^4He$ solid, correlation between the defects and the superelasticity and superfluidity is yet to be established. Therefore, a deeper insight into the chemical nature of the surfaces is necessary for one to gain a consistent understanding of the origin for the 4S.

The localized energy densification makes the skin stiffer and the densely- and tightly-trapped bonding charges polarize nonbonding electrons, if exist, to form locked skin dipoles. In addition, the sp-orbit hybridization of F, O, N, or C upon reacting with solid atoms generates nonbonding lone pairs or unpaired edge electrons that induce dipoles directing into the open end of a surface. Such a Coulomb repulsion between the negatively charged skins of the contacting objects not only lowers the effective contacting force and the friction but also prevents charge from being exchanged between the counterparts of the contact. Being similar to magnetic levitation, such Coulomb repulsion provides the force driving the 4S.

### 10.5.2 BOLS-NEP Transition Mechanism

The BOLS-NEP notion defined the skin supersolidity for water [3] and normal solids [98], as featured in Table 10.2. The presence or absence of the nonbonding electrons discriminates a hydrophobic solid skin from that of hydrophilic skin. Undercoordination induced bond contraction and bonding charge entrapment are common to all solid skin, but the presence and absence of nonbonding electrons makes a great difference. The combination of the supersolid skin of water and the hydrophobic solid skin ensures the 4S phenomena, otherwise, the otherwise.

**Table 10.2** Skin supersolidity of water and its counterpart for hydrophobicity and hydrophilicity defined by the BOLS-NEP premise[a]

| | | Water | Regular solid | |
|---|---|---|---|---|
| | | Hydrophobic | Hydrophobic | Hydrophilic |
| Bond length | $\Delta d$ | $\Delta d_H < 0$; $\Delta d_L > 0$ | <0 | |
| Bond energy | $\Delta E$ | $\Delta E_H > 0$; $\Delta E_L < 0$ | >0 | |
| Mass density | $\Delta \rho$ | <0 | >0 | |
| Melting point (atomic cohesive energy) | $\Delta T_m$ | >0 | <0 | |
| Elasticity (local energy density) | $\Delta Y$ | >0 | | |
| Quantum entrapment (core level shift) | $\Delta E_{1s}$ | >0 | | |
| Polarization of nonbonding electrons | $\Delta P$ | >0 | | 0 |

[a]Radiation using UV light and plasma removes the nonbonding electrons temporarily and hence annihilates the polarization, causing hydrophobicity–hydrophilicity transition

The BOLS-NEP notion also claims the responsibility not only for the hydrophobicity–hydrophilicity transition but also for the roughness enhancement. Alternative UV or plasma radiation can peel off the skin dipoles and hence switch the hydrophobicity off. Ageing recovers the skin dipoles and switches on the hydrophobicity. Nanoscaled roughness reduces the effective atomic CN, which enhances the quantum entrapment and nonbonding electron polarization and hence augments the initial hydrophobicity and hydrophilicity of the solid contacts, as Wenzel's notion expects.

The 4S phenomena have little to do with surface energy but the elasticity and electronic repulsivity. Furthermore, the classical definition of surface energy being required for cutting a sold into two halves per unit area is still subject to confusion. The energy density gain per unit volume in the skin region or the cohesive energy remnant per atom in the skin is meaningfully practical quantities. Energy density determines the elasticity and the cohesive energy dictates the thermal stability [99].

Figure 10.14a shows the theoretically predicted curvature ($K^{-1}$) dependence of the skin charge density, elasticity, and potential trap depth of the outermost shell of a spherical dot. The elasticity corresponds to binding energy density and the potential trap depth is proportional to bond energy [98]. The volume average correspond to the size dependence of the elastic modulus such as ZnO [30] and the core-level shift of nanostructures [100].

As illustrated in Fig. 10.14b, the drop and the wall surface are likely charged (green dots) and repelling each other, which ensures water skin not only high elasticity but also electric repulsivity—under compression stress. The droplet will lose its viscosity and becomes frictionless unless the surface dipoles are removed. Such a system runs in a way more like a "maglev train".

The superhydrophobicity phenomenon can be explained from the viewpoints of surface chemistry, energy and charge density enhancement. If the air pockets beneath a droplet on a sinusoidal substrate are open to the atmosphere, the superhydrophobic state can exist only when the substrate is hydrophobic, and that the geometric

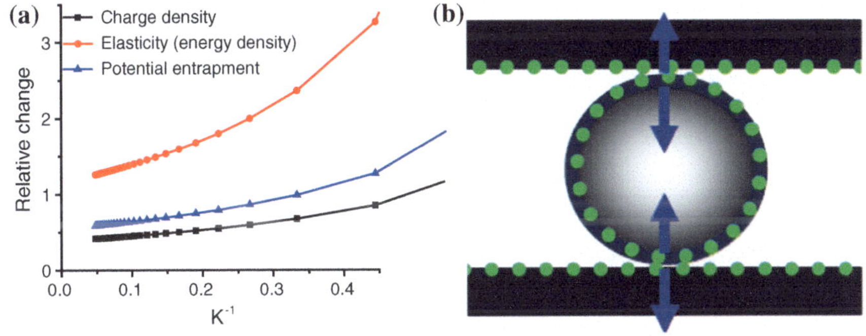


**Fig. 10.14** **a** Curvature ($K^{-1}$) dependence of the skin charge density, elasticity (energy density), and potential trap depth of the outermost shell of a spherical dot. **b** A water droplet of supersolid skin flowing through a nanochannel is subject to electrosteatic repulsion. The smaller the droplet is, the higher the supersolidity will be [98]

parameters of the microstructure have a great influence on the wetting behavior. Being similar to the superfluidity, polarization of the surface or the presence of lone pair electrons happens to both the fluidic drop and the material. The charged surface repels the ambient charged particles, such as water molecules, to result in superhydrophobicity. The UV radiation removes the polarized charges and the dark storage recovers the surface dipoles, being the same as the surface magnetism of noble metal clusters and the dilute magnetism of oxide nanostructures [101, 102].

BOLS-NEP notion combines both the Wenzel and Cassie–Baxter models from skin polarization point of view. Atomic undercoordination becomes more pronounced when the curvature of the proxy is increased, which enhances the entrapment-polarization defined by the BOLS-NEP notion [86] and then the Wenzel effect occurs. Air pockets will form underneath water droplets if the rough skin is hydrophobic. If the hydrophilic skin is roughens, it will be even hydrophilic and no air pockets will present. Atomic undercoordination induced local entrapment is global yet the polarization is subjective, which is why Wenzel's model works. Pt, Co and graphite skin show entrapment dominance while Cu, Ag, Au, Rh, W, Mo, and graphite point defects demonstrate polarization dominance; most oxides, nitrides, fluorides skins and defects are polarization dominance because of nonbonding electron pairs [31].

The 4S occurrences result from the reduction of the friction force ($f_r = \mu N$ with $\mu$ being the friction coefficient and N the contacting force). The lowering of the $f_r$ will reduce the process of friction or the extent of phonon and electron excitation. One surprising fact is that these 4S effects share a general identity of non-sticky and frictionless motion—with lowered effective contacting pressure and reduced friction coefficient. Skins for both the water and the substrate must be hydrophobic to ensure the superhydrophobicity at working.

The 4S phenomena must share a common elastic and repulsive origin in addition to the energetic and geometric descriptions of the existing models. Considerations from the perspectives of surface roughness, air pocket, and surface energy seem insufficient because the chemistry and the charge identities do alter at the surface skin up to two interatomic spacings [103]. In particular, the hydrophobicity–hydrophilicity recycling effect caused by UV irradiation and the subsequent dark aging is beyond the scope of Cassie's law and the PT mechanism of air pockets dominance. Furthermore, the superhydrophobicity of alkanes, oils, fats, wax, and the greasy and organic substances is independent of the surface roughness.

### 10.5.3 Hydrophobicity–Hydrophilicity Transition

Superhydrophobic materials have surfaces that are extremely difficult to wet, with water contact angles in excess of 150° or even greater, see Fig. 10.3. Surfaces with ultra-hydrophobicity have aroused much interest with their potential applications in self-cleaning coatings, microfluidics, and biocompatible materials and so on.

Many physical chemical processes, such as adsorption, lubrication, adhesion, dispersion, friction, etc., are closely related to the wettability of materials surfaces [104, 105]. Examples of hydrophobic molecules include alkanes, oils, fats, wax, and greasy and organic substances with C, N, O, or F as the key constituent element.

#### 10.5.3.1 UV Radiation

What is even more amazing is that the hydrophobic surface can switch reversibly between superhydrophobicity and superhydrophilicity when the solid surface is subject to UV radiation [106] which results in electron–hole pair creation [107]. After being stored in the dark over an extended period, the hydrophilicity is once again lost. The dipoles can be also demolished by thermal excitation, or excessively applied compression due to ionization or $sp^3$ orbital de-hybridization.

The UV radiation with excitation energy around 3.0 eV could break chemical bonds and ionize surface atoms, which could turn the hydrophobic surface to be hydrophilic, as it has widely been observed. $Ar^+$ sputtering the surface is expected to have the same function of removing dipole or monopole temporarily. If the polarized electrons were removed, the 4S characteristics would be lost. Aging of the specimen will recover the surface charges. The UV radiation reversing effect is the same as that observed in the surface magnetism of noble metal clusters and the dilute magnetism of oxide nanostructures [101, 102, 108]. Figures 10.15 and 10.16 show the hydrophobic–hydrophilic transition of ZnO [109] and graphene [110].

However, aging the samples in the ambient will recover the $sp^3$-orbital hybridization and the dipoles as well. Surface bias to a certain extent may also cause the depletion of the locked charges though this expectation is subject to verification. Overloaded pressure in the dry sliding will overcome the Coulomb repulsion, as the energy dissipation by phonon and electron excitation could occur under compression. On the other hand, a sufficiently large difference in the electro-affinity between the contact media, chemical bond may form under a certain conditions such as heating, compression, or electric ecitation, the interface will be adherent.

#### 10.5.3.2 Plasma Sputtering

If water bonds directly to the substrate, or exchange interaction occurs between water molecules and substrate atoms, hydrophilicity takes place, which is subject to conditions of crystal growth-lattice matching [111] and chemical conditions [112]. When encapsulated in hydrophilic nanopores [113, 114], or wetted in hydrophilic topological configurations [115], water molecules perform in an opposite way and melt at temperatures below the bulk $T_m$.

Water maintains its high lubricity under the normal pressure of 1.7 MPa at pulling when confined between silica plates, as shown in Fig. 10.17 [116], with the

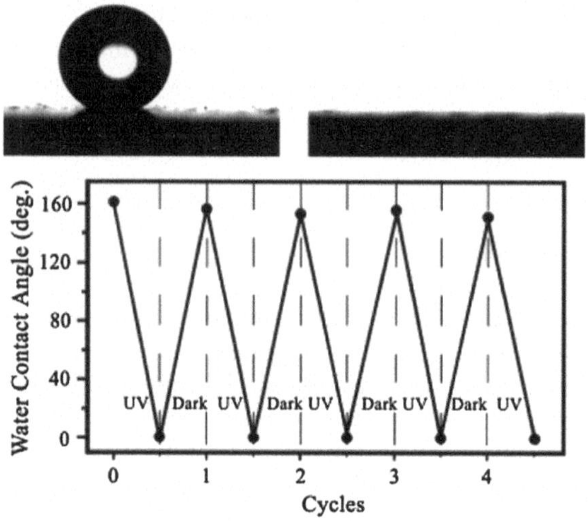


**Fig. 10.15** Photographs of water droplet shape on the aligned ZnO nanorod films before (*left*) and after (*right*) 365 nm UV illumination for 2 h cycle. **b** Reversible superhydrophobic–superhydrophilic transition of the as—prepared films under the alternation of UV irradiation and dark storage (Reprinted with permission from [109].)

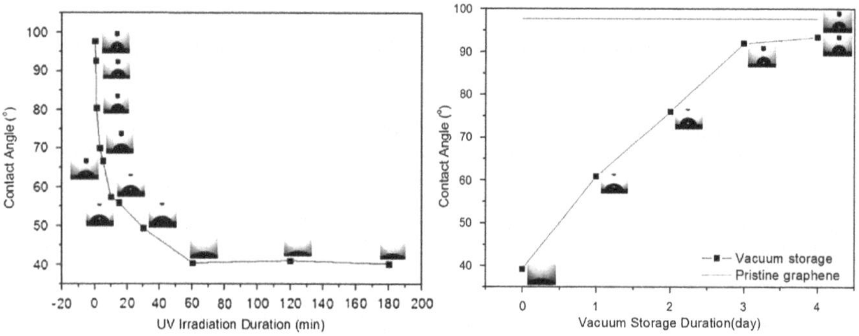


**Fig. 10.16** Graphene hydrophobicity–hydrophilicity conversion under UV irradiation and vacuum storage (Reprinted with permission from [110].)

characteristics $\omega_H$ of 3450 $cm^{-1}$ for the $H_2O/SiO_2$ hydrophobic supersolid skin that is elastic, polarized and frictionless with an air gap in between.

However, removal of the polarization skin by water vapor plasma sputtering not only raises the interface shear viscosity starting at 0.4 MPa but also recovers the $\omega_H$ from the skin characteristic 3450 $cm^{-1}$ to the bulk frequency of 3200 $cm^{-1}$ [116]. These observations indicate that the interface between silica and water is indeed hydrophobic but plasma sputtering alters it by removing the polarized electrons of silica skin.

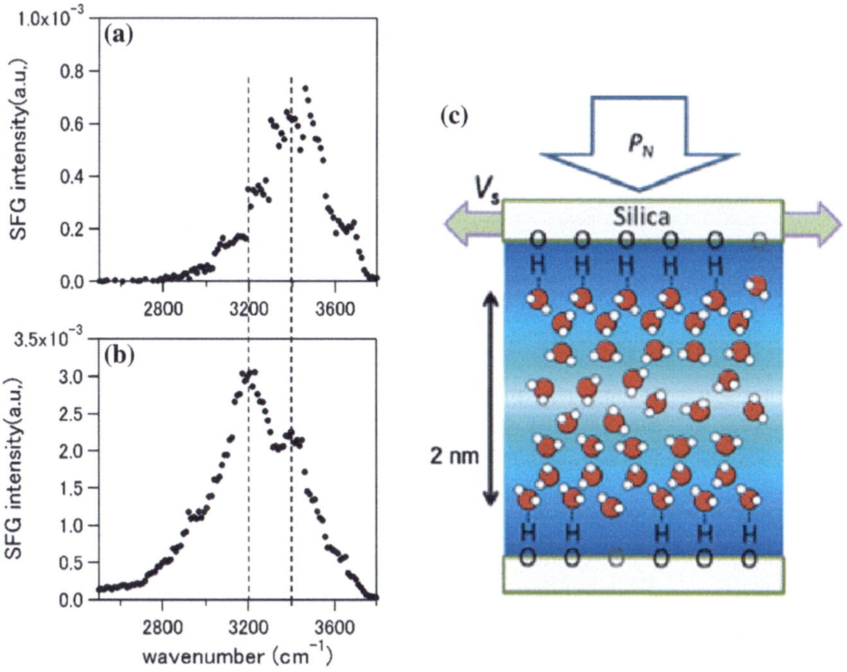


**Fig. 10.17** SFG vibrational spectra obtained from water on **a** untreated and **b** plasma-treated silica surfaces. Plasma sputtering removes the $SiO_2$ skin dipoles and switches the skin from hydrophobic to hydrophilic. $\omega_H$ = 3450 and 3200 cm$^{-1}$ are characteristics for the supersolid skin and the bulk water, respectively. The shear viscosity remains stable at $P_N$ = 1.7 MPa starts to increase at 0.4 MPa normal compression when the $SiO_2$ changes from hydrophobic to hydrophilic (Reprinted with permission from [116].)

These hydrophobic–hydrophilic transition of ZnO [109], graphene [110], and $SiO_2$ is beyond the description of the Wenzel–Cassie–Baster's law. However, observations comply with the presently proposed hydrophobicity and the hydrophobic–hydrophilic transition mechanism. Elastic Coulomb repulsion between the supersolid skin of liquid water and the polarized solid substrate dominate the 4S. Removal and recovering of the polarized states by UV or plasma irradiation switch off and on the hydrophobicity. Air gap formation between the liquid and solid is a consequences of the interface Coulomb repulsion.

## *10.5.4 Microchannel: How Does a Double Layer Form?*

### 10.5.4.1 Skin BOLS-NEP Notion

It is within expectation that the narrower the channel diameter is, the faster the flow of the fluid will be in microchannels [94, 117], because of the curvature-enhanced

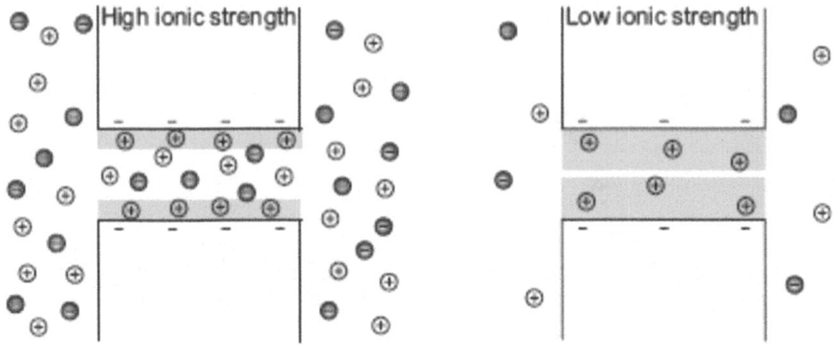


**Fig. 10.18** Electrical double layer (EDL), shaded in gray, at high ionic strength, it is thin, allowing co-ions and counter-ions to pass through the nanochannel. At low ionic strength, the EDL thickness increases, resulting in a counter-ion-selective nanochannels (Reprinted with permission from [91].)

supersolidity of the water droplet interacting with hydrophobic carbon nanotubes or micrchannels, which elaborate why the rate of the pressure-driven water flow through carbon nanotubes is orders higher in magnitude and faster than is predicted from conventional fluid-flow theory [93].

The transport of fluid in and around nanometer-sized objects with at least one characteristic dimension below 100 nm enables the occurrence of phenomena that are impossible at bigger length scales [91]. Nanofluids have significantly greater thermal and mass conductivity compared to their base fluids [92]. The surface-to-volume ratio increases with miniaturization of both the fluid and the channel. This high ratio in nanochannels results in surface-charge-governed transport, which allows for ion separation and is described by an electrokinetic theory of electrical double layer (EDL) scheme, as illustrated in Fig. 10.18.

According to the present developed premise, the EDL forms in the microchannel following the BOLS-NEP mechanism, as illustrated in Fig. 10.14b. Atomic/molecular undercoordination induced skin polarization of the nonbonding electrons of both the fluid and the channel provide the electrostatic force preventing the interface contacting and thus the fluid skin supersolidity makes the fluid flow in channel frictionlessly.

#### 10.5.4.2 Nanofluid Travelling in Microchannels

The structure of the water-hydrophobic interface is of fundamental interest for the understanding of biological and colloidal systems. Due to the perturbation of the

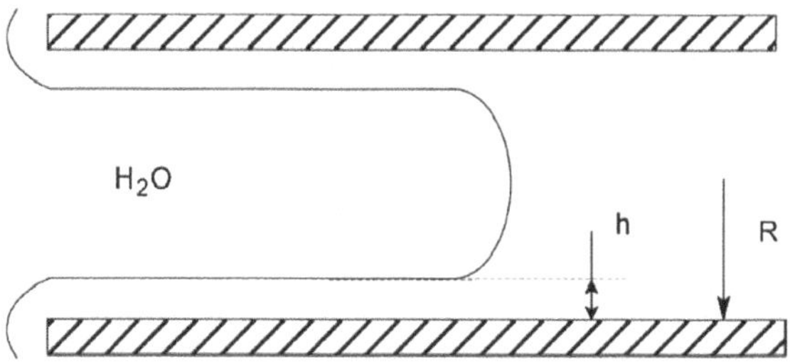


**Fig. 10.19** Water (non-wetting fluid) in a hydrophobic pore of R radius) is separated from the walls by a h = 0.6 nm thin layer of vapor/low-density film (wetting fluid) (Reprinted with permission from [79].)

hydrogen bonding, the structure of water near a hydrophobic surface is quite different from the bulk; in particular, there is an air gap between the channel and the fluid because of the electrostatic repulsion between supersolid skin dipoles of the fluid and the hydrophobic channel.

Helmy et al. [79] investigated intrusion of water in hydrophobic nanochannels (Fig. 10.19), using the water porosimetry method. They demonstrated that classical (macroscopic) theory fails to describe wetting at nanoscale because of the presence of an h = 0.6 nm thin vapor (low density) layer separating water and the hydrophobic surface.

## 10.6 Summary

Molecular undercoordination induced O:H–O bond relaxation and the associated binding electron entrapment and the nonbonding electron dual polarization clarify the anomalous behaviour of the skin supersolidity of liquid water. Agreement between numerical calculations and experimental observations verified the following:

(1) Undercoordination-induced O:H–O relaxation and associated dual polarizationresults in the supersolid phase that is elastic, hydrophobic, thermally more stable, and less dense, which dictates the unusual behaviour of water molecules at the boundary of the O:H–O networks or in the nanoscale droplet.
(2) H–O bond contraction densifies and entraps the core and bonding electrons; H–O bond stiffening shifts positively the O 1s energy, the $\omega_H$ and the $T_m$ of molecular clusters, surface skins, and ultrathin films of water.
(3) The combination of the supersolid skins for both water and the solid substrate with or without presence of polarization determines the 4S phenomena.
(4) Removal and recover of the skin dipoles control the hydrophobic–hydrophilic transition.

# References


1. Reporter, Water way to go: the unlucky ant trapped in a raindrop grave. Dailymail http://www.dailymail.co.uk/news/article-1371416/Photographer-Adam-Gormley-captures-ant-trapped-raindrop.html (2011)
2. Y. Huang, X. Zhang, Z. Ma, Y. Zhou, W. Zheng, J. Zhou, C.Q. Sun, Hydrogen-bond relaxation dynamics: resolving mysteries of water ice. Coord. Chem. Rev. **285**, 109–165 (2015)
3. X. Zhang, Y. Huang, Z. Ma, Y. Zhou, W. Zheng, J. Zhou, C.Q. Sun, A common supersolid skin covering both water and ice. PCCP **16**(42), 22987–22994 (2014)
4. C.Q. Sun, X. Zhang, J. Zhou, Y. Huang, Y. Zhou, W. Zheng, Density, elasticity, and stability anomalies of water molecules with fewer than four neighbors. J. Phys. Chem. Lett. **4**, 2565–2570 (2013)
5. Y. Huang, X. Zhang, Z. Ma, G. Zhou, Y. Gong, C.Q. Sun, Potential paths for the hydrogen-bond relaxing with $(H_2O)N$ cluster size. J. Phys. Chem. C **119**(29), 16962–16971 (2015)
6. X. Zhang, P. Sun, Y. Huang, Z. Ma, X. Liu, J. Zhou, W. Zheng, C.Q. Sun, Water nanodroplet thermodynamics: quasi-solid phase-boundary dispersivity. J. Phys. Chem. B **119** (16), 5265–5269 (2015)
7. J. Cooper, R. Dooley, IAPWS release on surface tension of ordinary water substance. International Association for the Properties of Water and Steam (IAPWS), Charlotte, NC, vol 2 (1994)
8. M. Sophocleous, Understanding and explaining surface tension and capillarity: an introduction to fundamental physics for water professionals. Hydrogeol. J. **18**(4), 811–821 (2010)
9. X.F. Gao, L. Jiang, Water-repellent legs of water striders. Nature **432**(7013), 36–36 (2004)
10. P.B. Miranda, L. Xu, Y.R. Shen, M. Salmeron, Icelike water monolayer adsorbed on mica at room temperature. Phys. Rev. Lett. **81**(26), 5876–5879 (1998)
11. A. Michaelides, K. Morgenstern, Ice nanoclusters at hydrophobic metal surfaces. Nat. Mater. **6**(8), 597–601 (2007)
12. E. Pennisi, Water's tough skin. Science **343**(6176), 1194–1197 (2014)
13. E. Martines, K. Seunarine, H. Morgan, N. Gadegaard, C.D.W. Wilkinson, M.O. Riehle, Superhydrophobicity and superhydrophilicity of regular nanopatterns. Nano Lett. **5**(10), 2097–2103 (2005)
14. C.Q. Sun, Y. Sun, Y.G. Ni, X. Zhang, J.S. Pan, X.H. Wang, J. Zhou, L.T. Li, W.T. Zheng, S. S. Yu, L.K. Pan, Z. Sun, Coulomb repulsion at the nanometer-sized contact: a force driving superhydrophobicity, superfluidity, superlubricity, and supersolidity. J. Phys. Chem. C **113** (46), 20009–20019 (2009)
15. T. Young, An essay on the cohesion of fluids. Philos. Trans. R. Soc. London, 65–87 (1805)
16. N.K. Adam, Use of the term 'Young's equation' for contact angles. Nature **180**, 809–810 (1957)
17. F. Mugele, J.-C. Baret, Electrowetting: from basics to applications. J. Phys.: Condens. Matter **17**(28), R705 (2005)
18. G. Whyman, E. Bormashenko, T. Stein, The rigorous derivation of Young, Cassie-Baxter and Wenzel equations and the analysis of the contact angle hysteresis phenomenon. Chem. Phys. Lett. **450**(4–6), 355–359 (2008)
19. A.B.D. Cassie, S. Baxter, Wettability of porous surfaces. Trans. Faraday Soc. **40**, 0546–0550 (1944)
20. F. Baldessari, Electrokinetics in nanochannels—Part I. Electric double layer overlap and channel-to-well equilibrium. J. Colloid Interface Sci. **325**(2), 526–538 (2008)
21. L. Prandtl, Mind model of the kinetic theory of solid bodies. Z. Ang. Math. Mech. **8**, 85–106 (1928)
22. G.A. Tomlinson, Molecular cohesion. Philos. Mag. **6**(37), 695 (1928)

23. A. Socoliuc, R. Bennewitz, E. Gnecco, E. Meyer, Transition from stick-slip to continuous sliding in atomic friction: entering a new regime of ultralow friction. Phys. Rev. Lett. **92**(13), 134301 (2004)
24. C.Q. Sun, Relaxation of the chemical bond. Springer Series in Chemical Physics 108. vol. 108 (Springer, Heidelberg, 2014), 807 pp
25. C. Lee, C.H. Choi, C.J. Kim, Structured surfaces for a giant liquid slip. Phys. Rev. Lett. **101** (6), 064501 (2008)
26. G.P. Fang, W. Li, X.F. Wang, G.J. Qiao, Droplet motion on designed microtextured superhydrophobic surfaces with tunable wettability. Langmuir **24**(20), 11651–11660 (2008)
27. W. Li, G.P. Fang, Y.F. Lij, G.J. Qiao, Anisotropic wetting behavior arising from superhydrophobic surfaces: parallel grooved structure. J. Phys. Chem. B **112**(24), 7234–7243 (2008)
28. X.P. Zheng, H.P. Zhao, L.T. Gao, J.L. Liu, S.W. Yu, X.Q. Feng, Elasticity-driven droplet movement on a microbeam with gradient stiffness: a biomimetic self-propelling mechanism. J. Colloid Interface Sci. **323**(1), 133–140 (2008)
29. Q.S. Zheng, Y. Yu, X.Q. Feng, The role of adaptive-deformation of water strider leg in its walking on water. J. Adhes. Sci. Technol. **23**(3), 493–501 (2009)
30. M. Zhao, W.T. Zheng, J.C. Li, Z. Wen, M.X. Gu, C.Q. Sun, Atomistic origin, temperature dependence, and responsibilities of surface energetics: an extended broken-bond rule. Phys. Rev. B **75**(8), 085427 (2007)
31. X.J. Liu, M.L. Bo, X. Zhang, L.T. Li, Y.G. Nie, H. TIan, Y. Sun, S. Xu, Y. Wang, W. Zheng, C.Q. Sun, Coordination-resolved electron spectrometrics. Chem. Rev. **115**(14), 6746–6810 (2015)
32. X. Zhang, T. Yan, Y. Huang, Z. Ma, X. Liu, B. Zou, C.Q. Sun, Mediating relaxation and polarization of hydrogen-bonds in water by NaCl salting and heating. PCCP **16**(45), 24666–24671 (2014)
33. K.R. Wilson, R.D. Schaller, D.T. Co, R.J. Saykally, B.S. Rude, T. Catalano, J.D. Bozek, Surface relaxation in liquid water and methanol studied by x-ray absorption spectroscopy. J. Chem. Phys. **117**(16), 7738–7744 (2002)
34. Y.I. Tarasevich, State and structure of water in vicinity of hydrophobic surfaces. Colloid J. **73**(2), 257–266 (2011)
35. B.H. Chai, H. Yoo, G.H. Pollack, Effect of radiant energy on near-surface water. J. Phys. Chem. B **113**(42), 13953–13958 (2009)
36. N.I. Hammer, J.W. Shin, J.M. Headrick, E.G. Diken, J.R. Roscioli, G.H. Weddle, M.A. Johnson, How do small water clusters bind an excess electron? Science **306**(5696), 675–679 (2004)
37. O. Marsalek, F. Uhlig, T. Frigato, B. Schmidt, P. Jungwirth, Dynamics of electron localization in warm versus cold water clusters. Phys. Rev. Lett. **105**(4), 043002 (2010)
38. S. Liu, J. Luo, G. Xie, D. Guo, Effect of surface charge on water film nanoconfined between hydrophilic solid surfaces. J. Appl. Phys. **105**(12), 124301–4 (2009)
39. K.R. Siefermann, Y. Liu, E. Lugovoy, O. Link, M. Faubel, U. Buck, B. Winter, B. Abel, Binding energies, lifetimes and implications of bulk and interface solvated electrons in water. Nat. Chem. **2**, 274–279 (2010)
40. D.H. Paik, I.R. Lee, D.S. Yang, J.S. Baskin, A.H. Zewail, Electrons in finite-sized water cavities: hydration dynamics observed in real time. Science **306**(5696), 672–5 (2004)
41. J.R.R. Verlet, A.E. Bragg, A. Kammrath, O. Cheshnovsky, D.M. Neumark, Observation of large water-cluster anions with surface-bound excess electrons. Science **307**(5706), 93–96 (2005)
42. R. Vacha, O. Marsalek, A.P. Willard, D.J. Bonthuis, R.R. Netz, P. Jungwirth, Charge transfer between water molecules as the possible origin of the observed charging at the surface of pure water. J. Phys. Chem. Lett. **3**(1), 107–111 (2012)
43. F. Baletto, C. Cavazzoni, S. Scandolo, Surface trapped excess electrons on ice. Phys. Rev. Lett. **95**(17), 176801 (2005)

44. L. Turi, W.S. Sheu, P.J. Rossky, Characterization of excess electrons in water-cluster anions by quantum simulations. Science **309**(5736), 914–917 (2005)
45. M. Abu-Samha, K.J. Borve, M. Winkler, J. Harnes, L.J. Saethre, A. Lindblad, H. Bergersen, G. Ohrwall, O. Bjorneholm, S. Svensson, The local structure of small water clusters: imprints on the core-level photoelectron spectrum. J. Phys. B **42**(5), 055201 (2009)
46. K. Nishizawa, N. Kurahashi, K. Sekiguchi, T. Mizuno, Y. Ogi, T. Horio, M. Oura, N. Kosugi, T. Suzuki, High-resolution soft X-ray photoelectron spectroscopy of liquid water. PCCP **13**, 413–417 (2011)
47. Y. Huang, X. Zhang, Z. Ma, Y. Zhou, J. Zhou, W. Zheng, C.Q. Sun, Size, separation, structure order, and mass density of molecules packing in water and ice. Sci. Rep. **3**, 3005 (2013)
48. S.A. Harich, D.W.H. Hwang, X. Yang, J.J. Lin, X. Yang, R.N. Dixon, Photodissociation of H2O at 121.6 nm: a state-to-state dynamical picture. J. Chem. Phys. **113**(22), 10073–10090 (2000)
49. U. Buck, F. Huisken, Infrared spectroscopy of size-selected water and methanol clusters. Chem. Rev. **100**(11), 3863–3890 (2000)
50. K.E. Otto, Z. Xue, P. Zielke, M.A. Suhm, The Raman spectrum of isolated water clusters. Phys. Chem. Chem. Phys. (2014). doi:10.1039/c3cp54272f
51. T.F. Kahan, J.P. Reid, D.J. Donaldson, Spectroscopic probes of the quasi-liquid layer on ice. J. Phys. Chem. A **111**(43), 11006–11012 (2007)
52. J. Ceponkus, P. Uvdal, B. Nelander, Water tetramer, pentamer, and hexamer in inert matrices. J. Phys. Chem. A **116**(20), 4842–50 (2012)
53. Y.R. Shen, V. Ostroverkhov, Sum-frequency vibrational spectroscopy on water interfaces: polar orientation of water molecules at interfaces. Chem. Rev. **106**(4), 1140–1154 (2006)
54. V. Buch, S. Bauerecker, J.P. Devlin, U. Buck, J.K. Kazimirski, Solid water clusters in the size range of tens-thousands of $H_2O$: a combined computational/spectroscopic outlook. Int. Rev. Phys. Chem. **23**(3), 375–433 (2004)
55. P.C. Cross, J. Burnham, P.A. Leighton, The Raman spectrum and the structure of water. J. Am. Chem. Soc. **59**, 1134–1147 (1937)
56. M. Sulpizi, M. Salanne, M. Sprik, M.-P. Gaigeot, Vibrational sum frequency generation spectroscopy of the water liquid–vapor interface from density functional theory-based molecular dynamics simulations. J, Phys. Chem. Lett. **4**(1), 83–87 (2012)
57. C.Q. Sun, X. Zhang, X. Fu, W. Zheng, J.-L. Kuo, Y. Zhou, Z. Shen, J. Zhou, Density and phonon-stiffness anomalies of water and ice in the full temperature range. J. Phys. Chem. Lett. **4**, 3238–3244 (2013)
58. U. Bergmann, A. Di Cicco, P. Wernet, E. Principi, P. Glatzel, A. Nilsson, Nearest-neighbor oxygen distances in liquid water and ice observed by x-ray Raman based extended x-ray absorption fine structure. J. Chem. Phys. **127**(17), 174504 (2007)
59. F. Mallamace, M. Broccio, C. Corsaro, A. Faraone, D. Majolino, V. Venuti, L. Liu, C.Y. Mou, S.H. Chen, Evidence of the existence of the low-density liquid phase in supercooled, confined water. Proc. Natl. Acad. Sci. U.S.A. **104**(2), 424–8 (2007)
60. J. Li, Y.X. Li, X. Yu, W.J. Ye, C.Q. Sun, Local bond average for the thermally driven elastic softening of solid specimens. J. Phys. D-Appl. Phys. **42**(4), 045406 (2009)
61. M.J. Holmes, N.G. Parker, M.J.W. Povey, Temperature dependence of bulk viscosity in water using acoustic spectroscopy. J. Phys: Conf. Ser. **269**, 012011 (2011)
62. D. Xu, K.M. Liechti, K. Ravi-Chandar, Mechanical probing of icelike water monolayers. Langmuir **25**(22), 12870–3 (2009)
63. K.B. Jinesh, J.W.M. Frenken, Experimental evidence for ice formation at room temperature. Phys. Rev. Lett. **101**(3), 036101 (2008)
64. C. Wang, H. Lu, Z. Wang, P. Xiu, B. Zhou, G. Zuo, R. Wan, J. Hu, H. Fang, Stable liquid water droplet on a water monolayer formed at room temperature on ionic model substrates. Phys. Rev. Lett. **103**(13), 137801–137804 (2009)

65. M. James, T.A. Darwish, S. Ciampi, S.O. Sylvester, Z.M. Zhang, A. Ng, J.J. Gooding, T.L. Hanley, Nanoscale condensation of water on self-assembled monolayers. Soft Matter **7**(11), 5309–5318 (2011)
66. T. Ishiyama, H. Takahashi, A. Morita, Origin of vibrational spectroscopic response at ice surface. J. Phys. Chem. Lett. **3**, 3001–3006 (2012)
67. H. Qiu, W. Guo, Electromelting of confined monolayer ice. Phys. Rev. Lett. **110**(19), 195701 (2013)
68. B. Winter, E.F. Aziz, U. Hergenhahn, M. Faubel, I.V. Hertel, Hydrogen bonds in liquid water studied by photoelectron spectroscopy. J. Chem. Phys. **126**(12), 124504 (2007)
69. L. Li, Y. Kazoe, K. Mawatari, Y. Sugii, T. Kitamori, Viscosity and wetting property of water confined in extended nanospace simultaneously measured from highly-pressurized meniscus motion. J. Phys. Chem. Lett. 2447–2452 (2012)
70. L. Vrbka, P. Jungwirth, Homogeneous freezing of water starts in the subsurface. J. Phys. Chem. B **110**(37), 18126–18129 (2006)
71. D. Donadio, P. Raiteri, M. Parrinello, Topological defects and bulk melting of hexagonal ice. J. Phys. Chem. B **109**(12), 5421–5424 (2005)
72. S. Strazdaite, J. Versluis, E.H. Backus, H.J. Bakker, Enhanced ordering of water at hydrophobic surfaces. J. Chem. Phys. **140**(5), 054711 (2014)
73. S.R. Friedman, M. Khalil, P. Taborek, Wetting transition in water. Phys. Rev. Lett. **111**(22), 226101 (2013)
74. D.P. Singh, J.P. Singh, Delayed freezing of water droplet on silver nanocolumnar thin film. Appl. Phys. Lett. **102**(24), 243112 (2013)
75. Q.T. Fu, E.J. Liu, P. Wilson, Z. Chen, Ice nucleation behaviour on sol-gel coatings with different surface energy and roughness. Phys. Chem. Chem. Phys. **17**(33), 21492–500 (2015)
76. Y.B. Fan, X. Chen, L.J. Yang, P.S. Cremer, Y.Q. Gao, On the structure of water at the aqueous/air interface. J. Phys. Chem. B **113**(34), 11672–11679 (2009)
77. C. Huang, K.T. Wikfeldt, D. Nordlund, U. Bergmann, T. McQueen, J. Sellberg, L.G.M. Pettersson, A. Nilsson, Wide-angle X-ray diffraction and molecular dynamics study of medium-range order in ambient and hot water. PCCP **13**(44), 19997–20007 (2011)
78. A. Uysal, M. Chu, B. Stripe, A. Timalsina, S. Chattopadhyay, C.M. Schlepütz, T.J. Marks, P. Dutta, What x rays can tell us about the interfacial profile of water near hydrophobic surfaces. Phys. Rev. B **88**(3), 035431 (2013)
79. R. Helmy, Y. Kazakevich, C.Y. Ni, A.Y. Fadeev, Wetting in hydrophobic nanochannels: a challenge of classical capillarity. J. Am. Chem. Soc. **127**(36), 12446–12447 (2005)
80. M. Mezger, H. Reichert, S. Schoder, J. Okasinski, H. Schroder, H. Dosch, D. Palms, J. Ralston, V. Honkimaki, High-resolution in situ x-ray study of the hydrophobic gap at the water-octadecyl-trichlorosilane interface. Proc. Natl. Acad. Sci. U S A **103**(49), 18401–4 (2006)
81. D.T. Limmer, A.P. Willard, P. Madden, D. Chandler, Hydration of metal surfaces can be dynamically heterogeneous and hydrophobic. Proc. Natl. Acad. Sci. **110**(11), 4200–4205 (2013)
82. C. Antonini, I. Bernagozzi, S. Jung, D. Poulikakos, M. Marengo, Water drops dancing on ice: how sublimation leads to drop rebound. Phys. Rev. Lett. **111**(1), 014501 (2013)
83. J.G. Leidenfrost, *De Aquae Communis Nonnullis Qualitatibus Tractatus* (Duisburg, 1756)
84. D. Richard, C. Clanet, D. Quere, Surface phenomena: contact time of a bouncing drop. Nature **417**(6891), 811–811 (2002)
85. T.M. Schutzius, S. Jung, T. Maitra, G. Graeber, M. Köhme, D. Poulikakos, Spontaneous droplet trampolining on rigid superhydrophobic surfaces. Nature **527**(7576), 82–85 (2015)
86. C.Q. Sun, Dominance of broken bonds and nonbonding electrons at the nanoscale. Nanoscale **2**(10), 1930–1961 (2010)
87. M.T. Suter, P.U. Andersson, J.B. Pettersson, Surface properties of water ice at 150–191 K studied by elastic helium scattering. J. Chem. Phys. **125**(17), 174704 (2006)

88. S.T. van der Post, C.S. Hsieh, M. Okuno, Y. Nagata, H.J. Bakker, M. Bonn, J. Hunger, Strong frequency dependence of vibrational relaxation in bulk and surface water reveals sub-picosecond structural heterogeneity. Nat Commun **6**, 8384 (2015)
89. G. Zhao, Q. Tan, L. Xiang, D. Cai, H. Zeng, H. Yi, Z. Ni, Y. Chen, Structure and properties of water film adsorbed on mica surfaces. J. Chem. Phys. **143**(10), 104705 (2015)
90. J.N. Israelachvili, R.M. Pashley, Molecular layering of water at surfaces and origin of repulsive hydration forces. Nature **306**(5940), 249–250 (1983)
91. R.B. Schoch, J.Y. Han, P. Renaud, Transport phenomena in nanofluidics. Rev. Mod. Phys. **80**(3), 839–883 (2008)
92. K.F.V. Wong, T. Kurma, Transport properties of alumina nanofluids. Nanotechnology **19**, **345702**(34), 345702 (2008)
93. J.A. Thomas, A.J.H. McGaughey, Reassessing fast water transport through carbon nanotubes. Nano Lett. **8**(9), 2788–2793 (2008)
94. M. Whitby, L. Cagnon, M. Thanou, N. Quirke, Enhanced fluid flow through nanoscale carbon pipes. Nano Lett. **8**(9), 2632–2637 (2008)
95. M. Majumder, N. Chopra, R. Andrews, B.J. Hinds, Nanoscale hydrodynamics—enhanced flow in carbon nanotubes. Nature **438**(7064), 44–44 (2005)
96. H.G. Park, Y. Jung, Carbon nanofluidics of rapid water transport for energy applications. Chem. Soc. Rev.: doi:10.1039/C3CS60253B (2014)
97. Q.Z. Yuan, Y.P. Zhao, Hydroelectric voltage generation based on water-filled single-walled carbon nanotubes. J. Am. Chem. Soc. **131**(18), 6374–6376 (2009)
98. C.Q. Sun, Thermo-mechanical behavior of low-dimensional systems: the local bond average approach. Prog. Mater. Sci. **54**(2), 179–307 (2009)
99. M.W. Zhao, R.Q. Zhang, Y.Y. Xia, C. Song, S.T. Lee, Faceted silicon nanotubes: structure, energetic, and passivation effects. J. Phys. Chem. C **111**(3), 1234–1238 (2007)
100. C.Q. Sun, Surface and nanosolid core-level shift: impact of atomic coordination-number imperfection. Phys. Rev. B **69**(4), 045105 (2004)
101. E. Roduner, Size matters: why nanomaterials are different. Chem. Soc. Rev. **35**(7), 583–592 (2006)
102. J.M.D. Coey, Dilute magnetic oxides. Curr. Opin. Solid State Mater. Sci. **10**(2), 83–92 (2006)
103. F. Matsui, T. Matsushita, Y. Kato, M. Hashimoto, K. Inaji, F.Z. Guo, H. Daimon, Atomic-layer resolved magnetic and electronic structure analysis of Ni thin film on a Cu(001) surface by diffraction spectroscopy. Phys. Rev. Lett. **100**(20), 207201 (2008)
104. W. Li, A. Amirfazli, Superhydrophobic surfaces: adhesive strongly to water? Adv. Mater. **19** (21), 3421–3422 (2007)
105. X.M. Li, D. Reinhoudt, M. Crego-Calama, What do we need for a superhydrophobic surface? A review on the recent progress in the preparation of superhydrophobic surfaces. Chem. Soc. Rev. **36**(8), 1350–1368 (2007)
106. G. Caputo, R. Cingolani, P.D. Cozzoli, A. Athanassiou, Wettability conversion of colloidal $TiO_2$ nanocrystal thin films with UV-switchable hydrophilicity. PCCP **11**, 3692–3700 (2009)
107. R.-D. Sun, A. Nakajima, A. Fujishima, T. Watanabe, K. Hashimoto, Photoinduced surface wettability conversion of ZnO and $TiO_2$ thin films. J. Phys. Chem. B **105**(10), 1984–1990 (2001)
108. X. Liu, M. Bauer, H. Bertagnolli, E. Roduner, J. van Slageren, F. Phillipp, Structure and magnetization of small monodisperse platinum clusters. Phys. Rev. Lett. **97**(25), 253401 (2006)
109. X. Feng, L. Feng, M. Jin, J. Zhai, L. Jiang, D. Zhu, Reversible super-hydrophobicity to super-hydrophilicity transition of aligned ZnO nanorod films. J. Am. Chem. Soc. **126**(1), 62–63 (2004)
110. Z. Xu, Z. Ao, D. Chu, A. Younis, C.M. Li, S. Li, Reversible hydrophobic to hydrophilic transition in graphene via water splitting induced by UV irradiation. Sci. Rep. **4**, 6450 (2014)

111. C. Zhu, H. Li, Y. Huang, X.C. Zeng, S. Meng, Microscopic insight into surface wetting: relations between interfacial water structure and the underlying lattice constant. Phys. Rev. Lett. **110**(12), 126101 (2013)
112. J. Liu, C. Wang, P. Guo, G. Shi, H. Fang, Linear relationship between water wetting behavior and microscopic interactions of super-hydrophilic surfaces. J. Chem. Phys. **139**(23), 234703 (2013)
113. F.G. Alabarse, J. Haines, O. Cambon, C. Levelut, D. Bourgogne, A. Haidoux, D. Granier, B. Coasne, Freezing of water confined at the nanoscale. Phys. Rev. Lett. **109**(3), 035701 (2012)
114. E.B. Moore, E. de la Llave, K. Welke, D.A. Scherlis, V. Molinero, Freezing, melting and structure of ice in a hydrophilic nanopore. PCCP **12**(16), 4124–4134 (2010)
115. Q. Yuan, Y.P. Zhao, Topology-dominated dynamic wetting of the precursor chain in a hydrophilic interior corner. Proc. Phys. Soc. London, Sect. A **468**(2138), 310–322 (2011)
116. M. Kasuya, M. Hino, H. Yamada, M. Mizukami, H. Mori, S. Kajita, T. Ohmori, A. Suzuki, K. Kurihara, Characterization of water confined between silica surfaces using the resonance shear measurement. J. Phys. Chem. C **117**(26), 13540–13546 (2013)
117. X.C. Qin, Q.Z. Yuan, Y.P. Zhao, S.B. Xie, Z.F. Liu, Measurement of the rate of water translocation through carbon nanotubes. Nano Lett. **11**(5), 2173–2177 (2011)

# Chapter 11
# Mpemba Paradox

- *Mpemba effect integrates the energy "emission–conduction–dissipation" dynamics of the hydrogen bond in the "source–path–drain" cycle system.*
- *O:H–O bond memory entitles water to emit energy at a rate proportional to its initial storage.*
- *Water skin supersolidity favors outward heat diffusion by raising the local thermal diffusivity.*
- *Non-adiabatic "source–drain" interface enables rapid heat dissipation, but convection, evaporation, frost, supercooling, and solutes contribute insignificantly.*

**Abstract** Numerical reproduction of observations confirms that water skin supersolidity enhances the local thermal diffusivity favoring heat diffusing outwardly in the liquid path. Analysis of experimental database reveals that O:H–O bond possesses memory to emit energy at a rate depending on its initial storage. Unlike other usual materials that lengthen and soften all bonds when they are absorbing thermal energy, water performs abnormally at heating to lengthen the O:H nonbond and shorten the H–O covalent bond through interoxygen Coulomb coupling. Cooling does oppositely to release energy, like releasing a coupled pair of bungees with full recoverability, at a rate of history dependence. Being sensitive to the source volume, skin radiation, and the drain temperature, Mpemba effect proceeds only in the strictly non-adiabatic 'source-path-drain' cycling system for the heat "emission-conduction-dissipation" dynamics with a relaxation time that drops exponentially with the rise of the initial temperature of the liquid source.

## 11.1 Challenge: Why Does Warm Water Freeze Quickly?

The Mpemba effect [1–5] is the assertion that hot water freezes quicker than its cold, even though it must pass through the same lower temperature on the way to freezing. Figure 11.1 shows numerical reproduction of the measured (insets) initial-temperature $\theta_i$ dependence of the thermal relaxation $\theta(\theta_i, t)$ profile and the temperature difference $\Delta\theta(\theta_i, t)$ between the skin and the bulk of liquid water under identical experimental conditions (purity, volume, drain temperature, etc.) [6], which demonstrate the following:



C.Q. Sun and Y. Sun, *The Attribute of Water*, Springer Series in Chemical Physics 113, DOI 10.1007/978-981-10-0180-2_11

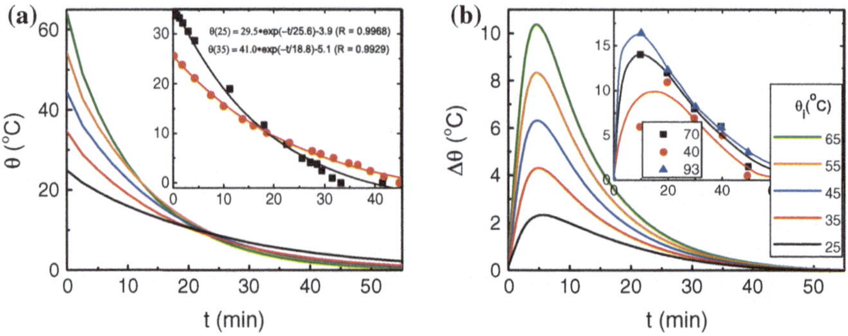


**Fig. 11.1** Numerical reproduction [7] of the measured (*insets*) initial-temperature and time dependence of **a** the $\theta(\theta_i, t)$ [6] and **b** the skin-bulk temperature difference $\Delta\theta(\theta_i, t)$ of water at cooling. *Inset* **a** shows cooling and freezing of 30 ml deionized water at $\theta_i$ = 25 and 35 °C in a glass beaker without cover or being mixed using magnetic stirring (Reprinted with permission from [7].)

(1) The liquid temperature $\theta$ drops exponentially with cooling time ($t$) until the transition from water to ice with a relaxation time $\tau$ that drops as $\theta_i$ is increased.
(2) Water skin is warmer than sites inside the liquid, and the skin of hotter water is even warmer throughout the course of cooling.
(3) Repeating this experiment is infrequent because of water's high sensitivity to the environment and experimental conditions.
(4) Quantitative reproduction of the observations was lacking despite numerous intuitive explanations with hypothetic factors.

## 11.2 Clarification: Hydrogen Bond Memory and Skin Supersolidity

Numerical reproduction of observations in Fig. 11.1 and experimental evidence of the O:H–O bond memory suggest the following mechanisms [7–9], see Fig. 11.2:

(1) Mpemba effect integrates the heat "emission–conduction–dissipation" dynamics in the "source–path–drain" cycle system [10].
(2) O:H–O bond disparity and O–O repulsivity enable its memory to emit energy at a rate proportional to its initial storage.
(3) Water skin supersolidity and heating elevate the local thermal diffusivity mainly by density reduction, favoring outward heat diffusion [8].
(4) The Mpemba effect proceeds only at the strictly non-adiabatic 'source–drain' interface.

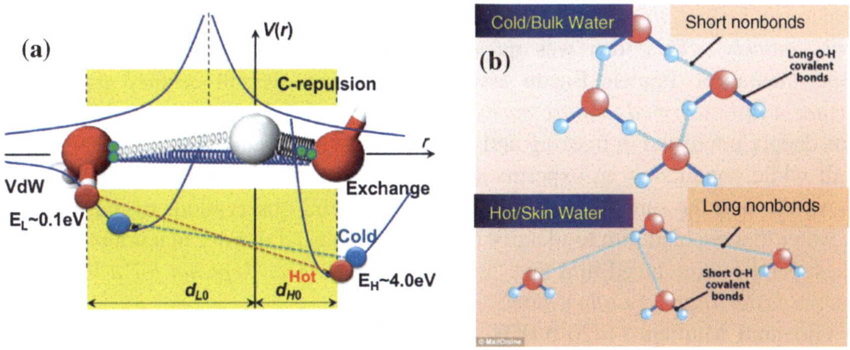


**Fig. 11.2** **a** O:H–O bond asymmetrical and short-range potentials, being analogous to springs. Potentials include the O:H nonbond vdW-like interaction ($E_L \sim 0.1$ eV), the H–O bond exchange interaction ($E_H \sim 4.0$ eV), and the O–O Coulomb repulsion [11]. Heating and molecular undercoordination dislocate O atoms to the left by different amounts with respect to the $H^+$ origin. O ions dislocate along the O:H–O bond potentials from hotter (*red line linked spheres*, labeled 'hot') to colder state (*blue line linked spheres*, labeled 'cold'). **b** Heating and molecular undercoordination shrinks molecular size ($d_H$) but enlarges the separations between molecules, lowering the mass density down to 75 % of the ideal unity at 4 °C. (Reprinted with permission from [7])

## 11.3 Historical Background

### *11.3.1 Mpemba Paradox*

The ability of hot water freezing faster than its cold seems to counter the intuitive that hot water must first become cold water and then freezes, and therefore, the time required for this process will always delay its freezing relative to cold water. Newton's law of cooling indicates that the rate at which a body cools is proportional to the temperature difference between the object and its surroundings. As it is infrequently observed and disobeys the Newton's Law of cooling (1701) in thermodynamics, sometimes this phenomenon is called a great hoax or paradox, the Mpemba paradox, named after Erasto Mpemba, a 13-year old student at the Magamba Secondary School in eastern Tanzania. Erasto asserted it in 1963 during making ice cream. However, the reason for this occurrence remains as one of the biggest mysteries since 350 B.C. when Aristotle firstly noted this effect.

#### 11.3.1.1 Historical Record

This paradox has attracted many inspiring minds. Aristotle (384–322 B.C.), Giovanni Marliani (1420–1483), Francis Bacon (1561–1626); René Descartes (1596–1660) all claimed that hot water does freeze more quickly. In his *Meteorologica* published in 350 B.C. [1], Aristotle wrote that "*if water has been previously heated, this contributes to the rapidity with which it freezes, for it cools more quickly*". This idea was questioned by the great medieval champion of experimentation. Francis Bacon asserted in the seventeenth century stated that "*water a little warmed is more easily frozen than that which is quite cold*". Bacon was deeply interested in freezing and refrigeration—he is said to have caught a fatal chill while conducting an experiment on preserving a chicken by stuffing it with snow. Around the same time, Descartes made careful observations of the freezing of water that enabled him to identify the liquid's unusual density maximum at 4 °C. These studies convinced him that "*water which has been kept hot for a long time freezes faster than any other sort*".

Giovanni Marliani (1420–1483) was an Italian physicist, doctor, philosopher, and astrologer. He was the first to empirically prove that hot water freezes faster than its cold in 1461. He used four ounces of unheated water and four ounces of boiled water, which he placed in similar containers outside on a cold winter's day. He eventually observed that the boiled water froze first; although he was unable to explain the mechanics of how it happened.

| | |
|---|---|
|  | *Aristotle (384–322 BC)* a Greek philosopher and scientist firstly found the Mpemba effect in 350 BC. (Free Wikimedia). |
|  | Francis Bacon (1561–1626), an English philosopher, statesman, scientist, jurist, orator, essayist, and author, wrote that "*slightly tepid water freezes more easily than that which is utterly cold*". |

(continued)

(continued)

| | |
|---|---|
|  | Rene Descartes (1596–1650), a French philosopher, mathematician and writer, also tried to solve the problem in 1637 and throughout his years. He wrote in his *Discourse on the Method*, *"One can see by experience that water that has been kept on a fire for a long time freezes faster than other, the reason being that those of its particles that are least able to stop being evaporate while the water is being heated."* (free Wikimedia). |

**Fig. 11.3** One can make snow or frozen fog by pouring boiling or hot water into cold ambient, –25 °C for instance [12]

Now a days, people are often use this phenomenon to make ice cubes by putting water in the sun before freezing. A snapshot of a video clip in Fig. 11.3 shows the making of snow or frozen fog from spreading hot water at −25 °C temperature [12].

#### 11.3.1.2 Mpemba's Experiment

Erasto Mpemba, a secondary—school student in Tanzania, may have been unaware of the historical claims, also observed the same in 1963. To make ice cream for a school project, he was told to boil milk mixture and then let it cool before putting it in the refrigerator. But, fearful of losing his place, Mpemba put his mixture in the fridge while it was still hot. He found that it frozed before the other, cooled mixtures.

However, many scientists find it hard to accept such a seemingly counterintuitive phenomenon. The problem is that the effect is frustratingly hard to reproduce—sometimes it appears, and sometimes not. In fact, no one has agreed exactly how the experiments should be conducted in the first place. Even if the Mpemba effect is real—if hot water can sometimes freeze more quickly than cold—it is not clear whether the explanation would be trivial or illuminating.

Monwhea Jeng [4] believes that scientists are much more likely to react with disbelief than lay people when they first hear about the phenomenon. That is because scientists know why it "cannot" be right. Indeed, when Mpemba learned about Newton's law of cooling a few years after making his discovery and asked his teacher how this could be reconciled with his observations, his teacher replied, "all I can say is that is Mpemba's physics and not the universal physics."

Fortunately, Mpemba was not deterred by this cruel put-down, and he went on to carry out further experiments of his own. When local physics professor Denis Osborne of University College in Dares Salaam visited the school, Mpemba seized the chance to ask for an explanation for his findings. Osborne had none, but he was less skeptical than Mpemba's teacher and wisely concluded that "it is dangerous to pass judgment on what can and cannot be". Osborne then repeated the experiments, and confirmed that Mpemba is right.

Erasto Mpemba (left) and Professor Denis Osborne (right; 1932–2014) presented at the presentation in London (Courtesy of Ben Gurr/The Times, 2013).

Mpemba and Osborne placed 70 ml samples of water in 100 ml beakers in the ice box of a domestic refrigerator on a sheet of polystyrene foam. They recorded the time for *freezing to start* was longest with an initial temperature of 25 °C and that it was much less at around 90 °C. They ruled out loss of liquid volume by *evaporation* as a significant factor and the effect of *dissolved air* as well. In their setup most heat loss was found to be from the *liquid skin* [2]. In 1969, Osborne wrote about the work with Mpemba [2] and published it in the journal *Physics Education*. Coincidentally, George Kell [13] reported the same phenomenon in the *American Journal of Physics*.

These reports confirmed that the Mpemba effect was true. Kell suggested that "a car should not be washed with hot water because the water will freeze on it more quickly than cold water will, or that a skating rink should be flooded with hot water because it will freeze more quickly". Mpemba, meanwhile, pointed out that

Tanzanian ice-cream makers routinely froze their mixtures while they were hot, because that was a faster method. When Mpemba's work was described in an article in *New Scientist* in 1969, it prompted a rush of anecdotes about food-freezing practices and hot-water pipes freezing while cold ones did not.

#### 11.3.1.3 Subsequent Hunting

A number of scientists have investigated Mpemba's claim, but their results remain inconclusive. In 1977, Jearl Walker [14] reported in *Scientific American* that he had observed the time it took a beaker of water to cool to 0 °C from different initial temperatures under various conditions. These tests provided some clarification of the effect. Walker could reproduce most of his results with large deviations in some of them.

Jeng [4] suggests that "there exists a set of initial parameters, and a pair of temperatures, such that given two bodies of water identical in these parameters, and differing only in their initial temperatures, the hot one will freeze sooner." There are many such parameters that could affect the rate of freezing, the most obvious including the volume and type of water used, the size and shape of the containers, and the temperature of the fridge. This suggestion presents a significant challenge for the experimentalist, who in principle would have to set up a vast multidimensional array of experiments involving containers with different sizes and shapes, while independently varying the masses and gas content of the water and the refrigeration method used, in order to test for the effect.

Despite the continuing uncertainties surrounding the effect, Pablo Debenedetti, a physicist at Princeton University and a specialist in phase transitions of water, is happy to believe Mpemba's account. "I do not see any reason to doubt observations showing that under some circumstances hot water can freeze faster than cold water."

Such contradictions continue to make the Mpemba effect as puzzling as ever. Knight is happy to leave it that way, because he thinks that attempts to clarify it would demand too much effort for little return. But Jeng is more positive. He says that despite the complexity of the effect, the experiments needed to study it can be carried out by undergraduates and high-school students—so long as they are planned carefully. As well as thinking about exactly how to heat the water and the kind of thermometer that should be used, experimentalists should also consider precise details of the environment surrounding the container. "It can make a difference whether the water is in the middle of an empty freezer, or jammed between a frozen pizza and a frost-covered tub of ice cream," he says.

James Brownridge (1937–, a radiation safety officer at the State University of New York) [15, 16] had spent 10 years and conducted more than 20 experiments to examine all possible factors. He suggested that *supercooling* is the dominant factor.

Philip Ball [17] overviewed in 2006 the history of the Mpemba effect and noted in *Physics World*, "Even if the Mpemba effect is real—it is not clear whether the explanation would be trivial or illuminating." He pointed out that investigations of the phenomenon need to control a large number of initial parameters (including type and initial temperature of the water, dissolved gas and other impurities, and size, shape and material of the container, and temperature of the refrigerator) and need to settle on a particular method of establishing the time of freezing, all of which might affect the presence or absence of the Mpemba effect. The required vast multidimensional array of experiments might explain why the effect is not yet understood.

#### 11.3.1.4 RSC 2012 Contest

In 2012, the Royal Society of Chemistry held a competition calling for papers offering explanations to the Mpemba effect. More than 22,000 people entered and Erasto Mpemba himself announced Nikola Bregović [6] who works at the Laboratory of Physical Chemistry in the Department of Chemistry of the University of Zagreb, as the winner. Bregović suggests that *convection* and *supercooling* were the reasons for the effect and provides a qualitative explanation:

> The statement by Brownridge, 'Hot water will freeze before cooler water only when the cooler water supercools, and then, only if the nucleation temperature of the cooler water is several degrees lower than that of the hot water. Heating water may lower, raise or not change the spontaneous freezing temperature,' summarizes in great part the conclusions that may be drawn from almost all the data I have collected myself and others presented earlier. However, the effect of convection, which enhances the probability of warmer-water freezing should be emphasized in order to express a more complete explanation of the effect. The fact that this effect is not fully resolved to this day, was an indication to me that fundamental problems lie underneath it, but still I did not expect to find that water could behave in such a different manner under so similar conditions. Once again this small, simple molecule amazes and intrigues us with its magic.

In August 2012, Osborne described the following about his work with Mpemba: "In line with his question made in front of his school staff and peers, we tested and found that hot water in Pyrex beakers on polystyrene foam in a domestic freezer froze before cooler samples. We attributed this to *convection* creating a continuing hot top, noting that:

- If two systems are cooled, the water that starts hotter may freeze first, but we did not look for ice and measured the time as that until a thermocouple in the water read 0 °C.
- A graph of 'time to start freezing' against initial temperature showed that the water starting at about 26 °C took longest to freeze; water starting at 60 °C took twice as long as water starting at 90 °C.
- Thermocouples near top and bottom showed a temperature gradient in the water. A hot starter kept a hot top while its lower levels were cooler than for the cool starter (due to convention effect).
- An oil film on the water surface delayed freezing for several hours, suggesting that without this film, most of the heat escaped from the top surface (thermal insulation effect).

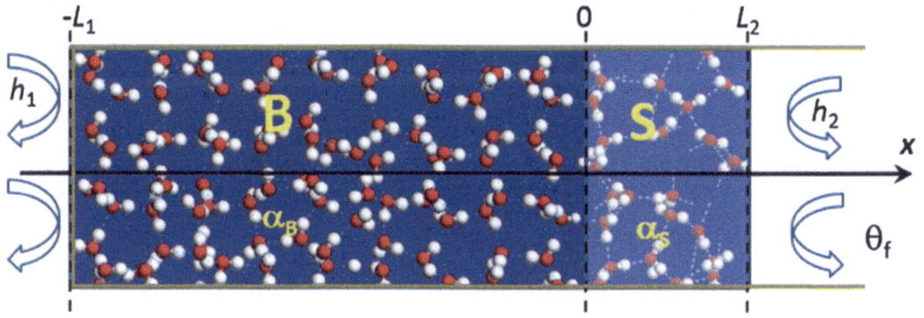


**Fig. 11.4** Water in the adiabatically walled, open-ended, one-dimensional tube cell at initial temperature $\theta_i$ is cooled in the drain of constant temperature $\theta_f$. The liquid source is divided along the $x$-axis into the bulk $B$ ($-L_1 = -9$ mm, 0) and the skin $S$ (0, $L_2 = 1$ mm), with respective thermal diffusivities $\alpha_B$ and $\alpha_S$. The mass densities are $\rho_B = 1$ and $\rho_S/\rho_B = 3/4$ g/cm$^3$ [27, 28] in the respective region. The point of $x = 0$ is the bulk-skin interface; $h_j$ is the heat transfer (radiation) coefficient at tube ends with absence ($j = 1$, *left-hand end*) and presence ($j = 2$, *right-hand end*) of the supersolid skin. (Reprinted with permission from [7])

- Changes in volume due to evaporation were small; the latent heat of vaporization for all the water to cool to 0 °C and start freezing accounted for less than 30 % of the cooling.
- We used recently boiled water for all the trials, making dissolved air an unlikely factor. We failed to check and report the ambient temperature in the freezer or its consistency during cooling. Lower ambient air temperatures might increase heat loss rates from the top surface, cause more rapid convection and increase the difference in freezing times.
- Different mechanisms may assume more importance in different situations. We gave one example, with Mpemba's initial discovery in mind, and we wrote: 'rapid cooling of a system that starts hot may be accelerated if it establishes thermal contact with the case of the freezer cabinet through melting the layer of ice and frost on which it rests'.

### 11.3.2 Latest Development

Sun and coworkers [7] examined the process in 2012 at a molecular level from the perspective of O:H–O bond cooperativity. They reproduced observations of Mpemba and Bregović by solving the Fourier thermo-fluid transportation dynamics using finite element method with involvement of skin effect and appropriate boundary conditions. They suggested that the Mpemba effect integrate the heat "emission–conduction–dissipation" dynamics proceeded in the "source–path–drain" cycle system. The dynamics of heat emission and conduction are intrinsically related to the O:H–O bond thermal and undercoordination relaxation dynamics and the heat dissipation is related to the extrinsic operation conditions [7, 11]. In particular, one cannot separate any process from the rest two in heat "emission–conduction–dissipation" dynamics whose relaxation time has little to do with fleeting lifetime of water molecules [18].

Sun and cowaorkers deposited their preliminary work in ArXiv Physics [9] in October 2012. This preliminary explanation attracted a flush of public attention and was hailed by numerous news and scientific media such as *Times, ArXiv Media, The Telegraph, Daily Mails, Chemical World, Chem Views, Physics Today, IOP Physics, Nature Chemistry,* etc. More quantitative solution was published in the Journal of *Physical Chemistry and Chemical Physics* in October 2014 [7].

One needs to note that the energy release relaxation time of the O:H–O bond in the Mpemba effect is in the *min* order, and the relaxation time varies with the initial temperature and experimental conditions[7]. On the other hand, as the inverses of the respective phonon frequencies, the phonon lifetimes are $\tau_L = v_L^{-1} = (2\pi c\omega_L)^{-1} \sim 3$ ns (for 200 $cm^{-1}$) and the $\tau_H = v_H^{-1} = (2\pi c\omega_H)^{-1} \sim 50$ ns(for 3000 $cm^{-1}$) order. The H–O phonon relaxation time, detected using ultrafast–IF, varies from 200 to 800 fs [19]. Therefore, the O:H–O bond heat release relaxation time is different from the fleeting times or the phonon life times.

### *11.3.3 Notes on Existing Explanations*

The following hypothetic factors have been employed to explain this paradox:

(1) *Evaporation* [20]: The evaporation of the warmer water reduces the mass of the water to be frozen. Evaporation is endothermic, meaning that the water mass is cooled by vapour carrying away the heat, but this alone probably does not account for the entirety of the effect. Pable Debenedetti, Professor of Chemical and Biological Engineering, Princeton University, point out that there could be at least one obvious explanation for it. If the containers are left open, the hot water will evaporate more quickly and its volume will decline compared with that of the cold water. With a smaller volume, the cooling of the hot water could then overtake that of the cold. That should be easy to test, according to Debenedetti, because the evaporation rate is proportional to the open area of the liquid surface. However, experiments confirmed that the mass loss is only 1.5 % or less when cooled from 75 to −40 °C [7].

(2) *Convection* [21–23]: Accelerating heat transfers. Reduction of water density below 4 °C (39 °F) tends to suppress the convection currents that cool the lower part of the liquid mass; the lower density of hot water would reduce this effect, perhaps sustaining the more rapid initial cooling. Higher convection in the warmer water may also spread ice crystals around faster. Numerical examination revealed however, that contribution of the convection velocity to the intersecting temperatures of two $\theta(\theta_i, t)$ relaxation curves is insignificant [7].

(3) *Frost* [4, 16]: Frost insulates thermal dissipation. The lower temperature water will tend to freeze from the top, reducing further heat loss by radiation and convection, while the warmer water will tend to freeze from the bottom and sides because of water convection. This is disputed as there are experiments that account for this factor.

(4) *Solutes* [25]: The effects of calcium carbonate, magnesium carbonate among others. The Mpemba effect can be observed in deionized water without any solutes being involved.
(5) *Thermal conductivity*: The container of hotter liquid may melt through a layer of frost that is acting as an insulator under the container (frost is an insulator), allowing the container to come into direct contact with a much colder lower layer that the frost formed on (ice, refrigeration coils, etc.). The container now rests on a much colder surface (or one better at removing heat, such as refrigeration coils) than the originally colder water, and so cools far faster from this point on.
(6) *Dissolved gases* [4]: Cold water can contain more dissolved gases than hot water, which may somehow change the properties of the water with respect to convection currents, a proposition that has some experimental support but no theoretical explanation. Hot water generally holds less dissolved gas than cold, which means that two samples that differ only in their initial temperature would not contain "identical" substances. Debenedetti points out that tiny bubbles of gas can provide nucleation sites where ice crystals start to form. In principle, this might be expected to make ice formation easier in cold water, contrary to the Mpemba effect. But Debenedetti says that the solubility of nonpolar gases such as nitrogen or methane do not necessarily vary smoothly with temperature, so there could be temperature ranges within which the hotter water contains more dissolved gas. Experiments to pinpoint these influences would require the water to be thoroughly degassed. The effects of other dissolved impurities could be even harder to probe: for example, one could divide the water up into tiny droplets in an oil–water emulsion so that most of them are too small to contain any impurity particles.
(7) *Supercooling* [16, 17, 23]: When placed in a freezing environment, cool water supercools more than hot water in the same environment, thus solidifying slower than hot water. However, supercooling tends to be less significant where there are particles that act as nuclei for ice crystals, thus precipitating rapid freezing. Then there is the role of chance, since the nucleation of ice in freezing water depends on enough water molecules coming together to form the core of an ice crystal that can then grow indefinitely. The further the water is below freezing point, the more likely this is to happen. But because it can take some time for ice crystals to nucleate, water can often be "supercooled" such that it remains liquid well below freezing. Random impurities in the liquid, such as specks of dust, can, however, increase the rate of nucleation and suppress supercooling.
In 1995 German physicist David Auerbach at the Max Planck Institute for Fluid Dynamics in Göttingen looked at the role of supercooling in the Mpemba effect. David [24] described how the effect can be observed in samples in glass beakers placed into a liquid cooling bath. In all cases the water supercools, reaching a temperature of typically −6 °C to −18 °C before spontaneously freezing. Considerable random variation was observed in the time required for spontaneous freezing to start and in some cases this resulted in the water which started

off hotter (partially) freezing first. What he found only made things more complicated. He observed that hot water froze at a higher temperature than cold and therefore in a sense froze "first". However, the cold water took less time to reach its supercooled state and so seemed to freeze "faster". To add to the confusion, earlier researchers had reported the opposite: that initially hot water could be supercooled to lower temperatures than cold water. In 1948 Noah Dorsey of the US National Bureau of Standards argued that this is because heating expels impurity particles that acted as nucleation sites for ice. It has been claimed that this effect leads to hot-water pipes bursting more readily than cold, since deeper supercooling leads to ice fingers that advance right across the pipe and block the flow, while freezing nearer to 0 °C just produces a sheath of ice on the pipe surfaces with an open channel in the centre. Experimentally derived crossing temperatures at θ > 0 °C (Fig. 11.1a) involves no supercooling effect.

However, all above factors are extrinsic and intuitive to the energy 'emission–conduction–dissipation' dynamics in the 'source–path–drain' cycle system in which the Mpemba paradox takes place. Focusing on the nature and relaxation dynamics of the O:H–O bond [26] with quantitative solution is necesary as the O:H-O bond relaxation is the primary constituent of the liquid source and the path in this event.

## 11.4 Numerical Resolution: Skin Supersolidity

### *11.4.1 Fourier Thermal–Fluid Transport Dynamics*

Figure 11.4 illustrates an adiabatically–walled, open-ended, one-dimensional tube cell for solving the thermal fluid transport problem using the finite element method [7]. Water at an initial temperature $\theta_i$ in the cell is divided along the $x$-axis into two regions: the bulk and the skin. The tube is cooled in a drain of constant temperature $\theta_f$ that is subject to variation to allow it to be examined for sensitivity.

The time dependent temperature gradient at any site $x$ follows the Fourier transport function and meets the initial and boundary conditions:

$$
\begin{aligned}
&\frac{\partial\theta(x)}{\partial t} = \nabla\cdot(\alpha(\theta(x),x)\nabla\theta(x)) - v\cdot\nabla\theta(x) \\
&\begin{pmatrix} \alpha(\theta,x) = \frac{\kappa_B(\theta,x)}{\rho_B(\theta,x)C_{pB}(\theta,x)} \times \begin{cases} 1 & (Bulk) \\ \approx \rho_B/\rho_S(=4/3) & (Skin), \end{cases} \\ v_S = v_B = 10^{-4}(m/s) \end{pmatrix} \\
&\begin{cases} \theta = \theta_i & (t=0) \\ \theta(0^-) = \theta(0^+);\theta_x(0^-) = \theta_x(0^+) & (x=0) \\ h_i(\theta_f - \theta) \pm \kappa_i\theta_x = 0 & (x = -l_1; l_2). \end{cases}
\end{aligned}
\qquad (11.1)
$$

The first term in the right-hand side describes thermal diffusion and the second term describes thermal convection in the Fourier equation. The $\alpha$ is the thermal

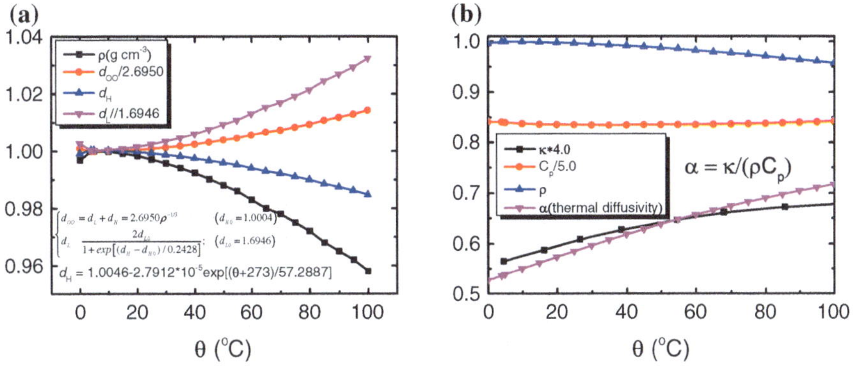


**Fig. 11.5** Temperature dependence of **a** the mas density $\rho(\theta)$, H–O length $d_H(\theta)$, O:H length $d_L(\theta)$, O–O distance $d_{OO}(\theta)$ formulated by the *inset* [28], and, **b** thermal conductivity $\kappa(\theta)$, specific heat $C_p(\theta)$, and thermal diffusivity $\alpha(\theta)$ of liquid water

diffusivity and the $v$ is the convection rate. The known temperature dependence of the thermal conductivity $\kappa(\theta)$, the mass density $\rho(\theta)$, and the specific heat under constant pressure $C_p(\theta)$, given in Fig. 11.5, determines the thermal diffusivity $\alpha_B$ of the bulk water. The skin supersolidity [27] contributes at least to the $\alpha_S$ in the form $\alpha_S(\theta) \approx 4/3\alpha_B(\theta)$, because the skin mass density 0.75 $\text{gcm}^{-3}$ is 3/4 times the standard density at 4 °C. $\alpha_S(\theta)$ is subject to optimization as the skin supersolidity may modify the $\kappa(\theta)/C_p(\theta)$ value in a yet unknown manner.

The boundary conditions represent that the $\theta$ and its gradient $\theta_x = \partial\theta/\partial x$ continue at the interface ($x = 0$) and the thermal flux $h(\theta_f - \theta)$ is conserved at each end for $t > 0$. The convection velocity $v$ takes the bulk value of $v_S = v_B = 10^{-4}$ m/s, or zero for examination. As the heat transfer (through radiation) coefficient $h_j$ depends linearly on the thermal conductivity $\kappa$ in the respective region [29], the standard value of $h_1/\kappa_B = h_2/\kappa_S = 30$ w/(m$^2$K) [30] is necessary for solving the problem. The $h_2/\kappa_S$ term contains heat reflection by the boundary. The ratio $h_2/h_1 > 1$ describes the possible effect of skin thermal radiation.

### *11.4.2 Convection, Diffusion, and Radiation*

The computer reads in the $\alpha_B(\theta)$ in Fig. 11.5 before each iteration of calculating the partitioned elemental cells. Besides the thermal diffusivity and the convection velocity in the Fourier equation, systematic examination of all possible parameters in the boundary conditions, shown in Figs. 11.6 and 11.7, revealed the following:

(1) Characterized by the intersecting temperature, the Mpemba effect happens only in the presence of the supersolid skin ($\alpha_S/\alpha_B > 1$).

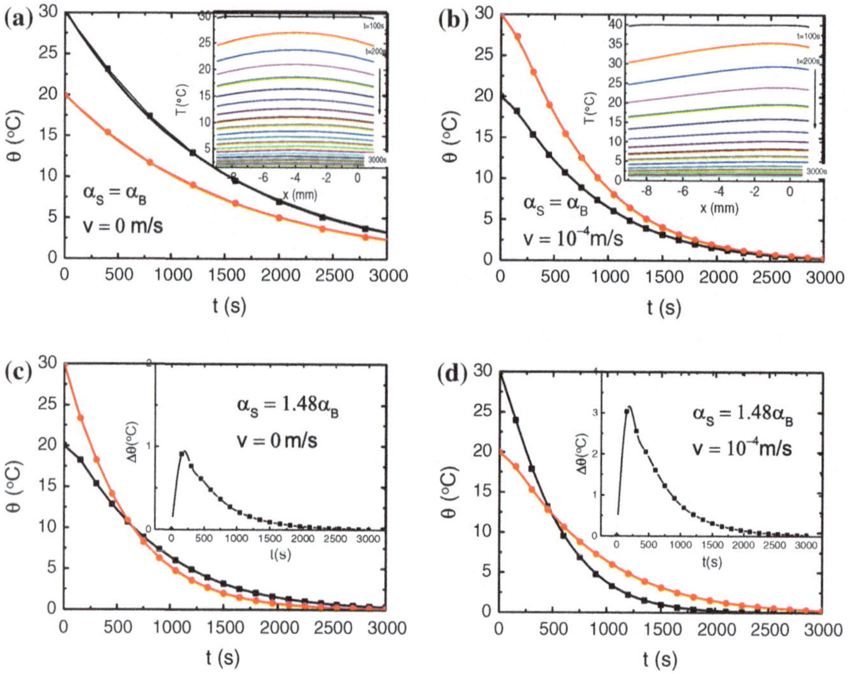


**Fig. 11.6** Thermal relaxation curves $\theta(\theta_i, t)$ at $x = 0$: **a**, **b** with supersolid skin absence ($\alpha_S/\alpha_B = 1$); and **c**, **d** with supersolid skin presence (optimized at $\alpha_S/\alpha_B = 1.48$); and **a**, **c** with thermal convection absence ($v_S = v_B = 0$); and **b**, **d** with thermal convection presence ($v_S = v_B = 10^{-4}$ m/s). The Mpemba effect is characterized by the crossing temperature which occurs only in the presence of the skin supersolidity, irrespective of the thermal convection. The *insets* in **a** and **b** show the time-dependent thermal field in the tube. Supplementing the skin supersolidity, convection only slightly raises $\Delta\theta$ and the crossing temperature—as the *insets* in **c** and **d** show. The $\Delta\theta$ may vary with the orientation of the tube put in the frige because of gravity and water thermal density (Reprinted with permission from [7].)

(2) Complementing skin supersolidity, thermal convection distinguishes only slightly difference temperature $\Delta\theta$ between the skin and the bulk, and raises negligibly the crossing temperature.
(3) The Mpemba effect is sensitive to the source volume, the $\alpha_S/\alpha_B$ ratio, the radiation rate $h_2/h_1$ and the drain temperature $\theta_f$.
(4) The bulk/skin thickness ($L_1$:$L_2$) ratio and the thermal convection have little effect on observations.

The insensitivity to the thickness ratio evidences the long-range response to perturbation due to molecular undercoordination. Increasing the liquid volume may annihilate the Mpemba effect because of the non-adiabatic process of heat dissipation. It is understandable that cooling a drop of water (1 mL) needs shorter time than cooling one cup of water (200 mL) at the same $\theta_i$ and under the same conditions. Higher skin radiation $h_2/h_1 > 1$ promotes the Mpemba effect. Therefore,

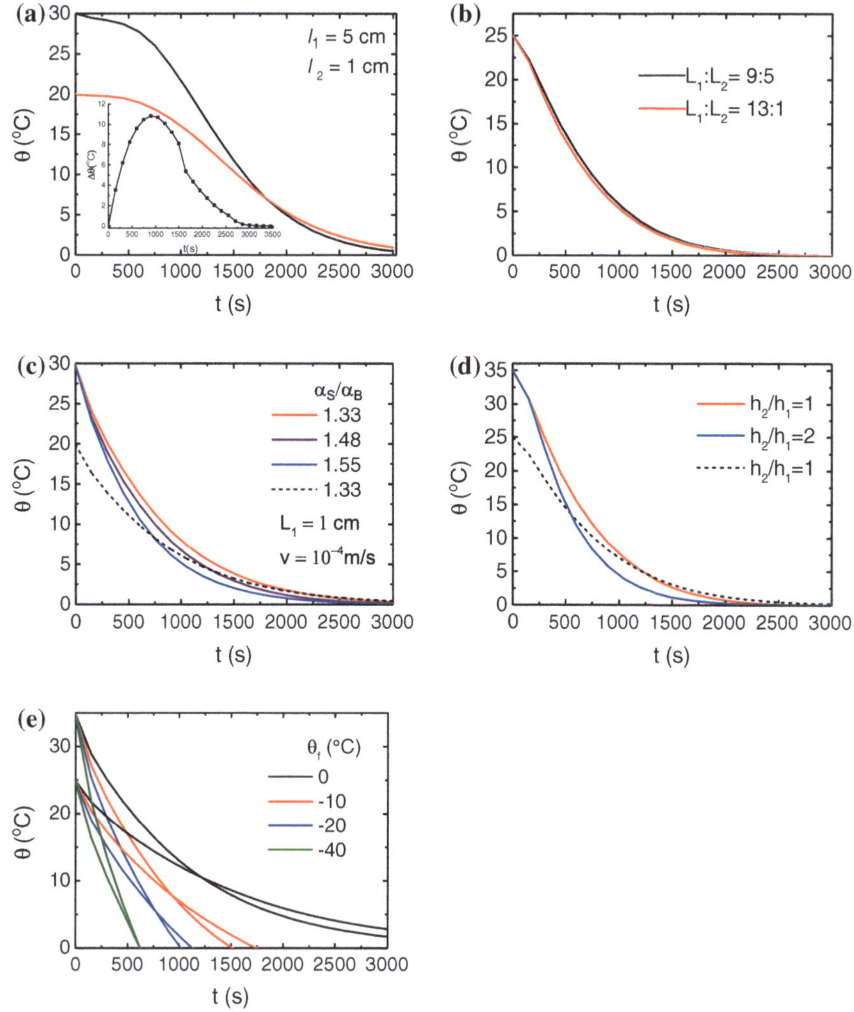


**Fig. 11.7** Sensitivity of the Mpemba effect (crossing temperature) to: **a** the source volume; **b** the bulk/skin thickness ratio ($L_1$:$L_2$); **c** the supersolidity ratio $\alpha_S/\alpha_B$; **d** the radiation rate $h_2/h_1$; and **e** the drain temperature $\theta_f$. Volume inflation (from 1 to 5 cm) in **a** prolongs the time until the crossing temperature is reached, and raises the skin temperature (see *inset*). **b** The $L_1$:$L_2$ ratio has little effect on the relaxation curve. Increasing **c** the $\alpha_S/\alpha_B$ and **d** the $h_2/h_1$ ratio promotes the Mpemba effect. **e** Lowering the $\theta_f$ shortens the time until the crossing temperature is reached. The sensitivity examination is conducted based on the conditions of $\alpha_S/\alpha_B = 1.48$, $v_S = v_B = 10^{-4}$ m/s, $\theta_f = 0$ °C, $L_1 = 10$ mm, $L_2 = 1$ mm, $h_1/\kappa_B = h_2/\kappa_S = 30$ w/(m$^2$K) unless indicated (Reprinted with permission from [7].)

conditions for the Mpemba effect are indeed very critical, which explains why it is not frequently observed. Figure 11.1 shows the numerical reproduction of the observed Mpemba attributes under the optimal conditions [2, 6].

## 11.5 Experimental Derivative: O:H–O Bond Memory

### *11.5.1 O:H–O Bond Relaxation Velocity*

The following formulates the decay curve $\theta(\theta_i, t)$ shown in Fig. 11.1a [6]:

$$\begin{cases} d\theta = -\tau_i^{-1}\theta dt & (decay\ function) \\ \tau_i^{-1} = \sum_j \tau_{ji}^{-1} & (relaxation\ time) \end{cases}, \tag{2}$$

where the $\theta_i$-dependent relaxation time $\tau_i$ is the sum of $\tau_{ji}$ over all possible *j*-th process of heat dissipation during cooling.

A combination of the measured $\theta(\theta_i, t)$ (Fig. 11.1a inset) and the $d_H(\theta)$ (Fig. 11.5) reveals that the O:H–O bond has memory without needing any assumption or approximation. This memory means that the O:H–O bond relaxation velocity is initial deformation or energy storage dependent. As the O:H nonbond and the H–O bond are correlated, the relaxation velocities of their lengths and energies are readily available, since $E_x = k_x(\Delta d_x)^2/2$ approximates the energy storage with the known $d_H$ velocity. For simplicity and conciseness, focus will be given on the instantaneous velocity of the $d_H(\theta)$ and the $\theta(\theta_i, t)$ delay curve during relaxation:

$$\begin{cases} \theta(\theta_i, t) = \theta_i \exp(-t/\tau_i) + \theta_0 \\ d\theta/dt = -\tau_i^{-1}\theta \end{cases}$$

and,

$$\begin{cases} d_H(\theta) = 1.0042 - 2.7912 \times 10^{-5} \exp[(\theta + 273)/57.2887] \\ d[d_H(\theta)]/d\theta = \Delta[d_H(\theta)]/57.2887 \end{cases}$$

where:

$$\Delta(d_H(\theta)) = -2.7912 \times 10^{-5} \exp[(\theta + 273)/57.2887].$$

Combining both slopes immediately yields the linear velocity for $d_H(\theta)$ relaxation at cooling.

$$\frac{d(d_H(\theta))}{dt} = \frac{d(d_H(\theta))}{d\theta}\frac{d\theta}{dt} = -\tau_i^{-1}\theta\frac{\Delta(d_H(\theta))}{57.2887}, \tag{3}$$

Figure 11.8b plots the $\theta_i$-dependence of the $d_H$ linear velocity, which confirms that the O:H–O bond indeed possesses memory. Although passing through the same temperature on the way to freezing, the initially shorter H–O bond at higher temperature remains highly active compared to its behavior otherwise when they meet on the way to freezing.

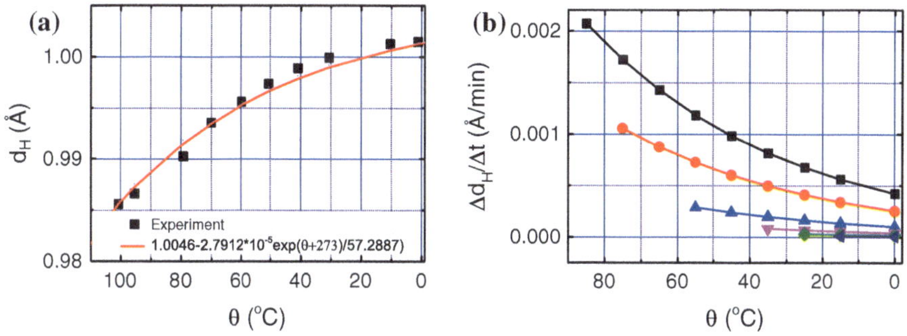


**Fig. 11.8** The **a** measured (*scattered data*) and simulated (*solid line*) $d_H(\theta)$ decay and **b** the experimentally derived $\theta_i$ (corresponds to the starting point of each line) dependence of the $d_H$ velocity at cooling. The velocity of the initially shorter H–O bond at higher $\theta_i$ remain always higher than those initially longer ones at lower $\theta_i$ values when they meet (Reprinted with permission from [7].)

### 11.5.2 *Relaxation Time Versus Initial Energy Storage*

Solving the decay function (3) yields the relaxation time $\tau_i(t_i, \theta_i, \theta_f)$:

$$\tau_i = -t_i\left[Ln\left(\frac{\theta_f + b_i}{\theta_i + b_i}\right)\right]^{-1}. \tag{4}$$

An offset of the $\theta_f$ (= 0 °C) and the $\theta_i$ by a constant $b_i$ is necessary to ensure $\theta_f + b_i \geq 0$ in the solution ($b_i$ = 5 was taken with reference to the fitted data in Fig. 11.1a. A combination of the known $t_i$, $\theta_i$ and $\theta_f$, given in Fig. 11.9a (scattered data) yields the respective lifetime $\tau_i$, shown by the solid line. The $\tau_i$ drops exponentially with the increase of the $\theta_i$ (Fig. 11.9a), or with the increase of initial

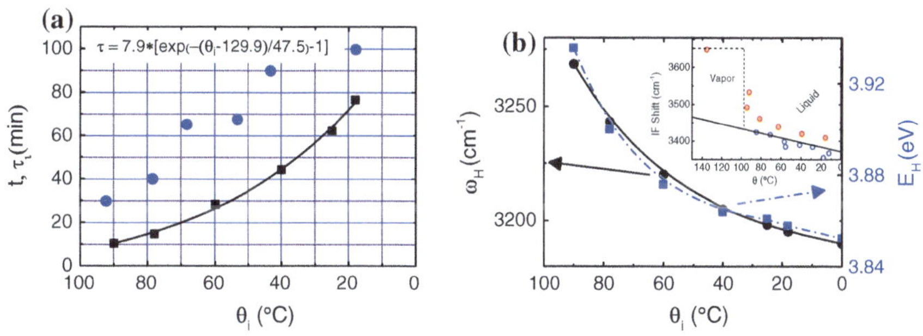


**Fig. 11.9** Initial temperature $\theta_i$ dependence of **a** the cooling time $t$ (*scattered circles*), relaxation time $\tau_i$ (*fitted solid line*), and **b** the measured initial energy $E_H$ (*solid black line*) and vibration frequency $\omega_H$ (*broken blue line*) for liquid water [31]. *Inset* in b shows the $\omega_H$ peak shift with temperature of the liquid water [31] (Reprinted with permission from [7].) (color online)

energy storage, or the initial vibration frequency measurements [31]. For instance, making ice using 78 °C water takes 40 min with a 15 min relaxation time; making ice from 18 °C water takes 100 min with a relaxation time of 75 min. The $\tau_i$ also drops with the increase of initial energy storage, or the initial vibration frequency [31], as Fig. 11.9b shows.

## 11.6 Heat 'Emission–Conduction–Dissipation' Cycling Dynamics

### *11.6.1 Source and Path: Heat Emission and Conduction*

Figure 11.2a illustrates the cooperative relaxation of the O:H–O bond in water under thermal cycling. An interplay of the O:H nonbond potential, the H–O exchange interaction, the O–O Coulomb repulsion, heating dislocates O atoms in liquid water to the left along the respective potential paths [32].

Generally, heating stores energy in a substance by stretching all bonds involved. However, heating excitation stores energy in water by lengthening the O:H nonbond. The O:H expansion weakens the Coulomb interaction, which shortens the H–O bond by shifting the $O^{2-}$ towards the $H^+$ (red line linked spheres in Fig. 11.2a are in the hot state). Cooling does the opposite (blue line liked spheres), analogous to suddenly releasing a pair of coupled, highly deformed springs, one of which is stretched and the other is compressed. This relaxation emits energy at a rate that depends on the deformation history (i.e., how much they were stretched or compressed). Energy storage and emission of the entire O:H–O bond occurs mainly through H–O relaxation, since $E_L$ (about 0.1 eV) is only 2.5 % of $E_H$ (about 4.0 eV) [27].

### *11.6.2 Source–Drain Interface: Non-Adiabatic Heat Dissipation*

Mpemba effect happens only in circumstances where the water temperature drops abruptly from $\theta_i$ to $\theta_f$ at the source–drain interface [7]. Examination has indicated that the Mpemba crossing temperature is sensitive to the volume of the liquid source (Fig. 11.7a). Larger liquid volumes may prevent this effect from happening by heat–dissipation hindering. As confirmed by Brownridge [16], any spatial temperature decay between the source and the drain could prevent the Mpemba effect. Examples of such decay include sealing the tube ends, an oil film covering, a vacuum isolating the source from the cold drain chamber, muffin-tin-like containers making in one copper plate, or multiple sources contributing to a limited fridge volume. Conducting experiments under identical conditions is necessary to minimize artifacts such as radiation, source/drain volume ratio, exposure area, container material, etc.

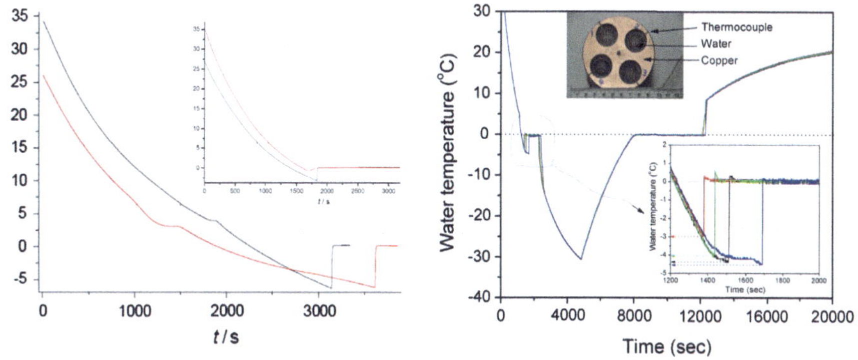


**Fig. 11.10** Cooling and freezing of 30 ml deionized water in icebox without stir mixing. The *inset* in beaker with magnetic stir mixing [6]. Four specimens of tap hard water in a copper container designed to maintain identical cooling conditions for each specimen [15]

Figure 11.10 shows the $\theta(\theta_i, t)$ profiles obtained under different conditions, which suggests that the O:H–O bond is very sensitive to stimulus during relaxation. Conditions for the Mpemba effect are indeed very critical, which explains why it is not frequently observed.

### *11.6.3 Other Factors: Supercooling and Evaporating*

Water molecules with fewer than four nearest neighbors, such as those that form the skin, a monolayer film, or a droplet on a hydrophobic surface or hydrophobically confined are subject to stretching dispersion of the quasi-solid phase boundaries. The phase boundary dispersion results in the melting point elevation and the freezing point depression. Approximately, the $E_H$ determines the critical temperature for melting and the $E_L$ dominates freezing [32]. The critical temperature elevation/depression is different from the process of superheating/superheating. Supercooling is associated with the initially longer O:H bond at higher temperature, which contributes positively to the liner velocity of O:H–O bond relaxation with a thermal momentum at cooling. Supercooling of colder water can never happen because of the relatively stiffer O:H nonbond with higher $\Theta_{DL}$ value. Therefore, the colder water reacts more slowly to the relaxation at freezing than bonds in the warmer water, because of the lower momentum of relaxation–memory effect.

The involvement of ionic solutes or impurities [33, 34] mediates Coulomb coupling because of the alternation of charge quantities and ion volumes [35]. Salting shifts the H–O phonon positively in the same way as heating [36–38] to weaken the Coulomb repulsion. Salting or impurities are expected to raise the velocity of heat emission at cooling but in Mpemba's observation, the only variable is temperature. Mass loss due to evaporation [3] has no effect on the relaxation rate of the O:H–O bond. The mass loss for samples at slightly different temperature is negligible.

## 11.7 Summary

Reproduction of observations revealed the following:

(1) Mpemba effect integrates the heat "emission–conduction–dissipation "dynamics in the "source–path–drain" cycle system as a whole. One cannot separate these processes and treat them independently.
(2) O:H–O bonds possess memory whose thermal relaxation defines intrinsically the rate of energy emission. Heating stores energy in water by O:H–O bond deformation. The H–O bond is shorter and stiffer in hotter water than its cold. Cooling does the opposite, emitting energy with a thermal momentum that is history-dependent.
(3) Heating enhances the skin supersolidity that elevates the skin thermal diffusivity with a critical ratio of $\alpha_S/\alpha_B \geq \rho_B/\rho_S = 4/3$. Convection alone produces no Mpemba effect.
(4) Mpemba effect proceeds only in the highly nonadiabatic "source-drain" interface to ensure immediate energy dissipation. The Mpemba crossing temperature is sensitive to the volumc of liquid source being cooled, the drain temperature and skin radiation.
(5) The Mpemba effect takes place with a characteristic relaxation time that drops exponentially with increased initial temperature, or with initial energy storage in the O:H–O bond.

## Appendix: Featured News

Content of this section was featured by *ArXiv Editor Picks, Times, Telegraph, Daily Mail, Nature Chemistry, Physics Today, IOP News, AIN News, Sing Chow Daily*, etc. The following is one of the records.

**The Telegraph**

HOME » NEWS » SCIENCE » SCIENCE NEWS


Have scientists worked out why hot water freezes faster than cold water?
Scientists claim to have solved why hot water appears to freeze faster than cold water
By Richard Gray, Science Correspondent
1:18PM GMT 01 Nov 2013

Scientists in Singapore claim to have worked out why hot water freezes faster than cold water Photo: ALAMY By Richard Gray, Science Correspondent

It is a phenomenon that has baffled the world's brightest minds since the time of Aristotle.

Now a team of physicists believe they may have solved the centuries old mystery of why hot water freezes faster than cold water.

Known as the Mpemba effect, water behaves unlike most other liquids by freezing into a solid more rapidly from a heated state than from room temperature.

Scientists have suggested dozens of theories for why this may occur, but none have been able to satisfactorily explain this strange physical property.

A team of physicists at the Nanyang Technological University in Singapore have now published what they believe may be the solution.

They claim that the explanation lies in the unusual interaction between the molecules of water.

Each water molecule is bound to its neighbor through a highly charged electromagnetic bond known as a "hydrogen bond".

It is this that produces surface tension in water and also gives it a higher than expected boiling point compared to other liquids.

However, Dr Sun Changqing and Dr Xi Zhang from Nanyang Technological University, argue this also determines the way water molecules store and release energy.

They argue that the rate at which energy is released varies with the initial state of the water and so calculate that hot water is able to release energy faster when it is placed into a freezer.

Dr Changqing said: "The processes and the rate of energy release from water vary intrinsically with the initial energy state of the sources."

The Mpemba effect is named after a Tanzanian student called Erasto Mpemba, who observed that hot ice cream mix froze before the cold mix.

Together with a physics professor at University College at Dar es Salaam, he published a paper in 1969 that showed equal volumes of boiling water and cold in similar containers would freeze at different times, with the hot water freezing first.

Similar observations have been described in the past, however, by Aristotle, Francis Bacon and Rene Descartes.

The effect can also have some real world implications, such as whether to use boiling water to defrost the windscreen of your car on a winter's day and whether hot water pipes are more prone to freezing than cold ones.

Some people deny that the effect exists at all and is in fact an artefact of experimental procedure, but others claim to have shown it using carefully controlled experiments.

There are a number of theories for might cause this, including that evaporation of hot water means there is less water to freeze. Another theory suggests that dissolved gasses in the water are released in hot water and so make it more viscous.

Last year the Royal Society of Chemistry offered a £1,000 prize to anyone who could explain how the Mpemba effect worked. Nikola Bregovic, a chemistry research assistant at the University of Zagreb, was announced as the winner for the prize earlier this year.

He conducted experiments using beakers of water in his laboratory and his resulting paper suggested that the effect of convection was probably responsible.

He said that convection currents set up in the warm water cause it to cool more rapidly. However, Dr Changqing and Dr Zhang have attempted to explain the effect further by examining the process at a molecular level.

Last week they published a paper in the journal Scientific Reports showing how water molecules arrange themselves when forming ice. They also published a paper on arXiv Chemical Physics that explained the Mpemba effect.

They say the interaction between the hydrogen bonds and the stronger bonds that hold the hydrogen and oxygen atoms in each molecule together, known as covalent bonds, is what causes the effect.

Normally when a liquid is heated, the covalent bonds between atoms stretch and store energy.

The scientists argue that in water, the hydrogen bonds produce an unusual effect that causes the covalent bonds to shorten and store energy when heated.

This they say leads to the bonds to release their energy in an exponential way compared to the initial amount stored when they are cooled in a freezer.

So hot water will lose more energy faster than cool water.

Dr Changqing said: "Heating stores energy by shortening and stiffening the H–O covalent bond. "Cooling in a refrigerator, the H–O bond releases its energy at a rate that depends exponentially on the initially stored energy, and therefore, Mpemba effect happens."

The Royal Society of Chemistry received more than 22,000 responses to its call for a solution to the Mpemba effect and it is still receiving theories despite the competition closing a year ago.

Mr Bregovic, who was judged to have developed the best solution by a panel of experts a conference at Imperial College London last year, said: "This small simple molecule amazes and intrigues us with its magic."

Aeneas Wiener, from Imperial College who helped to judge the competition, added: "The new paper demonstrates that even though a phenomenon seems simple, delving deeper reveals even more complexity and that is certainly worth looking at.

"We hope it'll inspire young people to pursue scientific studies."

Dr Denis Osborne, a lecturer at University College in Dar es Salaam who published the paper with Mr Mpemba on the effect they had observed, said: "Several different mechanisms may cause or contribute to an Mpemba effect.

"What the authors describe as a property of H-O bonding may be one of these."

## References


1. Aristotle, Meteorology 350 B.C.E, http://classics.mit.edu/Aristotle/meteorology.1.i.html
2. E.B. Mpemba, D.G. Osborne, Cool? Phys. Educ. **14**, 410–413 (1979)
3. D. Auerbach, Supercooling and the Mpemba effect—when hot-water freezes quicker than cold. Am. J. Phys. **63**(10), 882–885 (1995)
4. M. Jeng, The Mpemba effect: when can hot water freeze faster than cold? Am. J. Phys. **74**(6), 514 (2006)
5. C.A. Knight, The Mpemba effect: the freezing times of hot and cold water. Am. J. Phys. **64**(5), 524–16731 (1996)
6. N. Bregović, Mpemba effect from a viewpoint of an experimental physical chemist (2012), http://www.rsc.org/images/nikola-bregovic-entry_tcm18–225169.pdf
7. X. Zhang, Y. Huang, Z. Ma, Y. Zhou, J. Zhou, W. Zheng, Q. Jiang, C.Q. Sun, Hydrogen-bond memory and water-skin supersolidity resolving the Mpemba paradox. Phys. Chem. Chem. Phys. **16**(42), 22995–23002 (2014)
8. X. Zhang, Y. Huang, Z. Ma, Y. Zhou, W. Zheng, J. Zhou, C.Q. Sun, A common supersolid skin covering both water and ice. Phys. Chem. Chem. Phys. **16**(42), 22987–22994 (2014)
9. X. Zhang, Y. Huang, Z. Ma, C.Q. Sun, O:H–O bond anomalous relaxation resolving Mpemba paradox (2013), http://arxiv.org/abs/1310.6514
10. C.Q. Sun, Behind the Mpemba paradox. Temperature, **2**(1), 38–39 (2015)
11. Y. Huang, X. Zhang, Z. Ma, Y. Zhou, W. Zheng, J. Zhou, C.Q. Sun, Hydrogen-bond relaxation dynamics: resolving mysteries of water ice. Coord. Chem. Rev. **285**, 109–165 (2015)
12. Turning boiling or hot water into snow at −13° F (−25 °C) (2013)
13. G.S. Kell, The freezing of hot and cold water. Am. J. Phys. **37**, 564–565 (1969)

14. J. Walker, Hot water freezes faster than cold water. Why does It do so? Sci. Am. **237**(3), 246–257
15. J.D. Brownridge, A search for the Mpemba effect: when hot water freezes faster then cold water (2010), http://arxiv.org/ftp/arxiv/papers/1003/1003.3185.pdf
16. J.D. Brownridge, When does hot water freeze faster then cold water? A search for the Mpemba effect. Am. J. Phys. **79**(1), 78 (2011)
17. P. Ball, Does hot water freeze first. Phys. World **19**(4), 19–21 (2006)
18. M. Chaplin, Water structure and science, http://www.lsbu.ac.uk/water/
19. S.T. van der Post, C.S. Hsieh, M. Okuno, Y. Nagata, H.J. Bakker, M. Bonn, J. Hunger, Strong frequency dependence of vibrational relaxation in bulk and surface water reveals sub-picosecond structural heterogeneity. Nat. Commun. **6**, 8384 (2015)
20. M. Vynnycky, S.L. Mitchell, Evaporative cooling and the Mpemba effect. Heat Mass Transf. **46**(8–9), 881–890 (2010)
21. H. Heffner, The Mpemba effect (2001), http://www.mtaonline.net/~hheffner/Mpemba.pdf
22. M. Vynnycky, N. Maeno, Axisymmetric natural convection-driven evaporation of hot water and the Mpemba effect. Int. J. Heat Mass Transf. **55**(23–24), 7297–7311 (2012)
23. M. Vynnycky, S. Kimura, Can natural convection alone explain the Mpemba effect? Int. J. Heat Mass Transf. **80**, 243–255 (2015)
24. D. Auerbach, Supercooling and the Mpemba effect: when hot water freezes quicker than cold. Am. J. Phys. **63**(10), 882–885 (1995)
25. J.I. Katz, When hot water freezes before cold. Am. J. Phys. **77**(1), 27–29 (2009)
26. L.B. Kier, C.K. Cheng, Effect of initial temperature on water aggregation at a cold surface. Chem. Biodivers. **10**(1), 138–143 (2013)
27. C.Q. Sun, X. Zhang, J. Zhou, Y. Huang, Y. Zhou, W. Zheng, Density, elasticity, and stability anomalies of water molecules with fewer than four neighbors. J. Phys. Chem. Lett. **4**, 2565–2570 (2013)
28. Y. Huang, X. Zhang, Z. Ma, Y. Zhou, J. Zhou, W. Zheng, C.Q. Sun, Size, separation, structure order, and mass density of molecules packing in water and ice. Sci. Rep. **3**, 3005 (2013)
29. J.R. Welty, C.E. Wicks, R.E. Wilson, G.L. Rorrer, *Fundamentals of Momentum, Heat and Mass transfer* (John Wiley and Sons, 2007)
30. Water thermal properties—the engineering toolbox [online], http://www.engineeringtoolbox.com/water-thermal-properties-d_162.html
31. P.C. Cross, J. Burnham, P.A. Leighton, The Raman spectrum and the structure of water. J. Am. Chem. Soc. **59**, 1134–1147 (1937)
32. C.Q. Sun, X. Zhang, W.T. Zheng, Hidden force opposing ice compression. Chem. Sci. **3**, 1455–1460 (2012)
33. M. Freeman, Cooler still. Phys. Educ. **14**, 417–421 (1979)
34. B. Wojciechowski, Freezing of aqueous solutions containing gases. Cryst. Res. Technol. **23**, 843–848 (1988)
35. J.D. Smith, R.J. Saykally, P.L. Geissler, The effects of dissolved halide anions on hydrogen bonding in liquid water. J. Am. Chem. Soc. **129**, 13847–13856 (2007)
36. Q. Sun, Raman spectroscopic study of the effects of dissolved NaCl on water structure. Vib. Spectrosc. **62**, 110–114 (2012)
37. S. Park, M.D. Fayer, Hydrogen bond dynamics in aqueous NaBr solutions. Proc. Natl. Acad. Sci. U.S.A. **104**(43), 16731–16738 (2007)
38. X. Zhang, T. Yan, Y. Huang, Z. Ma, X. Liu, B. Zou, C.Q. Sun, Mediating relaxation and polarization of hydrogen-bonds in water by NaCl salting and heating. Phys. Chem. Chem. Phys. **16**(45), 24666–24671 (2014)

# Chapter 12
# Aqueous Solutions: Quantum Specification

- *An HX-type acid releases an $H^+$ that bonds to a $H_2O$ to form a hydronium ($H_3O^+$, $NH_3$-like tetrahedron with one lone pair). The $H_3O^+$ interacts with one of its four $H_2O$ neighbors through the O–H↔H–O anti-HB, which breaks the HB network in a point-by-point manner, quantum fagilation taking place.*
- *The YHO-type base dissolves into the $Y^+$ and the hydroxide ($HO^-$, HF-like tetrahedron with three lone pairs) that interacts with one $H_2O$ neighbor through the "O:↔:O" super-HB, which compresses the rest O:H nonbonds and lengthens the H–O bonds substantially, releasing burning heat.*
- *In the acid base-adduct, solutions, the $X^-$ and $Y^+$ create each an electronic field that aligns, clusters, polarizes and stretches its surrounding $H_2O$ molecular dipoles radially, elongating and polarizing the O:H–O bond forming the supersolid hydration shells.*
- *Salt electrification has the same effect of molecular undercoordination on the O:H–O bond forming the supersolid hydration shells. relaxation and polarization throughout the volume and hence elevates the hydrophobicity, lubricity, molecular dynamics, skin stress, reactivity, solubility, and viscoelasticity of the solution.*

**Abstract** Phonon spectrometrics and contact-angle measurements revealed the essentiality of anti-HB (hydrogen bond), super-HB, and electrified-HB representing molecular interactions in the Lewis acid, base, and adduct (salt) solutions, respectively. Hydronium creation ($H_3O$ tetrahedron with one lone pair) in acid solution results in the H↔H anti-HB that breaks the O:H–O bond network, diluting blood flow for instance; hydroxide (OH tetrahedron with three lone pairs) leads to the O:↔:O super-HB that serves as a point compressor to the hydration network, releasing heat by softening the H–O bond at hydrating. Salt ions create each an electric field that aligns, clusters, polarizes, and stretches water molecular dipoles in a supersolid hydration-shell manner. The electrification-induced phonon relaxation disperses the quasisolid phase boundary outwardly and hence lowers the freezing temperature and raises the melting point. O:H–O bond electrification also raise the viscosity, skin stress, H–O phonon lifetime, but depresses the order of fluctuation and the coefficient of molecular rotation and self-diffusion in the hydration shells. The extent of electrification is molecular site, solute concentration and type dependent, following the Hofmeister series.

## 12.1 Challenge: Why Is Salted Water so Special?

Aqueous saline solutions are capable of changing their surface stress and salting out or salting in proteins, which is ubiquitously important to many chemical and biological processes [1]. However, the mechanics behind the fascinations remains



C.Q. Sun and Y. Sun, *The Attribute of Water*, Springer Series in Chemical Physics 113, DOI 10.1007/978-981-10-0180-2_12

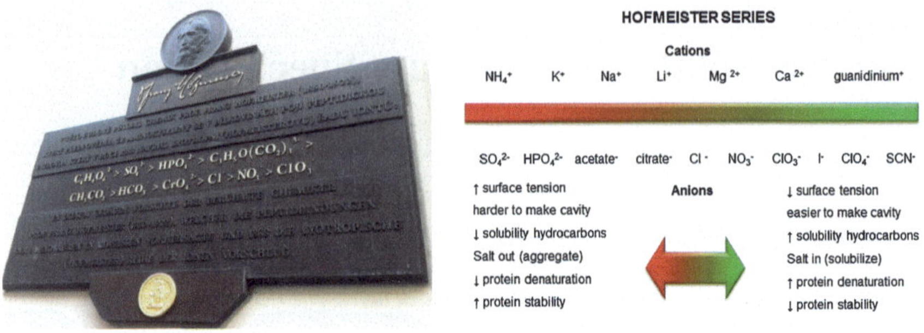


**Fig. 12.1** A commemorative plaque at the Medical Faculty of the Charles University in Prague (in Czech and German) reading as [4]: "*Professor Franz Hofmeister (1850–1922), who carried out research in this building, predicted that amino acids in proteins are connected by a peptide bond and, in 1888, derived the lyotropic (Hofmeister) series of ions*," as shown on the right (Reprinted with permission from [3].)

unclear despite intensive investigations carried out since 1888 when Hofmeister [2] firstly discovered this effect and ranked the series according to the extent of such abilities (see Fig. 12.1) [3]. Key concerns include:

(1) How many factors are involved in the Hofmeister series?
(2) How can one diagnose and correlate the functionalities of the solute and the solvent attributes?
(3) How do ions interact with water and biomolecules—direct or indirect interactions with or without dispersion forces?
(4) How do solute type and concentration mediate the O:H–O bond segmental length and energy at different sites, and the skin stress and solubility of the solution?

## 12.2 Clarification: O:H–O Bond Electrification

Besides the involvement of the supersolid skin and the temperature of operation [5], salt addition resolves the following features of water solvent:

(1) Solute ions create each a radial electric field that aligns, clusters, polarizes, and stretches $H_2O$ molecules in a core-shell fashion with an association of solvent O:H–O bond cooperative relaxation and nonbonding electron polarization (see Fig. 12.2).
(2) Ion size, concentration, charge sign and quantity, and $H_2O$ dipole self-screening determine the extent of O:H–O bond electrification.

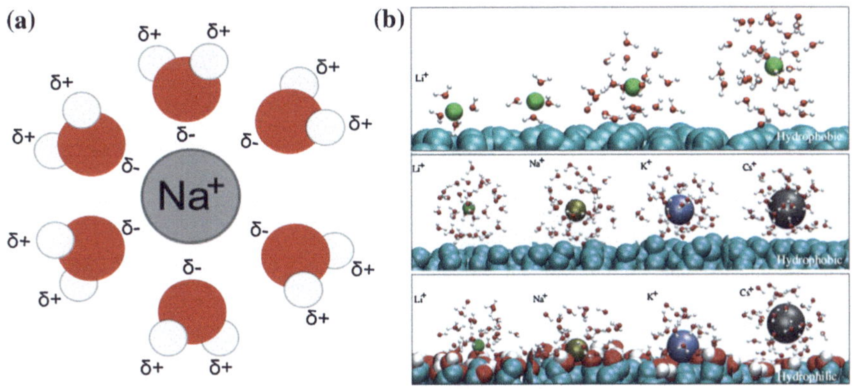


**Fig. 12.2** **a** Serving as a point charge source, the solute ion aligns, clusters, and stretches water molecules via electrostatic attraction and O–O repulsion. **b** The ion centered clusters interact with the hydrophobic and hydrophilic surfaces of proteins in different manners. Solute type, concentration, ionic radius, charge sign and quantity determine the number of $H_2O$ molecules in the first and the subsequent hydration shells (Reprinted with permission from [6].)

(3) Ionic electrification effects the same to heating, stretching, and molecular undercoordination on the O:H–O phonon dynamics but it enhances non-bonding electron polarization (Fig. 12.3).
(4) Such electrification disperses the quasisolid phase boundary, which determines the thermal stability, reactivity, skin stress and viscosity of the solution. The O:H peak offset from below (due to polarization) to above 500 $cm^{-1}$ indicates the repulsivity of the H↔H quantum fagilation that shortens and stiffens the O:H nonbond.

## 12.3 History Background

### *12.3.1 Wonders of Hofmeister Series*

Inorganic salts dissolve in water and form aqueous electrolyte solutions containing individually isolated or complex ions. Sea water and body fluid are well known electrolyte solutions that play essential roles in living bodies [8]. The behavior of aqueous ions has profound impact on biological molecules such as proteins and DNA, enzyme, and membrane, and thus, tremendous implications for health care and disease curing. Ions added in the form of bases, acids, salts, sugar, or buffer agents to protein solution are crucial to maintaining protein stability. Different ions are better or worse at preventing aggregation unfolding and self-association folding of DNA in activating or deactivating ion channeling.

Besides, aqueous solutions lubricate the chemical machinery at the molecular scale. Sweet solutions can be obtained upon heating mixtures of simple carbohydrates, urea and inorganic salts to moderate temperatures, to give new chiral media for organic reactions [9]. Salt can promote dissociating snow to improve the public

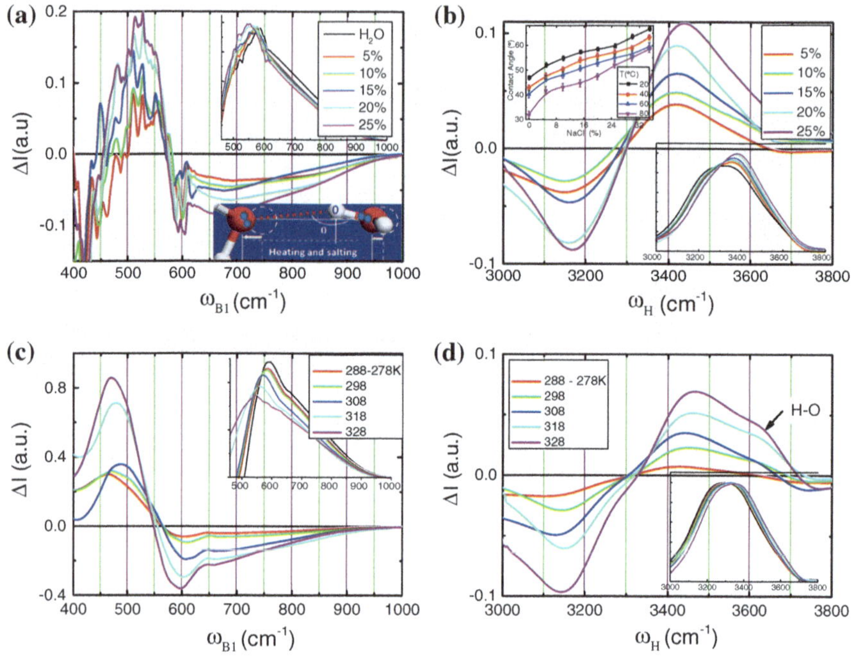


**Fig. 12.3** Differential IR spectra for **a**, **c** O:H–O bending ($\omega_{B1}$) and **b**, **d** H–O stretching ($\omega_H$) phonon relaxation as a function of (**a**, **b**) salinity (wt% NaCl) and (**c**, **d**) temperature with respect to the referential spectrum collected from deionized water at 278 K. The *insets* in (**a**, **b**) illustrate, respectively, the O:H–O cooperative relaxation and the contact angle (polarization degree) variation with salinity and temperature. The $\omega_H$ peak width and the $\omega_{B1}$ intensity fingerprint the order of fluctuation, polarization, $\omega_H$ relaxation time, and viscoelasticity of the solution. Heating softens but salting stiffens the ∠O:H–O bending vibration mode. Salting raises the molecular order with lowers IR absorption probability (Reprinted with permission from [7].)

traffic conditions upon a heavy snowing in cold zones. Therefore, sugar and salt addition could lower the dissociation energy and hence anti-icing takes place. The solubility of sugar increases with temperature but drops with pressure [10].

In 1880s, Franz Hofmeister dissolved gloppy egg-white proteins in solutions, duly noting that some, such as sulfate and fluoride, caused the proteins to precipitate readily, whereas others, such as iodide and isocyanate, did not. Hofmeister also found that ions varied in their effects on other fundamental properties of ionic solutions, such as their abilities to unfold proteins and affect surface stress.

Hofmeister noted that anions appear to have a larger effect than cations, and they are usually ordered as

$$F^- \approx SO_4^{2-} > HPO_4^{2-} > \text{acetate} > Cl^- > NO_3^- > Br^- > ClO_3^- > I^- > ClO_4^- > SCN^-$$

Cations are usually in the order of

$$NH_4^+ > K^+ > Na^+ > Li^+ > Mg^{2+} > Ca^{2+} > \text{guanidinium}$$

Randall and Failey [15–17] realized that the efficiency of common ions as salting-out agents follows these orders:

For anions,

$$OH^- > SO_4^{2-} > CO_3^{2-} > ClO_4^- > BrO_3^- > Cl^- \\ > CH_3COO^- > IO_3^- > IO_4^- > Br^- > I^- > NO_3^-;$$

For cations,

$$Na^+ > K^+ > Li^+ > Ba^{2+} > Rb^+ > Ca^{2+} > Ni^{2+} > Co^{2+} > Mg^{2+} > Fe^{2+} > Zn^{2+} \\ > Cs^+ > Mn^{2+} > Al^{3+} > Fe^{3+} > Cr^{3+} > NH_4^+ > H^+$$

Understanding the impact of ions on the properties of aqueous solutions and how these modified properties influence chemical and conformational dynamics remains an important and elusive objective for physical chemistry communities though numerous contributions [1, 4, 11–14] have updated comprehensively and timely the progress in understanding the Hofmeister series.

## *12.3.2 Known Facts and Mechanisms*

The effect of adding ions into solutions of nonelectrolytes is very complicated, due to different types of intermolecular interactions that involve ions, solvents, and the solute molecules [1, 18]. There are multiple mechanisms explaining the Hofmeister series from the perspective of interaction length scales and the ability of order and disorder making [8].

### 12.3.2.1 Hofmeister: Structure Maker and Breaker

Hofmeister [19] made a heuristic attempt at interpreting his observations, based on the theory of electrolytic dissociation. Hofmeister extended his studies to additional proteins and colloidal particles, and tried to connect the observed ordering of ions with their strength of hydration, denoted as the water-absorbing effects of salts. Hofmeister's explanation for the ionic ordering was eventually framed into the theory of structure-making and structure-breaking ions during 1930s–1950s [20–22]. Table 12.1 features the attributes of structure-maker and structure-breaker.

Zangi et al. [23] have shown how, even when the direct interactions of ions and solutes is taken into account, it can be hard to tell a simple story about ion solvation

**Table 12.1** Attributes of solute ion structure-making and structure-breaking

| | Structure maker (kosmotropes) | Structure breaker (chaotropes) |
|---|---|---|
| Solute size | Small | Large |
| Charge | High | Low |
| Functionality | • Enhance hydrogen bond order;<br>• Strengthen hydration;<br>• 'Steal' water from proteins to result in the salting-out effect;<br>• Precipitate proteins and prevent unfolding;<br>• Strengthen the hydrophobic interaction;<br>• Commonly used in protein purification through the use of ammonium sulfate precipitation | • Disrupt water structure order and denature DNA;<br>• Weakens hydrogen bonding;<br>• Increases the unfolding or denaturation of protein;<br>• Weaken the hydrophobic effect;<br>• Interact more strongly with the unfolded form of a protein than with its native form |
| Examples | $F^-$; $Mg_2^+$ | $I^-$; $NH_4^+$; $SCN^-$ |

Ions produce long-range effects on the structure of water, leading to changes in water's ability of letting proteins fall out of, or stay dissolved in, a solution

effects using single concept. They considered how ions interact with small hydrophobic particles of 0.5 nm across and found that ions with high charge density (q) induce salting-out by promoting stronger hydrophobic interactions that cause particle aggregation. But low-q ions could have either a salting-out or a salting-in effect, depending on their concentration. These effects are related to preferential absorption or exclusion of the ions at the particle surfaces, but not in any simple, monotonic fashion. High-q ions tend to be depleted at the surface of the hydrophobic particle clusters, but are tightly bound to water elsewhere, thereby decreasing the number of water molecules available for solvating the particles. Low-q ions are absorbed preferentially at the particle surfaces, and at high ionic concentrations this can lead to salting-in because the hydrophobic particles form clusters surrounded by ions in a micelle-like arrangement. At lower concentrations, the ions are unable to solubilize aggregation in this way, and salting-out occurs.

There are, however, at least two serious problems with the kosmotropes/chaotropes notion [4]. First, this explanation leaves the chemical details of the surfaces of proteins or other hydrated solutes out of the picture. Second, ions may affect only on their immediate hydration layers without long-range water ordering [24, 25] or on the subsequent hydrogen shells with long-distance ordering [26]. Furthermore, anions may prefer the location at the solvent-air interface with polarization nature [27]. Without considering the nature of the protein itself and the solvent-protein interaction, it is impossible to properly rationalize the regular ordering of ions, or the well-documented exceptions to Hofmeister behavior.

#### 12.3.2.2 Collins: Affinity Matching—Ion Specificity

To provide a qualitative rationale, Collins formulated an empirical "law of matching water affinities" in the 1990s [28, 29], stating that oppositely charged ions of comparable hydration free energies tend to pair in water. The rule takes cognizance of electrostatic interactions and systematizes (unquantified) hydration interaction between ions, and between ions and surface sites of opposite charge. Salt effects do not only depend on the individual ions, but also show interplay of the effects between these ions. Ion pairing, as a source of ion cooperativity has consequences in many aspects of aqueous solutions, such as activity coefficients, surface stress, and salt effects on proteins.

The currently popular view is that Hofmeister effects stem largely from the varying abilities of different salt ions to replace water at the nonpolar molecular or macroscopic surfaces. The surface potential difference and surface stress at an air-salt solution interface could explain how ions affect protein stability and solubility through indirect interactions at the protein-solution interface [30].

#### 12.3.2.3 Quantum Dispersion and Ion-Surface Induction

Models based solely on electrostatics cannot explain ion specific properties of electrolyte solutions without considering quantum mechanical and ion specific dispersion interactions of ions with other ions and with water molecules [31]. Calculations based on Collins's rule established dispersion interactions to ionic interactions, which explained several puzzling properties of electrolyte solutions [28]: the variation in solvation energy among ions of the same size, the small repulsion of iodide from the air–water interface, and the affinity of large ions for each other in water embodied in Collins's rules.

Conversely, Liu et al. [32] observed strong Hofmeister effect in $Ca^{2+}/Na^{+}$ exchange on a permanently charged surface over a wide range of ionic strengths and argued that their observations could not be attributed to dispersion forces, classical induction forces, ionic size, or hydration effects other than a new force active in the ion-surface induction interactions. The strength of this induction force was up to $10^4$ times that of the classical induction force, and could be comparable to the Coulomb force. Coulomb interaction, dispersion, and hydration effects appeared to be intertwined to affect the induction force. The presence of the observed strong non-classical induction force implied that energies of non-valence electrons of ions/atoms at the interface might be underestimated, and possibly just those underestimated energies of non-valence electrons determined Hofmeister effects.

Xie et al. [8] shows that the often neglected cation-anion cooperativity plays a very important role in the Hofmeister effects. Those increase hydrogen donor concentrations such as urea and salts with strongly solvated cations/weakly hydrated anions tend to dissolve protein backbone acting as secondary structure denaturants, whereas those lack of hydrogen donors but rich in acceptors have the opposite effect.

#### 12.3.2.4 Spectroscopic Approaches

A combination of MD simulations, spectroscopy and light-scattering measurements forms a powerful means to investigate Hofmeister ordering of salt ions at charged amino acid side-chains [33, 34]. It turns out that the negatively charged carboxylic acid side-chain groups prefer pairing with smaller, rather than larger, cations. This translates to protein surfaces being more easily 'poisoned' by sodium than by potassium [35], which may be one of the clues why cytosol is rich in $K^+$ but poor in $Na^+$. Similarly, the positively charged side-chain groups of basic amino acids pair more efficiently with smaller, rather than with larger, anions, leading to reversed Hofmeister ordering at these sites [36, 37]. Interactions with these charged groups can be comparable to—or even overwhelm—those with the backbone, which would then cause Hofmeister reversal for the whole peptide or protein.

The interaction between a solute ion and a water molecular dipole affects water's reorientation dynamics and the H–O ($\omega_H$) and D–O ($\omega_D$) stretching phonon frequencies in $H_2O + D_2O$ solvents. For instance, Smith et al. [38] probed the phonon spectra of HOD/$D_2O$ water with and without presence of 1 M KX (X = $F^-$, $Cl^-$, $Br^-$, and $I^-$) and found that larger anions with lower electronegativity stiffen the $\omega_H$ more than the otherwise and that the $F^-$ has almost no effect on phonon relaxation. Using ultrafast 2-dimensional infrared (2DIR) spectroscopy and MD simulations, Park and coworkers [39] found that 5 % NaBr addition to HOD in $H_2O$ shifts the D–O stretching frequency from 2509 to 2539 $cm^{-1}$ and the extent of shift varies with the relative number (8, 16, 32) of $H_2O$ molecules attached to a $Br^-$ ion. The fewer the relative number (higher concentration) the more significant of the shift. Furthermore, the D–O phonon frequency blueshift is associated with an increase of its relaxation time in the ultrafast IR spectroscopy. An increase of NaBr concentration increases the $\omega_D$ shift and the solution viscosity by lowering the translational motion dynamics of water molecules or the degree of molecular fluctuation. Besides, aqueous LiCl addition drops the supercooling (homogeneous freezing) temperature of the solution from 248 to 190 K [40].

Using Raman spectroscopy, Li et al. [41] studied the hydrogen bonded structure of water in the presence of sodium halide salts at various temperatures. They resolved the Raman band for the H–O stretching vibration to five Gaussians at 3051, 3233, 3393, 3511, and 3628 $cm^{-1}$ and assigned the two higher wavenumbers to water molecules with fewer than four hydrogen bonds and the three lower ones to water molecules with all four ice-like hydrogen bonds intact, mainly in view of the temperature dependence between 0 and 100 °C for pure water. They then showed that at 20 °C, $F^-$ does not affect the Raman spectrum appreciably, but $Cl^-$, $Br^-$, and $I^-$ ions do so in an increasing manner, in the direction of further breaking the ice-like hydrogen bonding. They proposed that halogenic ions have the same effect to heating on breaking water structure. Creation more free water molecules to turn the hydrogen bonds into halogenic ion–water hydrogen bonds.

Smith et al. [38] argued that the Raman $\omega_H$ in aqueous salt solutions are related to the degree of hydrogen bonding, instead of structure breaking. They contend that, based on measurements on 1 M potassium halide solutions with 14 mol% HOD in $D_2O$, the difference in the spectra of the salt solutions relative to water arises primarily from the electric fields rather than from rearrangement of the hydrogen bonds beyond the first hydration shell.

As shown in Fig. 12.3, the $\Delta\omega_H$ blueshift is always accompanied by the $\Delta\omega_L$ redshift and the cooperative $\Delta\omega_x$ shift depends merely on the O:H–O bond relaxation in length and energy regardless of the source of perturbation, heating or salting. The polarization and depolarization of the nonbonding electrons change the $\omega_H$ peak width and the intensity and the frequency of the $\omega_L$ peak [5]. Therefore, contributions of the non-valence electrons are significant [32] because of the H–O bond entrapment and nonbonding electron polarization, and the local field electrification is dominant more than rearrangement of water molecules [38].

#### 12.3.2.5 Viscosity Versus Relaxation Time

The viscosity of salt solutions is one of the most important macroscopic parameters often used to classify water soluble salts into structure making or structure breaking. The Jones–Dole empirical expression for the changes in viscosity with concentration for electrolyte solutions is popular [42],

$$\eta/\eta_0 - 1 = A\sqrt{c} + Bc$$

Coefficients A is related to the nobilities and ion–ion interactions between the solute ions and have constant values for alkali halide solutions. Coefficient B reflects the interactions between the ions and solvent molecules. $\eta_0$ is the viscosity of neat water. This empirical expression describes reasonably well the experimentally observed viscosity $\eta$ behavior for dilute solutions and is usually used for concentrations c below 0.1 M.

Nickolov and Miller [43] examined the $\Delta\omega_D$ in solutions of five alkali halide salts (KF, KI, NaI, CsF, and CsCl) of 4 wt% $D_2O$ in $H_2O$ mixtures using FTIR at the full concentration range. They correlated the $\Delta\omega_D$ to the solute characteristics of structure making ($B > 0$) and breaking ($B < 0$). They attributed the $\Delta\omega_D$ red shift at 2500 $cm^{-1}$ for KI, NaI, and CsCl to structure breaking but the blueshift of this peak for KF and CsF to structure making.

Omta et al. [24] examined the effects of ions on the orientation correlation time of water molecules in $Mg(ClO_4)_2$, $NaClO_4$, and $Na_2SO_4$ solutions by means of femtosecond pump-probe spectroscopy. They suggested that an addition of ions had no influence on the rotational dynamics of water molecules outside the first solvation shells of the ions and suggested that the presence of ions does not lead to an enhancement or a breakdown of the hydrogen-bond network in liquid water. However, supported by neutron diffraction [44] and viscosity measurements [30], Tielrooij et al. [26] showed that certain salts can change water reorientation dynamics at a longer distance or multiple shells. A combined terahertz and femtosecond IR spectroscopic study of water dynamics around different ions (specifically magnesium, lithium, sodium, and cesium cations, as well as sulfate, chloride, iodide, and perchlorate anions) revealed that the effect of ions and counterions on water can be strongly interdependent and non-additive, and in certain cases extends well beyond the first solvation shell of water molecules directly surrounding the ion.

As noted by Jungwirth and Cremer [4], further progress in understanding ion-specific effects in biological systems will require researchers to go beyond the simplifying concept of separate anionic and cationic Hofmeister series. What matters is not only the behavior of individual ions at the protein surface but, to varying extents, also interactions between the salt ions themselves, both near the protein and in the bulk aqueous solution. Such effects become operational at relatively high salt concentrations and are distinct from non-specific electrostatic interactions.

Consideration of the anomalous performance of water solvent from the perspective of ion electrification induced O:H–O relaxation and polarization would be necessary [1, 14, 45], which will be possible when the performance of the O:H–O bond under ion electrification is well understood.

### 12.3.3 Acid-Base Solutions

Another important issue is the concept of acid, base, and adduct. There are several acid-base theories describing the characteristics of the acids, bases, and adducts. Lewis [46] developed a theory of acids and bases that is based upon the sharing of electron pairs. A Lewis acid is a substance that can accept a pair of electrons from another atom to form a new bond. A Lewis base is a substance that can donate a pair of electrons to another atom to form a new bond. A simple example is the formation of the hydronium ion from a proton (no electrons) and water (has electron pairs to donate):

| | Acid | Base |
|---|---|---|
| | 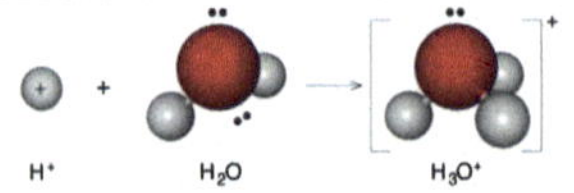 | 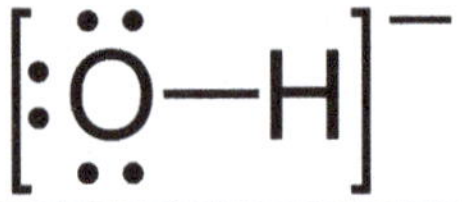 |
| Svante Arrhenius (1884) | Increases the concentration of $H^+$ in solution | Increase the concentration of $OH^-$ in solution |
| Brønsted–Lowry (1923) | Donate $H^+$ | Accept $H^+$ |
| Gilbert N. Lewis (1923) | Accept an electron pair to form a covalent bond | Donate an electron pair |
| Chang Q. Sun (2015) | H↔H anti-HB point breaker (quantum fragilation) | O:↔:O super-HB point compressor (quantum compressor) |
| | Acid + Base = Adduct: HI + NaOH = $Na^+$ + $I^-$ + $H_2O$ | $Na^+$ and $I^-$ ($Y^+$ and $X^-$) point polarizer |

| | |
|---|---|
|  | **Gilbert Newton Lewis** (October 23, 1875–March 23, 1946) was an American physical chemist known for the discovery of the covalent bond and his concept of electron pairs; his Lewis dot structures and other contributions to valence bond theory have shaped modern theories of chemical bonding. Lewis successfully contributed to thermodynamics, photochemistry, and isotope separation, and is also known for his concept of acids and bases. Though he was nominated 35 times, G.N. Lewis never won the Nobel Prize in Chemistry |
|  | **Svante August Arrhenius** (19 February 1859–2 October 1927) was a Swedish scientist, and one of the founders of the science of physical chemistry. He received the Nobel Prize for Chemistry in 1903 for his acid-base theory |
|  | **Johannes Nicolaus Brønsted** (February 22, 1879–December 17, 1947) was a Danish physical chemist. He earned his Ph.D. in 1908 from the University of Copenhagen and was immediately thereafter appointed professor at the same university. In 1906 he published the first paper on electron affinity, and, simultaneously with the Thomas M Lowry, he introduced the protonic theory of acid-base reactions in 1923 |
|  | **Thomas Martin Lowry** CBE FRS (26 October 1874–2 November 1936) was an English physical chemist who developed the Brønsted–Lowry acid–base theory simultaneously with and independently of Johannes Nicolaus Brønsted and was a founder-member and president of the Faraday Society |

The product is called adduct or complex. When one substance donates all the electrons for the bond the bond is called the coordinate covalent bond. In this example, $H^+$ is the Lewis acid and $H_2O$ is the Lewis base.

All metal cations are potential Lewis acids because their positive charge will readily attract electron pairs and they all have at least one empty orbital. The hydroxide ion ($HO^-$) is an excellent Lewis base and so it will bind readily to metal cations to give metal hydroxides. Some of these metal hydroxides are amphoteric. An amphoteric metal hydroxide can behave as a Bronsted base and react with a Bronsted acid, or it can behave as a Lewis acid and react with a Lewis base.

The Lewis acid-base convention also accounts for the fact that oxides of non-metals behave as acids. An example is carbon dioxide, which has acid behavior that can be understood in this notion. Because the oxygen atom in $CO_2$ is more electronegative than the carbon atom, the carbon atom will have a slight positive charge that will be attractive to the $OH^-$ group. The $CO_2$ molecule then acts as a Lewis acid accepting the donated electrons to form the hydroxyl group.

## 12.4 Quantitative Resolution

### *12.4.1 Dominating Factors*

In addition to the supersolid skin of the solution and the thermal effect of operation at different temperatures [5], three parameter domains, shown in Fig. 12.4, are involved in the Hofmeister effect—solute, solvent, and protein. The intradomain interaction and interdomain coupling determine the series but the response of the $H_2O$ solvent to the type and concentration of the solutes and the proteins plays the role of dominance.

However, it is impossible for one model to cover all factors and their coupling simultaneously without sorting the significance of these parameters. It would be necessary to sort these factors, as listed in Table 12.2, according to their domains

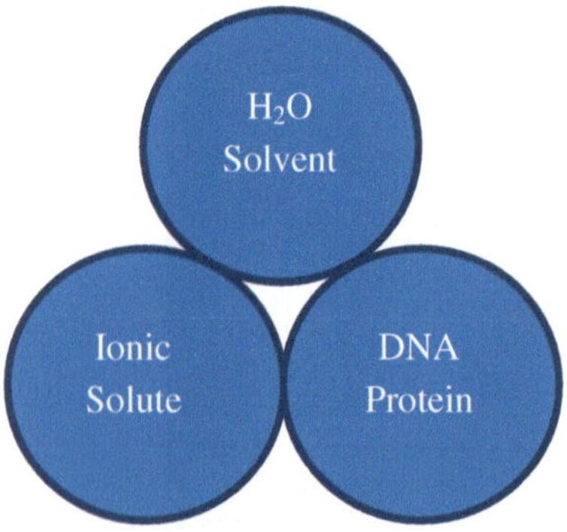


**Fig. 12.4** The intradomain and interdomain parameter interactions determine the Hofmeister effect with $H_2O$ playing the key role of matrix

**Table 12.2** Factors contributing to the Hofmeister series and acid-base solutions

| Domains | Factors |
|---|---|
| Ionic solute | Solute type, concentration, size, charge, electronegativity difference, contact ion pair (CIP) or separated ion pair (SIP) configuration, complex formation, the nature and the formation manner of the hydration shells, etc. |
| $H_2O$ solvent | O–O repulsion, O:H–O bond segmental disparity and its response to the fields of ions and proteins, relaxation in length and stiffness, polarization, molecular mobility, etc. |
| DNA and protein | Backbone and side chains, polarization, etc. |
| Solvent skin and solute hydration shell | Preference of ions occupancy, supersolidity, dipole orientation, ionic eletrification, quantum fragilation, quantum compression, etc. |
| $H_2O$-protein interface | Hydrophilic bonding, hydrophobic nonbonding, supersolidity, etc. |
| Diagnostics | Phonon frequency shift, dielectrics, skin stress, life time, chemical stability, polarizability, solubility, thermal stability, viscosity, etc. |

and interfaces and clarify one-by-one their functionalities at a time—two domains and their interface as a point of starting.

O:H–O bond cooperative relaxation and the associated polarization dictate the chemical reactivity, stress and solubility, and thermal stability of the hydration shells and the solution skins. Ions provide mainly the sources for electrification and polarization with tiny chance of direct bonding with charge sharing or transforming between the solute ions and $H^+$ proton or oxygen lone pairs.

Biomolecules also contain the O:H nonbonding lone pairs and dipole moments, which is also very sensitive to perturbation. Therefore, it is essential to focus firstly on the solute and solvent interdomain and intradomain interactions. O:H–O bond relaxation, bonding charge entrapment, nonbonding electron polarization, acid anti-HB quantum fragilation, base super-HB quantum compression, O:H nonbond switching on and off, could be major events happening in reaction under the electric field of ions. The strong H–O bond of 4.0 eV energy is hard to break without catalysis. There exists no electron exchange interaction between solutes and $H_2O$ or biomolecules but only induction and polarization.

Ionic electrification elongates the O:H–O bond and polarizes the nonbonding electrons of the solvent and the biomolecules. Such electrification correlates the skin stress, viscoelasticity, and the critical pressures and temperatures for phase transition of the solution and the hydrophobicity, solubility, and reactivity of the solute-air and solute-protein interfaces. The solvent-protein interaction may be hydrophobic (dominated by electrosteatic repulsion) or hydrophilic (hydrogen bond like formation through electrosteatic attraction) depending on the nature of the particular protein. Understanding interaction between solute and biomolecules would become easier when the solute-solvent-protein knowledge is adequately established.

## 12.4.2 *Salt Electrification—Point Polarizer*

### 12.4.2.1 Solute Ionic Electric Field

Water dissolves YX salts into the $Y^{n+}$ cations and $X^{n-}$ anions, see Fig. 12.5 for NaX instance. The $X^{n-}$ stays alone but the $Na^+$ may be alone or attached to a host $H_2O$ to form $OH_2Na$ by replacing an electron of their lone pairs or form a NaOH by replacing an $H^+$ though the former is unlikely. However, it is ready to justify which way the $Y^{n+}$ and $X^{n-}$ prefers simply by Raman scanning over the full frequency to identify if new phonon peaks present due to the complex formation. A complex formation will create a peak of phonon frequency that depends on the length and energy, and the reduced mass of the complex dimer.

Nevertheless, the $Y^{n+}$ and the $X^{n-}$ remains each a charge center of different diameters. The anions and cations create each a radial electric field directing from positive to negative. The $Y^{n+}$ and $X^{n-}$ can be separated each other but the distance varies not only with the salt concentration but also with the electronegativity difference between them, $\Delta\eta$. If the $\Delta\eta$ is too high, the $Y^{n+}$ and $X^{n-}$ tend to stay closer, which creates a dipole electric field around, such as NaF ($\Delta\eta = 4.0 - 0.9 = 3.1$),

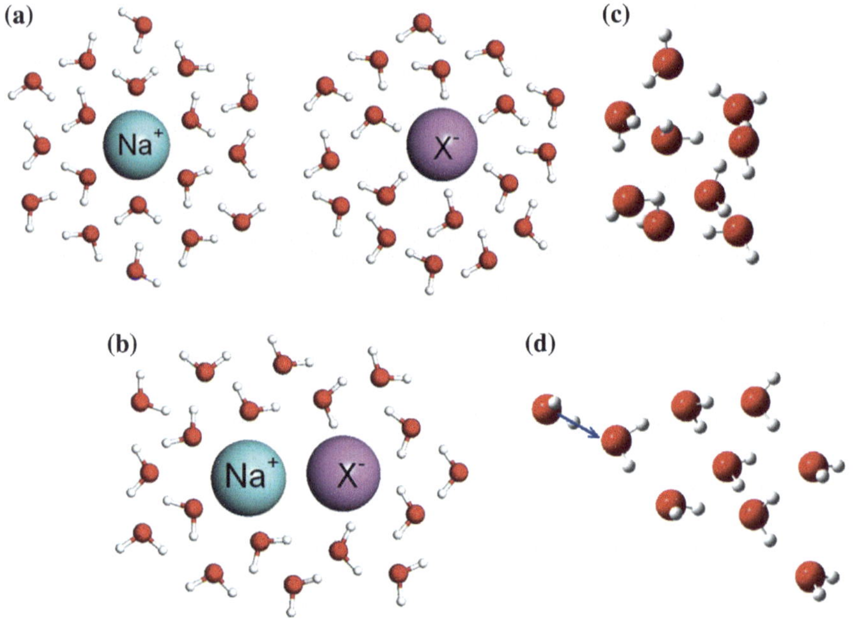


**Fig. 12.5** Schematic illustration of ionic hydration shells. Solvent-separated ion pair (SIP) forms in (**a**) at low concentrations and the relatively lower number (N) of water molecules per ion. Contacted ion pair (CIP) forms in (**b**) at higher concentrations with N less than 18 (≥3.0 M). The randomly ordered $H_2O$ molecules under (**c**) E = 0 field become ordered when the electric field (**d**) E = 0.024 V/m, for instance, is applied (Reprinted with permission from [48].)

which is regarded as contacted ion pair (CIP). The otherwise is called solvent-separated ion pair (SIP). The local field of the CIP should be much shorter than that of the SIP.

On the other hand, the inner filed of the $H_2O$ dipole will screen and partially weaken the ionic field. The resultant electric field determines the extent of O:H–O bond relaxation and polarization. For a single ion, the electric field should be long-range but for a mirror pairing ions, the distance of the electric field is relatively shorter, as discussed in Sect. 3.4.5. Therefore, both single and multiple molecular-layer hydration shells of different solutes are possible [24, 26].

#### 12.4.2.2 O:H–O Bond Elongation and Polarization

The electric field of the point charge will align, stretch, and polarize water molecules in the hydration shells. Molecules in the first hydration shell are preferably oriented with $H^+$ protons toward anions and the lone pairs of $O^{2-}$ toward cations without exchange interaction. Such water molecular electrification may extend to a second or higher number of hydration layers, depending on the intensity of the local ionic fields. Therefore, both long and short range inductive interactions should exist. Water molecular dipole stretching is equivalent to O:H–O bond elongation—the O:H nonbond becomes longer and the H–O bond shorter. The segmental disparity elongate the O:H more than the H–O contracts.

Therefore, ionic field electrification of $H_2O$ molecules reduces their sizes but elongates their separations in the hydration shell. Meanwhile, the electrification is associated with strong polarization of the nonbonding electrons and lowed degrees of molecular fluctuation and self-diffusion [49], which has the same effect to molecular undercoordination but it is subject to the extent of supersolidity. The supersolidity of the hydration shell and the solution skin is responsible for the longer H–O phonon life time, higher stress, higher viscoelasticity, higher solubility, higher thermal stability, but lower molecular dynamics and mass density.

The detectable properties of the solution are correlated by the O:H–O bond identities such as segmental length $d_x$ and energy $E_x$, the reduced segmental mass $\mu_x$, and the polarization [5, 50]. Therefore, solute electrification effects the same to heating, stress, and molecular undercoordination on modulating the interoxygen Coulomb interaction $V_C(r_O)$ and the associated phonon frequency shift $\Delta\omega_x$ that depends functionally on the segmental stiffness only without discriminating the nature of the sources as hidden variables [51, 52]:

$$\begin{cases} \Delta\omega_x \propto \sqrt{E_x/\mu_x}/d_x & \text{(Phonon frequency)} \\ V_C(r_O) = \frac{q_O^2}{4\pi\varepsilon_r\varepsilon_0 r_O} & \text{(O–O potential)} \end{cases} \quad (12.1)$$

There are three parameters in the $V_c(r_O)$ potential: the dielectric constant $\varepsilon_r$, O–O distance $r_O$, and the net charge on oxygen $q_O$. Conditioning by compressing, clustering, heating, and salting has the same effect on the $V_c(r_O)$ by different extent without charge exchanging.

Ionic electrification weakens the Coulomb interaction but enhances polarization through O:H–O stretching. O:H–O bond alignment and polarization add a local field opposing the solute electrification [39, 40, 53–56].

One can conveniently identify the factors dominating the Hofmeister effect by monitoring the phonon frequency shift, contact angle change, and variation of the critical conditions for phase transition such as critical temperatures for melting and freezing, critical pressures for ice formation, and critical time for sol-gel transition (called gelation). Contact angle variation indicates the extent of polarization, which affects the solution viscosity, and molecular dynamics.

### 12.4.3 Acid Anti-HB and Base Super-HB

Lewis firstly defined the acid and base in 1923 in terms of electron pair gain and loss when they are dissolved. However, molecular interactions in the aqueous solutions remain yet unattended. It is necessary to examine the solution molecular bonding with concepts of anti-hydrogen bond (anti-HB) (O–H↔H–O), super hydrogen bond (super-HB) (O:↔:O), and the electrified hydrogen bond (O:H–O) to describe molecular interactions in the Lewis solutions.

An HX type acid releases an $H^+$ proton that bonds to a $H_2O$ molecule to form a hydronium ($H_3O^+$, $NH_3$-like tetrahedron with one lone pair). The $H_3O^+$ reacts with one of its tetrahedrally-coordinated $H_2O$ neighbors through the O–H↔H–O interaction, which breaks the HB network in a point-by-point manner. However, the YHO type base releases the hydroxide ($HO^-$, HF-like tetrahedron with three lone pairs) that interacts with one $H_2O$ neighbor through the "O:↔:O" interaction that compresses the rest O:H nonbonds and lengthens the H–O bonds. Furthermore, in the acid, in the base, or in their combination, the $X^-$ and $Y^+$ create each an electronic field that aligns, clusters, polarizes and stretches its surrounding $H_2O$ molecular dipoles, elongating and polarizing the O:H–O bond. Table 12.3 illustrates the consequence of Lewis acid and base on the solution intermolecular bonding.

### 12.4.4 Salt Hydration Quantum Polarization

#### 12.4.4.1 Thermal Fluctuation and Salt Electrification

Phonon spectrometrics probes the stiffness variation of the O:H–O bond under stimulus. Molecular coordination environment splits a specific phonon peak into components corresponding to O:H–O bond in the bulk (3200 $cm^{-1}$), in the solution

**Table 12.3** Molecular bonding in the Lewis-Hofmeister solutions

| Solution | | | Reaction | Point controller | Property |
|---|---|---|---|---|---|
| Acid (pH < 7) | HX + $H_2\ddot{O}$: | ⇒ | $X^-$ + $H_3\ddot{O}^+$ (O–H↔H–O anti-HB and H↔H point breaker) | $H_3\ddot{O}^+$ forms a $NH_3$-like tetrahedron with one lone pair, interacting with one of its four $H_2O$ neighbors through the anti-HB that breaks the HB network point-by-point | Sour taste; capable of turning blue litmus red; corrosive, dilution, depressing viscosity and skin stress; relieving hypertension, etc. |
| Adduct (salt) | YX + $H_2\ddot{O}$: | | $X^-$ + $Y^+$ + $H_2\ddot{O}$: ($X^-$ and $Y^+$ point polarizer) | Electric fields of the $X^-$ and $Y^+$ align, polarize and stretch the O:H–O bond, whose extent depends on the ion size, solute concentration and type separation | Hofmeister series: stress and viscosity elevation; thermal stability; protein solubility; polarization; supersolid hydration shells; hypertension enhancement, etc. |
| Base (pH ≥ 7) (alkali metals) | YHO + $H_2\ddot{O}$: | | H: $\ddot{O}$:$^-$ + $Y^+$ + $H_2\ddot{O}$: (O:↔:O super-HB and point compressor) | H:$\ddot{O}$:$^-$ forms an H:$\ddot{F}$: like tetrahedron with three lone pairs. The O:↔:O super HB point compressor shortens the neighboring O:H nonbond and lengthens the H–O bond | Solution greasiness and slipperiness; burning heat release (H–O softening) |

X = F, Cl, Br, I; Y = Li, Na, K, Rb, Cs for the simple cases
The number of $H_2O$ molecules in the hydration shell varies with the size and charge quantity of the $Y^+$ and $X^-$ ions

skin and the hydration shell (3450 $cm^{-1}$), and to the H–O dangling bond free radicals (3600 $cm^{-1}$). The O:H–O bond segmental length and energy variation shifts the $\omega_x$ frequency. The molecular fluctuation order determines the FWHM of the specific component. The phonon abundance determines the integral of the specific component. The sum of the phonon abundance over all components for a particularly convoluted spectral peak conserves. The abundance loss and its gain may not be identical for a specific component but the loss and the gain of the convoluted peak is identical.

Figure 12.6 shows decomposition of the $\omega_H$ spectra for water at 298 and 338 K and 3 M/l NaI solution at 298 K using the bulk, skin, and H–O radical components. Figure 12.7 compares the site-resolved differential phonon spectra (DPS) and the respective $\omega_H$, FWHM, and phonon abundance for the heating and NaI salting effect.

The spectral decomposition reveals the following, see Table 12.4:

(1) Both heating and salting stiffen the $\omega_H$ for the bulk and the skin at different rate.
(2) Heating from 298 to 338 K softens the H–O radical $\omega_H$ from 3611 to 3608 $cm^{-1}$ with enhanced intensity, indicating higher degree of molecular thermal fluctuation.
(3) NaI salting at 3.0 M concentration stiffens the H–O radical $\omega_H$ from 3611 to 3625 $cm^{-1}$ with lowered intensity, resulting from the skin preferential occupancy of $I^-$ anions.
(4) The FWHM reduction of the DPS component indicates the lowered degree of molecular fluctuation, which happens to the skin by salting but the heating effect is opposite.
(5) Salting depresses the phonon abundance of the bulk and the H–O radicals but raises the skin abundance substantially as a result of hydration shell formation. The increased number of phonons having frequency higher than that of the skin arises from the ionic hydration shells.

Likewise, the IR DPS of deionized water and NaCl solution in Fig. 12.3 also show the same thermal and salinity trends [7]:

(1) O:H–O bonds in the bulk, hydration shells, and the H–O radical responds discriminatively to heating and salting in relaxation of stiffness, abundance, and structural order.
(2) Salting has the same effect of heating on stiffening the $\omega_H$ from 3150 to 3450 $cm^{-1}$ and softening the $\omega_{B1}$ ($\sim$600 $cm^{-1}$ corresponding to the ∠O:H–O bending mode) despite slight $\Delta\omega_{B1}$ difference [51, 59] at 25 % salting, which is consistent to the documented results [53, 60].
(3) Heating stiffens the $\omega_H$ gradually from 3200 to the same value of 3450 $cm^{-1}$ with presence of a shoulder at 3650 $cm^{-1}$ (correspond to 3610 $cm^{-1}$ in refection) absorption that corresponds to the dangling H–O bond radicals and gaseous monomers. Heating also depolarizes and enhances thermal fluctuation that widens the $\omega_H$ peak.
(4) Salting shifts the $\Delta\omega_{B1}$ from 600 to 530 $cm^{-1}$ but heating shifts it to 470 $cm^{-1}$ due to depolarization The low intensity of absorbance of the 530 $cm^{-1}$ peak indicates salting enhances polarization and structure ordering, which lowers the IR transmitancy; the absence of the 3610 $cm^{-1}$ feature indicates higher skin occupation probability of $Cl^{-1}$ anions, which annihilates the features of skin H–O bond radicals [27].
(5) Contact angle measurements confirm the expectation of salting polarization and heating depolarization [61]. The IR is more sensitive than the Raman to the bond bending.

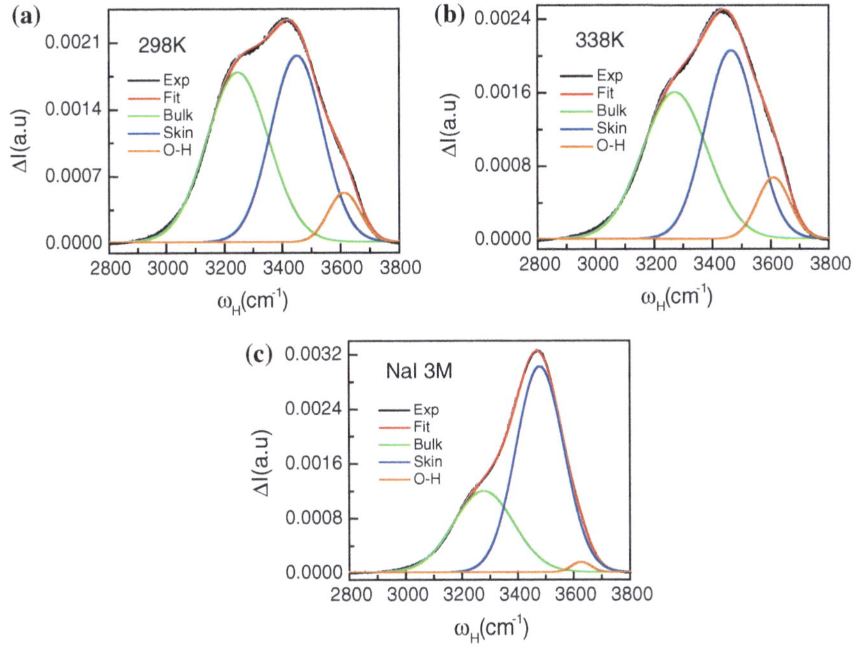


**Fig. 12.6** Raman $\omega_H$ decomposition of neat water at **a** 298 K and **b** 338 K and 3 M NaI solution at 298 K (Reprinted with permission from [57].)

#### 12.4.4.2 YI (Y = Na, K, Rb, Cs) Solute Type Dependence

The full-frequency Raman scans for aqueous YI solutions at 2.1 M concentration and 298 K temperature, as shown in Fig. 12.8a, revealed no extra features for YI compounds in the solution. Therefore, Y and I ions remain as isolated charge centers that polarize neighboring molecules without bond formation between the ions and water molecules. The segmental DPS in Fig. 12.8b, c revealed that YI solutes stiffen the H–O bond and meanwhile soften the O:H nonbond at extents that are insensitive to the type of $Y^+$ cations. While the phonon abundance loss is identical to its gain, the FWHM of the gains in both H–O and O:H regimes become narrow. The narrower FWHM indicates the depression of molecular dynamics in terms offluctuation or elevation of structural order.

Figures 12.9 and 12.10 compare the solute type and concentration dependence of the DPS for aqueous YI solutions. This set of data shows that the $\Delta\omega_x$ follows the series: $Y^+(\eta/R) = Na^+$ (0.9/0.98) > $K^+$ (0.8/1.33) > $Rb^+$ (0.8/1.49) > $Cs^+$ (0.8/1.65), but the polarizability ($\omega_L$ upward shift) seems in the order of slightly inverse. Except for the $Na^+$(0.9), all the rest $Y^+$ cations have the same electronegativity of 0.8. Smaller $Y^+$ ions have stronger effect on the $\omega_x$ shift, which means that the first hydration shell contains fewer $H_2O$ molecules so the polarization effect is more significant.

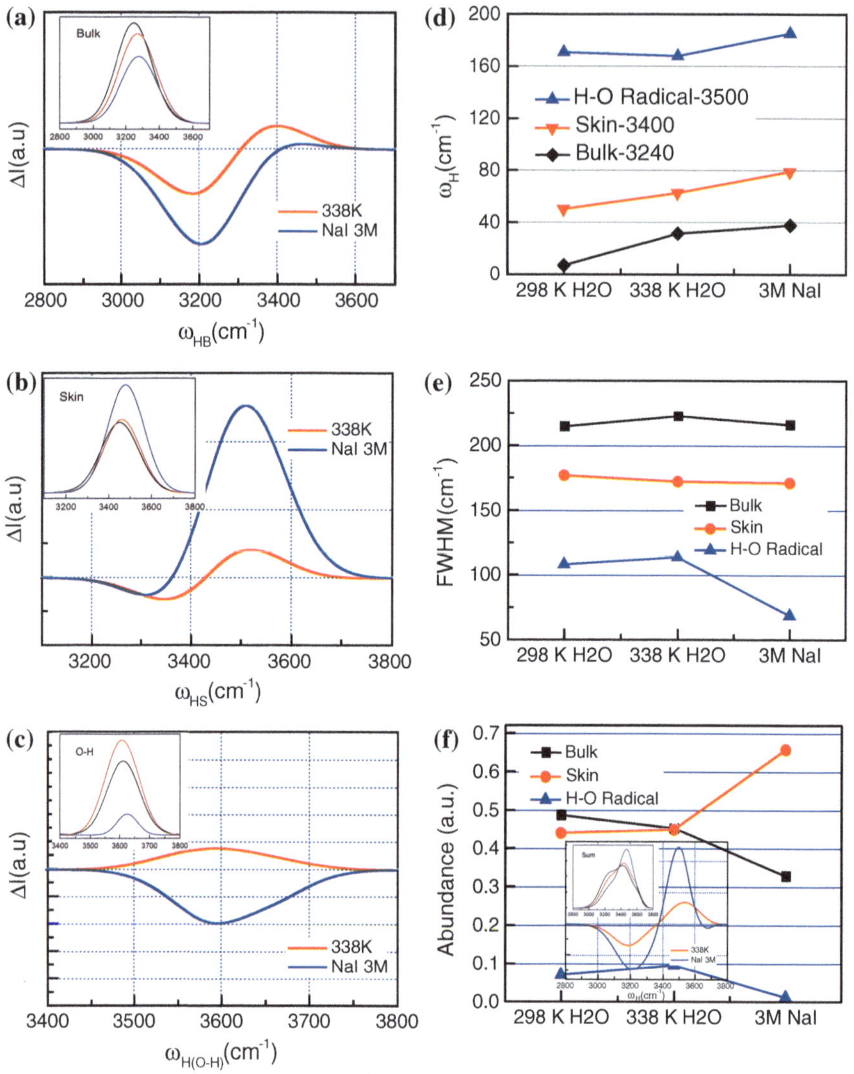


**Fig. 12.7** Molecular site-resolved DPS for 338 K water and 3 M NaI solution with reference to that for the 298 K water (**a**–**c**) and comparison of the site-resolved, **d** bond stiffness, **e** molecular structure order (inversely proportional to the FWHM), and **f** phonon abundance as a function of heating and salting. The *insets* show the heating and salting effect on the respective component and the *inset* (**f**) shows the convoluted DPS for the H–O phonon (Reprinted with permission from [58].)

Nevertheless, the $\omega_x$ is overall less sensitive to the types of $Y^+$ cations in the solutions compared with the $X^-$ anions in the NaX because of their compatible electronegativity and ionic radius, as indicated above.

**Table 12.4** Gaussian decomposition of the $\omega_H$ for $H_2O$ and 3M NaI solution measured at 298 K

| | Solution | Bulk | Skin/shell | H–O radical | Notes |
|---|---|---|---|---|---|
| $\omega_H$ ($cm^{-1}$) (bond stiffness) | 298 K $H_2O$ | 3246.91 | 3450.92 | 3611.30 | • Heating stiffens the skin and bulk H–O bonds but softens the H–O radical • NaI salting stiffens all H–O bonds |
| | 338 K $H_2O$ | 3271.38 | 3463.26 | 3608.42 | |
| | 3 M NaI | 3277.62 | 3479.57 | 3625.70 | |
| FWHM ($cm^{-1}$) (fluctuation order) | 298 K $H_2O$ | 214.48 | 177.52 | 108.48 | • Heating depresses the skin molecular dynamics but raises the molecular fluctuation in the bulk and the H–O radicals • NaI depresses the skin molecular order and the H–O radical |
| | 338 K $H_2O$ | 222.50 | 172.76 | 113.91 | |
| | 3 M NaI | 215.66 | 171.70 | 69.21 | |
| Peak area (abundance) | 298 K $H_2O$ | 0.49 | 0.44 | 0.07 | • Skin phonon abundance is thermally stable • NaI enriches the skin and shell phonons |
| | 338 K $H_2O$ | 0.45 | 0.45 | 0.10 | |
| | 3 M NaI | 0.33 | 0.66 | 0.01 | |

Reprinted with permission from [57]

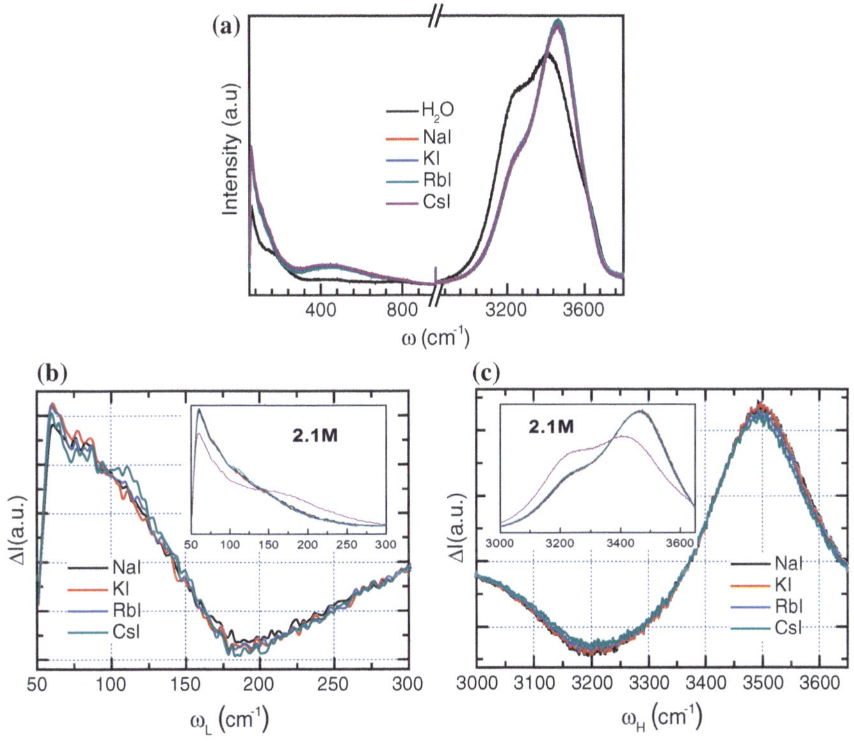


**Fig. 12.8** Raman probing of the 2.1 M YI (Y = Na, K, Rb, Cs) solutions at 298 K. The $\omega_x$ DPS shows that solute electrification offsets the $\omega_L$ and $\omega_H$ in opposite directions and raises the order of molecules in the hydration shells and solution skin (Reprinted with permission from [57].)

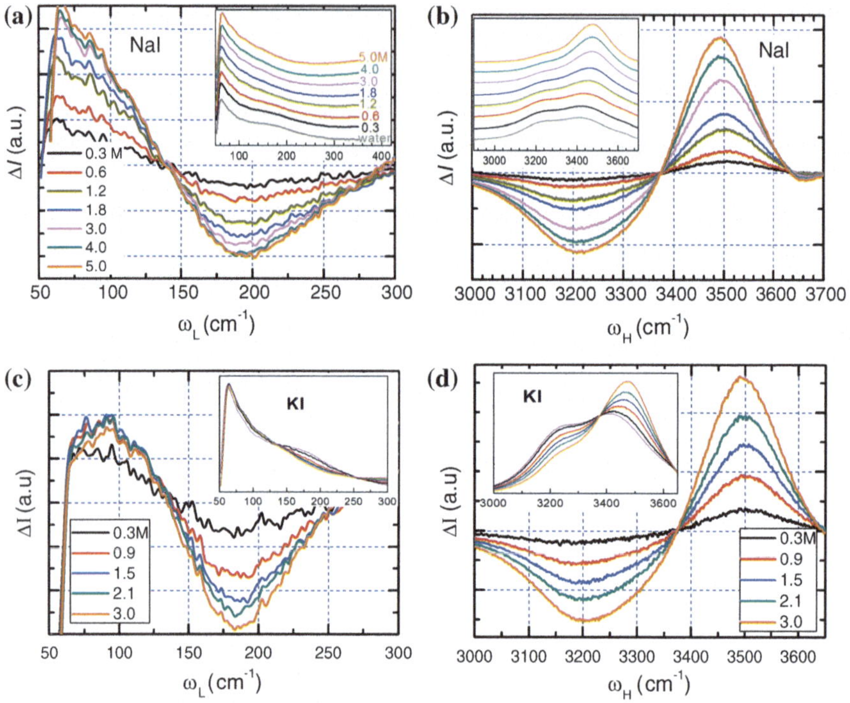


**Fig. 12.9** Concentration dependence of the segmental DPS for aqueous **a**, **b** NaI and **c**, **d** KI solutions at 298 K with respect to the referential spectrum collected from neat water at 298 K. Electrification shifts the $\omega_H/\omega_L$ from the reference of 3200/190 to 3500/70 cm$^{-1}$ and sharpens them by structural order elavation (Reprinted with permission from [58].)

Three Gaussian-component decomposition of the 298 K Raman spectra collected from neat $H_2O$ and 2.1 M YI solutions, shown in Fig. 12.11, revealed the molecular site-resolved O:H–O bond relaxation in terms of H–O bond stiffness, molecular structural order, and phonon abundance. Components at 3200, 3450 and 3610 cm$^{-1}$ correspond to the H–O stretching phonons in the bulk, skin, and the dangling radicals, respectively. As shown in Fig. 12.11 and summarized in Table 12.5, YI raises the skin/shell phonon abundance by rendering the abundance of the bulk and the H–O radicals. A further componential DPS gives more on-site information of the local O:H–O bond relaxation as shown in Fig. 12.12.

The componential DPS in Fig. 12.12 resolves minus difference among observations. The abundance increase of the solution skin and hydration shells renders the loss of the bulk component. The skin and the hydration shells are naturally the same, as molecular undercoordination and solute electrification derive the supersolidity of the less coordinated water molecules. The abundance loss and a 20 cm$^{-1}$ blueshift of H–O radical features compared with that of neat water evidence the preferential occupancy of $I^-$ anions at the solvent-air interface, which reduce the

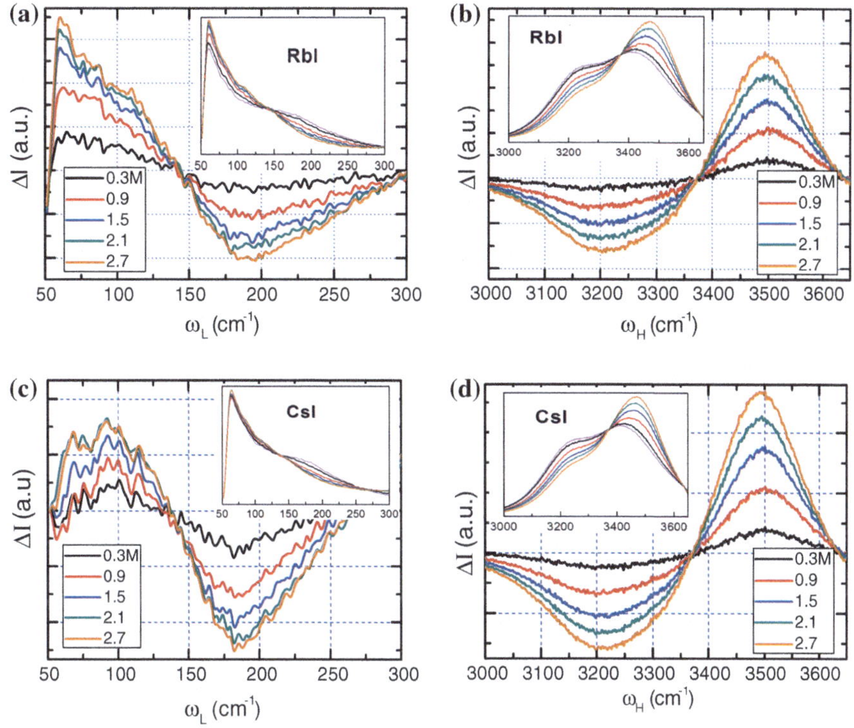


**Fig. 12.10** Concentration dependence of the $\omega_x$ DPS for aqueous **a**, **b** RbI and **c**, **d** CsI solutions. The maximal solubility of CsI is 2.7 M (Reprinted with permission from [62].)

number of the H–O radicals and strengthen the local electric field subjected by the H–O radicals, as the insets (b, c) illustrated [27, 63].

#### 12.4.4.3 NaX(X = F, Cl, Br, I) Solute Type Dependence

Figure 12.13 compares the $\omega_L$ and $\omega_H$ DPS and the full-frequency Raman spectra collected from deionized water and 3 M NaX solutions. They show the same trend of electrification on shifting the $\omega_L$ and $\omega_H$ in opposite directions. Figure 12.14 shows the convoluted DPS for the 0.9 M NaX with the same trend of phonon relaxation by slightly different amounts.

The room temperature DPS for 0.9 and 3.0 M aqueous NaX solutions shows consistently the same salting $\omega_x$ trend, which follows the Hofmeister series: $X^-$ (R/η) = $I^-$ (2.2/2.5) > $Br^-$ (1.96/2.8) > $Cl^-$ (1.81/3.0) > $F^-$ (1.33/4.0) ≈ 0. The electrification effect of a larger ion with lower electronegativity is more pronounced than others. The effect of NaF addition is too small to be resolvable. This tiny spectral difference evidences that the $Na^+$ and $F^-$ stay closer and explains why the maximal

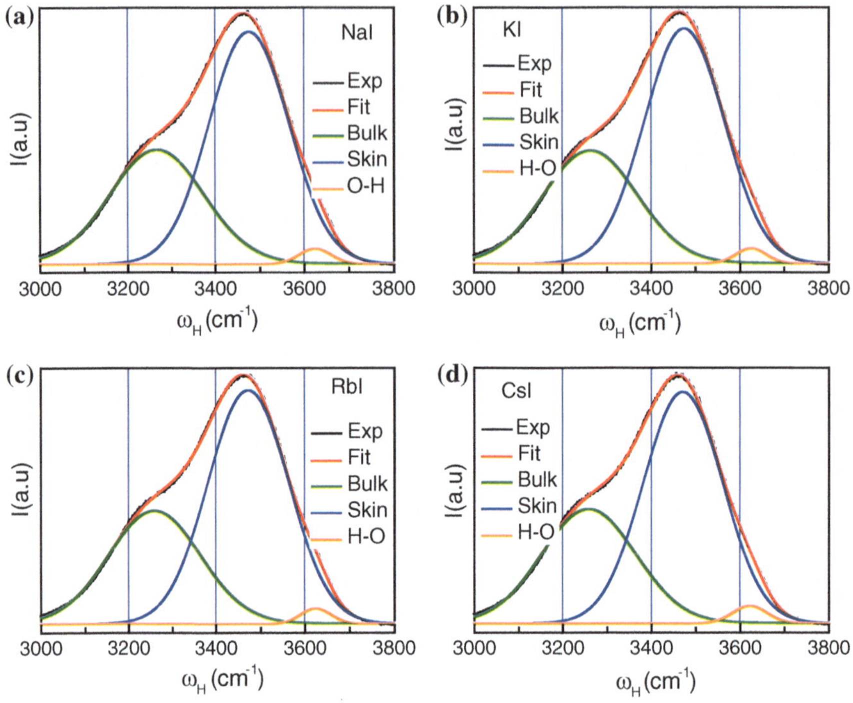


**Fig. 12.11** Decomposition of the Raman spectra for 2.1 M YI solutions (Reprinted with permission from [62].)

**Table 12.5** Gaussian decomposition of the $\omega_H$ for YI solutions measured at 298 K

| | | Bulk | Skin/shell | H–O | Notes |
|---|---|---|---|---|---|
| $\omega_H$ ($cm^{-1}$) (bond stiffness) | $H_2O$ | 3245.94 | 3450.55 | 3611.59 | • Solute electrification and molecular undercoordination stiffen the H–O bond and soften the O:H nonbond consequently • H–O stiffness is less sensitive to the type of $Y^+$ |
| | NaI | 3266.29 | 3476.39 | 3624.90 | |
| | KI | 3258.93 | 3475.16 | 3625.91 | |
| | RbI | 3258.94 | 3473.30 | 3625.25 | |
| | CsI | 3258.40 | 3471.45 | 3623.71 | |
| FWHM ($cm^{-1}$) (fluctuation order) | $H_2O$ | 217.27 | 177.05 | 107.14 | • Molecular undercoordination depresses molecular dynamics • Molecular dynamics is slightly sensitive to the $Y^+$ cations |
| | NaI | 216.04 | 179.71 | 72.19 | |
| | KI | 212.77 | 179.66 | 68.65 | |
| | RbI | 210.45 | 181.26 | 68.96 | |
| | CsI | 212.40 | 178.82 | 77.58 | |
| Peak area (abundance) | $H_2O$ | 0.50 | 0.43 | 0.07 | • Hydration-shell abundance gain renders the bulk loss • H–O radical abundance loss and stiffening evidence preferential occupancy of I anions |
| | NaI | 0.36 | 0.62 | 0.02 | |
| | KI | | | | |
| | RbI | | | | |
| | CsI | | | | |

Reprinted with permission from [62]

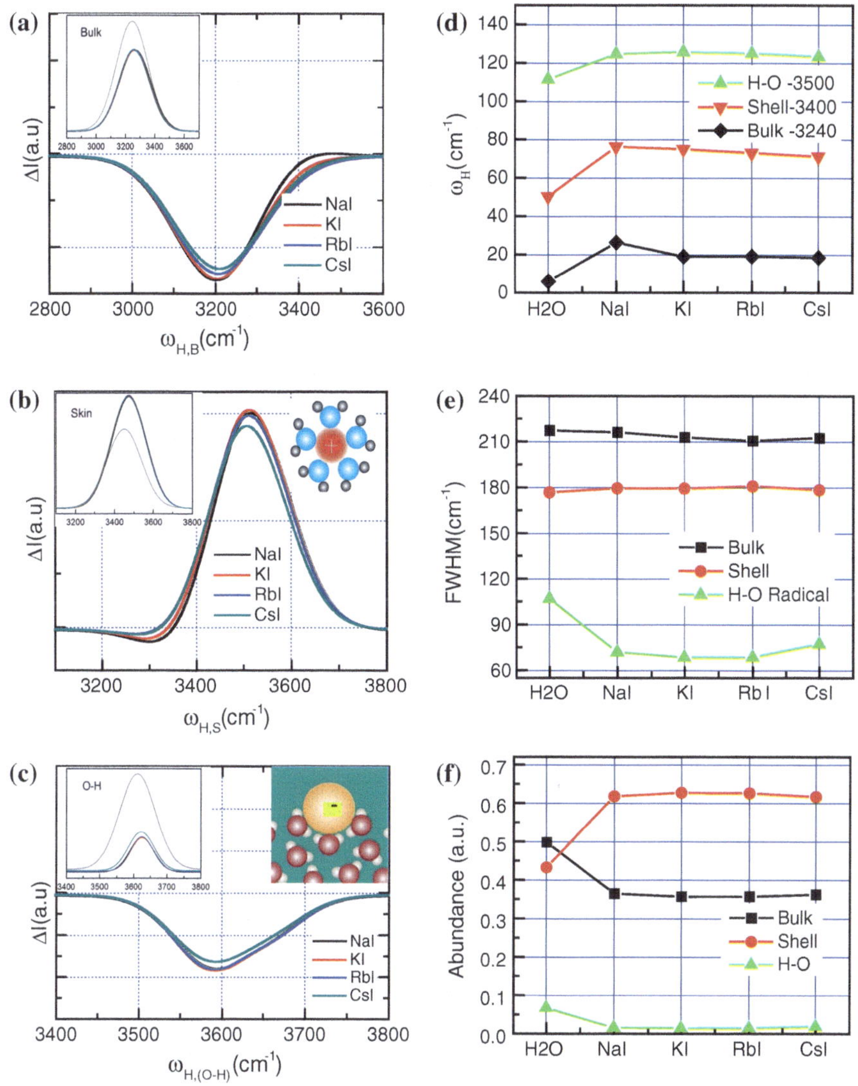


**Fig. 12.12** Site-resolved DPS (**a–c**) for 2.1 M YI solutions with respect to the 298 K water revealed that the $Y^+$ has the similar effect on the (**d–f**) H–O bond stiffness, molecular dynamics, and phonon abundance. *Insets* (**b**) and (**c**) illustrate the solute hydration shell of a cation or an anion and the skin preferential occupancy of anions [27, 63, 64] (Reprinted with permission from [62].)

solubility of NaF is only 0.9 M (n($H_2O$)/n(NaF) ≈ 60/1). The concentration of 3.0 M corresponds to situation that one salt molecule is surrounded by 18 $H_2O$ molecules.

One needs to note that the molecular fraction of 3 M solution is 1/19, 1/17, and 1/16 for NaCl, NaBr, and NaI, respectively. Precision can be improved by using the molecular ratio of different solutions instead of the number of Moles per liter.

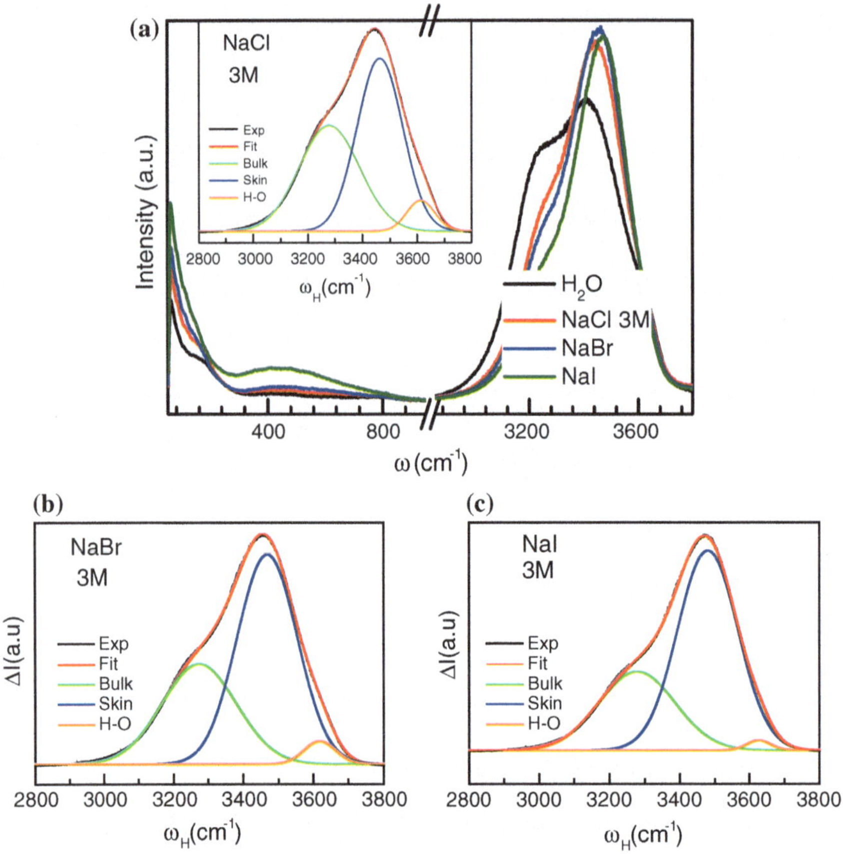


**Fig. 12.13** Raman probing of **a** the 3.0 M NaX (X = F (limited to 0.9 M), Cl, Br, I) solutions at 298 K and the Gaussian decomposition of the H–O phonon peak (Reprinted with permission from [57].)

Results in Fig. 12.15 and Table 12.6 revealed that the bulk and H–O radical undergo abundance loss and the hydration shells gain the abundance. Salting stiffens the bulk and radical H–O bond slightly in the order of $I^- > Br^- > Cl^-$ but the $Cl^+$ is more significant than others to the bulk.

#### 12.4.4.4 NaI Concentration Dependence

Figure 12.16 shows the full-frequency phonon spectroscopy of NaI solution as a function of solute concentration and the segmental DPS. The $\omega_H$ shift towards higher frequencies with abundance transition from the bulk water to hydration shells. Repeating the same iteration of DPS, one can obtain the concentration

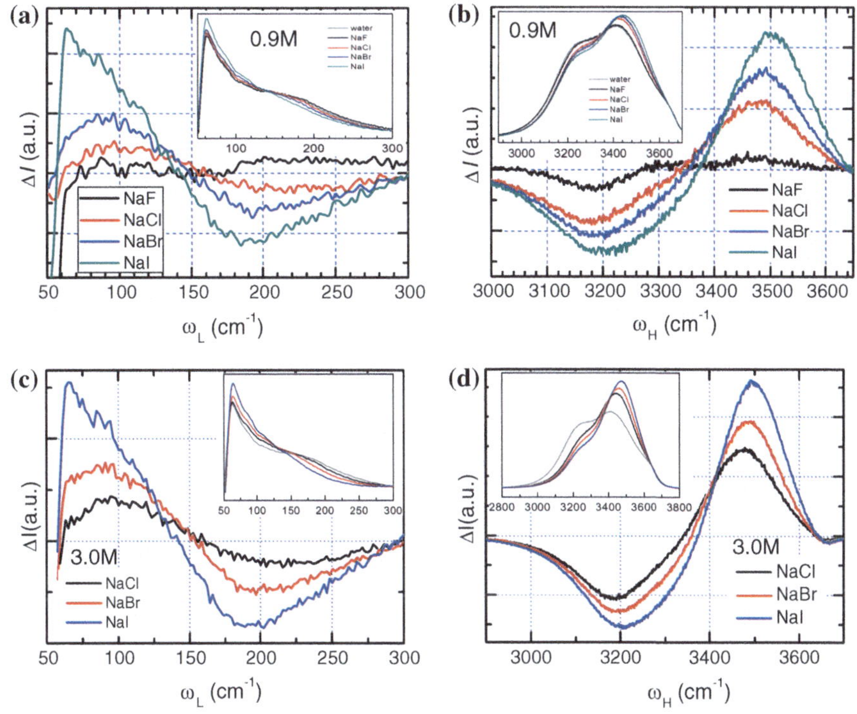


**Fig. 12.14** Room temperature DPS for **a**, **b** 0.9 M, **c**, **d** 3.0 M NaX solutions shows the same trend of solute electrification that offsets the $\omega_L$ and $\omega_H$ in opposite directions. *Insets* are the raw spectra (Reprinted with permission from [58].)

dependence of the bond stiffness, order of molecular fluctuation, and the abundance of the three components according to their molecular coordination numbers. Figure 12.17 shows the componential DPS and the site-resolved H–O bond stiffness, molecular fluctuation order, and phonon abundance.

Results in Fig. 12.17 and Table 12.7 show the following:

(1) H–O bond stiffness in the ionic hydration shells are independent of the concentration but it is overall stiffer than the H–O bond in the skin of neat water, which indicates that the supersolidity of hydration shells is higher than the skin of pure water.
(2) The order of molecular fluctuation and the abundance of the hydration shell increase with concentration, which means that the viscosity and the number ratio of molecules in the first hydration shells increase with concentration.
(3) The bulk and the H–O radicals are subject to net loss and the fraction of loss increases with NaI concentration. The hydration shells undergo a net gain in the abundance.

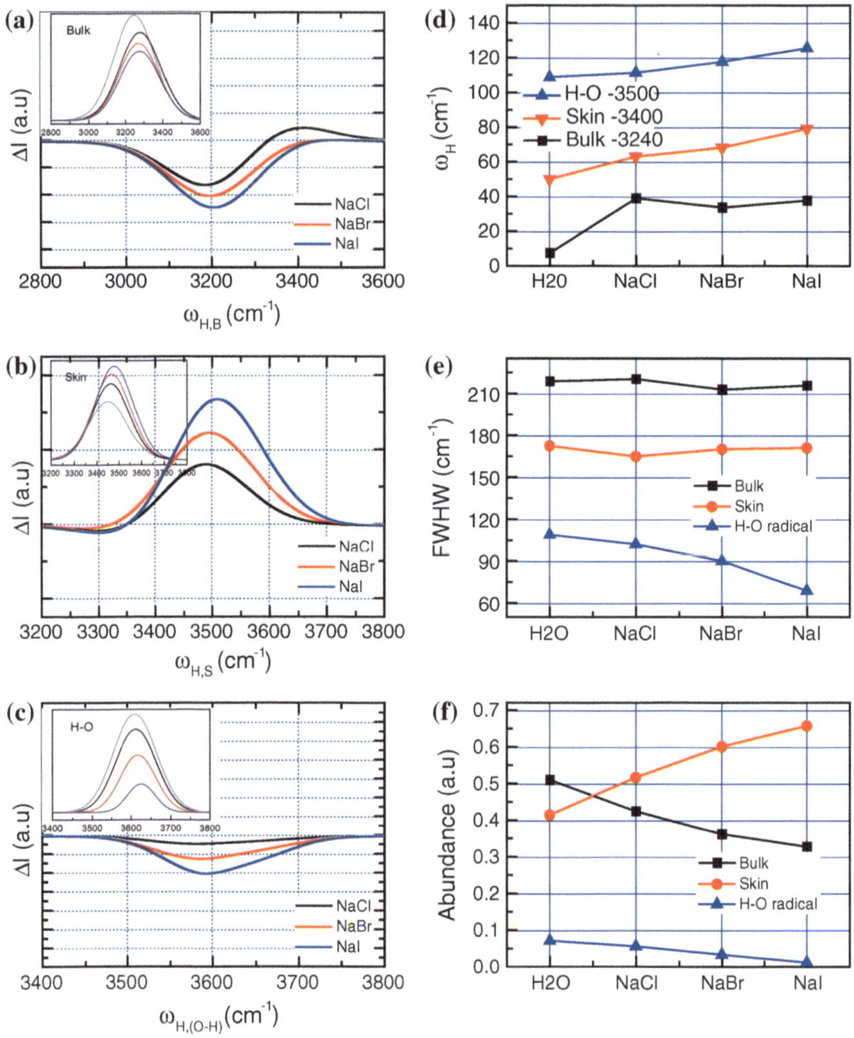


**Fig. 12.15** Molecular-site-resolved $\omega_H$ DPS for the bulk, hydration shell/skin, and the H–O radicals and the respective **d** bond stiffness, **e** molecular fluctuation order, and **f** phonon abundance for 3.0 M NaX solutions (Reprinted with permission from [62].)

(4) The bulk H–O undergoes slight blueshift as the number of ions increase, which belong to the high-order hydration shells.

(5) The $\omega_H$ for the H–O radials increases to saturation at 3 M and then drops to a certain value; the fluctuation order and abundance drop when the concentration increases. The blueshift and abundance drop indicates the skin preferential occupancy of the anions up to a saturation pint at 3 M.

**Table 12.6** Molecular-site-resolved liquid O:H–O bond thermal relaxation in terms of bond stiffness, order of fluctuation, and phonon abundance in 0.9 and 3.0 M NaX solutions

| | | 298 K | Bulk | Shell/Skin | H–O radical | Notes |
|---|---|---|---|---|---|---|
| $\omega_H$ ($cm^{-1}$) (bond stiffness) | 0.9 | $H_2O$ | 3247.48 | 3450.45 | 3609.08 | • H–O bond in the hydration shell is relatively stable<br>• H–O bond stiffness increases with solute concentration<br>• The sensitivity to a certain solute varies from site to site |
| | | NaF | 3259.71 | 3457.97 | 3612.54 | |
| | | NaCl | 3256.58 | 3456.16 | 3611.16 | |
| | | NaBr | 3256.56 | 3459.41 | 3612.06 | |
| | | NaI | 3254.87 | 3465.77 | 3614.61 | |
| | 3.0 | NaCl | 3279.02 | 3463.41 | 3611.64 | |
| | | NaBr | 3273.64 | 3468.50 | 3617.87 | |
| | | NaI | 3277.70 | 3479.65 | 3625.77 | |
| FWHM ($cm^{-1}$) (fluctuation order) | 0.9 | $H_2O$ | 218.67 | 173.16 | 109.51 | • Molecules in the hydration shell and in the bulk are relatively stable<br>• H–O radicals become stiffer but less mobile |
| | | NaF | 231.40 | 170.15 | 104.65 | |
| | | NaCl | 219.30 | 172.28 | 106.25 | |
| | | NaBr | 219.43 | 174.37 | 103.70 | |
| | | NaI | 218.08 | 180.47 | 98.02 | |
| | 3.0 | NaCl | 220.28 | 165.57 | 102.70 | |
| | | NaBr | 212.83 | 170.68 | 90.40 | |
| | | NaI | 215.66 | 171.70 | 69.21 | |
| Peak integral (abundance) | 0.9 | $H_2O$ | 0.51 | 0.42 | 0.07 | • Phonon abundance transfers from the bulk to the shell<br>• H–O radical abundance loss with blueshift due to the preferential skin occupancy of the $X^-$ anions |
| | | NaF | 0.55 | 0.39 | 0.07 | |
| | | NaCl | 0.48 | 0.45 | 0.06 | |
| | | NaBr | 0.46 | 0.48 | 0.06 | |
| | | NaI | 0.44 | 0.52 | 0.04 | |
| | 3.0 | NaCl | 0.42 | 0.52 | 0.06 | |
| | | NaBr | 0.36 | 0.60 | 0.03 | |
| | | NaI | 0.33 | 0.66 | 0.01 | |

Reprinted with permission from [62]

#### 12.4.4.5 NaI Hydration Shell Thermal Stability

Figure 12.18 shows the full-frequency Raman spectra of NaI salt solutions of 0.1 mol ratio (equivalent of 6 M) as a function of temperature in the range of 278 and 368 K. The higher spectral intensity particular at low frequencies indicates the high reflectivity due to the structure order increase.

The convoluted DPS in Fig. 12.19 shows that the $\omega_x$ undergoes opposite shift as a result of H–O contraction and O:H elongation. NaI addition turns the bulk water into the hydration shells surrounding the ionic solutes. The FWHM for the hydration shell is narrower than that of the bulk water as a result of the molecular dynamics drop in the hydration shells. Heating enhances the solute

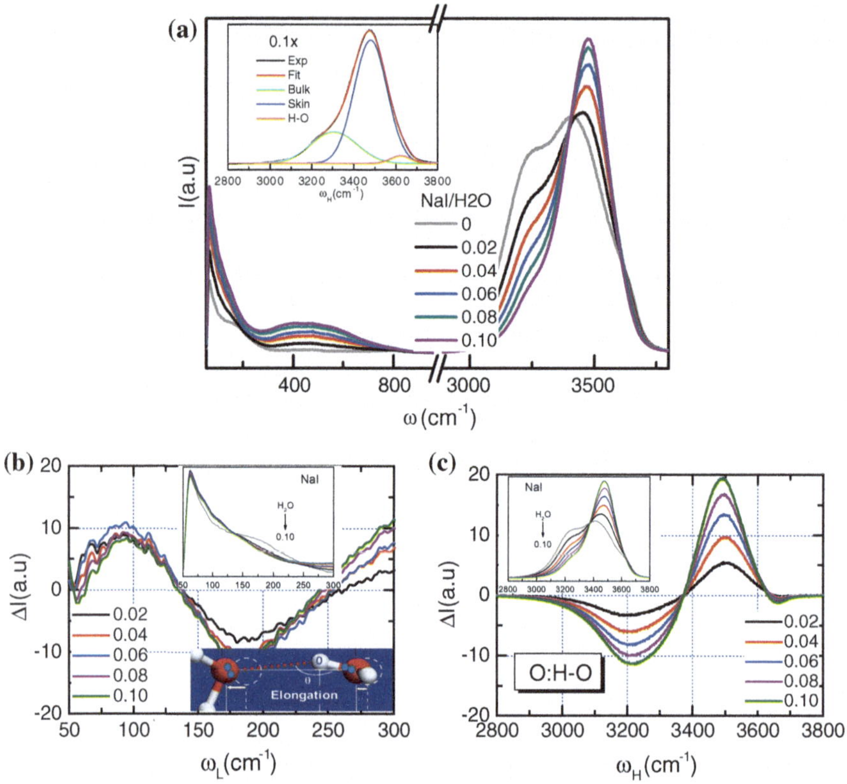


**Fig. 12.16** **a** Full-frequency Raman spectra for NaI solution as a function of mole concentration and **b**, **c** the segmental DPS (Reprinted with permission from [62].)

electrification effect. Figure 12.20 shows the decomposition of the $\omega_H$ into the bulk, shell/skin, and H–O radical components, as a function of joint heating and salting effect.

Table 12.8 and Fig. 12.21 shows the molecular site and temperature resolved O: H–O bond relaxation in terms of H–O bond stiffness, structural order, and phonon abundance in comparison to that for deionized water. While the abundance of the hydration shells undergo net gain and the H–O radicals net loss, the bulk matrix evolves to asymmetrical gain and loss at 318 K and above due to the heating enhanced blue shift of the bulk phonons. However, the spectral change of the H–O radicals and the bulk is very tiny. While the bulk of water and the solution undergo $\omega_H$ heating blueshift, the skin and the shell is extraordinarily stable due to their supersolidity nature. The change of the H–O radical is almost negligible.

## 12.4.5 *Acid Hydration—Quantum Fragilation*

### 12.4.5.1 HX Acid Solutions (X = Cl, Br, I)

Figure 12.22 displays the full-frequency and the DPS Raman spectra for 0.1 HX solutions with respect to the spectrum of deionized water measured at 298 K. Features around 500 $cm^{-1}$ shows order enhancement of the ∠O:H–O angle. The inset shows the $\omega_x$ decomposition into the bulk, hydration shells/solution skin and the H–O dangling bond radicals. The effect of ionic electrification on the O:H–O bond relaxation follows the same order in salts X = I > Br > Cl. Compared with the spectra for salts that the $\omega_L$ shifts from 190 to 75 $cm^{-1}$ but the acids shift the $\omega_L$ from 190 to 120 $cm^{-1}$, which indicate H↔H quantum fragilation that compresses the O:H nonbond in the matrix.

It is convenient to use the molar ratio for direct comparison of the extent and the effect of solute electrification on the O:H–O bond relaxation and polarization. The $H_2O$/HX number ratio is (1.0 + x)/(0.47x) = 147, 42, and 21 for x = 0.015, 0.05, and 0.1. The 0.47 is the concentration of the initially commercially available solution.

Figure 12.22a inset and Fig. 12.23 decompose the $\omega_H$ spectral peak into three components, which derives the molecular-site and solute-type resolved $\omega_H$ stiffness, order of molecular fluctuation, and the phonon abundance. The componential DPS in Fig. 12.24a–c displays the evolution of the phonon abundance for each component, As summarized in Table 12.9 and Fig. 12.24d–f, acid solutions effect differently on the O:H–O bond in the bulk, hydration shells and the H–O dangling bond radicals.

Results in Table 12.9 and Fig. 12.24 reveal the following:

(1) Acid has the same effect of salt on O:H–O bond relaxation at a site-resolved sensitivity but the $\omega_H$ phonon abundsnce is lower than that of the same type salt of the same concentration because the $H^+$ does not form a polarizer but a H↔H breaker.
(2) The extent of the $\omega_H$ blue shift for the hydration shell and the H–O radical follow the Hofmeister series order X = I > Br > Cl but the bulk $\omega_H$ shifts in opposite direction. The H–O radical $\omega_H$ is less insensitive to the type of acids.
(3) The fluctuation order (the inverse of the FWHM) of the hydration shell is similar to the skin of water but the HCl and HBr improve the hydration-shell order only slightly. However, the bulk structure order drops substantially with respect to water and the HCl lowers the structure order more than HBr and HI.
(4) The H–O radical drops in abundance with slight blue shift in the order of I > Br > Cl, which indicate the preferential occupancy of larger anions, being the same to salts.
(5) Observations indicate that the stiffness and molecular order of the hydration shells are relatively more stable than the bulk.

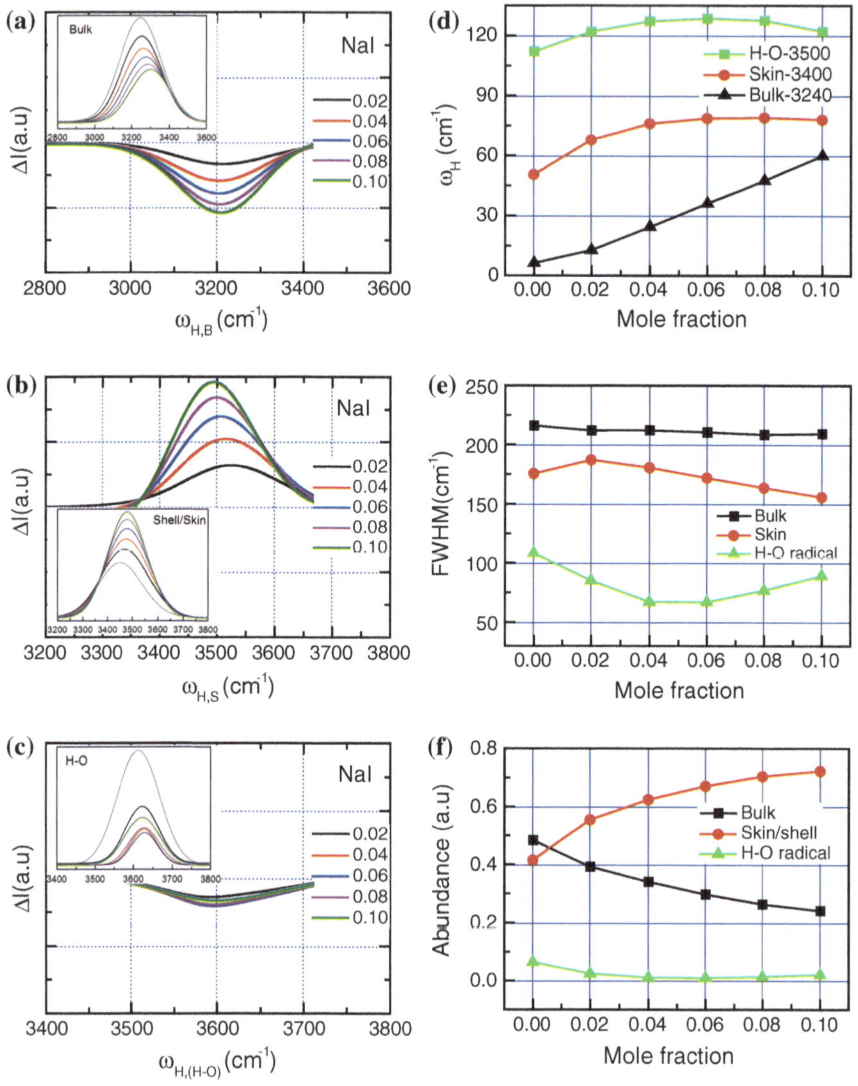


**Fig. 12.17** NaI solute concentration and molecular site dependence of the $\omega_H$ DPS for the bulk, hydration shell/skin, and the H–O radicals and the respective **d** bond stiffness, **e** molecular fluctuation dynamics, and **f** phonon abundance (Reprinted with permission from [62].)

#### 12.4.5.2 HI Solute Concentration

Figure 12.25a shows the full-frequency Raman spectra for the HI solution as a function of mole ratio and deionized water measured at 298 K. The inset shows the $\omega_H$ decomposition into the bulk, hydration shell, and H–O radical components. The convoluted DPS in Fig. 12.25b, c showed that the acid ionic electrification effect

**Table 12.7** NaI concentration dependence and the molecular-site-resolved liquid O:H–O bond thermal relaxation in terms of bond stiffness, order of fluctuation, and phonon abundance

| 298 K | NaI | Bulk | Shell/Skin | H–O radical | |
|---|---|---|---|---|---|
| $\omega_H$ (cm$^{-1}$) (bond stiffness) | 0 | 3246.29 | 3450.83 | 3612.53 | • Solute electrification has the same effect of molecular undercoordination on stiffening the H–O bond • Bulk H–O stiffness increases linearly with concentration • Shell and radical H–O bond stiffness saturates at 3 M and then recover slightly |
| | 0.02 | 3252.78 | 3468.21 | 3622.42 | |
| | 0.04 | 3264.35 | 3476.40 | 3627.56 | |
| | 0.06 | 3276.08 | 3479.10 | 3628.89 | |
| | 0.08 | 3287.45 | 3479.39 | 3627.93 | |
| | 0.10 | 3299.86 | 3478.41 | 3622.56 | |
| FWHM (cm$^{-1}$) (fluctuation order) | 0 | 216.01 | 176.26 | 108.93 | • Molecular dynamics is solute concentration and molecular site sensitive • Solute depresses the molecular dynamics in the shell and the bulk at different rates |
| | 0.02 | 211.93 | 187.78 | 85.76 | |
| | 0.04 | 212.06 | 181.31 | 67.67 | |
| | 0.06 | 210.35 | 172.55 | 67.45 | |
| | 0.08 | 208.26 | 164.10 | 77.42 | |
| | 0.10 | 209.10 | 156.45 | 89.68 | |
| Peak integral (a.u.) (abundance) | 0 | 0.4833 | 0.4180 | 0.0678 | • Shell gains abundance at the expense of bulk abundance loss • H–O radical abundance changes inversely with its phonon frequency |
| | 0.02 | 0.3919 | 0.5578 | 0.0266 | |
| | 0.04 | 0.3400 | 0.6272 | 0.0131 | |
| | 0.06 | 0.2971 | 0.6734 | 0.0119 | |
| | 0.08 | 0.2628 | 0.7074 | 0.0158 | |
| | 0.10 | 0.2406 | 0.7252 | 0.0233 | |

Reprinted with permission from [62]

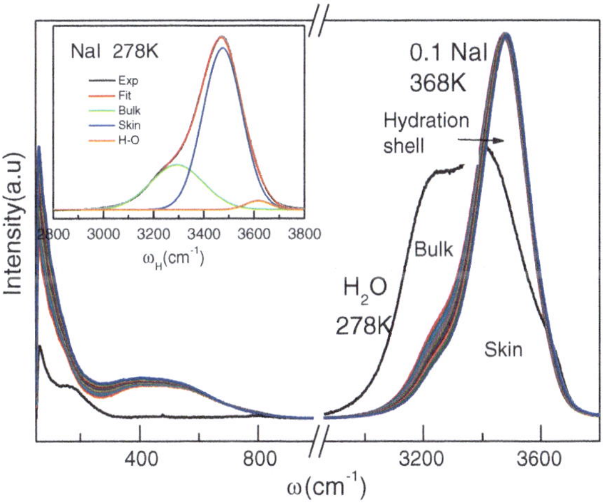


**Fig. 12.18** Temperature dependence of the full-frequency Raman spectra for 0.1 NaI solution compared with neat water. Higher structure order increases the photon reflectivity in measurements (Reprinted with permission from [65].)

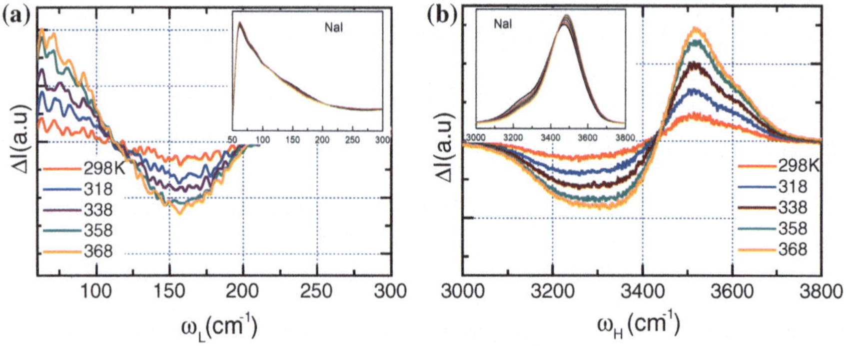


**Fig. 12.19** Convoluted **a**, **b** $\omega_x$ DPS for 0.1 NaI solutions as a function of temperature with respect to the solution at 278 K (Reprinted with permission from [65].)

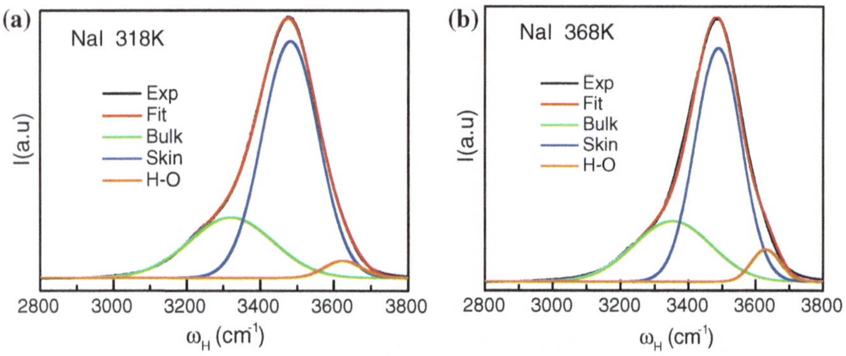


**Fig. 12.20** The best-fit decomposition of the $\omega_H$ for 0.1 NaI solutions at typical temperatures. Heating evolves the bulk component into hydration shells (Reprinted with permission from [65].)

is proportional to the concentration, and that the $\omega_L$ evolves from 190 to 120 $cm^{-1}$ rather than 75 $cm^{-1}$ by salting or heating, because of the H↔H fragilation repulsive effect.

Figure 12.25a inset and Fig. 12.26 decompose the $\omega_H$ peak into three components, which derives the molecular site and HI concentration dependence of the local H–O bond stiffness, molecular structural order, and the phonon abundance. Figure 12.27 and Table 12.10 summarizes the detailed information showing that acids effect differently on the O:H–O bond in the bulk, hydration shells, and the H–O dangling bond radicals.

Results in Table 12.10 and Fig. 12.27 reveal the following:

(1) The $\omega_H$ undergoes blueshift with extent increasing with concentration but the $\omega_H$ shifts less in the hydration shell and H–O radicals, which indicates the higher stability of the hydration shells than the bulk.

**Table 12.8** Molecular-site and temperature resolved O:H–O bond relaxation dynamics in terms of H–O bond stiffness, molecular fluctuation, and phonon abundance

| | | Bulk | Skin | H–O radical | Notes |
|---|---|---|---|---|---|
| $\omega_H$ ($cm^{-1}$) | 278 K water | 3239.00 | 3442.60 | 3603.04 | • The H–O stiffness in the hydration shell is thermally more stable than that in the water skin<br>• The bulk H–O stiffness increases abruptly at 320 K due to the joint effect |
| | 278 K NaI | 3294.60 | 3476.01 | 3618.68 | |
| | 298 | 3307.40 | 3478.89 | 3621.23 | |
| | 318 | 3320.36 | 3481.56 | 3623.76 | |
| | 328 | 3333.84 | 3483.34 | 3624.68 | |
| | 338 | 3330.64 | 3481.86 | 3625.97 | |
| | 355 | 3340.64 | 3485.04 | 3628.62 | |
| | 368 | 3352.52 | 3488.04 | 3629.68 | |
| FWHM ($cm^{-1}$) | 278 K water | 217.10 | 167.46 | 121.84 | • Molecules in the hydration shells are less mobile than water skin and insensitive to temperature<br>• At 320 K and above, the bulk molecules become more active but the H–O less mobile |
| | 278 K NaI | 211.92 | 156.95 | 100.22 | |
| | 298 | 219.97 | 155.00 | 98.70 | |
| | 318 | 226.72 | 152.75 | 99.62 | |
| | 328 | 236.44 | 150.65 | 102.28 | |
| | 338 | 250.36 | 170.36 | 104.36 | |
| | 358 | 259.87 | 167.87 | 107.87 | |
| | 368 | 269.17 | 162.17 | 100.17 | |
| Abundance | 278 K water | 0.532 | 0.385 | 0.083 | • At 320 K and above, the shell abundance losses to the bulk due to thermal fluctuation dominance<br>• The abundance of H–O radicals drops slightly when heated |
| | 278 K NaI | 0.267 | 0.708 | 0.025 | |
| | 298 | 0.266 | 0.705 | 0.029 | |
| | 318 | 0.266 | 0.701 | 0.033 | |
| | 328 | 0.283 | 0.681 | 0.036 | |
| | 338 | 0.249 | 0.695 | 0.056 | |
| | 358 | 0.262 | 0.683 | 0.055 | |
| | 368 | 0.284 | 0.660 | 0.056 | |

Reprinted with permission form [65]

(2) Molecular structure order in the hydration shell is the same to the skin of water, which is insensitive to the acid concentration. The H–O radical becomes more ordered and the bulk H–O bond less than the bulk water.
(3) The H–O radical drops in abundance with slight blue shift when the concentration increases, which indicates the preferential occupancy of larger anions, being the same to salts. The bulk H–O bond becomes shorter with decreased structural order compared with neat water or the solution skin.
(4) The stiffness and molecular order of the hydration shells are relatively more stable and insensitive to HI concentration.

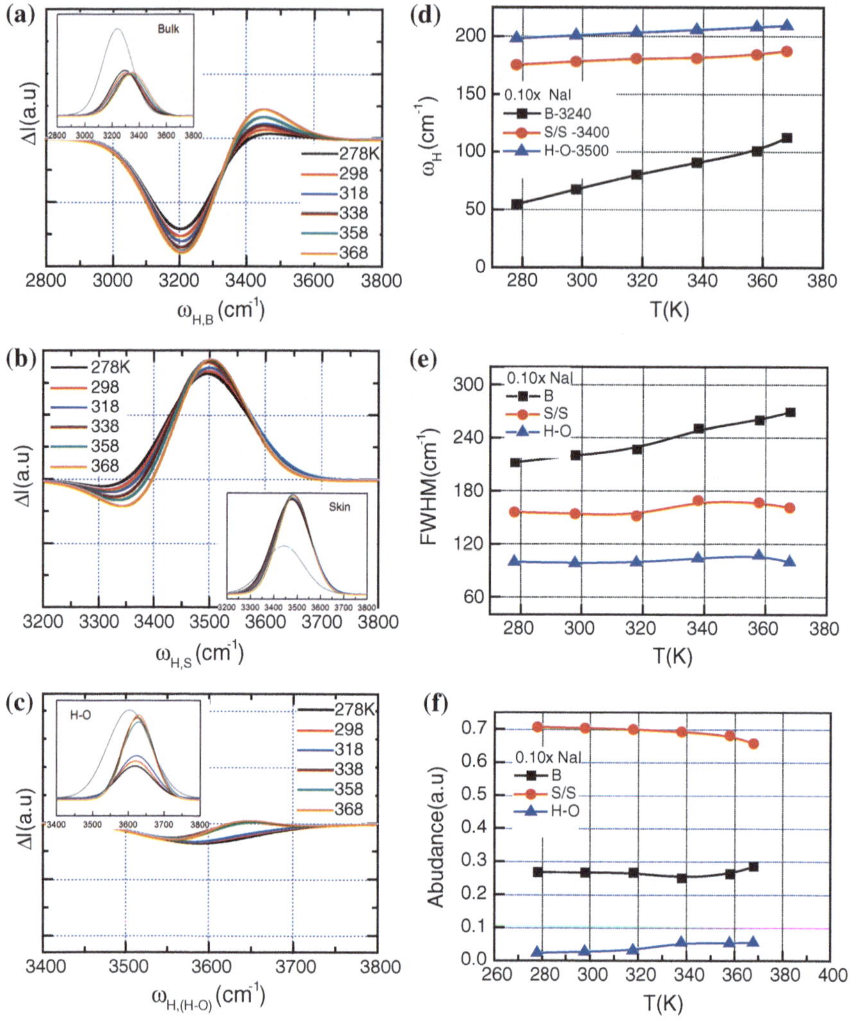


**Fig. 12.21** Molecular-site and temperature dependent phonon relaxation dynamics (**a–c**) in terms of **d** H–O bond stiffness, **e** molecular fluctuation order, and **f** phonon abundance. *B* represents the bulk, *S/S* the skin/shell, and H–O the dangling bond radicals (Reprinted with permission form [65].)

### 12.4.5.3 HI Hydration-Shell Thermal Stability

The full-frequency Raman spectra in Fig. 12.28 for the 0.1 HI solution show the same trend of heating O:H–O bond relaxation to that of NaI solution despite its ability of depolarization. Acid shifts the $\omega_x$ in opposite direction—H–O contraction and O:H elongation. Heating enhances the acid effect on phonon relaxation, as confirmed in the convoluted DPS in Fig. 12.30. Compared with NaI, the $\omega_H$ DPS

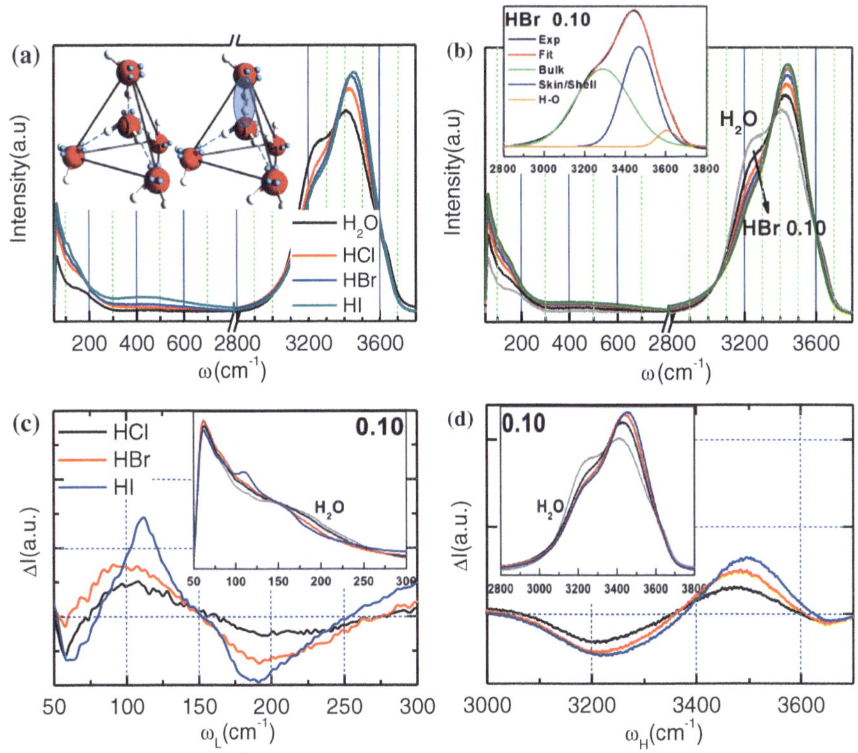


**Fig. 12.22** Full-frequency Raman spectra for **a** $H_2O$ and 0.1 (mole fraction) HX solutions and **b** concentration dependence of the HBr solution at 298 K. (c, d) Segmental DPS of 0.1 HX solutions. An replacement of the central $H_2O$ with a $H_3O$ hydronium in the inset (**a**) results in the H↔H point breaker (shaded) fragilating acid solutions, inset (**b**) decomposes the $\omega_H$ into three components and insets (**c, d**) the normalized segmental spectra. (Reprinted with permission from [66].)

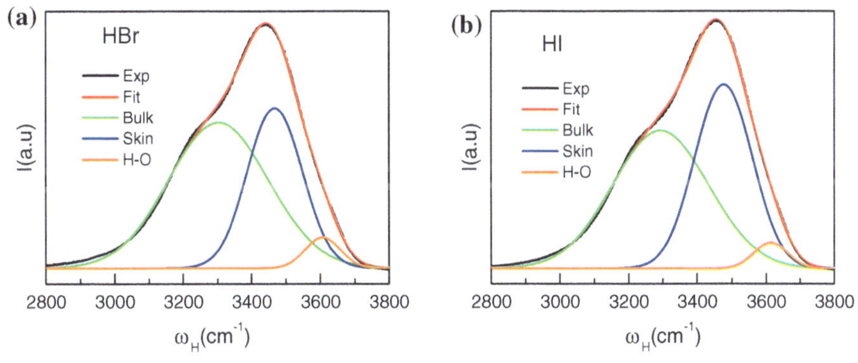


**Fig. 12.23** Gaussian decomposition of the 298 K Raman spectra for **a** HBr and **b** HI solutions of 0.1 molar concentration (Reprinted with permission from [66].)

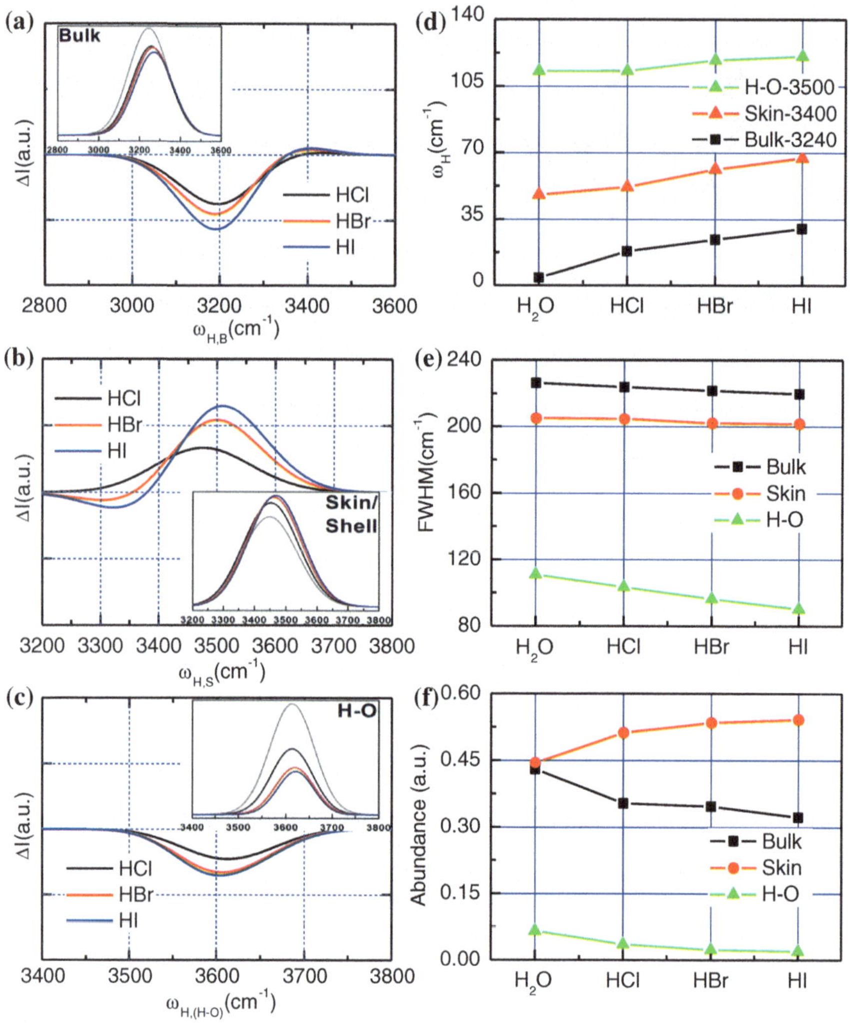


**Fig. 12.24** The **a** bulk, **b** skin/shell, **c** H–O radical componential DPS for 0.1x HX acid solutions and their molecular-site-resolved **d** H–O bond stiffness, **e** order of molecular structure and **f** phonon abundance at 298 K (Reprinted with permission from [66].)

for the HI is less asymmetrical as the HI raises the structure order to a lower extent than the NaI of the same concentration under the same temperature (Fig. 12.29).

The best-fit of the $\omega_H$ spectra in Fig. 12.30 shows that the bulk component for the HI solution is relatively abundant than that of the NaI solution, which indicates the smaller hydration shells of the HI ions under the same concentration and temperature. However, from the full-frequency Raman spectra and the convoluted DPS, one can hardly tell the difference between the functionalities of the HI and the

**Table 12.9** Molecular-site-resolved O:H–O bond relaxation in terms of H–O bond stiffness, order of molecular fluctuation and phonon abundance in 0.1 molar ratio HX solutions compared with deionized $H_2O$ measured at 298 K temperature

| | 0.10 | Bulk | Skin | H–O radical |
|---|---|---|---|---|
| Frequency ($cm^{-1}$) | $H_2O$ | 3244.13 | 3447.98 | 3613.22 |
| | HCl | 3258.08 | 3452.13 | 3613.30 |
| | HBr | 3264.34 | 3461.41 | 3618.92 |
| | HI | 3270.02 | 3467.31 | 3620.88 |
| FWHM ($cm^{-1}$) | $H_2O$ | 226.27 | 205.17 | 111.17 |
| | HCl | 223.77 | 204.84 | 103.73 |
| | HBr | 221.65 | 202.30 | 96.64 |
| | HI | 219.67 | 201.85 | 90.61 |
| Abundance (a.u.) | $H_2O$ | 0.43 | 0.4448 | 0.0666 |
| | HCl | 0.3533 | 0.5131 | 0.0368 |
| | HBr | 0.3460 | 0.5357 | 0.0247 |
| | HI | 0.3218 | 0.5429 | 0.0212 |

NaI in solutions but their polarizability. What differenciates the similarity is the H↔H anti-HB quantum fragilation in the acid solution.

Figure 12.31 and Table 12.11 summarize the site and temperature resolved O:H–O bond relaxation in the HI solutions compared with that of deionized water. The H–O radical in the solution undergoes heating blueshift but in the deionized water it is subject to redshift, as a result of heating fluctuation dominance in neat water and screening dominance in the solution because of the $I^-$ preferential occupancy in the skin though the radical's abundance is negligibly small, the shell $\omega_H$ undergoes slight blueshift in both water and HI solution but the molecular order in both are extremely stable.

#### 12.4.5.4 Acid Solution Anti-HB Formation

One can assume that the $I^-$ in NaI and HI has the same concentration dependence of electrification and the $Na^+$ and $H^+$ have different effect of electrification. The relative number of $H_2O$ molecules being electrified by the ions meets the criterion: $1 = N_I/N_I > x_{Na} > x_H \geq 0$, because of the ionic radius difference. Therefore, $(1 + x_{Na})/(1 + x_H) = A_{NaI}/A_{NI}$, where A is the phonon abundance (number of electrified $H_2O$ molecules) that is proportional to the integral of the $\omega_H$ gain in Fig. 12.16c for NaI and Fig. 12.25c for HI at the same mole concentration. Figure 12.32 shows the concentration dependence of the phonon abundance. At c = 0.02 and 0.10, $(1 + x_{Na})/(1 + x_H) = 1.48–1.5$ and 1.97–2.0, respectively, which indicates the following:

(1) The $N_I$ increases with concentration and reaches its maximum of 0.10 (red).
(2) The $N_{Na}$ increases with concentration linearly in a slope of 1.39 (blue).
(3) The N ~ 0; the $N_I$ increases with concentration nonlinearly.

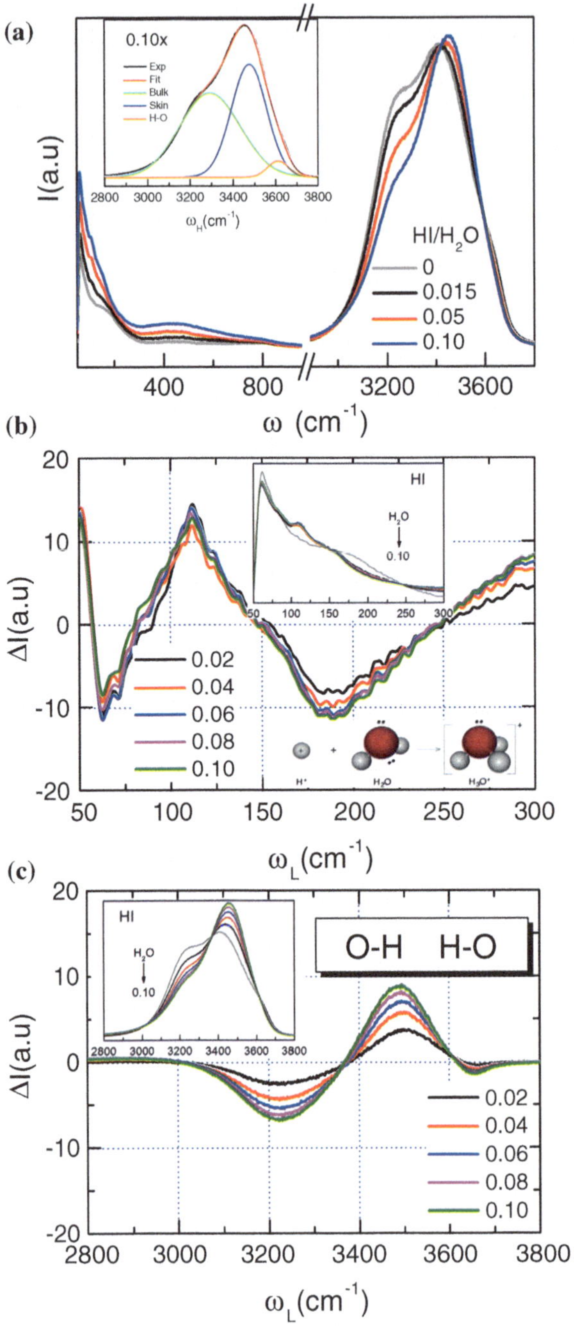


**Fig. 12.25** **a** Full-frequency Raman scan and **b**, **c** the $\omega_x$ DPS spectra for the HI acid solutions, measured at 298 K. The DPS shows the extent of the electrification is proportional to concentration. *Inset* is the Gaussian decomposition of 0.10 HI solutions. $I^-$ electrification and the hydronium O–H↔H–O anti-hydrogen bond determine the spectral features. The H↔H breaks the H-bond network in a point-by-point manner, diluting blood flow for instance (Reprinted with permission from [66].)

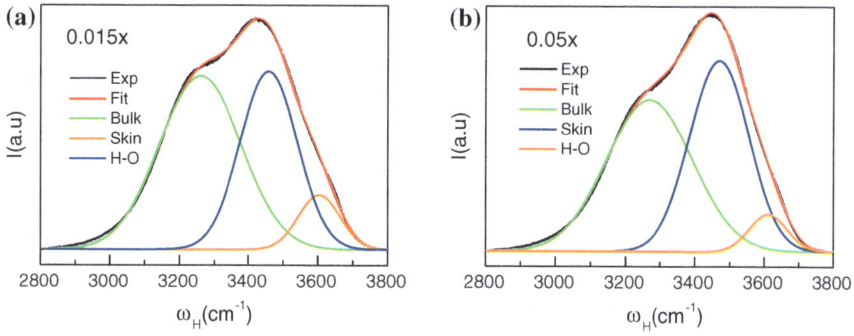


**Fig. 12.26** Gaussian decomposition of HI solution at 298 K (Reprinted with permission from [66].)

Therefore, the $H^+$ has no such effect of electrification but bonds to a $H_2O$ molecule to form the $H_3O^+$ hydronium, which interacts with one of its four coordination neighbors through the O-H↔H-O anti-HB.

In the acid solution, the X- polaization retains, the H↔H has two effects. One is the fragilation of the hydration network and the other is the H↔H repulsion. The X-polarization stiffens the H-O bond and softens the O:H nonbond. The H↔H compresses the neigbouring O:H nonbonds raising its phonon frequency, as observed.

(see phonon abundance in Table 12.12).

## *12.4.6 Base Hydration—Quantum Compression*

The full-frequency Raman spectra in Fig. 12.33a–d show consistently that the YOH (Y = Li, Na, K) solutes depress indeed the $\omega_H$ because of the H-O bond elongation. The $\omega_x$ relaxation is insensitive to the cation type Y = Li, Na, and K of the same concentration. The segmental DPS shown in Fig. 12.33e and f refine the observations, which reveal that the $\omega_H$ shifts from 3150–3450 to 2500–3150 $cm^{-1}$ and that the $\omega_L$ shifts from below to above the 250 $cm^{-1}$. The low concentration $\Delta\omega_L$ reproduces the $\Delta\omega_L$ feature of compressed water. The sharp peak at 3610 $cm^{-1}$ characterizes the H-O bond of the hydroxide. Comparison of Fig. 12.34c and d verified the expectation on the quantum compression of the O:H-O bond. The quantum compression is more significant than the applied mechanical pressure on shortening the O:H-O bond of the liquid water and aqueous solutions and even ice.

The O:↔:O in NaOH serves as a point source of extremely high pressure that shortens the O:H nonbond and polarization and meanwhile lengthens and softens the H–O bond. The H–O bond in the hydronium shows the spectral feature at 3600 $cm^{-1}$ being identical to that of the H–O free radicals. Fig. 12.34 shows the segmental DPS of the NaOH solution with respect to the deionized water under compression. The H–O abundance of the $H_3O^+$ increases with concentration and the $\omega_H$ redshift substantially more than that of the deionized water subjecting mechnical pressure up to

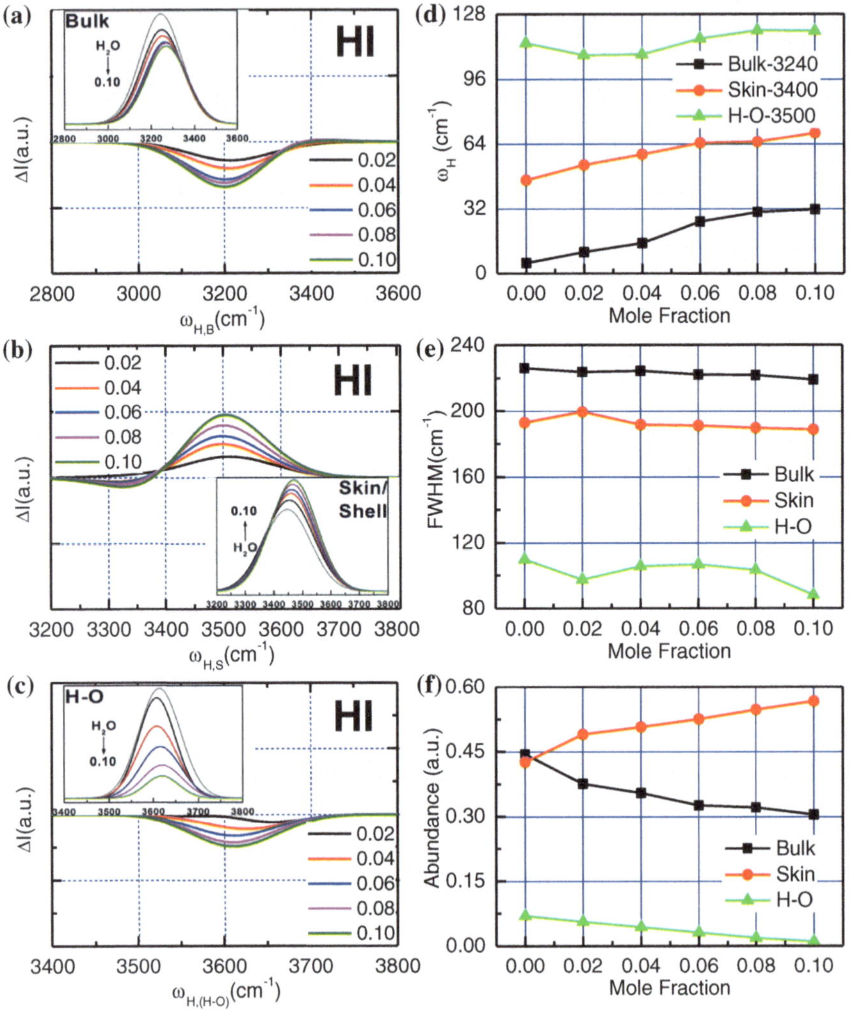


**Fig. 12.27** The **a** bulk, **b** skin, **c** H–O radical component DPS for HI acid solutions and their molecular-site-resolved, **d** H–O bond stiffness, **e** molecular structural order and **f** phonon abundance at 298 K (Reprinted with permission from [66].)

1.33 GPa. The point compressor shortens the O:H bond and stiffens the $\omega_L$ phonon and meanwhile lengthens the H-O bond and softens its phonon from 3500 to 2500 $cm^{-1}$. The effect of $Na^+$ electrification is annihilated by the O:↔:O compression. The concentration dependent DPS in Fig. 12.35 further clarify our expectations on the effect of the HB network compressors. The phonon abundance is unambiguously proportional to the number of the bonds contributed to relaxation. However, the polarization by $Y^+$ cations could hardly be resolved from the spectrometrics because of its annihilation by the quantum compression. With the known data of ($d_H$, $E_H$, $\omega_H$)

**Table 12.10** Molecular-site and concentration resolved O:H–O bond relaxation in terms of H-O bond stiffness, molecular structural order, and phonon abundance in the HI solution compared with deionized $H_2O$ measured at 298 K temperature

| 298 K | HI | Bulk | Shell | H–O radical | Notes |
|---|---|---|---|---|---|
| $\omega_H$ ($cm^{-1}$) Bond stiffness | 0 | 3245.23 | 3446.29 | 3613.96 | • $\omega_H$ increases proportionally to the concentration but the less-coordinated bond is less sensitive |
| | 0.02 | 3250.66 | 3453.74 | 3608.04 | |
| | 0.04 | 3255.06 | 3459.01 | 3608.42 | |
| | 0.06 | 3265.60 | 3464.61 | 3616.18 | |
| | 0.08 | 3270.26 | 3465.23 | 3620.25 | |
| | 0.10 | 3271.62 | 3469.67 | 3620.00 | |
| FWHM ($cm^{-1}$) Fluctuation order | 0 | 226.07 | 193.19 | 110.06 | • Shell molecules are insensitive to concentration but bulk molecules become more active |
| | 0.02 | 223.69 | 199.75 | 97.76 | |
| | 0.04 | 224.40 | 191.90 | 105.89 | |
| | 0.06 | 222.14 | 191.33 | 106.93 | |
| | 0.08 | 221.75 | 189.89 | 103.56 | |
| | 0.10 | 219.05 | 188.99 | 88.44 | |
| Abundance | 0 | 0.4441 | 0.4268 | 0.0708 | • The shell and bulk phonon abundance is stable |
| | 0.02 | 0.3755 | 0.4910 | 0.0574 | |
| | 0.04 | 0.3543 | 0.5081 | 0.0449 | |
| | 0.06 | 0.3253 | 0.5265 | 0.0324 | |
| | 0.08 | 0.3207 | 0.5485 | 0.0201 | |
| | 0.10 | 0.3042 | 0.5682 | 0.0115 | |

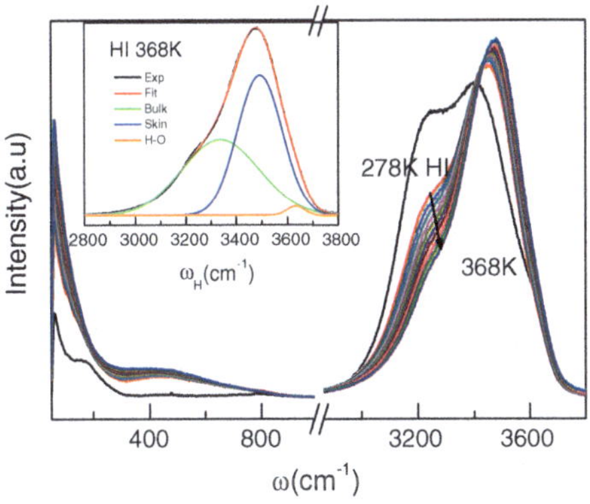


**Fig. 12.28** Full-frequency DPS and Raman spectra for 0.1 HI solution heating from 278–368 K (Reprinted with permission from [65].)

= (0.10 nm, 4.0 eV, 3200 $cm^{-1}$)$_{Bulk}$, we can estimate the compressed H-O bond energy (0.105 nm, 2.70 eV, 2500 $cm^{-1}$)$_{Base}$ by assuming the $d_H \sim 0.105$ nm:

$$E_2 = \frac{E_1 d_{H2}^2}{d_{H1}^2}\left(\frac{\omega_1}{\omega_2}\right)^{-2} = 4.0 \times (1.05/1.0)^2 \left(\frac{2500}{3200}\right)^2 \approx 2.70\,ev$$

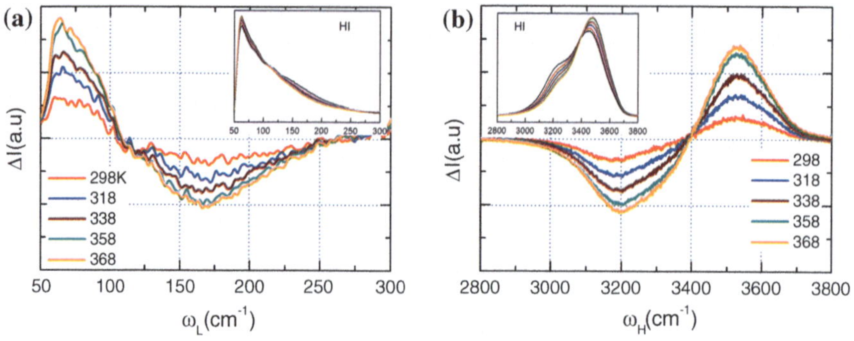


**Fig. 12.29** Convoluted $\omega_x$ DPS for the 0.1 HI solution at heating that stiffens the $\omega_H$ and softens the $\omega_L$ (Reprinted with permission from [65].)

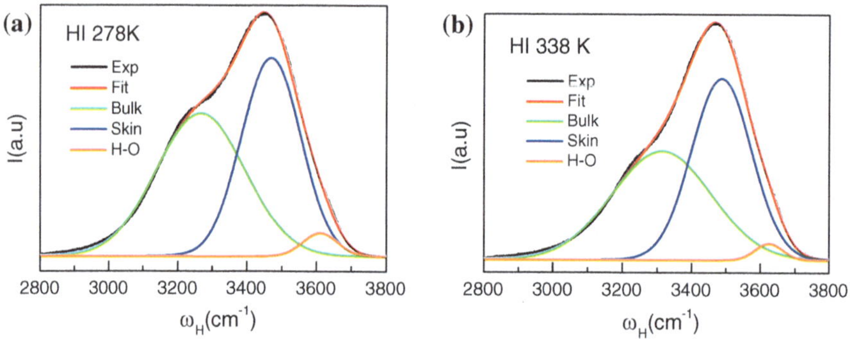


**Fig. 12.30** The best-fit decomposition (**a**, **b**) of the $\omega_H$ for 0.1 HI solution at typical temperatures (Reprinted with permission from [65].)

## *12.4.7 Methanol, Ethanol, and Glycerol Solutions*

One feels drunk accompanied with dizziness, blush, chest stuffiness, and sensation of a lump or discomfort in the throat when drinking liquor (methanol) because of the presence of OH- hydroxide and lone pairs that form thus important functional groups for Methanol, Ethanol, and glycerol and many drugs as well, like the Artemisinin treating malaria received 2015 Nobel Prize for medicine. However, it is unclear how the OH- interacts with solvent water and functionalizes the HB network? A phonon spectrometrics examination may help answering this question.

Figure 12.36 shows the Raman spectra for the methanol, ethanol, and glycerol and their concentration dependent solutions. Sharp peaks centered at 2900±100 $cm^{-1}$ correspond to the undercoordinated C-C bond vibration in liquid state. This peak whose abundance increases with solute concentration approaches to two-fold of the primary G mode for carbon dimer 1510 $cm^{-1}$. These molecules dissolve in water with additional lone pairs and C-H dangling bonds, both of which interact with

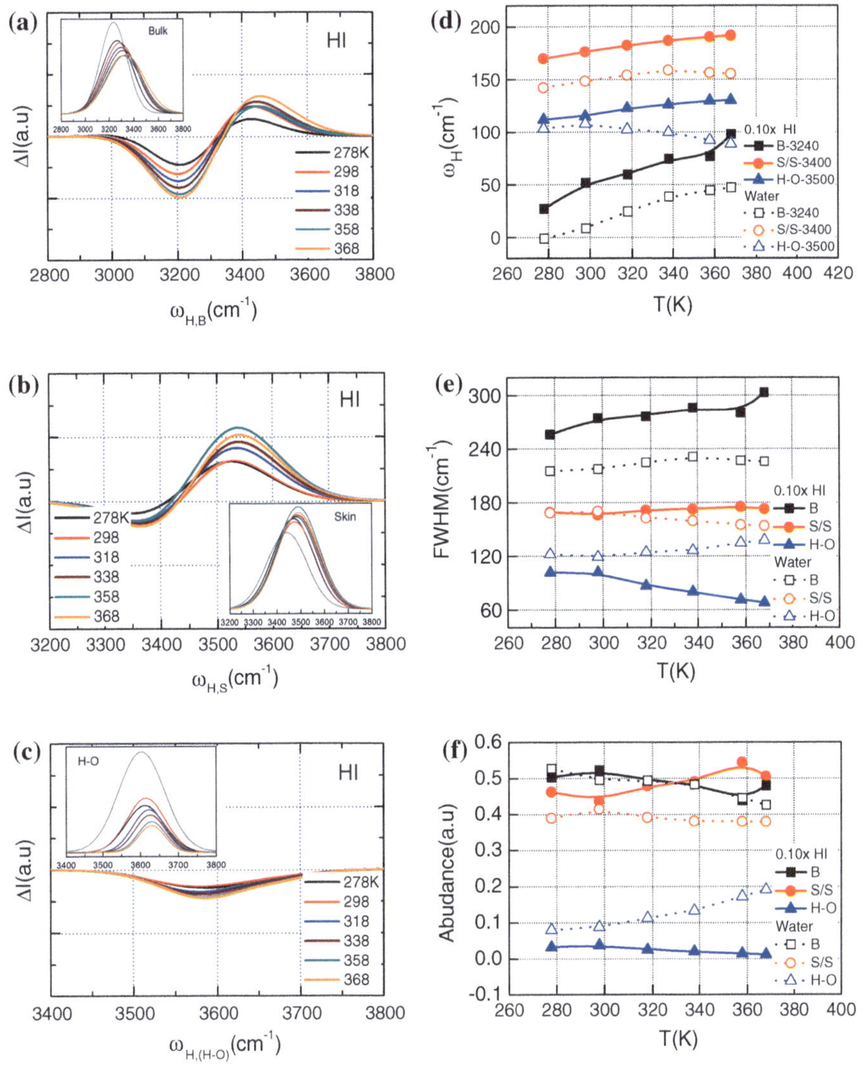


**Fig. 12.31** Componential DPS (**a–c**) and site-resolved O:H–O bond thermal relaxation dynamics in the **a** H–O bond stiffness, **b** molecular order, and **c** phonon abundance of HI solution of 0.1 mole fraction in comparison with those of neat water (Reprinted with permission from [65].)

water molecules through the nonbonds having possible polarization, H↔H fragilation, and O:↔:O compression effects.

The segmental DPS in Fig. 12.37 confirm the expectation of the compression ($\omega_H$ redshift) and polarization ($\omega_L$ redshift) effect but insignificant fragilation effect. Observations explain why these chemicals have lower melting point and why they can serve as lubricant.

**Table 12.11** Molecular-site and temperature resolved O:H–O bond relaxation dynamics in 0.1 HI/$H_2O$ solutions

| | | Bulk | Skin | H–O radical | Notes |
|---|---|---|---|---|---|
| 0 | 278 K water | 3239.00 | 3442.60 | 3603.04 | • Heating stiffens all the H–O bonds in acid solution but softens the H–O radical of the deionized water |
| | 278 K HI | 3267.08 | 3470.32 | 3612.28 | |
| | 298 | 3291.48 | 3476.80 | 3615.39 | |
| | 318 | 3299.52 | 3482.78 | 3623.00 | |
| | 338 | 3314.31 | 3487.46 | 3626.33 | |
| | 358 | 3316.41 | 3490.89 | 3630.27 | |
| | 368 | 3337.89 | 3492.22 | 3630.39 | |
| FWHM ($cm^{-1}$) | 278 K water | 217.10 | 167.46 | 121.84 | • Joint heating and aciding stabilizes the mobility of the hydration shells • Heating solution stiffens the H–O radicals |
| | 278 K HI | 255.77 | 169.89 | 101.84 | |
| | 298 | 274.11 | 166.66 | 102.15 | |
| | 318 | 276.39 | 172.21 | 86.95 | |
| | 338 | 286.07 | 173.65 | 80.58 | |
| | 358 | 280.28 | 176.15 | 71.64 | |
| | 368 | 302.97 | 173.80 | 68.45 | |
| Abundance | 278 K water | 0.532 | 0.385 | 0.083 | • H–O radical abundance drops in heated solution but increases in heated water • Skin abundance of the acid solution is less sensitive to that of NaI salt solution |
| | 278 K HI | 0.503 | 0.464 | 0.033 | |
| | 298 | 0.523 | 0.439 | 0.038 | |
| | 318 | 0.493 | 0.481 | 0.026 | |
| | 338 | 0.487 | 0.492 | 0.021 | |
| | 358 | 0.438 | 0.546 | 0.015 | |
| | 368 | 0.480 | 0.507 | 0.013 | |

Reprinted with permission from [65]

## *12.4.8 Polarization Depolarization*

### 12.4.8.1 Salt and Base Ionic Polarization

The degree of skin polarization determines the contact angle between the substrate and the solution droplet on it. The temperature, solute type and concentration dependence of the contact angle of NaCl and NaI solutions show in Fig. 12.35 consistently that heating lowers the surface stress by depolarization but salting raises it by polarization, though both heating and salting relax the $\omega_x$ phonon in the

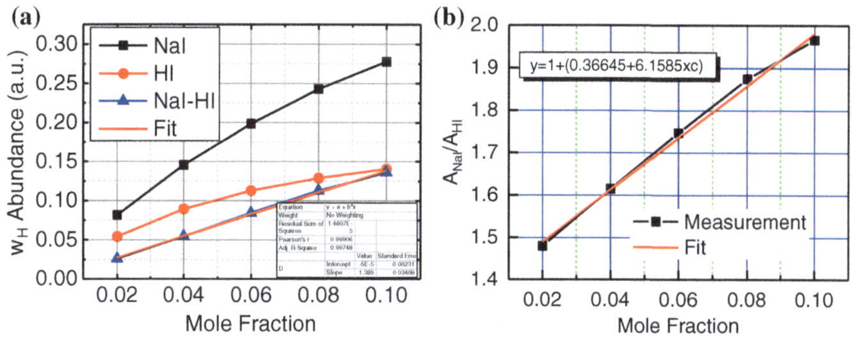


**Fig. 12.32** Concentration dependence of the **a** NaI/HI hydration-shell phonon abundance and **b** their ratio. The integral of the wH abundance gain results from the electrification of both the $Na^+$ and $I^-$ but the $H^+$ that forms the $H_3O^+$ in NI solution

**Table 12.12** Concentration dependence of the hydration-shell phonon $\omega_H$ abundance and its formulation

| Mole ratio (c) | $N_{H2O}/N_{YX}$ | $A_{NaI}$ (a.u) | $A_{HI}$ (a.u) | $A_{NaI}/A_{HI}$ |
|---|---|---|---|---|
| 0.02 | 49 | 0.0817 | 0.0552 | 1.4800 |
| 0.04 | 24 | 0.1456 | 0.0902 | 1.6145 |
| 0.06 | 16 | 0.1985 | 0.1136 | 1.7466 |
| 0.08 | 12 | 0.2429 | 0.1296 | 1.8742 |
| 0.10 | 9 | 0.2778 | 0.1413 | 1.9660 |

same manner. Therefore, bond relaxation and electronic dynamics are critical issues determining the performance of a solution.

Figure 12.36 compares the solute type and concentration dependence of the contact angle for (a) NaX (b) YI, and (c) YOH solutions. The contact angle generally increases with the solute type and concentration in the order of $X^- = I^- > Br^- > Cl^- > F^-$, $Y^+ = Na^+ > K^+ > Rb^+ > Cs^+$. The NaX constructs the surface stress by Y+ and $X^-$ polarization but the YOH and by $Y^+$ the O:↔:O compression. However, due to limited $Y^+$ ionic radius and its hydration shell size, the polarization effect of $Y^+$ can hardly be detected in the phonon spectrometrics.

#### 12.4.8.2 Acid Depolarization

Figure 12.37 shows the concentration dependence of the contact angles between HX solutions and glass substrate in comparison with the temperature dependence of the contact angle between deionized water and coper foil substrate. Acid concentration has the same effect of heating on solution skin depolarization, showing the

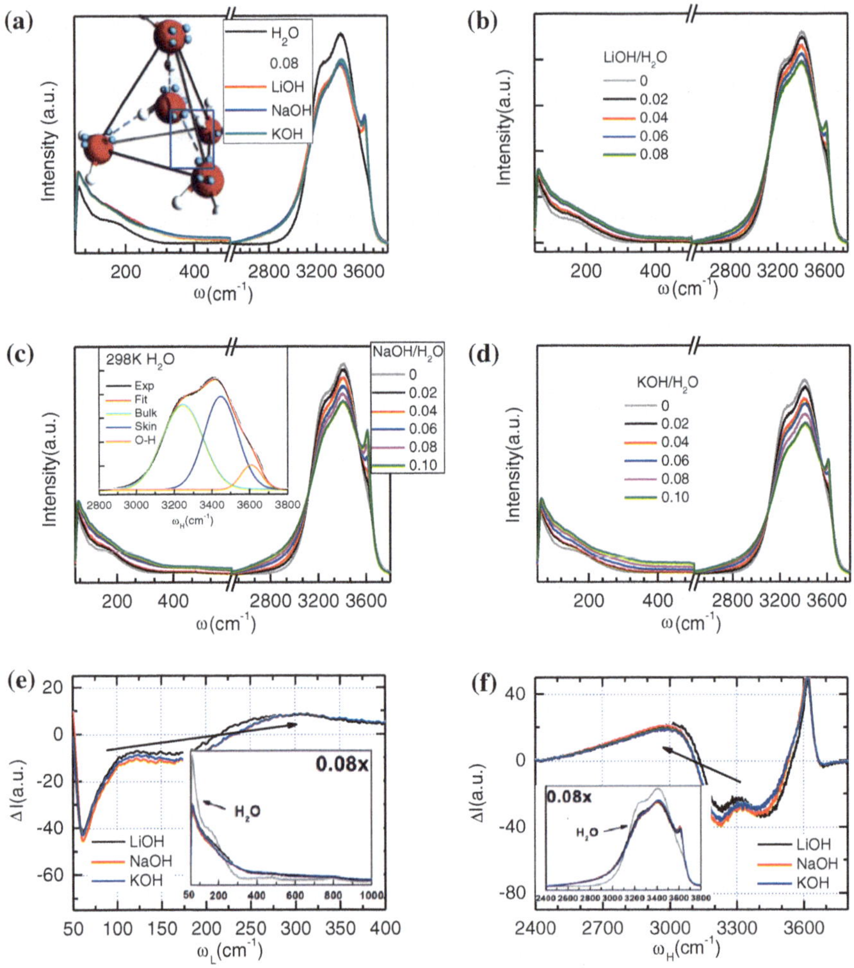


**Fig. 12.33** Full-frequency Raman spectra for (**a**) the 298 K 0.08 mole concentration YOH aqueous solutions and the concentration dependence of the (**b**) LiOH (saturates at 0.08), (**c**) NaOH and (**d**) KOH solutions. A replacement of the central $H_2O$ with a OH- in the unit cell (inset **a**) results the O:↔:O point compressor that offsets the phonon spectra. Inset (**c**) shows the bulk (3200), skin (3450), and H-O radical (3610 $cm^{-1}$) components of the $\omega_H$ peak for the 298 K deionized water

opposite trend of salting effect. The additional H-O bond of the $H_3O^+$ forms the O-H↔H-O, namely, the anti-HB, with one of its four $H_2O$ neighbors. The H↔H terminates the HB network at this particular point, which embritles the HB network as the $H^+$ ion does in metals, which is responsible for the acid solution corrosive and dilutive nature.

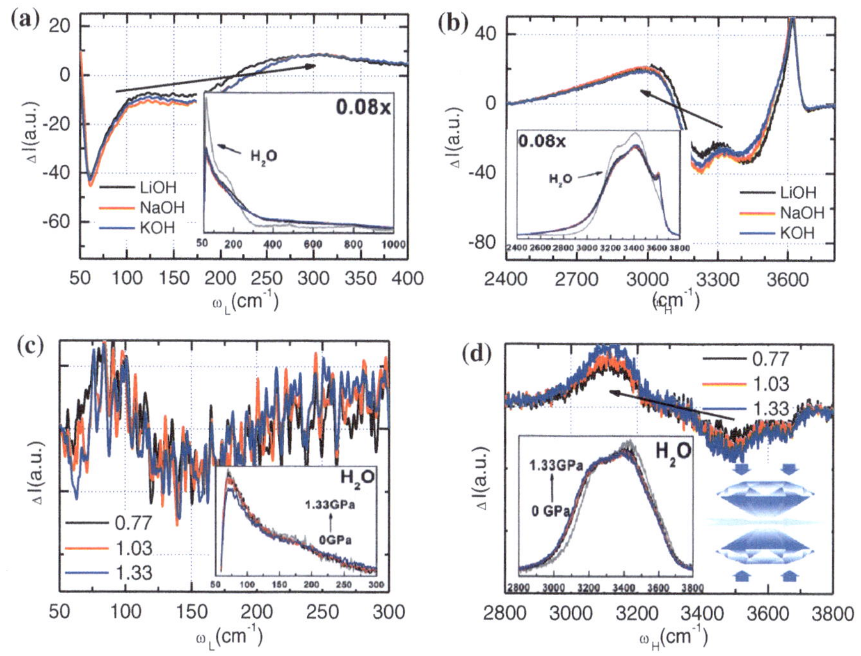


**Fig. 12.34** Comparison of the effects of (**a**, **b**) the internal O:↔:O quantum compression and (**c**, **d**) the external compression on the O:H-O phonon relaxation. Compression shortens and stiffens (**a**, **c**) the O:H nonbond phonon and meanwhile lengthens and softens (**b**, **d**) the H-O bond phonon

### 12.4.9 O:H–O Bond Length and Energy

With the known $E_x$ and $d_x$ for water bulk and skin [5], one can estimate the $d_x$ and $E_x$ for the O:H–O bond in the aqueous solutions by using the relation:

$$\omega_i \propto \frac{\sqrt{E_i/\mu_i}}{d_i}$$

$$E_i = \frac{E_{0i}}{(\omega_0 d_0)_i^2}\omega_i d_i = \begin{cases} 3.906 \times 10^{-7}\omega_H d_H & (\text{eV}) \\ 2.375 \times 10^{-3}\omega_L d_L & (\text{meV}) \end{cases}$$

Results confirms that ionic field electrification raise the $\Delta E_H > 0$ and reduces the $\Delta E_L < 0$ with respect to the bulk neat water. The associated polarization and viscosity are responsible for the solubility of biomolecules like DNA and proteins.

The $\omega_x$ peak positions give rise immediately the segmental length and energy when calibrated against the known bulk and skin features at different temperatures [5]. One can estimate the segmental length and energy with a certain error tolerance

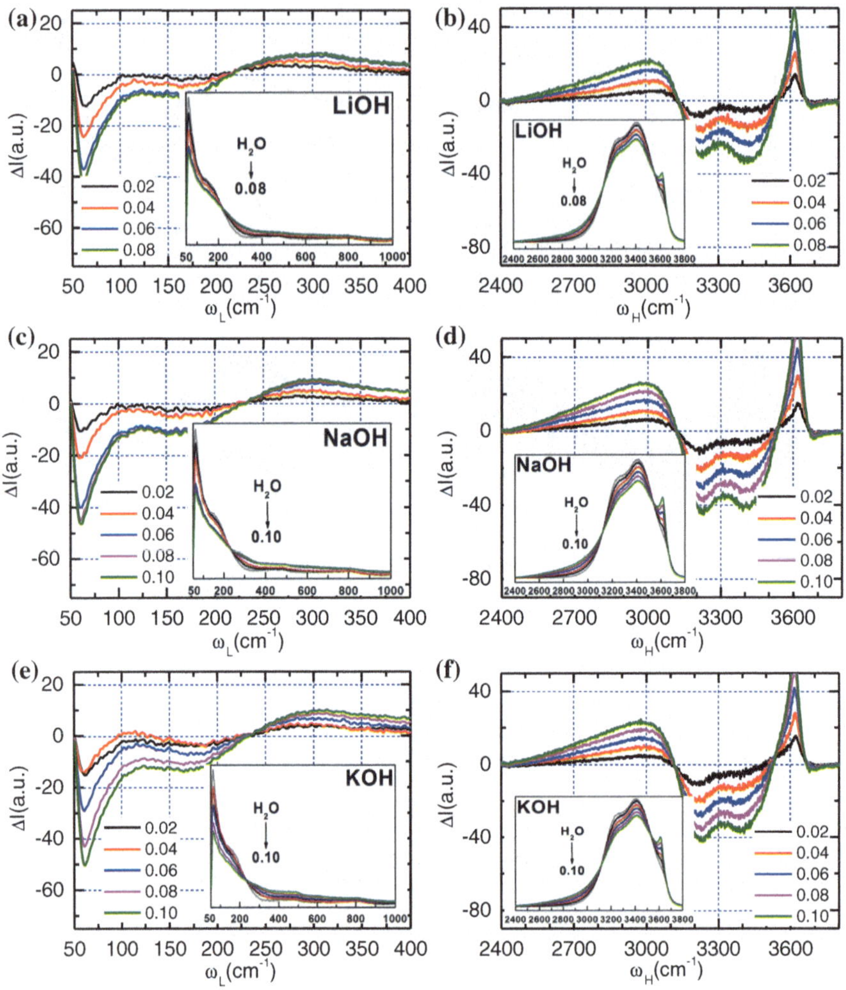


**Fig. 12.35** Concentration dependence of the DPS for (**a,b**) LiOH (saturates at 0.08), (**c,d**) NaOH and (**e,f**) KOH solutions. Insets show the normalized segmental spectra

for other aqueous solutions at different concentrations, with the universal relationships in Chap. 5.

Table 12.13 estimates the H–O and O:H length in NaX solution. The relaxation trend of segmental length, phonon frequency, and cohesive energy is sufficient for interpreting the functional activities of the solutions, including the solubility, surface stress, viscosity, and the critical temperatures for phase transition. For instance, one can elucidate that drinking salt water could raise blood pressure because O:H–O bond elongation and polarization. Salting not only promote snow melting on the

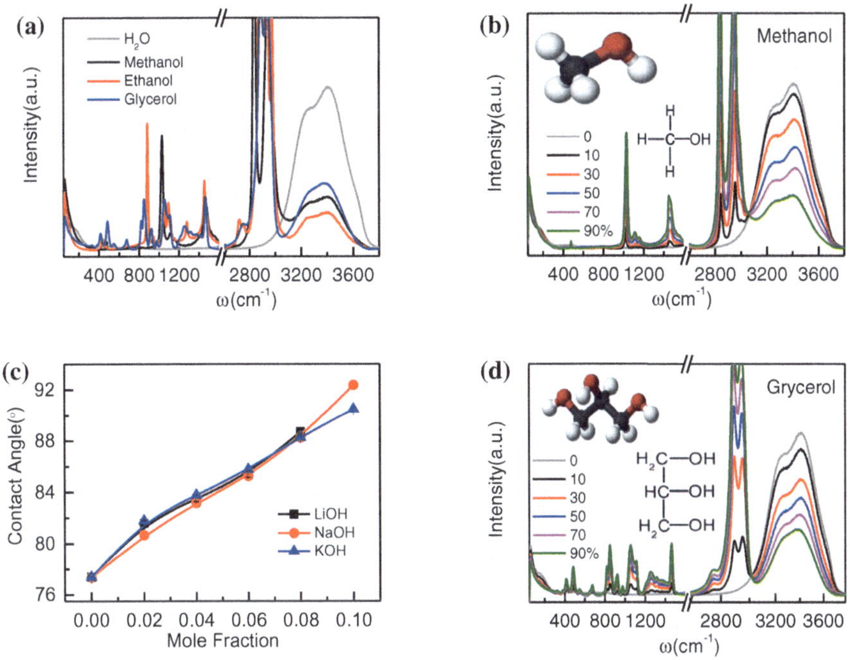


**Fig. 12.36** Full-frequency Raman spectra for the (**a**) type (pure) and concentration dependence of (**b**) Methanol, (**b**) Ethanol, and (**d**) Glycerol solutions. Insets show the respective molecular structures composed of H (white), carbon (*black*) and oxygen (*red*). Sharp peaks centered at 2900 ±100 $cm^{-1}$ correspond to the C-C bond vibration in liquid state

road but also accelerate sol-gel transition, called gelation, of colloid solution. Under these guidelines, one can find means to weaken the H–O bond for hydrogen generation and storage for green chemistry and energy management.

The consistency in the heating and salting effect on the $\omega_x$ relaxation proves that the $d_H$ contraction and $d_L$ elongation takes place in both situations. In the liquid phase, the lower specific heat of the O:H nonbond ensures its dominance of thermal expansion, which weakens the O–O Coulomb repulsion and then the H–O bond recovers to its initially short length without influence of Coulomb repulsion. Salting does the same by ionic field electrification that aligns and stretches water molecular dipoles. The extent of O:H–O bond elongation varies with temperature in the former and with the intensity of the electric field in the latter.

The difference between salting/basing and heating/acding is the polarization. Heating enhances molecular dynamics and depolarizes molecules; aciding breakes the network regulaly. Salting electrification does the opposites to polarize and stiffen the O:H–O bond, resulting in the slight upper shift of the $\omega_L$ and narrow the $\omega_x$ peaks; basing compresses bond network to stiffen the $\omega_L$ but soften the $\omega_H$ but enhances

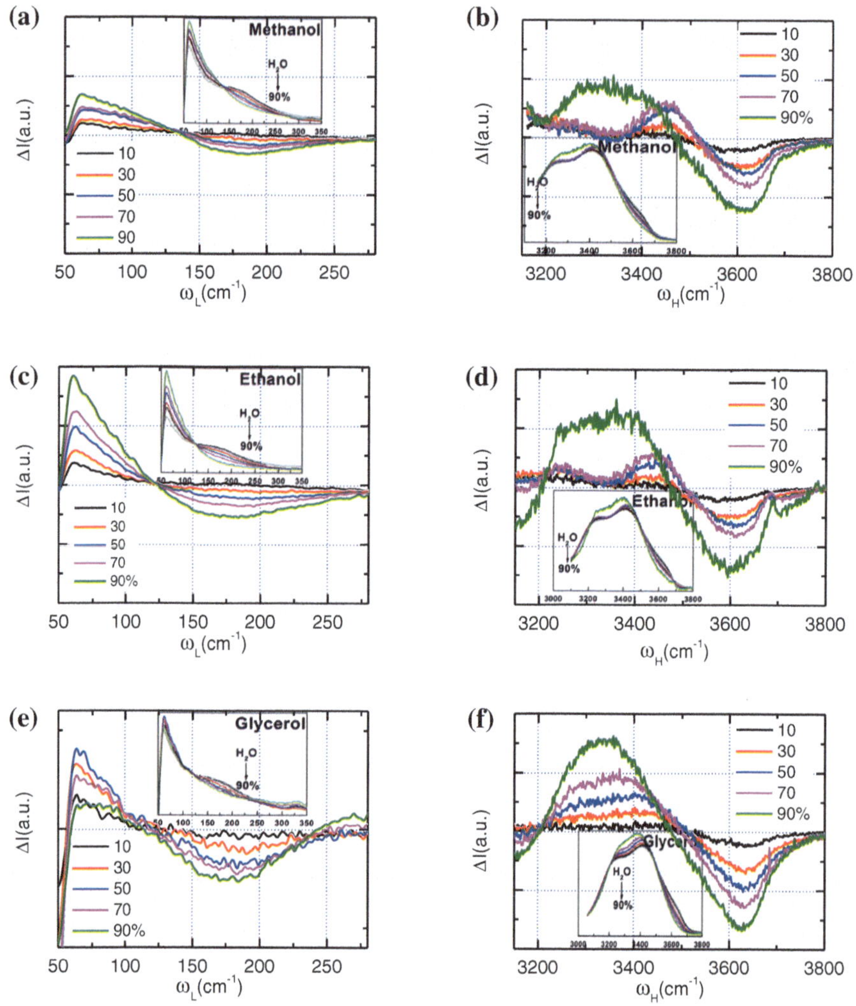


**Fig. 12.37** Quantum compression and polarization of the O:H-O bond by addition of (**a, b**) methanol (**c, d**) alcohol $C_2H_5OH$ (ethanol), and (**e, f**) glycerol ($C_3H_5(OH)_3$) in solutions. O:↔:O compression soften the $\omega_H$ from 3200–3500 to as low as 2500 $cm_{-1}$ and polarization softens the $\omega_L$ to 75 $cm^{-1}$. The annihilation of the 3600 $cm^{-1}$ peak indicates that the dangling H-O bond has found neigbours to for the regular O:H-O bonds

polarization. The polarization enhances the skin supersolidity in determining the hydrophobicity, viscoelasticity, and solubility. The ionic field electrification also disperses the quasisolid phase boundaries to lower the freezing point and raise the melting point of the solutions, being identical to the undercoordination effect.

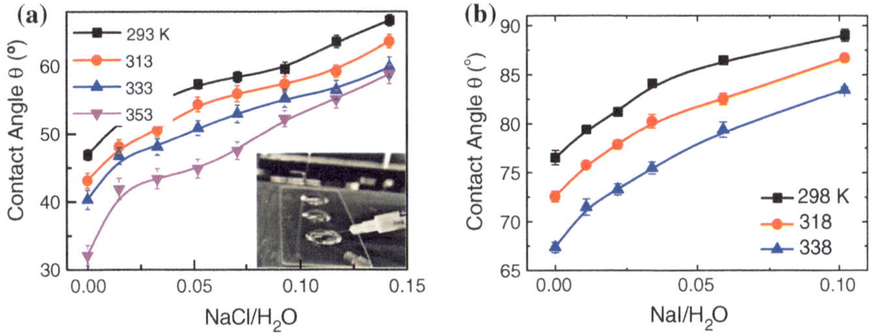


**Fig. 12.38** Concentration and temperature dependence of the contact angle for **a** NaCl and **b** NaI solutions. The *inset* is the contact-angle measurements set up of water on glass substrate. Results revealed that heating weakens while salting enhances the polarization that modulates the skin stress. NaI is more significant than NaCl on the solution skin polarization (Reprinted with permission from [7].)

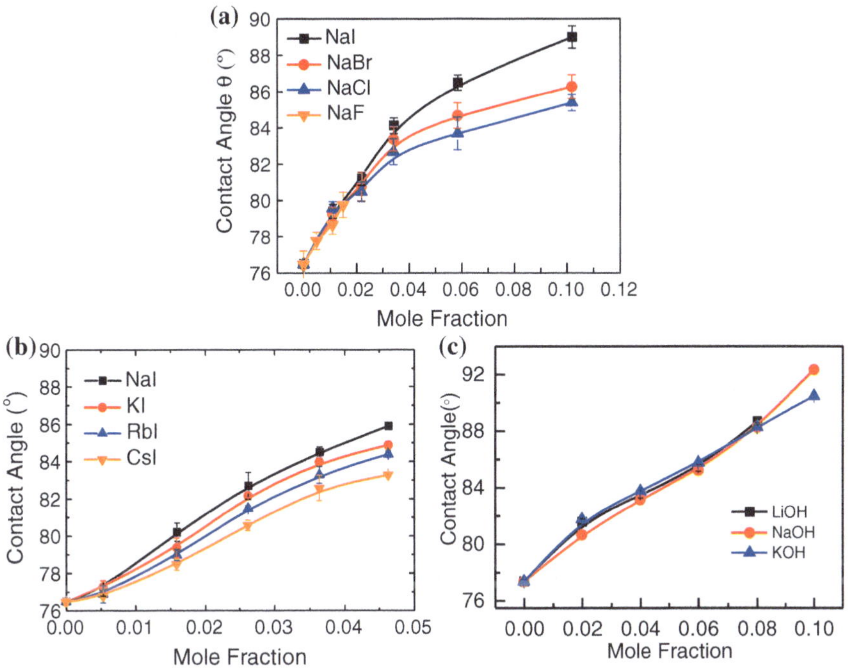


**Fig. 12.39** Concentration dependence of the contact angle for **a** NaX **b** YI, and **c** YOH solution on glass substrate measured at 298 K. The contact angle increases with concentration and with the solute type in the order X = I > Br > Cl > F (saturates at 0.9 M), Y = Na > K > Rb > Cs (saturates at 0.05) (Reprinted with permission from [69].)

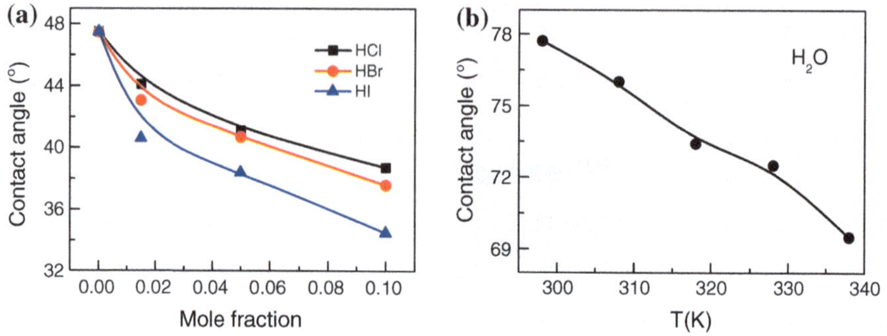


**Fig. 12.40** **a** Molar fraction concentration dependence of the contact angles between HX solutions and glass substrate under the ambient condition and **b** temperature dependence of the contact angle between deionized water and copper foil sub substrate (Reprinted with permission from [66].)

**Table 12.13** O:H–O bond identities in the neat water and in the NaX salt and HX acid solutions

| | $H_2O$ (277 K) | $H_2O$ (338 K) | $H_2O$ (skin) | NaX (shell) | HX (shell) |
|---|---|---|---|---|---|
| $\omega_L$ ($cm^{-1}$)[a] | 200 | Fluctuated | 75 | 65–90 | 120 |
| $\omega_H$ 0.9 M | 3150 | 3340 | 3450 | 3466 | 3459 |
| $\omega_H$ 3 M | – | – | – | 3480 | 3468 |
| $d_L$ (Å) | 1.70 | 1.7170 | 2.0757 | ≥1.95 | – |
| $d_H$ (Å) | 1.00 | 0.9945 | 0.8893 | ≤0.95 | – |
| $E_L$ (meV) | 95 | – | 59 | ≤59 | – |
| $E_H$ (eV) | 3.98 | 4.43 | 4.97 | ≥4.43 | – |
| O–O (Å) | 2.6950 [5] | 2.7115 [5] | 2.9650 [70] | ≥2.965 | – |

Change with respect to bulk water

All examined cases shows the electrification effect on the O:H–O elongation with slight difference of the extent

[a]Subject to polarization and fluctuation. Polarization raises the $\omega_L$ slightly upward and fluctuation widens the peak

## *12.4.10 Acid-Base-Salt Point Switchers*

Why salts represented by NaI, acids represented by HI, and bases represented by NaOH function differently in their solutions and in relaxing the O:H–O bond in terms of bond stiffness, structure order, and phonon abundance in the bulk, hydration shell, and the H–O dangling bond? The following defines the respective solution in terms of molecular anti-HB, super-HB, and polarized-HB proposed in the present volume:

(1) **Acid point breaker.** The $H^+$ released by acid bonds to the $H_2O$ in solution to form the hydronium $H_3O^+$ tetrahedron with one electron lone pair. The $H_3O^+$

interacts with one of its tetrahedrally coordinated $H_2O$ neighbors through the O–H↔H–O anti-HB.

(2) **Base point compressor.** The $OH^-$ dissolved from bases stands alone with three lone pairs, which interacts with one $H_2O$ through the O:↔:O super-HB. The super-HB provides extraordinarily high pressure to compress the rest O: H–O bond, which lowers the H–O bonding energy substantially from 3500 to 2500 $cm^{-1}$. It is clear why dissolving bases in water release burning heat as the H–O bond release energy when stretched by the point compressor.

(3) **Salt point polarizer.** The $Y^+$ and $X^-$ remain alone in the form of CIP or SIP to form each a point polarizer to align, cluster, polarize, and stretch the O:H–O bond in a supersolid hydration-shell manner. The electrification shortens and stiffens the H–O bond, shortens the H–O phonon time, and makes the solution viscoelastic, thermally more stable.

These observations thus verified expectations from the perspectives of O:H–O bond cooperative relaxation, potential modulation, and polarization. The known mechanism of $\omega_x$ thermal relaxation for liquid water [71, 72] and the mechanicm of ionic field electrification explain the salting effect—Hofmeister series—solute type (ionic radius and electronegativity difference) and concentration dependence, which determines the intensity of the local electric field that a molecule is experiencing. These observations may lead to the acid-base-salt redefination according to, respectively, their abilities of creating $H^+$ or hydroniums, hydroxide or addiitonal lone pairs, and point polarisors.

It is clear now why one is often sweating when drink soup with addition of vinegar and why one need to drink water if he eats too much salt in food. Salt solute polarizes water molecules and turns water molecules in the blood from fluid into supersolid state in a hydration shell form—strongly polarized and ice like! The acid hydration shell is much smaller than that of the salt, so acid will dilute the fluid making blood flowing easily. It might be helpful to our health to drink sourer than salty soups. Aspirin tends to dilute our blood preventing from being stroke due to blood thickening but the vessels become fragile if takes the Asprin too often.

It is also clear why the solution releases burning heat when the base is at hydrating. The $\omega_H$ transition from higher to lower frequencies, or softening the H-O bond, ejects energy in the form of Joule heat. The effect of compressor annihilates the effect of $Y^+$ polarizer, raising the lower phonon frequencies to 200 $cm^{-1}$ and above. Compared with the phonon spectra of deionized water under 1.3 GPa pressure, the pressure of the quantum compressor is much greater. The $O^{2\delta-}$ :↔: $O^{2\delta-}$ electrostatic repulsion is four times $(2\delta)^2/\delta^2$ that of the $H^{\delta+}$↔$H^{\delta+}$ and the lone pair interaction is softer than the H-O bond. Hence, the $O^{2\delta-}$:↔:$O^{2\delta-}$ makes the base solution greasy and more repulsive than the $H^{\delta-}$↔$H^{\delta-}$ acid breaker. The net charge quantity $\delta$ is about 0.63±0.2 electron but for convenience, we took the $\delta$ unity for simplifying discussion.

## 12.5 Summary

A combination of Raman spectrometrics and contact angle detection has enabled clarification and quantification of the acid, base, and salt ionic electrification on the molecular-site, solute concentration, solute type, and temperature resolved O:H–O bond relaxation dynamics in terms of H–O bond length and stiffness, order of molecular fluctuation, the abundance of phonon, and the polarizability of the solutions. Consistency in theoretical predictions, numerical derivatives, and experimental observations clarifies the following:

(1) HX-type acids dissolve into the $X^-$ and the $H^+$ and the $H^+$ bonds to a $H_2O$ to form a hydronium $H_3O^+$ with one lone pair. The $H_3O^+$ interacts with one of its four neighbors through the O–H↔H–O anti-HB. The H↔H serves as a point breaker that breaks the HB network, which lowers the skin stress, solution viscosity, and makes the solution corrosive, capable of diluting solutions like blood.
(2) YOH-type bases dissolve into the $Y^+$ and the $OH^-$ hydroxide with three lone pairs. The $OH^-$ interacts with one neighbor through the O:↔:O super-HB that serves as a point compressor, which shortens and stiffens the O:H nonbond and lengthens and softens the H–O bond substantially, easing H–O dissociation in applications.
(3) $Y^+$ and $X^-$ serve each as a point polarizer that shares the same effect of molecular undercoordination on O:H–O bond elongation and polarization, resulting in higher extent of supersolidity of the hydration shells.
(4) Electrification aligns, clusters, and stretches the solvent $H_2O$ molecules, reducing molecular size but enlarging molecular separations. The supersolid hydration shells are more thermally stable and viscoelastic than the supersolid skin of the deionized water, which thickens the human blood, detrimental to hypertension, for example.
(5) The quantum switchers form the keys not only to determining the drug-cell and soltion-prtien interactions but also to functionizing the bleeding, hypertension, stroke, signaling, messeging, and regularing in biological systems. Observations suggest definition of acidebase-salt solutions according the abilities of creating quantum breaker-compressor-polarizer in an aqueous solution.

## References

1. P. Lo Nostro, B.W. Ninham, Hofmeister phenomena: an update on ion specificity in biology. Chem. Rev. **112**(4), 2286–2322 (2012)
2. F. Hofmeister, Concerning regularities in the protein-precipitating effects of salts and the relationship of these effects to the physiological behaviour of salts. Arch. Exp. Pathol. Pharmacol. **24**, 247–260 (1888)
3. E.K. Wilson, Hofmeister still mystifies. Chem. Eng. News Arch. **90**(29), 42–43 (2012)

4. P. Jungwirth, P.S. Cremer, Beyond Hofmeister. Nat. Chem. **6**(4), 261–263 (2014)
5. Y. Huang, X. Zhang, Z. Ma, Y. Zhou, W. Zheng, J. Zhou, C.Q. Sun, Hydrogen-bond relaxation dynamics: resolving mysteries of water ice. Coord. Chem. Rev. **285**, 109–165 (2015)
6. N. Schwierz, D. Horinek, R.R. Netz, Anionic and cationic Hofmeister effects on hydrophobic and hydrophilic surfaces. Langmuir **29**(8), 2602–2614 (2013)
7. X. Zhang, T. Yan, Y. Huang, Z. Ma, X. Liu, B. Zou, C.Q. Sun, Mediating relaxation and polarization of hydrogen-bonds in water by NaCl salting and heating. PCCP **16**(45), 24666–24671 (2014)
8. W.J. Xie, Y.Q. Gao, A simple theory for the Hofmeister series. J. Phys. Chem. Lett. 4247–4252 (2013)
9. G. Imperato, E. Eibler, J. Niedermaier, B. Konig, Low-melting sugar-urea-salt mixtures as solvents for Diels-Alder reactions. Chem. Commun. **9**, 1170–1172 (2005)
10. M.D.A. Saldaña, V.H. Alvarez, A. Haldar, Solubility and physical properties of sugars in pressurized water. J. Chem. Thermodyn. **55**, 115–123 (2012)
11. A. Salis, B.W. Ninham, Models and mechanisms of Hofmeister effects in electrolyte solutions, and colloid and protein systems revisited. Chem. Soc. Rev. **43**(21), 7358–7377 (2014)
12. W.J. Xie, C.W. Liu, L.J. Yang, Y.Q. Gao, On the molecular mechanism of ion specific Hofmeister series. Sci. Chin. Chem. **57**(1), 36–47 (2014)
13. M.T. Record, E. Guinn, L. Pegram, M. Capp, Introductory lecture: interpreting and predicting Hofmeister salt ion and solute effects on biopolymer and model processes using the solute partitioning model. Faraday Discuss. **160**, 9–44 (2013)
14. C.M. Johnson, S. Baldelli, Vibrational sum frequency spectroscopy studies of the influence of solutes and phospholipids at vapor/water interfaces relevant to biological and environmental systems. Chem. Rev. **114**(17), 8416–8446 (2014)
15. M. Randall, C.F. Failey, The activity coefficient of gases in aqueous salt solutions. Chem. Rev. **4**(3), 271–284 (1927)
16. M. Randall, C.F. Failey, The activity coefficient of non-electrolytes in aqueous salt solutions from solubility measurements. The salting-out order of the ions. Chem. Rev. **4**(3), 285–290 (1927)
17. M. Randall, C.F. Failey, The activity coefficient of the undissociated part of weak electrolytes. Chem. Rev. **4**(3), 291–318 (1927)
18. D.F. Parsons, M. Boström, P.L. Nostro, B.W. Ninham, Hofmeister effects: interplay of hydration, nonelectrostatic potentials, and ion size. PCCP **13**(27), 12352–12367 (2011)
19. F. Hofmeister, Zur Lehre von der Wirkung der Salze. Arch. Exp. Pathol. Pharmakol. **25**(1), 1–30 (1888)
20. W.M. Cox, J.H. Wolfenden, *The Viscosity of Strong Electrolytes Measured by a Differential Method*, vol. 145 (1934), pp. 475–488
21. P. Ball, J.E. Hallsworth, Water structure and chaotropicity: their uses, abuses and biological implications. Phys. Chem. Chem. Phys. **17**(13), 8297–8305 (2015)
22. K.D. Collins, M.W. Washabaugh, The Hofmeister effect and the behaviour of water at interfaces. Q. Rev. Biophys. **18**(04), 323–422 (1985)
23. R. Zangi, B. Berne, Aggregation and dispersion of small hydrophobic particles in aqueous electrolyte solutions. J. Phys. Chem. B **110**(45), 22736–22741 (2006)
24. A.W. Omta, M.F. Kropman, S. Woutersen, H.J. Bakker, Negligible effect of ions on the hydrogen-bond structure in liquid water. Science **301**(5631), 347–349 (2003)
25. S. Funkner, G. Niehues, D.A. Schmidt, M. Heyden, G. Schwaab, K.M. Callahan, D.J. Tobias, M. Havenith, Watching the low-frequency motions in aqueous salt solutions: the terahertz vibrational signatures of hydrated ions. J. Am. Chem. Soc. **134**(2), 1030–1035 (2012)
26. K. Tielrooij, N. Garcia-Araez, M. Bonn, H. Bakker, Cooperativity in ion hydration. Science **328**(5981), 1006–1009 (2010)
27. Y. Levin, Polarizable ions at interfaces. Phys. Rev. Lett. **102**(14), 147803 (2009)

28. K.D. Collins, Why continuum electrostatics theories cannot explain biological structure, polyelectrolytes or ionic strength effects in ion–protein interactions. Biophys. Chem. **167**, 43–59 (2012)
29. K.D. Collins, Charge density-dependent strength of hydration and biological structure. Biophys. J. **72**(1), 65–76 (1997)
30. K.D. Collins, Ions from the Hofmeister series and osmolytes: effects on proteins in solution and in the crystallization process. Methods **34**(3), 300–311 (2004)
31. T.T. Duignan, D.F. Parsons, B.W. Ninham, Collins's rule, Hofmeister effects and ionic dispersion interactions. Chem. Phys. Lett. **608**, 55–59 (2014)
32. X. Liu, H. Li, R. Li, D. Xie, J. Ni, L. Wu, Strong non-classical induction forces in ion-surface interactions: General origin of Hofmeister effects. Sci. Rep. **4** (2014)
33. B. Hess, N.F.A. van der Vegt, Cation specific binding with protein surface charges. Proc. Natl. Acad. Sci. **106**(32), 13296–13300 (2009)
34. J.S. Uejio, C.P. Schwartz, A.M. Duffin, W.S. Drisdell, R.C. Cohen, R.J. Saykally, Characterization of selective binding of alkali cations with carboxylate by x-ray absorption spectroscopy of liquid microjets. Proc. Natl. Acad. Sci. **105**(19), 6809–6812 (2008)
35. L. Vrbka, J. Vondrášek, B. Jagoda-Cwiklik, R. Vácha, P. Jungwirth, Quantification and rationalization of the higher affinity of sodium over potassium to protein surfaces. Proc. Natl. Acad. Sci. **103**(42), 15440–15444 (2006)
36. J. Paterová, K.B. Rembert, J. Heyda, Y. Kurra, H.I. Okur, W.R. Liu, C. Hilty, P.S. Cremer, P. Jungwirth, Reversal of the Hofmeister series: specific ion effects on peptides. J. Phys. Chem. B **117**(27), 8150–8158 (2013)
37. J. Heyda, T. Hrobárik, P. Jungwirth, Ion-specific interactions between halides and basic amino acids in water†. J. Phys. Chem. A **113**(10), 1969–1975 (2009)
38. J.D. Smith, R.J. Saykally, P.L. Geissler, The effects of dissolved halide anions on hydrogen bonding in liquid water. J. Amer. Chem. Soc. **129**, 13847–13856 (2007)
39. S. Park, M.D. Fayer, Hydrogen bond dynamics in aqueous NaBr solutions. Proc. Natl. Acad. Sci. U.S.A. **104**(43), 16731–16738 (2007)
40. F. Aliotta, M. Pochylski, R. Ponterio, F. Saija, G. Salvato, C. Vasi, Structure of bulk water from Raman measurements of supercooled pure liquid and LiCl solutions. Phys. Rev. B **86** (13), 134301 (2012)
41. R. Li, Z. Jiang, F. Chen, H. Yang, Y. Guan, Hydrogen bonded structure of water and aqueous solutions of sodium halides: a Raman spectroscopic study. J. Mol. Struct. **707**(1–3), 83–88 (2004)
42. G. Jones, M. Dole, The viscosity of aqueous solutions of strong electrolytes with special reference to barium chloride. J. Am. Chem. Soc. **51**(10), 2950–2964 (1929)
43. Z.S. Nickolov, J. Miller, Water structure in aqueous solutions of alkali halide salts: FTIR spectroscopy of the OD stretching band. J. Colloid Interface Sci. **287**(2), 572–580 (2005)
44. R. Mancinelli, A. Botti, F. Bruni, M. Ricci, A. Soper, Hydration of sodium, potassium, and chloride ions in solution and the concept of structure maker/breaker. J. Phys. Chem. B **111** (48), 13570–13577 (2007)
45. X.P. Li, K. Huang, J.Y. Lin, Y.Z. Xu, H.Z. Liu, Hofmeister ion series and its mechanism of action on affecting the behavior of macromolecular solutes in aqueous solution. Prog. Chem. **26**(8), 1285–1291 (2014)
46. G.N. Lewis, Acids and bases. J. Franklin Inst. **226**(3), 293–313 (1938)
47. W.B. Jensen, *The Lewis Acid-Base Concepts: An Overview* (Wiley, 1980)
48. L.J. Bartolotti, D. Rai, A.D. Kulkarni, S.P. Gejji, R.K. Pathak, Water clusters $(H_2O)n$ [n = 9–20] in external electric fields: exotic OH stretching frequencies near breakdown. Comput. Theor. Chem. **1044**, 66–73 (2014)
49. L. Zhao, K. Ma, Z. Yang, Changes of water hydrogen bond network with different externalities. Int. J. Mol. Sci. **16**(4), 8454–8489 (2015)
50. C.Q. Sun, *Relaxation of the Chemical Bond*. Springer Series in Chemical Physics, vol. 108 (Springer, Heidelberg, 2014), 807 pp

51. X. Zhang, Y. Huang, Z. Ma, Y. Zhou, W. Zheng, J. Zhou, C.Q. Sun, A common supersolid skin covering both water and ice. PCCP **16**(42), 22987–22994 (2014)
52. Y. Huang, X. Zhang, Z. Ma, Y. Zhou, G. Zhou, C.Q. Sun, Hydrogen-bond asymmetric local potentials in compressed ice. J. Phys. Chem. B **117**(43), 13639–13645 (2013)
53. Q. Sun, Raman spectroscopic study of the effects of dissolved NaCl on water structure. Vib. Spectrosc. **62**, 110–114 (2012)
54. S. Park, M.B. Ji, K.J. Gaffney, Ligand exchange dynamics in aqueous solution studied with 2DIR spectroscopy. J. Phys. Chem. B **114**(19), 6693–6702 (2010)
55. S. Park, M. Odelius, K.J. Gaffney, Ultrafast dynamics of hydrogen bond exchange in aqueous ionic solutions. J. Phys. Chem. B **113**(22), 7825–7835 (2009)
56. K.J. Gaffney, M. Ji, M. Odelius, S. Park, Z. Sun, H-bond switching and ligand exchange dynamics in aqueous ionic solution. Chem. Phys. Lett. **504**(1–3), 1–6 (2011)
57. Y. Zhou, Y. Huang, Y. Gong, C.Q. Sun, Skin preferential occupancy of the $I^-$ anion in NaI-water solution. Communicated (2016)
58. Y. Zhou, Y. Huang, Y. Gong, C.Q. Sun, *Molecular site resolved O:H-O bond thermal relaxation in liquid water.* Communicated (2015)
59. T.F. Kahan, J.P. Reid, D.J. Donaldson, Spectroscopic probes of the quasi-liquid layer on ice. J. Phys. Chem. A **111**(43), 11006–11012 (2007)
60. M. Baumgartner, R.J. Bakker, Raman spectroscopy of pure H2O and NaCl-H2O containing synthetic fluid inclusions in quartz—a study of polarization effects. Mineral. Petrol. **95**(1–2), 1–15 (2008)
61. X. Zhang, C.Q. Sun, *Coulomb mediation of hydrogen-bond short-range interactions by programmable heating and salting,* http://arxiv.org/abs/1310.0893 (2013)
62. Y. Zhou, Y. Huang, Y. Gong, C.Q. Sun, O:H-O bond electrification in the aqueous YI solutrean (Y = Na, K, Rb, Cs). Communicated (2016)
63. D. Monroe, Focus: a surface attraction. Phys. Rev. Focus **24**, 25 (2009)
64. C.A. Silvera Batista, R.G. Larson, N.A. Kotov, Nonadditivity of nanoparticle interactions. Science **350**(6257) (2015)
65. Y. Zhou, C. Sun, Ultrahigh thermal stability of HI and NaI hydration shells. Communicated (2015)
66. Y. Zhou, C.Q. Sun, Molecular site, temperature, solute concentration, and solute type resolved O:H-O bond relaxation in HX acid solutions (X = I, Br, Cl). Communicated (2016)
67. X. Zhang, P. Sun, Y. Huang, T. Yan, Z. Ma, X. Liu, B. Zou, J. Zhou, W. Zheng, C.Q. Sun, Water's phase diagram: from the notion of thermodynamics to hydrogen-bond cooperativity. Prog. Solid State Chem. **43**, 71–81 (2015)
68. C.Q. Sun, X. Zhang, W.T. Zheng, Hidden force opposing ice compression. Chem. Sci. **3**, 1455–1460 (2012)
69. Y. Zhou, Y. Huang, Y. Gong, C.Q. Sun, O:H-O bond electrification by NaZ aqueous solutions ((Z = $HSO_4$, $NO_3$, SCN, $NO_3$). Communicated (2016)
70. K.R. Wilson, R.D. Schaller, D.T. Co, R.J. Saykally, B.S. Rude, T. Catalano, J.D. Bozek, Surface relaxation in liquid water and methanol studied by x-ray absorption spectroscopy. J. Chem. Phys. **117**(16), 7738–7744 (2002)
71. C.Q. Sun, X. Zhang, X. Fu, W. Zheng, J.-L. Kuo, Y. Zhou, Z. Shen, J. Zhou, Density and phonon-stiffness anomalies of water and ice in the full temperature range. J. Phys. Chem. Lett. **4**, 3238–3244 (2013)
72. X. Zhang, Y. Huang, Z. Ma, Y. Zhou, J. Zhou, W. Zheng, Q. Jiang, C.Q. Sun, Hydrogen-bond memory and water-skin supersolidity resolving the Mpemba paradox. PCCP **16**(42), 22995–23002 (2014)

# Chapter 13
# Hydration Shells and Water Skin

- *Solutes form hydration shells with an association of H–O bond contraction and subjective nonbonding electron polarization.*
- *HX acid solutions exhibit quantum fragilation dominance, which depresses the structural order and skin supersolidity.*
- *Salt solutions demonstrate polarization dominance, which raises the structural order and skin supersolidity.*
- *H–O phonon frequency shift correlates the phonon lifetime, structure order, skin stress, skin solubility, solute viscosity, and thermal stability proportionally.*

**Abstract** A solute forms a hydration shell by clustering water molecular dipoles surrounding it, which elongates the O:H–O bond in the shell and stiffens its H–O phonon and softens the O:H–O nonbond phonon by different extents. Polarization dominance of salt solutions raises the H–O phonon lifetime, molecular structural order, skin stress, solution viscosity, and thermal stability. Quantum fragmentsation of acid solutions weakens molecular structural order, skin stress, and the reflectivity of Raman photon and the transmittance of IR photons. The difference in electronegativity, electronic structure, and ionic size between $H^+$ and other alkali metals could be origin.

## 13.1 Challenge: Skin Stress Versus Molecular Lifetime

Figures 12.35, 12.36 and 12.37 compared the concentration trends of the contact angle and skin stress of YX salt, YOH base, and HX acid aqueous solutions. The contact angle and the stress of YX salt solutions are proportionally correlated following the Hofmeister series order. However, mechanism and correlation for the following issues remain challenging:

(1) How are the contact angle and stress correlated?
(2) What is the nature of the solute hydration shells?
(3) How are the H–O phonon frequency, lifetime, structure order, and the viscosity correlated?
(4) Why does $H^+$ cation differ from the $Na^+$ in determining the skin stress and structure order?


C.Q. Sun and Y. Sun, *The Attribute of Water*, Springer Series in Chemical Physics 113, DOI 10.1007/978-981-10-0180-2_13

## 13.2 Clarification: Acid Fragilation and Salt Polarization

The segmental $\omega_X$ DPS for the HX acid and NaX salt solutions, shown in Figs. 12.14 and 12.22, clarifies the following:

(1) Both acid and salt shorten the H–O bond and stiffen its phonon and elongate the O:H bond and soften the $\omega_L$ by different extents due to $X^-$ anion polarization.
(2) The H–O phonon lifetime is proportional to its frequency shift with a coefficient that is higher for the skin than it is for the bulk [3].
(3) The long-lived H–O phonon indicates higher structure order, higher solution viscosity, higher thermal stability, and lower molecular self-diffusivity.
(4) All these quantities are correlated to the O:H–O bond relaxation and nonbonding electron polarization and $H^+$ induced depolarization.

## 13.3 H–O Phonon Frequency Dependent Lifetime

The ultrafast IR absorption spectroscopy probes information about the phonon frequency shift and its relaxation time. A blueshift of the high-frequency phonon will be accompanied with elongation of the relaxation time [3], which enhances the viscosity of the solution. Figures 13.1 and 13.2 show the ultrafast IR spectra for NaBr and $NaClO_4$ solutions using HOD + $H_2O$ as solvent [4, 5]. Indeed, increasing the solute concentration not only stiffens the O–D phonon frequency but also lengthens the relaxation time because of the slowdown of the water molecules in the solution by polarization.

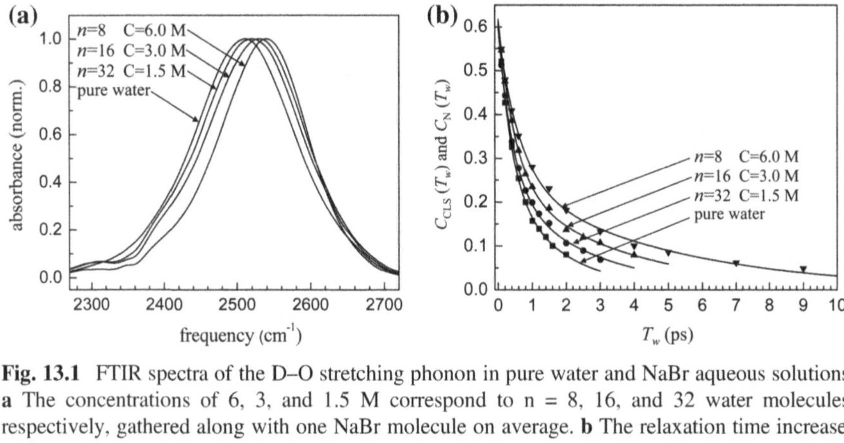


**Fig. 13.1** FTIR spectra of the D–O stretching phonon in pure water and NaBr aqueous solutions. **a** The concentrations of 6, 3, and 1.5 M correspond to n = 8, 16, and 32 water molecules, respectively, gathered along with one NaBr molecule on average. **b** The relaxation time increases with NaBr concentration by slowing the dynamics of water molecules (Reprinted with permission from [4].)

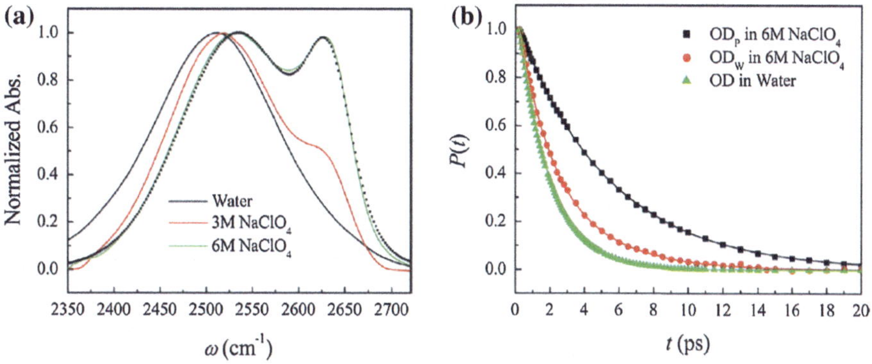


**Fig. 13.2** **a** FTIR spectra of the D–O stretch phonon in 3 and 6 M aqueous $NaClO_4$ solutions and **b** vibrational population decays of the D–O in two ensembles ($\omega_w = 2534$ and $\omega_P = 2633$ $cm^{-1}$ peaks) in the 6 M $NaClO_4$ aqueous solution. Stiffer phonons have longer relaxation time (Reprinted with permission from [5].)

## 13.4 Lifetime Versus Diffusivity

MD calculations [6] of the microstructure and lifetime of 1-butyl-3-methylimidazolium iodide/water solution ion associations and ion transport dynamics from bulk phase ionic liquid to the dilute solution revealed that, as shown in Fig. 13.3, the diffusion coefficient of ions and water molecules drops with the concentration of the solute, which is accompanied with longer solute rotational relaxation time. Solvent water molecular electrification raises the viscosity and depresses the mobility and diffusivity of both solute ions and $H_2O$ solvent molecules.

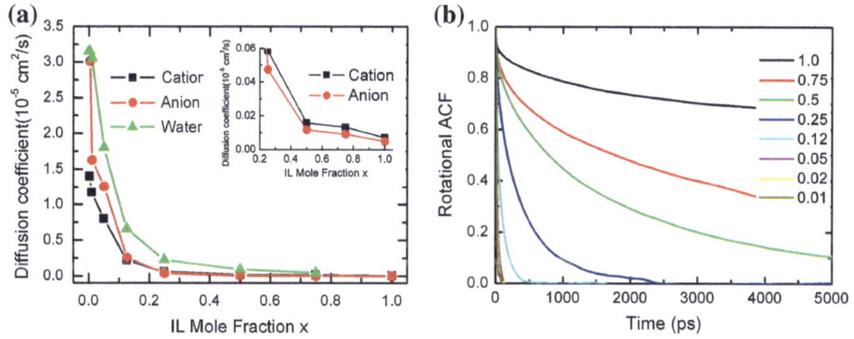


**Fig. 13.3** Imidazolium solute concentration dependence of **a** the diffusion coefficients of cations, anions and water molecules and **b** the solute rotational autocorrelation function as a function of time. The *inset* in (**a**) enlarges the diffusion coefficients of cation and anion in high fractions (Reprinted with permission from [6].)

The dynamics of O:H–O bonded network is pretty fast in pure water. Specifically, the spectral diffusion time for HDO/$D_2O$ system is less than 200 fs [7] and the vibrational lifetime in pure $H_2O$ is 200 fs [8]. The H–O phonon lifetime is frequency and molecular site sensitive; of the same frequency, the skin molecules are long-lived up to 700 fs [3] because of the skin supersolidity [9].

As the segmental phonon frequency depends only on its stiffness without discriminating the source of stimulus such as salting, heating, compressing or undercoordinating, one can define the ionic electrification effect on the dynamics of hydration shells by comparing the phonon relaxation dynamics with those of pure water in terms of the bond stiffness (frequency), abundance (peak area), and molecular dynamics (FWHM) of the characteristic peaks. The $\omega_H$ blueshift is associated with the elongation of its lifetime which indicates the enhancement of the viscosity of the solution because of the slowing of $H_2O$ molecular transitional and rotational motion. Higher order of molecular fluctuation weakens the intensity of Raman refection or IR transmission. The DPS shape and shift are consistent with the lifetime measurements, which fingerprint interaction fluctuation and electron polarization of $H_2O$ molecules in an aqueous solution.

## 13.5 Polarization Versus Skin Stress

The variation in phonon frequency, skin stress, and viscosity of aqueous solutions with molecular site, solute type and concentration, and temperature of operation provide direct and useful information on the functionality of solutions when interacting with biological molecules. One can thus conduct the desired biochemical process without needing interference and in providing physicochemical support. It infers that the particular concentration would be effective in tissue or cell culture process for biological research in factorization of the ionic or the molecular forces. Therefore, in a process of such factorization, alignment, relaxation and polarization of the O:H–O bond due to the fields of salt or acid ions or water molecules vis-a-vis cohesive and dispersive forces are readily characterized.

SFG, Raman, IR spectrometrics and skin stress investigations [2] revealed that the $\omega_H$ = 3450 $cm^{-1}$ peak intensity increases with the fraction of the aqueous hydrogen–halide (i.e., strong acid) and salt solutions at 298 K, as typically shown in Figs. 12.12 and 12.24. The spectra display no apparent frequency shift as it approaches the saturation $\omega_H$ value that corresponds to $d_H$ of 0.95 Å and $d_L$ of 1.95 Å. The phonon spectroscopy probes only the effect of the ionic field on the stiffness relaxation of the O:H–O bond without discriminating the nature of perturbation, which is the advantage of the spectroscopy that follows the principle of Fourier transition.

However, acids reduce but salts increase the skin stress and the contact angle. The extent of contact angle and skin stress changes with the concentration of the solutes, as shown in Figs. 12.35, 12.36 and 12.37. NaX salting polarization enhances the skin stress but HX acid depolarization/fragilation reduces the skin stress [10]. The lowered skin stress or structural order weakens the Raman phonon reflectivity and IR phonon absorptivity of the solution. The extent of electron polarization discriminates the stress from tensile (<0) to compressive (>0) though ions of both salts and acids form electric fields stretching the O:H–O bond.

## 13.6 Skin Stress Versus Viscoelasticity

Figures 13.4 and 13.5 compare the effect of Hofmeister series on the skin stress and viscosity of the alkali metal phosphate salt solutions as a function of solute concentration and temperature. Both skin stress and viscosity increase with solute concentrations but decrease with measurement temperature. For example, the skin stress of 0.1 M $K_3PO_4$ solution decreases from 72.82 to 72.11 mN/m when heated from 298 to 303 K; the stress of 0.3 M $K_3PO_4$ solution drops from 72.54 to 72.20 mN/m when heated from 298 K by 5 K. These observations agree with trends of salting and heating on the phonon frequency and contact angle relaxation [10]. These quantities are closely correlated.

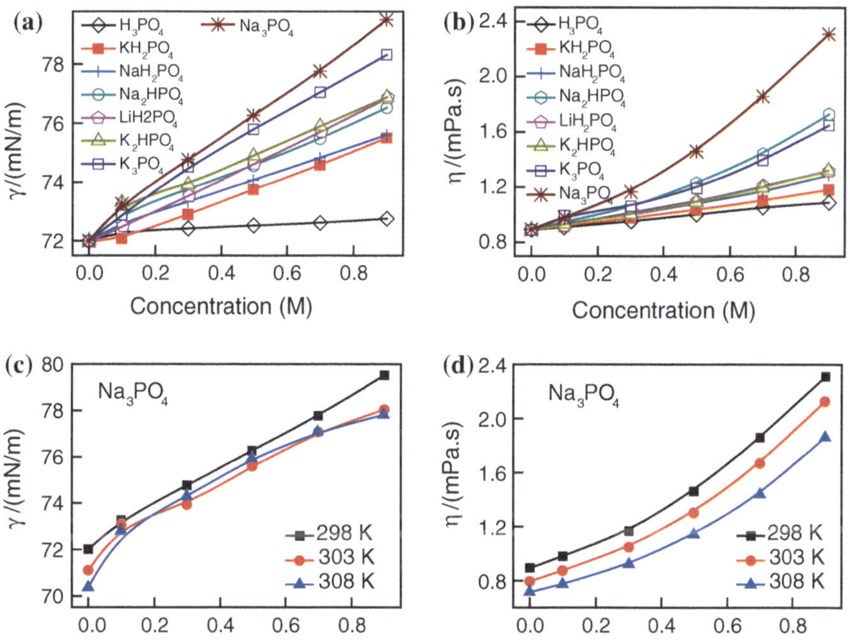


**Fig. 13.4** Solute type and concentration dependence of the **a** skin stress γ and **b** viscosity η of alkali metal phosphate salts solutions and the temperature dependence for $Na_3PO_4$ (**c**, **d**) (Reprinted with permission from [11].)

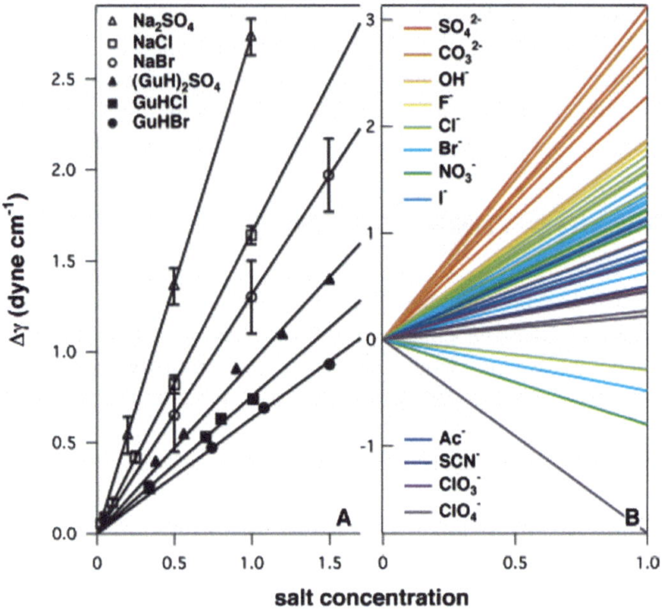


**Fig. 13.5** Skin stress for selected Hofmeister salts. *Open symbols* represent sodium salts, while the corresponding *filled symbol* denotes the guanidinium salt of the same anion [12]. The symbols for the various anions are as follows: sulfate (*triangle*), chloride (*square*), and bromide (*circle*) (Reprinted with permission from [13].)

## 13.7 Insight Extension

### *13.7.1 Water on Mars*

We may extend these observations to the understanding of supersolid saline water on Mars [14]. Salt solutes make their hydration shells and the solution skin to be supersolid that is viscous, lower freezing temperature and higher melting point. The electrification stretches and polarizes the O:H–O bond with elevated $\omega_H$ and depressed $\omega_L$ phonon frequencies that not only disperse the quasisolid phase boundary but also create the supersolidity of the skin and the hydration shells. For instance, sodium perchlorate can lower the freezing point of water by up to 40 K, whereas magnesium perchlorate and magnesium chlorate can depress the freezing point even more, by up to 70 K.

### *13.7.2 Hydration and Hypertension*

Conversely, sweet solutions are obtained upon heating mixtures of carbohydrates, urea and inorganic salts to moderate temperatures, to give new chiral media for organic reactions [15]. The solubility of sugar increases with temperature but drops with pressure [16]. Salt assists the solid/quasisolid transition of snow and ice by ionic electrification that softens the O:H nonbond and stiffens the H–O bond. Therefore, sugars and salts share some common mechanism for anti-icing and de-icing.

Effect of heating, sugaring, and salting on the O:H–O bond relaxation and the skin stress and $\omega_x$ shift enhancement may explain why hot water is a better cleaning a detergent than cold water. Soaps, sugars and detergents may functionalize in a similar manner to heating and salting in hydration cleaning. Thermal softening of the O:H bond enhances water molecular fluctuation and diffusivity, which reduces the skin stress and hence make hot water an active and better 'wetting agent' that can penetrate into pores and fissures rather than bridging them by skin stress of the viscos liquid. Soaps and detergents may further weaken the O:H nonbond and help the hydration cleaning process. Replacing $O^{2-}$ with ions of salts, acids or sugars, together with heating, weaken the Coulomb repulsion between charged ions. According to the present understanding, reduced repulsion weakens the O:H bond and heating depolarization weakens it further.

Salt, sugar, and heating have the same trend of O:H–O bond relaxation but different in polarization and depolarization. If you are suffering from hypertension, medical Doctor will suggest you minimize the intake of salt or sugar. However, one hardly explain why it is so from molecular point of view. It is now clear that electrification by salting and sugar intaking will reduces the size but increases the separation between water molecules by polarization in the blood cells, which raises the viscosity of the blood and consume more power for the heart to pump the blood flowing in vessels.

Likewise, salted and sugared water increase the skin stress and viscosity of the hydration shells by solute ionic electrification –O:H–O bond elongation and polarization. Therefore, salted soils, being harmful to plant growth, or preventing plants from completely absorbing healthy water for fertilizing.

## 13.8 Summary

A combination of Raman spectrometrics, contact angle detection, molecular relaxation life time, skin stress and viscosity measurements, has enabled clarification and quantification of the ionic electrification on the molecular-site-resolved the O:H–O bond segmental length and stiffness, order of molecular fluctuation and the abundance of phonon in the supersolid hydration shells in aqueous solutions. Consistency in theoretical predictions, numerical derivatives, and experimental observations clarifies the following:

(1) Electrification of the O:H–O bond raises the $\omega_H$ frequency, molecular life time, diffusivity, skin stress, solubility, and viscosity of the aqueous solutions collaboratively.

(2) Anions may or not occupy preferentially at the solution-air interface, which depends on the type of the solute. The preference of anion interface occupancy reduces the number but further shortens the nearby H–O radicals.
(3) O:H–O bond electrification is molecular site, solute type and concentration dependent. The supersolid hydration shells are very critical to the Hofmeister effect.
(4) The polarization dominance of NaX represented salts and the depolarization/fragilation dominance of HX acids discriminates the structural order and viscosity of the hydration shells and skins of their solutions.

## References


1. Y. Zhou, C.Q. Sun, Molecular site, temperature, solute concentration, and solute type resolved O:H-O bond relaxation in HX acid solutions (X = I, Br, Cl). Communicated (2016)
2. L.M. Levering, M.R. Sierra-Hernández, H.C. Allen, Observation of hydronium ions at the air–aqueous acid interface: vibrational spectroscopic studies of aqueous HCl, HBr, and HI. J. Phys. Chem. C **111**(25), 8814–8826 (2007)
3. S.T. van der Post, C.S. Hsieh, M. Okuno, Y. Nagata, H.J. Bakker, M. Bonn, J. Hunger, Strong frequency dependence of vibrational relaxation in bulk and surface water reveals sub-picosecond structural heterogeneity. Nat. Commun. **6**, 8384 (2015)
4. S. Park, M.D. Fayer, Hydrogen bond dynamics in aqueous NaBr solutions. Proc. Natl. Acad. Sci. U.S.A. **104**(43), 16731–16738 (2007)
5. S. Park, M. Odelius, K.J. Gaffney, Ultrafast dynamics of hydrogen bond exchange in aqueous ionic solutions. J. Phys. Chem. B **113**(22), 7825–7835 (2009)
6. M. Sha, H. Dong, F. Luo, Z. Tang, G. Zhu, G. Wu, Dilute or concentrated electrolyte solutions? Insight from ionic liquid/water electrolytes. J. Phys. Chem. Lett. 3713–3720 (2015)
7. C. Fecko, J. Eaves, J. Loparo, A. Tokmakoff, P. Geissler, Ultrafast hydrogen-bond dynamics in the infrared spectroscopy of water. Science **301**(5640), 1698–1702 (2003)
8. A. Lock, S. Woutersen, H. Bakker, Ultrafast energy equilibration in hydrogen-bonded liquids. J. Phys. Chem. A **105**(8), 1238–1243 (2001)
9. Y. Huang, X. Zhang, Z. Ma, Y. Zhou, W. Zheng, J. Zhou, C.Q. Sun, Hydrogen-bond relaxation dynamics: resolving mysteries of water ice. Coord. Chem. Rev. **285**, 109–165 (2015)
10. X. Zhang, T. Yan, Y. Huang, Z. Ma, X. Liu, B. Zou, C.Q. Sun, Mediating relaxation and polarization of hydrogen-bonds in water by NaCl salting and heating. PCCP. **16**(45), 24666–24671 (2014)
11. R.K. Ameta, M. Singh, Surface tension, viscosity, apparent molal volume, activation viscous flow energy and entropic changes of water + alkali metal phosphates at T = (298.15, 303.15, 308.15) K. J. Mol. Liq. **203**, 29–38 (2015)
12. A. Kumar, Aqueous guanidinium salts: Part II. Isopiestic osmotic coefficients of guanidinium sulphate and viscosity and surface tension of guanidinium chloride, bromide, acetate, perchlorate and sulphate solutions at 298.15 K. Fluid Phase Equilib. **180**(1–2), 195–204 (2001)
13. L.M. Pegram, M.T. Record, Hofmeister salt effects on surface tension arise from partitioning of anions and cations between bulk water and the air–water interface. J. Phys. Chem. B **111** (19), 5411–5417 (2007)

14. L. Ojha, M.B. Wilhelm, S.L. Murchie, A.S. McEwen, J.J. Wray, J. Hanley, M. Masse, M. Chojnacki, Spectral evidence for hydrated salts in recurring slope lineae on Mars. Nat. Geosci. (2015). doi:10.1038/ngeo2546
15. G. Imperato, E. Eibler, J. Niedermaier, B. Konig, Low-melting sugar-urea-salt mixtures as solvents for Diels-Alder reactions. Chem. Commun. (9), 1170–1172 (2005)
16. M.D.A. Saldaña, V.H. Alvarez, A. Haldar, Solubility and physical properties of sugars in pressurized water. J. Chem. Thermodyn. **55**, 115–123 (2012)

# Chapter 14
# Aqueous Solution Phase Transition

- *Solvent O:H–O phonon relaxation by electrification mediates the $T_C$ for phase transition by dispersing the quasisolid phase boundaries.*
- *The $\Delta E_L$ loss depresses the $T_N$ for ice/quasisolid transition; the $\Delta E_H$ gain elevates the $T_m$ for liquid/quasisolid transition.*
- *Liquid/VI, VI/VII, and VII/XI phase transition at constant $T_C$ requires excessive $\Delta P_C$ to recover the electrification-deformed H–O bond; mechanical impulsion raises but ionic electrification depresses the freezing temperature of the quasisolid (supercooled) water by modifying the O:H energy.*
- *Colloidal gelation at constant $T_C$ and $P_C$ takes time that follows the Hofmeister series but the concentration dependence is less conclusive because of the involvement of other kinds of ions.*

**Abstract** Solute ionic electrification of the O:H–O bond modulates significantly the critical pressures, temperatures, and gelation times for transiting aqueous solution into solid by dispersing the boundaries of the quasisolid phase. High-pressure in situ Raman spectrometrics revealed that transiting NaX solutions into ice VI and then into ice VII phase requires higher excessive pressures at 298 K temperature. The $\Delta P_C$ varies in the order of Hofmeister series: X = I > Br > Cl > F ~ 0. Meanwhile, salting st iffens the $\omega_H$ and elongates the $d_{OO}$ throughout the course of compressure when transiting phase VII to phase X at even higher pressure. Recovering the electrification-shortened H–O bond needs excessive energy for the same sequence of phase transitions. Concentration dependence of the NaI solution indicates a different mechanism from that of solution type but it is similar to heating on the Liquid-VI-VII phase transition dynamics.

## 14.1 Challenge: Salt Solution Phase Transition

Ions of salts or other impurities can mediate the critical temperatures, critical pressures, and gelation times for phase transition of their solutions, with unclear mechanisms:

(1) Salting promotes snow ice melting and shortens the time for solution gelation (solidification), see Fig. 14.1.
(2) The ambient water forms ice VI under ~1 GPa pressure and then turns into ice VI and then VII at elevated pressure ($P_C$). Salting raises the $P_C$ by an amount that varies with solution type and concentration following the Hofmeister series order [1].



C.Q. Sun and Y. Sun, *The Attribute of Water*, Springer Series in Chemical Physics 113, DOI 10.1007/978-981-10-0180-2_14

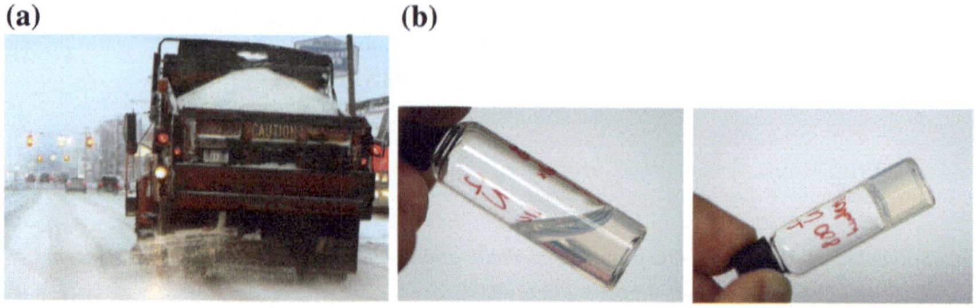

**Fig. 14.1** **a** Salt anti-icing. If live in a place that gets snow, you will have noticed trucks spreading salt over the roads in order to prevent ice slippery by promoting ice melting (Avon Barksdale, Feb 2014). **b** Transforming the macroscopic colloidal fluid initially flows (**b**, *left*) into gel solid (**b**, *right*) takes gelation time that varies with the type and concentration of the solute (Reprinted with permission from [4].)

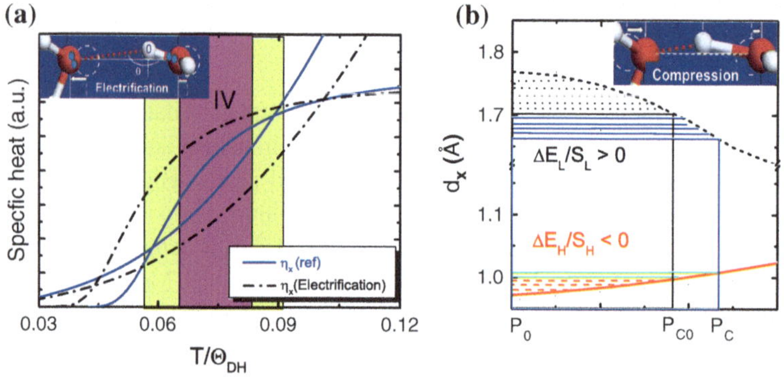


**Fig. 14.2** **a** Electrification disperses the quasisolid phase boundaries by O:H–O bond elongation, which raises the $T_m$ and depresses the $T_N$ of the solution. **b** Compressing an aqueous solution into ice requires excessive energy $\Delta E_H$ (<0, *green shaded*) and $\Delta E_L$ (>0, *blue shaded*) to recover the initially electrification-deformed H–O and O:H by raising the critical pressure from $P_{C0}$ to $P_C$. *Insets* (**a**, **b**) illustrate the electrification elongation and mechanical compression of the O:H–O bond, respectively (Reprinted with permission from [5].) (color online)

(3) At a certain critical concentration the Liquid-VI and VI-VII transitions emerge into the Liquid-VII transition [2].

(4) Collision of flying salt particles with, and then impinge into, supercooled water droplet modulate the freezing behavior of the droplet [3].

## 14.2 Clarification: Quasisolid Phase-Boundary Dispersion

Figure 14.2 clarifies the mechanisms for the phase transition behavior of aqueous solutions:

(1) O:H–O bond relaxation by ionic field electrification disperses the quasisolid phase boundaries whose dispersion extent depends on the solute type and concentration.
(2) H–O bond energy gain raises the $T_m$ for Liquid-Quasisolid transition and the O:H energy loss depresses the $T_N$ for quasisolid-solid transition.
(3) Liquid-VI, VI-VII, and VII/X transitions require excessive energies to recover the electrification-shortened H–O bond by raising the $P_C$ at constant $T_C$ for mechanical icing.
(4) O:H energy loss depresses the $T_N$ for snow-quasisolid transition and the gelation time for quasisolid solution-solid gel transition but the latter is more complicated because of the activation of other types of interactions.

## 14.3 Quantitative Resolution

### *14.3.1 Principle for Solution Phase Transition*

The following formulates the critical temperature $T_{xC}$ as a function of the segmental volume, $v_x = s_x d_x$, critical pressure $P_C$, and the O:H–O bond segmental cohesive energy at the reference equilibrium $E_{xo}$ for the neat water and the $E_x$ for an aqueous solution (see energy diagram in Fig. 14.2b). Ionic electrification shortens the H–O bond and lengthens the O:H nonbond. Compression recovers the deformation and then stores energy into the O:H–O bond to a certain extent at which phase transition takes place.

$$T_C \propto \sum_{L,H} E_{xC} = \begin{cases} \sum_{L,H} \left( E_{x0} - s_x \int_{P_0}^{P_{C0}} p \frac{dd_x}{dp} dp \right) (H_2O) \\ \sum_{L,H} \left( E_x - s_x \int_{P_0}^{P_C} p \frac{dd_x}{dp} dp \right) (\text{Solution}) \end{cases} \quad (14.1)$$

The $E_{xC}$ is the segmental energy required for phase transition. The $\Delta E_x = E_x - E_{x0}$ is the segmental deformation energy caused by electrification. Electrification results in $\Delta E_H > 0$ and $\Delta E_L < 0$ because of the H–O bond stiffening and the O:H nonbond softening. The change of the cross-sectional area $s_x$ of the specific segment of $d_x$ length is assumed insignificant at relaxation. Phase transition of aqueous solution takes place in situations of constant $T_C$, constant $P_C$, or constant $T_C$ and $P_C$ but different gelation time $t_{gel}$, as addressed in the subsequent sections.

## 14.3.2 Mechanical Icing of Ambient Solutions

### 14.3.2.1 Formulation

Compression turns the neat water and a solution into ice of the same temperature—room-temperature mechanical icing. In this situation, the $T_C$ remains the same for both neat water and solution, $\Delta T_C = 0$. The same type of phase transition proceeds at different pressures for the neat water and an aqueous solution. Thus, (14.1) evolves into

$$\Delta E_x - s_x\left(\int_{P_0}^{P_C} p\frac{dd_x}{dp}dp - \int_{P_0}^{P_{C0}} p\frac{dd_x}{dp}dp\right) = 0$$

The known pressure trends of the segmental length $d_x$ and the observed $P_C$ elevation constrain the transition for an aqueous solution [6, 7],

$$\frac{dd_L}{dp} < 0, \frac{dd_H}{dp} > 0, \text{ and } \quad P_C > P_{C0}$$

This expression defines uniquely the segmental deformation energy due to electrification,

$$\Delta E_x = s_x\left(\int_{P_0}^{P_C} p\frac{dd_x}{dp}dp - \int_{P_0}^{P_{C0}} p\frac{dd_x}{dp}dp\right)\begin{cases} > 0 & (x = H) \\ < 0 & (x = L) \end{cases}$$

It is thus justified that ionic electrification raises the $\Delta E_H$ by H–O shortening and depresses the $\Delta E_L$ by O:H elongation at a constant temperature. If $P_C < P_{C0}$, the $\Delta E_L$ loss dictates the phase transition, otherwise, the $\Delta E_H$ dictates. If the $s_x$ and the $dd_x/dp$ are exactly known, one can ideally determine the $\Delta E_x$. However, the $dd_x/dp$ for the solution is different from that for the neat water because of their different initial equilibrium lengths though the pressure trend retains.

### 14.3.2.2 NaX (F, Cl, Br, I) Resolved $P_C$

Figure 14.3a shows that transition from Liquid to ice VI and then into ice VII take place at different $P_C$ for 0.9 M NaX solutions at $T_C \equiv 298$ K, along the pressure path moving across the Liquid, ice VI, and ice VII phases, as (b) indicated.

The pressure and solute type dependence of the $\Delta\omega_x$ of the 0.9 M aqueous NaX solutions, shown in Figs. 14.4 and 14.5, reveal the following [7]:

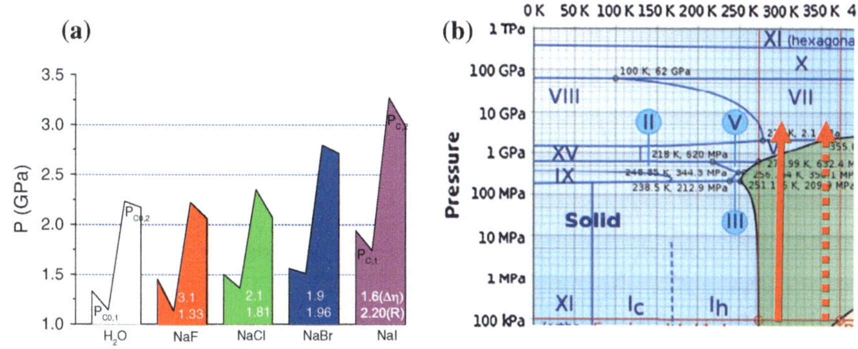


**Fig. 14.3** Solute type resolved critical pressures $P_{C,1}$ and $P_{C,2}$ for the Liquid-VI and VI-VII phase transition at room temperature. Denoted values are the electronegativity difference ($\Delta\eta$) and $X^-$ ionic radius (R). The *vertical red line* in the phase diagram **b** shows the pressure path of Raman spectrometrics crossing the Liquid, VI, and VII phases and their boundaries at 298 K. The *broken red line* indicates the same pressure path crossing the Liquid and VII phases only at elevated temperature (Reprinted with permission from [5].) (color online)

(1) Compression shortens the O:H nonbond and stiffens its $\omega_L$ phonon but lengthens the H–O bond and softens its $\omega_H$ phonon over the full pressure range except for crossing the VI/VII phase boundary where both O:H and H–O contract abruptly and abnormally. This observation agrees with the prediction on the phase boundary of constant pressure, where both segments undergo contraction [8].
(2) Zones from lower to higher pressures correspond to the Liquid, ice VI, and ice VII phases and their boundaries, as the phase diagram Fig. 14.3b indicates.
(3) At transition, the gauged pressure drops, resulting from geometric reconstruction that weakens the O–O repulsion during transition.
(4) Most strikingly, the transition pressures increases with the $X^-$ anion size or with the drop of its electronegativity, following Hofmeister series: $X^-$(R/η) = $I^-$ (2.20/2.5) > $Br^-$ (1.96/2.8) > $Cl^-$ (1.81/3.0) > $F^-$ (1.33/4.0) ~ 0.

Table 14.1 features the electrification and compression effect on the phonon frequency shift and the transition pressures of aqueous 0.9 M NaX solutions. The solute type dependence of the $\Delta P_C$ verified the prediction that ionic electrification deforms the O:H–O bond and that the H–O contraction dictates this series of Liquid-VI and VI-VII transition.

#### 14.3.2.3 NaI Concentration Dependent $P_C$

NaI is most sensitive in the Hofmeister series, which is then chosen to examine the concentration dependence of the critical $P_C$. Figure 14.6 compares the critical pressures for the NaI solution icing as a function of concentration. The Raman spectra shown in Figs. 14.7 and 14.8 reveal the following:

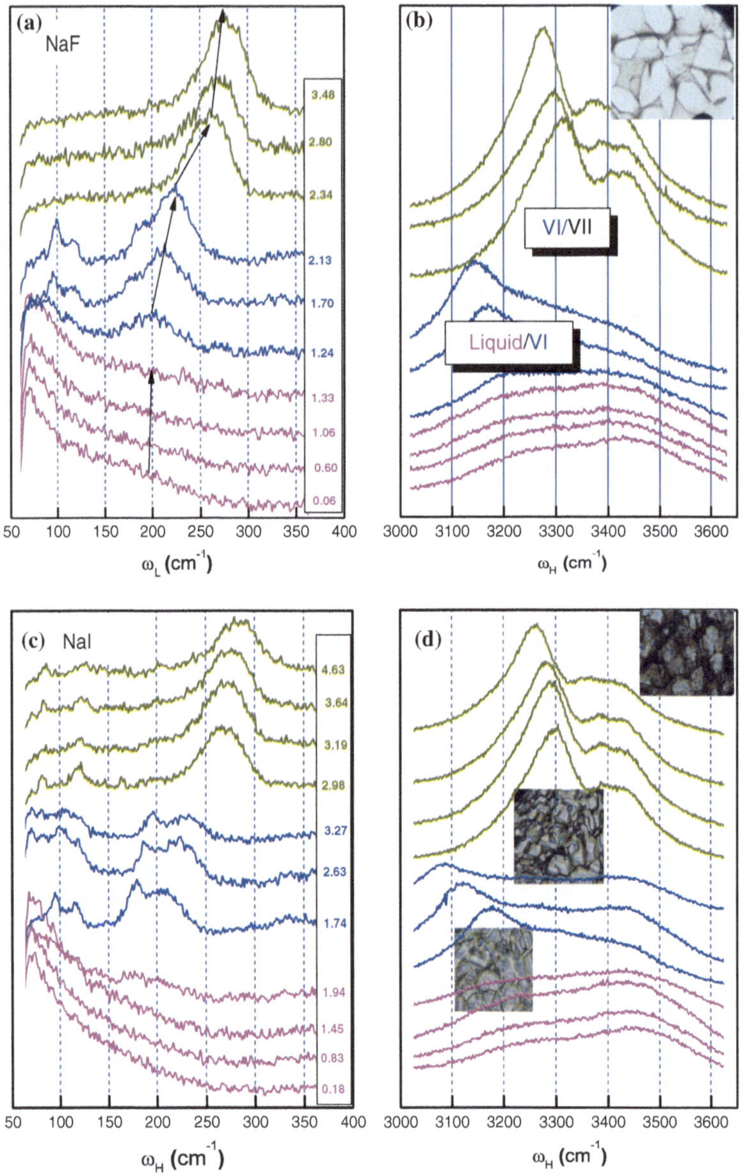


**Fig. 14.4** O:H–O phonon cooperative relaxation of the ambient (298 K) aqueous 0.9 M **a**, **b** NaF and **c**, **d** NaI solutions under compression. *Insets* in (**b**) and (**d**) show the optical image of ice. Liquid-VI-VII phase transitions take place ideally at the same $\omega_x$ but at critical pressures varying with solute type (Reprinted with permission from [5].)

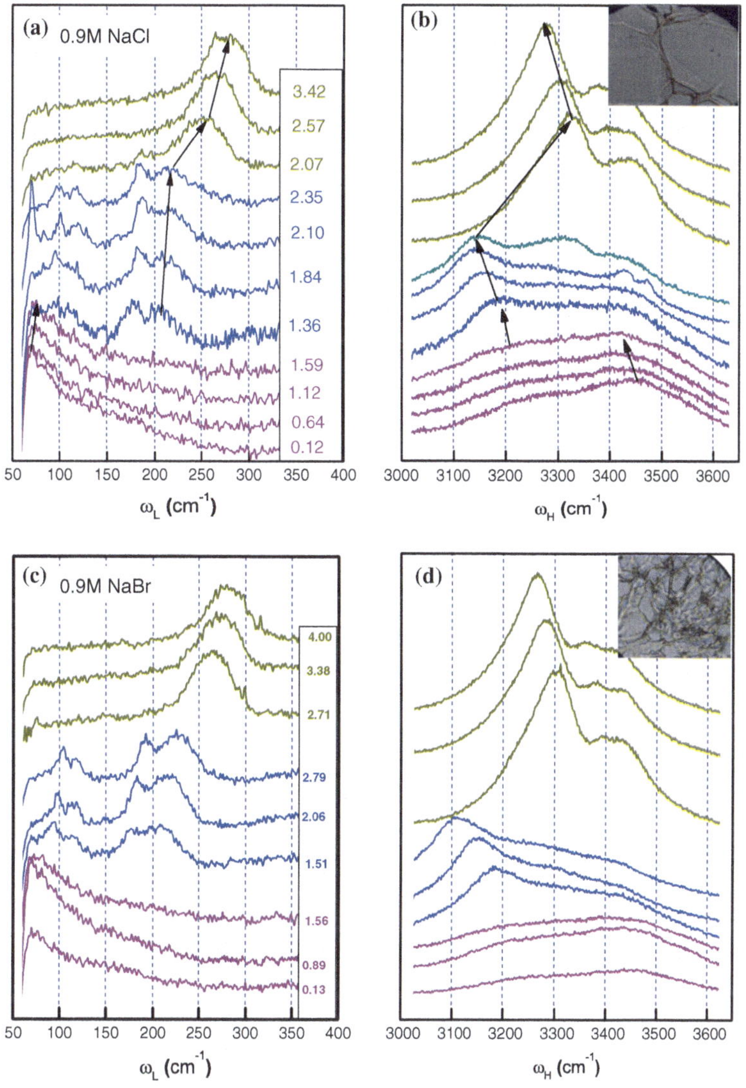


**Fig. 14.5** O:H–O phonon cooperative relaxation of the ambient (298 K) aqueous 0.9 M **a**, **b** NaCl and **c**, **d** NaBr solution under compression. *Insets* in (**b**) and (**d**) show the optical image of ice (Reprinted with permission from [5].)

(1) The $P_{C,1}$ for liquid-VI transition increases with the concentration till its maximum at 3.0 GPa and 6 M concentration.
(2) Variation of solution concentration changes the $P_{C,2}$ at the VI-VII phase boundary insignificantly, agreeing with the phase boundary in the phase diagram, which is in contrast to the solution type trend of the $P_{C,2}$, presented in Fig. 14.3, where the $P_{C,2}$ varies with solute type.

**Table 14.1** O:H–O bond relaxation dynamics of the salted ambient water transiting from Liquid to VI and VII ice under compression

| | | $H_2O$ | NaF | NaCl | NaBr | NaI |
|---|---|---|---|---|---|---|
| $\Delta\eta$ ($\eta_{Na} = 0.9$) | | – | 3.1 | 2.1 | 1.9 | 1.6 |
| R ($R_{Na+} = 0.98$ Å) | | – | 1.33 | 1.81 | 1.96 | 2.20 |
| Liquid VI VII | $\Delta\omega_L$ | >0 ($\Delta d_L < 0$) | | | | |
| | $\Delta\omega_H$ | <0 ($\Delta d_H > 0$) | | | | |
| Liquid → VI | $P_{C1}$ | 1.33 → 1.14 | 1.33 → 1.24 | 1.59 → 1.36 | 1.56 → 1.51 | 1.94 → 1.74 |
| | $\Delta\omega_L$ | >0 ($\Delta d_L < 0$) | | | | |
| | $\Delta\omega_H$ | <0 ($\Delta d_H > 0$) | | | | |
| VI → VII | $P_{C2}$ | 2.23 → 2.17 | 2.13 → 2.34 | 2.35 → 2.07 | 2.79 → 2.71 | 3.27 → 2.98 |
| | $\Delta\omega_L$ | >0 ($\Delta d_L < 0$) | | | | |
| | $\Delta\omega_H$ | >0 ($\Delta d_H < 0$) | | | | |

$\eta$ is the elemental electronegativity

Reprinted with permission from [5]

The critical pressures $P_{C1}$ and $P_{C2}$ varies with the type of salts following Hofmeister series. Larger radius and lower electronegativity raises the critical pressures

The pressure drop at transition indicates the weakening of the interoxygen Coulomb repulsion

Both the O:H and the H–O contract at VI-VII transition indicates spontaneous H–O contraction is involved

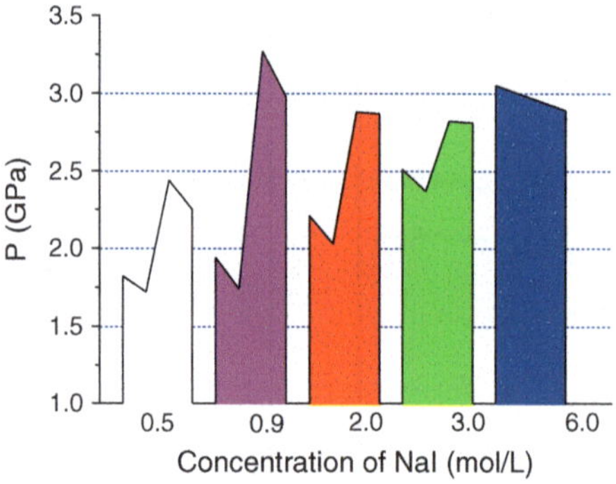


**Fig. 14.6** NaI concentration dependence of the critical pressures $P_{C,1}$ and $P_{C,2}$ for Liquid-VI and VI-VII phase transition. The $P_{C,2}$-$P_{C,1}$ separation reduction till zero at 6.0 M and the nearly constant $P_{C,2}$ evidence the equivalence of salting and heating on H–O bond contraction. Compression of 6 M solution is equivalent to compress 350 K neat water that undergoes only Liquid-VII phase transition, see the phase diagram in Fig. 14.3b (Reprinted with permission from [2].)

(3) Concentration increase is equivalent to heating, which shifts the pressure path towards higher temperatures, such as 360 K, see Fig. 14.3b. The $P_{C,1}$ goes up along the Liquid-VI boundary while the $P_{C,2}$ remains almost constant. At a certain critical concentration of 6 M, the $P_{C,1}$ and $P_{C,2}$ merge into the Liquid-VII boundary, resulting in the situation shown in Fig. 14.6.

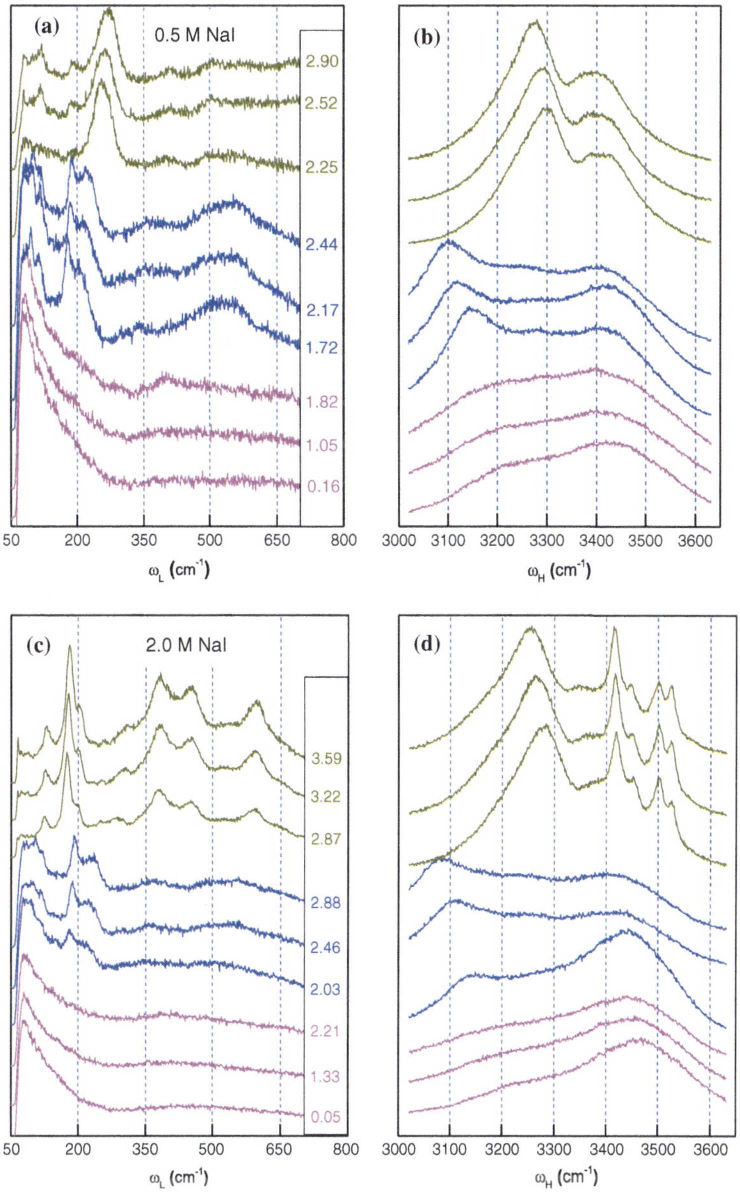


**Fig. 14.7** NaI concentration dependence of the O:H–O phonon frequency shift and the critical pressures at 298 K (Reprinted with permission from [2].)

(4) However, the function of the solute type of the same concentration is different. Solute type determines the extent of initial electrification of the O:H–O bond. Both $P_{C,1}$ and $P_{C,2}$ change with the initial H–O bond energy storage that follows the order of Hofmeister series.

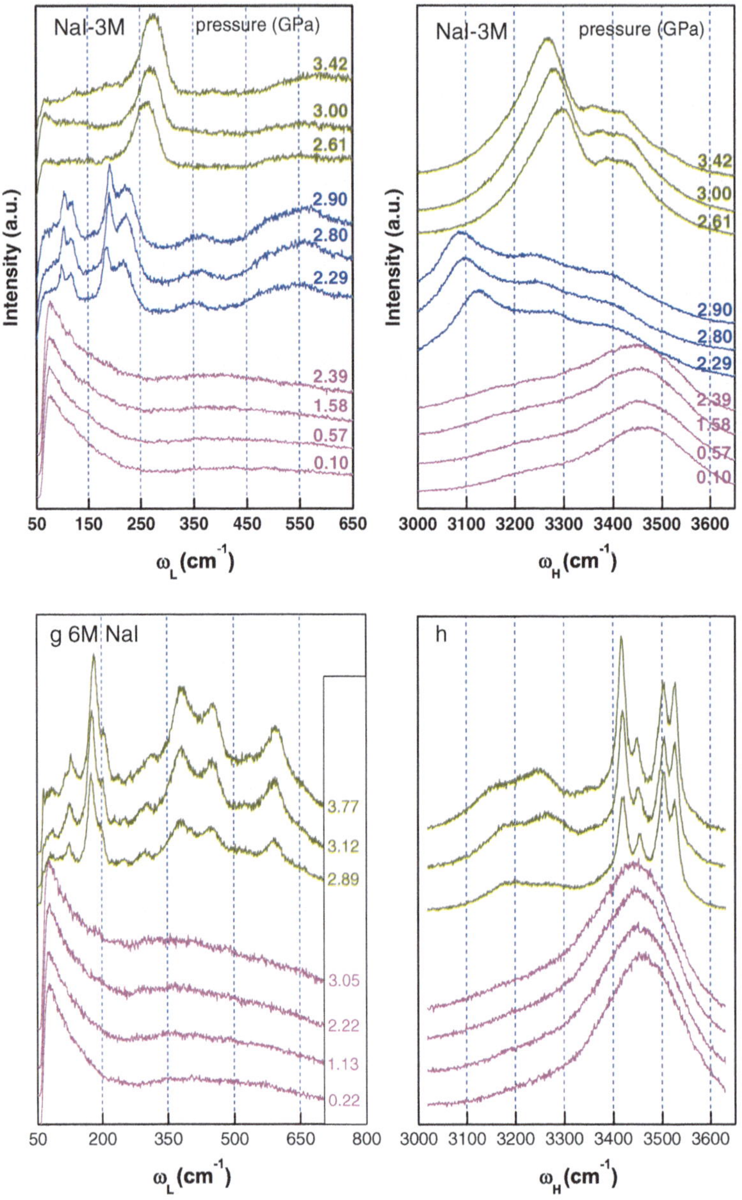


**Fig. 14.8** NaI concentration dependence of the O:H–O phonon frequency shift and the critical pressures at 298 K (Reprinted with permission from [2].)

Figures 14.7 and 14.8 present the NaI concentration dependence of the $\omega_x$ Raman spectra. Results show consistently that compression shortens the O:H nonbond and stiffens its phonons but the H–O bond responds to pressure oppositely throughout the pressure range unless at the phased boundaries. At higher concentrations, 3 and 6 M, the skin 3450 $cm^{-1}$ mode are relatively more active in responding to pressure, which further evidence the preferential skin occupancy of the $I^-$ anion that enhances the local electric filed [9]. The abrupt pressure shift at transition indicates the weakening of the O–O repulsion due to geometrical structure transition.

One must note that, the O:H–O bond is very sensitive to the environment such as pressure holding time, temperature, and chemical reaction, the measurement may not be readily reproducible, but the trends of measurements and the physical origin remain unchanged.

It is expected that the critical pressures for transiting the room temperature base solutions into ice should drop as the quantum compression enhances the mechanical compression on elongating the H-O bond whose cohesive energy dictates the $T_C$ for transition. Further investigation would be very interesting.

#### 14.3.2.4 VII-XI phase transition

Raman scattering measurements up to 130 GPa of ice VII containing NaCl or LiCl impurities probed that the transition pressure to the symmetric phase ice X is shifted up by about 30 GPa, even at small salt concentrations. Figure 6 shows the pressure dependence of the $\omega_H$ phonon relaxation and $d_{OO}$ distance of neat water and salt solution ices transiting from phase XII to phase X. Salt impurity not only stiffens the $\omega_H$ but also lengthens the O-O distance throughout the course of compression transiting from phase VII to phase X [19, 20].

### *14.3.3 Salt Impact Freezing*

#### 14.3.3.1 Effect of Ion Impulsion and Immersion

It is amazing that freezing more likely when the catalyst impinges upon the air −water interface. Much of the research in contact freezing has been motivated by its possible relevance for ice initiation in Earth's atmosphere [3] and the snow ice-quasisolid transition by salting. If a soluble substance is dissolved in water, both the melting and freezing temperatures are intuitively reduced. Addition of an insoluble substance will increase the characteristic freezing temperature of the solution.

Early work [10] showed that silver iodide, sand, and clay can trigger freezing at a higher temperature in the contact mode than in the immersion mode; however, salt and sugar contact can initiate freezing at −11 and −13.5 °C, respectively. Niehaus and Cantrell [3] demonstrated that collisions of six soluble substances (KCl, KI, NaCl, NaI, NaOH, and KOH) with moderately supercooled water trigger freezing at temperatures within 10 ns time scale, as shown in Table 14.2, rather than at −34 °C

**Table 14.2** Threshold temperatures for the initial freezing ($T_0$), 80 % freezing ($T_{80}$), and the eutectic temperature ($t_{eutectic}$) of the alkali salts tested for contact freezing activity [3]

| Salt | $T_0$ (−°C) | $T_{80\ \%}$ (−°C) | $T_{eutecic}$ (−°C) |
|---|---|---|---|
| NaI | 7 | 13 | 31.5 |
| KI | 8 | 12 | 23.2 |
| NaOH | 11 | 15 | 28.0 |
| KOH | 11 | 15 | 62.8 |
| NaCl | 12 | 15 | 21.2 |
| KCl | 12 | 13 | 10.8 |

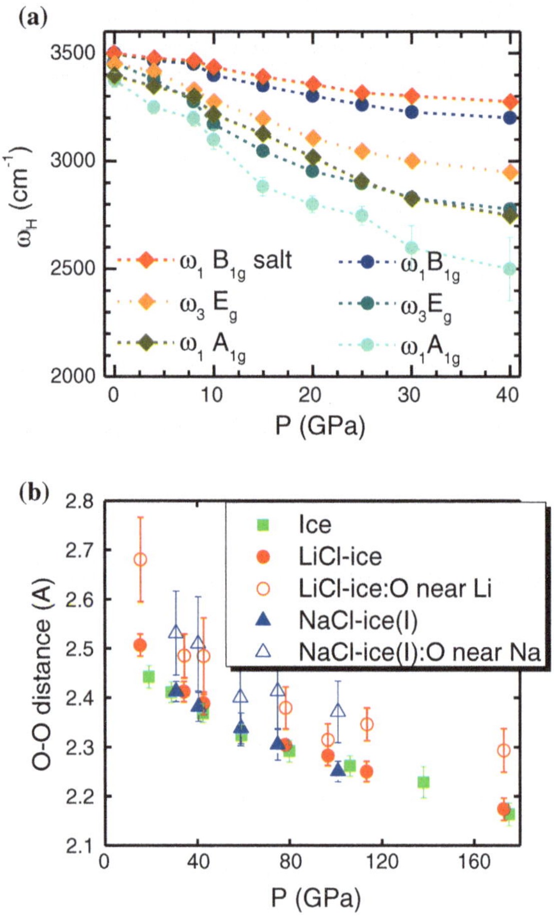


**Fig. 14.9** **a** Raman measurements of the OH stretching mode ($\omega_3$ Eg) in pure and LiCl ices (8 % mole concentration) and **b** the mean O-O distance in pure ices (*squares*), in LiCl ices (*solid circles*), and in NaCl ices (*solid triangles*). Open circles and triangles correspond to the mean O-O distances near the cation. (Reprinted with permission from [19, 20])

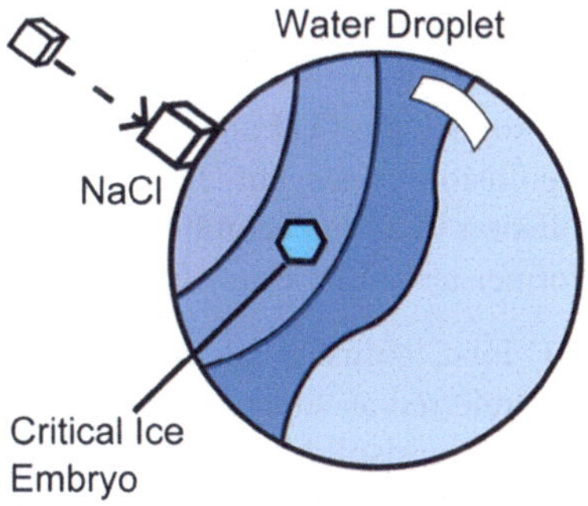


**Fig. 14.10** Schematic illustration of salt contact freezing. The collision of a flying ionic particle (NaCl) can raise the freezing temperature of the supercooled water droplet (Reprinted with permission from [3].)

for homogeneous ice nucleation of supercooled water. They examined the effect of particle size and density, impact velocity ($\sim$3 m/s), and collision kinetic energy on the temperature of freezing and suggested that the freezing behavior depends on the collision itself, as the freezing catalysts would depress the freezing point upon dissolving into the bulk, as illustrated in Fig. 14.10. In other words, the contact freezing nuclei need not be effective as immersion mode (immerse the salt into the solution). The kinetic energy from a mechanical impulse at contact is identified as the key factor in reducing the energy barrier for nucleation, enhancing the probability of a phase transition.

Available mechanisms for contact freezing include subcritical ice embryos adsorbed to the surface of incoming particles [11], a momentary reduction in the free-energy barrier between water and ice as a result of the heat of wetting [12], and intrinsic reduction in the free energy barrier at a three phase contact line [13, 14]. Knollenberg [15] proposed yet another mechanism for contact nucleation involving soluble substances, having recognized that most of the salts present in the atmosphere are endothermic upon dissolution.

Salt impinging upon a water surface induces cooling in the surrounding liquid as heat is absorbed as the bonds within the salt are broken and the resulting ions hydrated. If the water is cooled below the eutectic point for the water-salt system, freezing is possible with the solid salt as a substrate. The eutectic point in a phase diagram indicates the chemical composition and temperature corresponding to the lowest melting point of a mixture of components. Alternatively, water may be cooled below its homogeneous freezing limit, inducing freezing before the ions from the dissolving crystal have diffused into the region that has been supercooled to that point.

#### 14.3.3.2 Impulsion and Electrification

Besides the mechanical perturbation, the solute electrification plays its role in both contact and immersion modes to modulate the freezing point of the solution. The O:H bond energy determines the critical $T_N$ for transforming the quasisolid supercooling phase to homogenous ice, or the inverse. Mechanical collision and ionic electrification relax the O:H–O bond oppositely. Collision effects the same to compression to shortens and stiffen the O:H nonbond momently; Salting elongates and softens the O:H nonbond. These sequential events modulate the $T_N$ oppositely —collision raises but electrification depresses the $T_N$. In fact, the salt particle collision freezing combines discrete steps of mechanical icing [8] and the Hofmeister solute electrification [16]:

(1) First, collision of the flying salt particle with the quasisolid droplet (often referred as supercooling) provides impulse that promotes local transformation of quasisolid water into ice. This process is momently or temporally pressure dependence.
(2) Second, when the salt particle is dissolved, ion electrification effect comes into play, which lowers the freezing temperature and raises the melting temperature. This process is solute type and concentration dependence.
(3) The competition of mechanical collision and ion electrification matters the temperature of freezing and the sequence order of these two stimuli matters the freezing differently.
(4) The H–O bond energy dictates the mechanical freezing of aqueous solution but the O:H nonbond energy governs the solute impinge quasisolid-ice transition of water.

### 14.3.4 Solute Type Resolved $T_C$ Under Constant $P_C$

In this situation of $\Delta P_C = 0$, (14.1) turns into $\Delta T_C \propto \Delta E_H + \Delta E_L$ under the constant ambient pressure. Electrification deformed O:H–O bond segmental energy determines the change of the $T_C$. If $\Delta T_C > 0$, $\Delta E_H$ gain elevates the $T_C$, otherwise, $\Delta E_L$ loss dictates the transition. Therefore, it is not surprising that salt spreading on snow and ice depresses the $\Delta E_L$, which lowers the $T_N$. So salt promotes heavy snow transiting from solid into the gel-like quasisolid on the road first.

Another example is that the $T_C$ for the quasisolid sol transiting into the solid gel of methylcellulose (MC) aqueous solutions characterized by the specific heat $C_p$ peaks and enthalpies changing with its solute type and concentration. As shown in Fig. 14.11, NaCl results in the salt-out and NaI leads to salt-in of the MC [17, 18]. Increasing NaCl concentration lowers the $C_p$ peak temperature and raises the peak intensity, but NaI has the opposite effect. The peak corresponds to the transition between the sol and the gel of the substance. NaSCN has the same salt-in effect as NaI at a slight higher rate of the linear concentration dependence. Other NaZ salts show the salt-out effect, with the slopes in the order $NaNO_3 < NaBr < NaCl < NaSO_4 < Na_3PO_4$. The effect of salt-in and salt-out on the thermal behavior of the MC solution shares the common mechanism in the Hofmeister series. Salt type and concentration modulate the extent of O:H–O bond segmental relaxation and hence the $T_C$ under the ambient pressure.

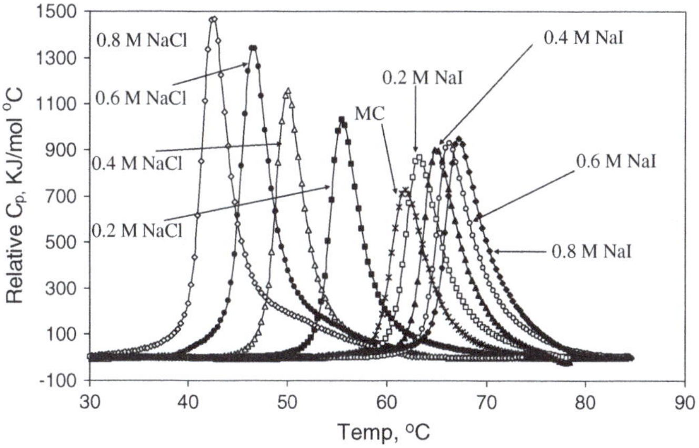


**Fig. 14.11** Salt type and concentration (M) dependence of the heat capacity $C_p$ peaks for methylcellulose solutions shows that NaCl depresses but NaI elevates the temperature for gelation (Reprinted with permission from [17].)

## 14.3.5 *Gelation Time Under Identical $P_C$ and $T_C$*

The gelation time $\Delta t_C$ is defined as that required to turn the macroscopic samples from quasisolid solution to gel solid under the ambient pressure $P_C$ and temperature $T_C$ that serves at the $T_N$. The time $\Delta t_{gel}$ for transition from colloid fluid to gel quasisolid depends on the $\Delta E_L$ change, which is extremely sensitive to the extent of electrification. Gelation takes longer time if the $\Delta E_L$ drops more, otherwise, it takes shorter time.

The solute type and concentration dependence of the gelation time of colloid silica shown in Fig. 14.12 indicates that a progressive shortening of gelation time

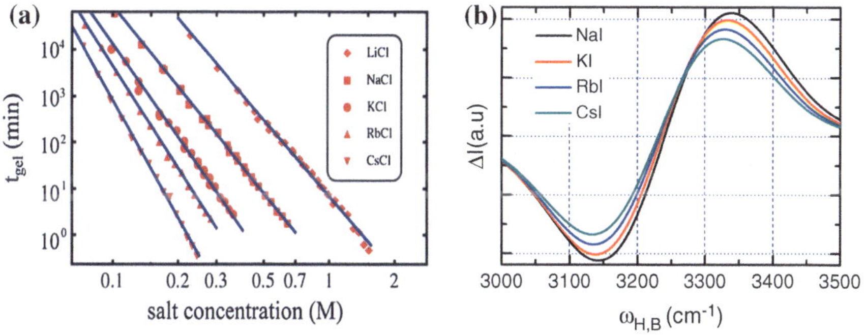


**Fig. 14.12** **a** YCl type and concentration dependence of the gelation time of colloid silica—larger $Y^+$ ions and higher concentrations shorten the gelation time. **b** The differential $\Delta\omega_H$ spectra indicate that larger ion shifts less the $\omega_x$ or deforms less the O:H–O bond to have lower O:H–O bond initial deformation energy (Reprinted with permission from [4].)

switching YCl samples with the same concentration [4]. The gelation time also drops with concentration of the same solute. The electrification depresses the $T_N$ from below toward the ambient temperature. The shortest gelation time of the CsCl solution and the least Raman shift, see Fig. 14.12b indicates its actual $T_N$ is closer than those of other solutions to the ambient temperature.

This solute type resolved gelation time agrees with the trend of Raman $\omega_H$ shift, as shown in Fig. 14.12b, follows the Hofmeister series [9]: $Li^+ > Na^+ > K^+ > Rb^+ > Cs^+$ ($\Delta t_{gel} \propto \Delta\omega_x \propto 1/\Delta R_+$). However, the concentration dependence of the $t_{gel}$ and $\omega_x$ disobeys what have observed from salted water: $\Delta t_{gel} \propto \Delta C \propto 1/\Delta\omega_x$. Salted water demonstrates that $\Delta\omega_x \propto \Delta C$. Larger $Y^+$ cation shortens the $t_{gel}$ and shifts less the $\omega_H$; higher $Y^+$concentration shortens the $t_{gel}$ but shifts more the $\omega_H$. These observations are inconsistent.

It is worthy of noting that salting effect on the $T_C$ or the $t_{gel}$ for a colloidal solution is different from that of neat water because the activation of other chemicals in the colloid, which makes the situation more complicated and sensitive such as the concentration effect on the $t_{gel}$ of silica colloid gelation [4] and the salt-out of NaCl and salt-in of NaI [17] in the methylcellulose solutions. Therefore, this situation is less conclusive.

## 14.4 Summary

Raman spectrometrics has enabled clarification of the Hofmeister effect on the critical pressures, critical temperatures, and gelation (quasisolid/solid transition) times for phase transition of aqueous solutions. Consistency in theoretical predictions, numerical derivatives, and experimental observations indicates the following:

(1) Electrification disperses the boundaries of the quasisolid phase to depress the $T_N$ for quasisolid-solid transition and elevates the $T_m$ for quasisolid–liquid transition, which is opposite to the effect of mechanical compression.
(2) Phase transition requires additional energy to recovers the electrification distorted solvent O:H–O bond by modulating the critical $T_C$, $P_C$ or $t_{gel}$ with extents depending on the solute type and concentration.
(3) Solute type and concentration affect the critical pressures for the mechanical icing of solution. Concentration increase is equivalent to heating; solute type effect on initial O:H–O bond distortion, which follows the Hofmeister series.
(4) The abrupt relaxation of phonons at Liquid/VI and VI/VII boundaries results from crystal structure change that weakens the O–O repulsion.
(5) Colloid gelation time is much more complicated and sensitive because the activation of other type solutes.
(6) Mechanical impulsion and ionic electrification contribute oppositely to mediating the freezing temperatures of quasisolid water.

# References

1. C.Q. Sun, Hofmeister pressures for liquid/VI and VI/VII phase transition. Communicated
2. Q. Zeng, Y. Zhou, W. Kai, B. Zou, C.Q. Sun, NaI concentration resolved critical pressures for the solution-ice VI-ice VII transition at room temperature. Communicated (2015)
3. J. Niehaus, W. Cantrell, Contact freezing of water by salts. J. Phys. Chem. Lett. (2015)
4. M. van der Linden, B.O. Conchúir, E. Spigone, A. Niranjan, A. Zaccone, P. Cicuta, Microscopic origin of the Hofmeister effect in gelation kinetics of colloidal silica. J. Phys. Chem. Lett. 2881–2887 (2015)
5. Q. Zeng, Y. Zhou, W. Kai, B. Zou, C.Q. Sun, Room temperature icing of NaX solution by compression (X = F, Cl, Br, I). Communicated (2015)
6. C.Q. Sun, X. Zhang, W.T. Zheng, Hidden force opposing ice compression. Chem. Sci. **3**, 1455–1460 (2012)
7. Q. Zeng, Y. Zhou, W. Kai, B. Zou, C.Q. Sun, Solution room temperature phase precipitation under compression. Communicated (2015)
8. X. Zhang, P. Sun, Y. Huang, T. Yan, Z. Ma, X. Liu, B. Zou, J. Zhou, W. Zheng, C.Q. Sun, Water's phase diagram: from the notion of thermodynamics to hydrogen-bond cooperativity. Prog. Solid State Chem. **43**, 71–81 (2015)
9. Y. Zhou, Y. Huang, Y. Gong, C.Q. Sun, Skin preferential occupancy of the $I^-$ anion in NaI-water solution. Communicated (2016)
10. N.R. Gokhale, J.D. Spengler, Freezing of freely suspended, supercooled water drops by contact nucleation. J. Appl. Meteorol. **11**(1), 157–160 (1972)
11. W.A. Cooper, A possible mechanism for contact nucleation. J. Atmos. Sci. **31**(7), 1832–1837 (1974)
12. N. Fukuta, A study of the mechanism of contact ice nucleation. J. Atmos. Sci. **32**(8), 1597–1603 (1975)
13. C. Gurganus, A.B. Kostinski, R.A. Shaw, Fast imaging of freezing drops: no preference for nucleation at the contact line. J. Phys. Chem. Lett. **2**(12), 1449–1454 (2011)
14. R.A. Shaw, A.J. Durant, Y. Mi, Heterogeneous surface crystallization observed in undercooled water. J. Phys. Chem. B **109**(20), 9865–9868 (2005)
15. R.G. Knollenberg, A laboratory study of the local cooling resulting from the dissolution of soluble ice nuclei having endothermic heats of solution. J. Atmos. Sci. **26**(1), 115–124 (1969)
16. X. Zhang, T. Yan, Y. Huang, Z. Ma, X. Liu, B. Zou, C.Q. Sun, Mediating relaxation and polarization of hydrogen-bonds in water by NaCl salting and heating. PCCP **16**(45), 24666–24671 (2014)
17. Y. Xu, L. Li, P. Zheng, Y.C. Lam, X. Hu, Controllable gelation of methylcellulose by a salt mixture. Langmuir **20**(15), 6134–6138 (2004)
18. Y. Xu, C. Wang, K. Tam, L. Li, Salt-assisted and salt-suppressed sol-gel transitions of methylcellulose in water. Langmuir **20**(3), 646–652 (2004)
19. L.E. Bove, R. Gaal, Z. Raza, A.-A. Ludl, S. Klotz, A.M. Saitta, A.F. Goncharov, P. Gillet, Effect of salt on the H-bond symmetrization in ice. Proceedings of the National Academy of Sciences **112**(27), 8216–8220 (2015)
20. Y. Bronstein, P. Depondt, L.E. Bove, R. Gaal, A.M. Saitta, F. Finocchi, Quantum versus classical protons in pure and salty ice under pressure. Phys. Rev. B **93**(2), 024104 (2016)

# Chapter 15
# Electrofreezing and Water Bridging

- *Capacitor effects the same to solute ions on O:H–O bond electrification in a long-range order.*
- *Quasisolid floating bridge forms since electrification raises the $T_m$ and depresses the $T_N$ by dispersing the quasisolid-phase boundary.*
- *Skin supersolidity further enhances the quasisolidity, which stiffens and stabilizes the plastic bridge.*
- *Solute ions destabilize the bridge as they create electric fields opposing that of the capacitor. Field of like charges depresses the thermal stability by inward dispersion of the quasisolid phase boundary.*

**Abstract** A combination of the electrification-induced quasisolidity and the undercoordination-induced skin supersolidity laid foundations for the water floating bridge, electrofreezing and electromelting of liquid water. Both electrification and molecular undercoordination disperse the quasisolid phase boundaries, which not only depresses the freezing temperature and molecular dynamics but also raises the melting point, H–O phonon lifetime, skin stress, and viscosity. The extent of quasisolidity is charge quantity and sign dependent. Aqueous solutions weaken the field of capacitors or charged particles, so aqueous solutions destabilize the floating bridge and wet faster soil particles.

## 15.1 Challenge: Why Does Water Form Bridge?

Water floating bridge forms between two containers of deionized water, placed on an insulator, connected by a thread, when a $10^6$ V/m field is applied. The bridge remains even when the containers are separated 3 cm apart. The typical configuration has a diameter of 1–3 mm remains stable for hours (Fig. 15.1). Floating bridge is also related to electrofreezing and electromelting of water droplet under various electric field. Mechanisms on the following issues remains unclear since 1893 when Sir William George Armstrong discovered the bridge [1].

(1) Why is the water bridge mechanically stiffer and thermally more stable than bulk water?
(2) How does the capacitor electric field modulate the intra- and intermolecular interactions?


C.Q. Sun and Y. Sun, *The Attribute of Water*, Springer Series in Chemical Physics 113, DOI 10.1007/978-981-10-0180-2_15

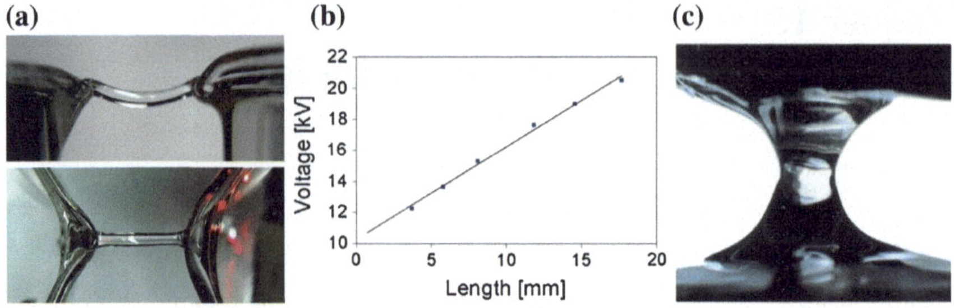


**Fig. 15.1** **a** *Front* and *top* views of a 14 mm water floating bridge formed between two beakers in a $10^6$ $Vm^{-1}$ field at room temperature. **b** DC voltage dependence (sloped at $\sim 0.6 \times 10^6$ V/m) of the water bridge length under a 0.5 mA constant current. **c** An 8 mm height paraffin column maintained by a $\sim 0.5 \times 10^6$ V/m electric field (Printed with permission from [2–4].)

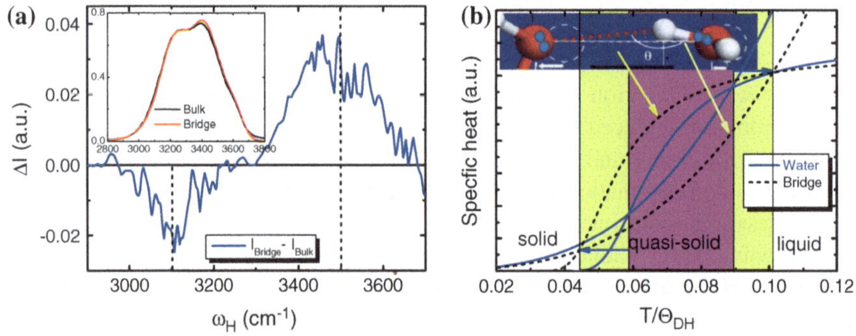


**Fig. 15.2** **a** Electrification stiffens the $\omega_H$ phonon from 3100 to 3500 $cm^{-1}$ associated with polarization (reprinted with permission from [5]). **b** The $\omega_x$ relaxation disperses the quasisolid phase boundary outwardly, which raises the $T_m$ and lowers the $T_N$ so the water floating bridge is in the rubber-like quasisolid state at the ambient temperatures (Reprinted with permission from [6, 7].)

(3) Why does salt solute electrification destabilize the bridge?
(4) How does additional charge modulate the temperatures of melting and freezing of water and ice?

## 15.2 Clarification: O:H–O Bond Ordered Electrification

Figure 15.2 illustrates the mechanisms for water floating bridge formation and electrofreezing and electromelting of water droplet:

(1) Capacitor has the same effect of salting but supplies a directional and long-range field, which aligns, elongates, and polarizes the $H_2O$ molecular dipoles. This electrification shortens the H–O bond and lengthens the O:H nonbond associated with polarization of the nonbonding electrons.

(2) Electrification raises the $T_m$ and depresses the $T_N$ by dispersing the quasisolid phase boundary through Eisenstein's relationship: $\hbar\omega_x = k\Theta_{Dx}$.
(3) Skin supersolidity due to molecular undercoordination enhances the quasisolidity of the bridge volume that is tough, viscous, stiff, and thermally stable.
(4) Ionic solutes destabilize the bridge by inserting electric fields opposing that of the capacitor; resultant fields of multiple sources mediate the freezing and melting of water droplets.

## 15.3 History Background: Armstrong Effect

Sir William George Armstrong reported the remarkable direct-current high-voltage experiment to the Literary and Philosophical Society of Newcastle [1] in 1893:

> Taking two wine glasses filled to the brim with chemically pure water, I connected the two glasses by a cotton thread coiled up in one glass, and having its shorter end dipped into the other glass. On turning on the current, the coiled thread was rapidly drawn out of the glass containing it, and the whole thread deposited in the other, leaving, for a few seconds, a rope of water suspended between the lips of the two glasses. This effect I attributed at that time to the existence of two water currents flowing in opposite directions and representing opposite electric currents, of which the one flowed within the other and carried the cotton with it. It required the full power of the machine to produce this effect, but, unfortunately when it went to London, and was fitted up in the lecture-room, I could not get the full power on account of the difficulty of effecting as good insulation in a room as in the outside air. I therefore failed in getting this result, after announcing that I could do it, and I daresay I got the credit of romancing.

William George Armstrong, FRS (26 November 1810–27 December 1900), an English industrialist and also an eminent engineer, scientist, inventor and philanthropist. Armstrong was knighted in 1859 after giving his gun patents to the government. In 1887, in Queen Victoria's golden jubilee year, he was raised to the peerage as Baron Armstrong of Cragside, becoming the first engineer—and the first scientist—to join the House of Lords. (Portrait by Mary Lemon Waller, Cragside, The Armstrong Collection (acquired through the National Land Fund and transferred to The National Trust in 1977), ©NTPL/Derrick E. Witty

Such "floating" liquid bridges resemble a small high-voltage fluid laboratory of their own, which enable the study of liquids in electric fields of kV/cm scale, even long time experiments like neutron or light scattering are feasible since the bridge keeps in a steady-state for hours. It is also an electro-chemical reactor where compounds are transported through the electrohydrodynamic flow, enabling the study of electrochemical reactions under potentials, which are otherwise not easily accessible. Last but not least the bridge provides the experimental biologist with the opportunity to expose living organisms such as bacteria to electric fields without killing them, but with a significant influence on their behavior, and possibly, even on their genome [8].

However, no record on the water floating bridge had been found till 1997 when Uhlig from the ETH Zürich reported a video of this experiment online. Then, in 2006, Fuchs and coworkers [9] started a systematic investigation of this phenomenon and now more research groups have been involved in this fascinating project both theoretically and experimentally.

## 15.4 Wonders of Water Electrification

### *15.4.1 Tylor Electric Cone-Jet-Spray*

The water floating bridge phenomenon is essentially an analogue of the Taylor cone and cone-jet formation [10–12], as illustrated in Fig. 15.3, which are responsible for electrosprays having important industrial applications. A Taylor cone is the formation of a cone shaped liquid surface due to high electric fields, which deforms further to emit a jet at higher field strength.

When a small volume of electrically conductive liquid is exposed to an electric field, the shape of liquid starts to deform from the shape caused by surface tension alone. As the voltage is increased the electric field effect becomes more prominent. As it approaches exerting a similar amount of force on the droplet as the surface tension does, a cone shape begins to form with convex sides and a rounded tip. This approaches the shape of a cone.

When a certain threshold voltage has been reached the slightly rounded tip inverts and emits a jet of liquid in the form of spraying fogs. This is called a cone-jet and is the beginning of the electrospraying process in which ions may be transferred to the gas phase. In order to achieve a stable cone-jet a slightly higher

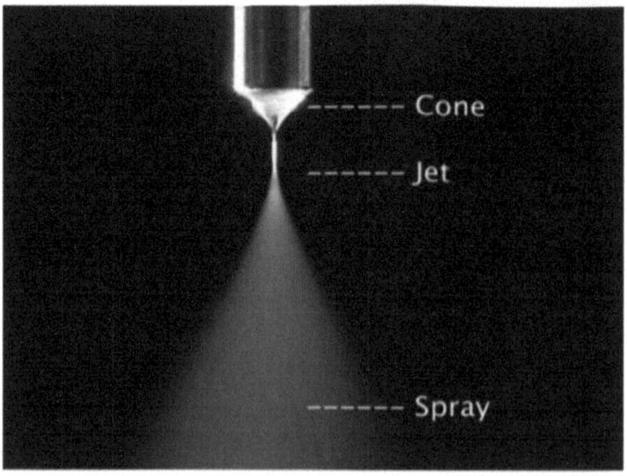


**Fig. 15.3** Tylor electro cone-jet-spray formation. An electric field is applied to the *nozzle* (public domain) [13]

than threshold voltage must be used. The shape of the spray can be adjusted by configuring the electric field.

This cone-jet phenomenon is understood in terms of the surface charge deforming the liquid toward a surface of equal electrical potential [14, 15]. According to the present understanding, the electric field aligns, stretches, polarizes, and then breaks the O:H–O bond at sufficiently high voltage, which breaks the skin stress and then emits jets whose spreading angle and velocity depends on the intensity and the configuration of the field.

### *15.4.2 Electric Freezing*

The freezing and melting temperatures of water and ice are very sensitive to any change of the electric field [16–19]. Dafour [20] reported in 1861 that an external electric field can raise the freezing temperature of water. In 1951, Rau [21] observed that supercooled water droplets instantly freezes when an electric field was applied between two bare electrodes in contact with water droplets. In the 1960s and 1970s, Pruppacher [22, 23] conducted experiments to show that freezing could be initiated by electrical discharges in water. However, discrepancy remain on such observations [24, 25], resulting from both the manner in which the electric field is applied and various secondary phenomena that can trigger ice nucleation.

The most common experiment to demonstrate electrofreezing consists of immersing bare metal electrodes in a small pool of water; an applied voltage sets up a volumetric electric field within the water. This also creates a small but finite current flow. Hozumi et al. [26] discovered that the freezing temperature depends on the electrode material, which suggests that surface reactions are important. The Joule heating due to current flow will work against nucleation; however, little efforts has been made to account for this current flow effect.

A daily example of electric freezing is that ice can form on the power cords under its radially outward electromagnetic field, see Fig. 15.4. Despite the cold

**Fig. 15.4** Ice formed on the power cables in snow blasting in January 2008 Southern China (from public domain)

weather, the electric field should play some yet unclear role in the preferential freezing on the cables. Ice formation on cables damages the power transport from the station to the end users. Considerable efforts have been made to find means preventing this blasting disaster. Therefore, understanding the mechanism of water bridge formation and electrofreezing is helpful.

## *15.4.3 Water Bridge*

### 15.4.3.1 Preparation and Characterisation

Elmar Fuchs and coworkers [3, 8, 9, 27–35] systematically investigated the water bridge using neutron diffraction, phonon spectroscopy, mass transport tracing, dielectric spectroscopy, and infrared thermography, etc. Ref [34] describes the visualized details on preparing water bridges and Fig. 15.1b shows the typical bias dependence of its stable length, which gives rise to critical electric field $\sim 0.6 \times 10^6$ V/m for stable bridge formation.

In a typical experiment, two 100 mL beakers are filled with deionized water to roughly 3 mm below the rims of the beakers, and exposed the water to 15 kV direct-current voltage with one beaker turning negative and the other positive. After building up electric charge, the water then spontaneously rises along the thread over the glass walls and forms a “water bridge” between them. When one beaker is slowly pushed away from the other, the structure remains. When the voltage rises to 25 kV, the structure can be pulled apart as far as 25 mm. If the thread is very short, then the force of the water may be strong enough to pull/push the thread from the positive glass into the negative glass.

The water generally travels from anode to cathode, but the mass flow direction may vary due to the different surface charge that builds up at the surface, which generates electrical shear stresses of different signs. The bridge breaks into droplets due to capillary action if separation between the beakers is too large, or the voltage is lower than the critical value.

However, an addition of salt ions or electrolytes dramatically lowers the stability. Electrolytes such as NaCl, NaOH, and $NH_4Cl$, modifies the ionization degree and reduces the bridge length but a suspension of the insolvable $Al_2O_3$ shows no such effect [8]. A stable water bridge can also be formed under high-frequency AC bias [5]. Furthermore, the bridge becomes less stable if the voltage is too high; a charged glass rod approaching will disrupt the bridge into water jets.

Vertical column of liquid can also form [36], pinned at each end between planar electrodes, and surrounded by a non-conducting, dielectric gas, as shown in Fig. 15.1c [4]. Although easy to reproduce, this watery connection demonstrates numerous fascinating phenomena as summarized in Table 15.1.

**Table 15.1** Summary of water bridge behavior [3, 4, 8, 9, 27–37]

| Quantity | Method | Attribute | Interpretation |
|---|---|---|---|
| Bridge dimension | Under $10^6$ V/m field across two containers of deionized water | Up to 3 cm long and 3 mm thick | Electrification, gravity, and surface tension |
| Temperature | IR thermography | Up to 60 °C before its breakdown | Joule heat due to current flow and resistance |
| Stability | | Several hours | |
| Mass transport | Water level and trace particles; laser Doppler anemometer | From anode to cathode and then the other way around | |
| Structure anisotropy Polarization | Laser polarizer [38] Neutron scattering [28] Optical birefringence [33] | Layered structure with rotating outer shell, exhibiting polarization effect and enhanced laser scattering Slight alignment along the bridge | Electrically enforced birefringence, commonly known as the electro-optical Kerr effect |
| Density gradient | Optical method [30] | 7 % drops from the edge to the middle of the bridge | Formation of nano and microbubbles [30] |
| Network structure | XRD [39, 40] | Bulk structure remains | |
| Young's modulus | Bending profile calculation [41]: $Y = mgl^3/(12\pi r^4 s)$(l: length; r: radius, s: flexure) | 10–24 MPa | Equivalent to rubber |
| H–O phonon stiffness | Raman and IR [5, 32, 42] | Shift from 3100 to 3500 $cm^{-1}$ | Polarization shell; H–O bond shortening and stiffening [6] |
| Dissolvable electrolyte (NaCl, NaOH, and $NH_4Cl$, etc.) [43] | | Lowers the stability and shortens the bridge | Ionic fields opposing capacitor electrification |
| Insolvable $Al_2O_3$ [38] | | No effect | Electrification less |
| Alternative liquids (methanol, glycerol, amyl alcohol, etc.) [40] | | Bridge forms under proper condition | O:H–O bond like formation and O:H compression by O:↔:O quantum compressors |

#### 15.4.3.2 Anisotropy or Isotropy

Spectroscopic, optical, and neutron scattering studies suggest that the bridge forms because of the formation of anisotropic chains of water molecules along the electric field in the liquid. Neutron scattering experiments on a floating bridge of $D_2O$ [29]

suggest an increase in scattered intensity at $Q \leq 2$ Å$^{-1}$ being associated with the presence of nanobubbles or some other forms of nanoscale objects, which lowers the density of the bridge. An artifact due to $H_2O$ contamination in the $D_2O$ sample with insignificant anisotropy results in the angular distribution of the scattered neutron intensity [28]. This anisotropy observed in the neutron and polarized light scattering suggests a preferred orientation of the hydrogen bonded water molecules along the electric field, at approximately $10^6$ V/m field. MD simulations also suggest that molecules alignment along the electric field but it occurs around $10^9$ V/m or higher electric field.

Using high-energy X-ray diffraction from a series of floating water bridges as a function of applied voltage, bridge length, and position within the bridge, Skinner et al. [39] demonstrated, shown in Fig. 15.5, however, that water molecules do not exhibit any significant preferred orientation along the electric field. The only structural changes observed were those due to heating, and these effects are the same to that for bulk water.

Dielectric measurements revealed insignificant difference in the microscopic hydrogen-bond structure between the bridge and the bulk water. Unfortunately, MD simulations can hardly elucidate the molecular-scale properties of the water bridge, as the number of water molecules that can be included in these simulations is limited. MD simulations do show effects of electric fields on water clusters, but these calculations concerned electric fields that are $10^3$ times higher than those needed to form the water bridge [8].

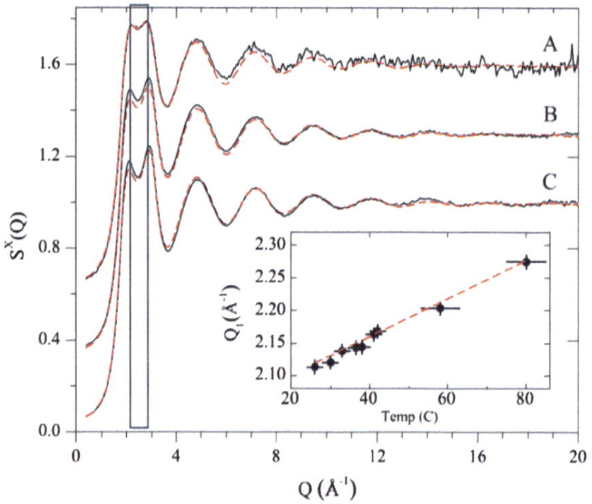


**Fig. 15.5** XRD structure factors for water bridges under conditions of **a** 58 °C, 17 kV, 9.4 mm long (1.81 × $10^6$ V/m); **b** 42 °C, 15 kV, 7 mm long (2.14 × $10^6$ V/m); and **c** 26 °C, 15.5 kV, 7 mm long (2.21 × $10^6$ V/m). The *red dashed lines* are normal deionized water measured under the same conditions without electric field. The *inset* shows the temperature dependence of the first—maximum against the trend for normal water (*red dashed line*) (Reprinted with permission from [39].) (color online)

#### 15.4.3.3 Long H–O Phonon Life—High Viscosity

Two concepts are important for the life times. One is the molecular population life time that is related to the O:H nonbond interaction and the other is the thermalization time related to the H–O bond vibration frequency. The H–O phonon relaxation time is proportional to its frequency and it is higher in the skin than in the bulk of the same frequency [44].

Figure 15.6 shows the H–O stretch vibration of an HDO molecule in the HDO: $D_2O$ water bridge probed using Ultrafast (pump–probe) laser spectroscopy [8, 32]. Results show that the molecular lifetime determined by O:H interaction of the bridge (~630 ± 50 fs) is between that for the 0 °C bulk water (~740 ± 40 fs) and that for the 0 °C ice I (384 ± 16 fs).

However, the thermalization dynamics associated with the H–O bond vibrational relaxation in the bridge is unusually slower (~1.5 ± 0.4 ps) than it is in the bulk (~0.25 ± 0.90 ps). Conversely, thermalization is much faster in ice than it is in liquid water. Therefore, it takes longer time for the excitation energy dissipating from one molecule to the next in the bridge. Compared with the lifetime of molecules in the Hofmeister salts, the longer H–O phonon life time indicates higher order of water molecules in the bridge that is much more viscous than bulk water.

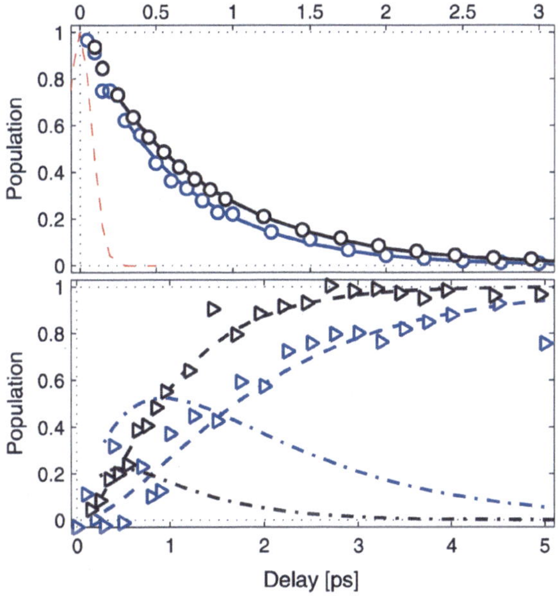


**Fig. 15.6** Molecular fluctuation order of the bridge—life time. *Top panel*: excited state population dynamics (*open circles*) for the water bridge (*blue*) and for the bulk reference (*black*). The *solid lines* fit to the kinetic model. The *dashed red line* represents the correlation profile of the pump and probe pulses. *Bottom panel*: thermalization dynamics of the water bridge (*in blue*) and the reference sample (*black*) following the relaxation of the H–O stretch mode. The *dash–dotted* and *dashed lines* follow from the fit to the model (Reprinted with permission from [32].) (color online)

#### 15.4.3.4 Rubber Like Elasticity

Teschke et al. [45] estimated the elastic modulus by assuming that the charged layer is plastic and not liquid using the following method. If an elongation is produced by the weight of the water mass (m) in a cylinder of length (l), radius (r) and produced flexure (s), the Young's modulus (Y) is given [30, 31]

$$Y = \frac{mgl^3}{12\pi s\left(r_{\text{out}}^4 - r_{\text{int}}^4\right)} \tag{15.1}$$

Because the ionized molecules are mostly restricted to the bridge surface, the expression of bending of a hollow tube with external radius $r_{out}$ and internal radius $r_{in}$ was employed. This high-voltage formed structure thickness is difficult to measure. Thus, Teschke et al. plotted the Young´s modulus as a function of $r_{in}$, in Fig. 15.7, and they observed that a substantial increase in the bulk modulus value for internal radii larger than 0.5 mm ($r_{out}$ = 0.56 mm). This observation leads to conclusion that the plastic structure has a thickness smaller than 0.50 nm. The bulk modulus for this thickness is 85 kPa, which is substantially higher than the value 0.63 kPa calculated for a full cylinder [41]. Table 15.2 features the calculated Young's modulus of water bridges using (15.1) for the full cylinder without skin,

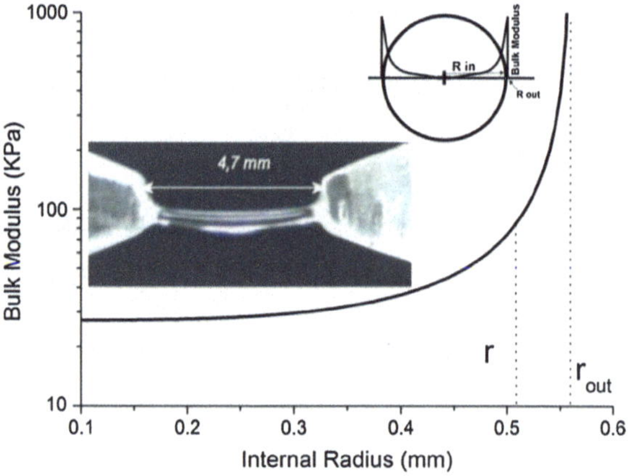


**Fig. 15.7** The calculated bulk modulus of the bridge as a function of the bridge outer layer thickness. The *arrow* indicates the point of the probable inner layer radius of 50 μm (Reprinted with permission from [45].)

**Table 15.2** Electric field dependence of the flexures and the bulk modulus (M) of Water Bridges

| Sample | l (cm) | r (mm) | s (mm) | E (MV/m) | M (MPa) |
|---|---|---|---|---|---|
| 1 | 0.59 | 0.38 | 0.42 | 0.20 | 0.63 |
| 2 | 0.90 | 0.56 | 0.46 | 0.21 | 2.16 |
| 3 | 1.38 | 0.72 | 0.68 | 0.22 | 6.19 |
| 4 | 1.23 | 0.41 | 0.36 | 0.25 | 13.10 |
| 5 | 0.60 | 0.30 | 0.02 | 0.43 | 24.03 |

Reprinted with permission from [41]

which shows that the elastic modulus and the flexure depend sensitively on the applied electric field.

This layer moves in a constant velocity along the longitudinal bridge. For a 10 μA current, $r_{out}$ = 0.56 mm, and a wall thickness of 0.5 mm, the corresponding current density is ~5.6 mA/cm$^2$. They suggested that the wall thickness, the electric field intensity inside the plastic bridge and the induced dipole moment describe a possible scenario for the operation of the floating bridge, but the stiffening effect of the supersolid skin was ignored.

## 15.4.4 Known Mechanisms

### 15.4.4.1 Classical Electrohydrodynamic Scheme

On the macroscopic scale, action of electrohydrodynamic forces caused by electric field gradients counteracts the gravity. This mechanism explains several observations of a horizontal water bridge in terms of the fluid motion in electric fields.

From the perspective of two-phase structure for water, the externally applied field aligns the "coherent domains" to form super-domains along the water bridge. This effect can be considered an electrical analogue to the domains in ferromagnetic materials when a magnetic field is applied, where order in the microscopic range suddenly creates a macroscopic effect. For this reason, Widom et al. [31] considered water to be an "electric ferrofluid", and they compared the water bridge's rheology to that of superfluids, and concluded that "...considering water as an electric ferrofluid subject to high electric fields allows for structures that are more than just a bit unusual...".

In order to understand this property, Widom et al. [31] showed that the forces responsible for holding up the water bridge stem from the pressure tensor of Maxwell electric field in dielectric polar fluids. The water bridge is viewed as a flexible cable as an electric field induced tension sufficiently large so as to explain its support. The need for deionized water in the experiment is evidently due prohibiting conductivity effects from masking the insulating dielectric effects. The Maxwell pressure tensor inside the water bridge is denoted by ($g_{ij}$ is volume element of the strained liquid and ε is the dielectric constant):

$$P_{ij} = Pg_{ij} - \frac{\varepsilon}{4\pi} E_i E_j$$

Therefrom they derived the catenary shape of the bridge, and therefore, the tension arises from long ordered chains of low entropy aligned coherent dipolar domains in the polar liquid approximating a ferrofluid. According to Widom et al. [31], Marın and Lohse [37], and Saija et al. [4], the high dielectric permittivity of water is the reason for a stable bridge being formed. Woisetschlager et al. [38] showed that a horizontal bridge of any dielectric fluid of polar molecules with sufficiently low conductivity is stable by the action of electrohydrodynamic forces caused by electric field gradients counteracting gravity.

#### 15.4.4.2 Quantum Mechanical Approach

Water can be described by quantum mechanics at the molecular scale. From the quantum mechanical point of view [42, 46], an electric field would stretch the intermolecular hydrogen bonds in the water network, eventually breaking the three dimensional morphologies to form linear, branched, or netlike structures, resulting in dipolar water monomers aligning along the field axis which coincides with the water bridge axis. However, the calculated field strength necessary in order to achieve such chains are considerably higher than the ones applied in the water bridge experiment.

However, MD simulations on small cold water clusters in relatively weak electric fields approximately $10^5$ V/m do not show any substantial structural changes but an increased, O:H related, molecular vibrational amplitude and molecular reorientation. Other MD simulations have suggested that it is the polarization of hydrogen bonds that induces the formation of the water bridge above a threshold field of $1.2 \times 10^9$ V/m. There is a conceptual gap at mesoscopic scale which is bridged by a number of theories including quantum mechanical entanglement and coherent structures in water [8].

#### 15.4.4.3 Force Equilibrium Considerations

Typically, there are two views on the forces equilibrating the bridge. One is the tension along the bridge caused by the electric field within the dielectric material [31]. The tension due to the electric field in a dielectric medium is given by:

$$T_{DE} = \varepsilon_0(\varepsilon_r - 1)E^2 A, \tag{15.2}$$

where $A = \pi D^2/4$ is the cross-sectional area of the bridge, $\varepsilon_r$ is the relative permittivity of water, and $\varepsilon_0$ is the permittivity of the vacuum. If a tension $T_{DE}$ is acting on a curved bridge of curvature $\xi$, the vertical force produced per unit length of the bridge is $\xi T_{DE}$, while the gravitational force per unit length is $A\rho g$. Thus the ratio of the dielectric force and the gravitational force is:

$$R_{DE} = \frac{\varepsilon_0(\varepsilon_r - 1)E^2 \xi}{\rho g}. \tag{15.3}$$

Water molecules assume a preferred orientation along the electric field within the floating water bridge [8].

The second mechanism suggests that the force holding the bridge is the surface tension and the electric field that avoids the breaking up the bridge into small droplets and maintains its stability. According to Aerov [47], the electric tension along the bridge is zero. A stable equilibrium for the bridge's surface is reached when the axial electric field is strong enough to counteract distortions caused by surface tension. The tension caused by surface tension is the sum of the tension on the sides (γl) and the repulsion caused by the pressure jump at the surface (−γl/2):

$$T_{ST} = \frac{\gamma l}{2}, \quad (15.4)$$

where l = πD is the perimeter of the cross-section of the bridge. According to this notion, the ratio between the upward surface-tension force and the gravitational force is then:

$$R_{ST} = \frac{2\gamma\xi}{\rho g D}. \quad (15.5)$$

The opinion [47] that the ordinary surface tension rather than the electric hydrodynamic force holds the bridge together conflicts with the observation [32] that an increased H–bond strength in the water bridge. The special properties of the bridges are attributed to performance of water molecules in the surface layers [39]. However, Teschke et al. [45] suggested that the charged clusters travel to the bridge surface and modify its structural properties to form a plastic layer with a ~85 kPa stiffness, which is responsible for the plasticity of the bridge.

#### 15.4.4.4 The Fourth Phase—Exclusion Zone

Although the phenomenon needs to be studied further, the scientific community agrees that polarization at the water surface when a high tangent electrical field is applied is responsible for the extraordinary stability of the system. Pollack [48] suggested that this bridge is made of an $H_3O^+$ lattice or the exclusion zone of water. The exclusion zone water is gel like, excludes charge and micro-organisms, and absorbs all sorts of energy. But it is unclear how the electric field generates the exclusion zone that is always associated with the hydrophilic contacting interface according to Pollack.

## 15.5 Quantitative Resolution

### *15.5.1 O:H–O Bond Deformation and Polarization*

Applying an electric field to deionized water by two separated electrodes is equivalent to put the liquid between two plates of a capacitor with some leaking of current. The electric field between the electrodes will align, stretch, and polarize water molecular dipoles in a long-range order, which effects the same to solute ions that create each a short-range electric field. The fields of ions are less-ordered but the field of a capacitor is long-range and directional.

If the electrodes are applied to aqueous solutions, fields of solute ions will align in such a way that oppose the field of the capacitor as the cation-anion dipoles will align against the capacitor field. The resultant field will be weaker than either alone. If the capacitor field is sufficiently high, ions will separate and flow in their respective direction, which destroy the water bridge cable.

Electrification of molecular dipoles by any form of electric fields will lengthen and polarize the O:H–O bond by elongating the O:H nonbond and shortening the H–O bond associated with phonon frequency relaxation. This relaxation will stiffen the H–O phonon and soften the O:H phonon, which will raise the $T_m$ and depress the $T_N$ by dispersing the boundary of the quasisolid phase. The polarization will raises the viscosity and the molecular and H–O phonon relaxation times, and therefore, the bridge will be plastic and stable and the molecular mobility turns to be lower.

Figure 15.8 shows a typical example for the alignment, elongation, and structure opening of $(H_2O)_N$ clusters [49]. Subjected to the $sp^3$-orbital hybridization and the applied electric field, $H_2O$ molecules remain with adjusted orientations, O:H length, and the polarization, as demonstrated in Fig. 15.9. At the critical electric field, the O:H nonbond breaks, which laid foundation of the Tylor cone-jet-spray formation, see Fig. 15.3.

Figure 15.10 shows the contour plots for the cluster size and electric field dependence of the molecular dipole moment and the optimal binding energy for the clusters. Figure 15.11 shows the typical IR spectra for the H–O stretching phonons. Indeed, electrification shortens the H–O bond and stiffens its phonon, which is associated with O:H nonbond lengthening and softening—higher vibration amplitude and lower frequency. Despite some discrepancies due to geometric configurations of the sized clusters, observations [49–51] agree with the expectation of the present electrification mechanism—electric field aligns, elongates, and polarizes the O:H–O bond.

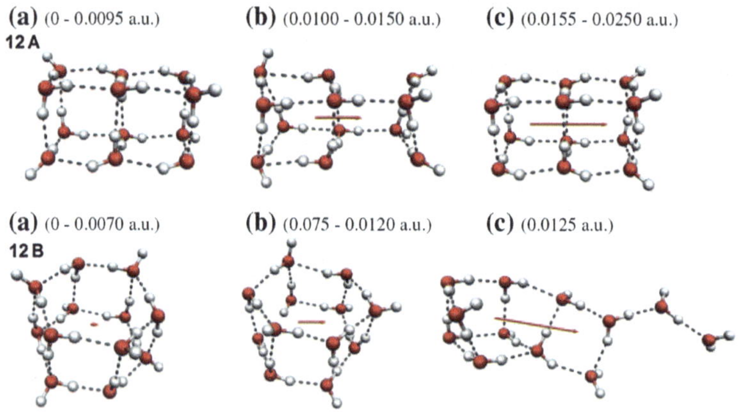


**Fig. 15.8** Typical samples of the energetically stable $(H_2O)_{12}$ conformers in the order of increasing energy, at zero field, evolving with applied electric field in the direction of the dipole moment (An atomic unit of electric field is 51.42 eV/Å. Reprinted with permission from [49].)

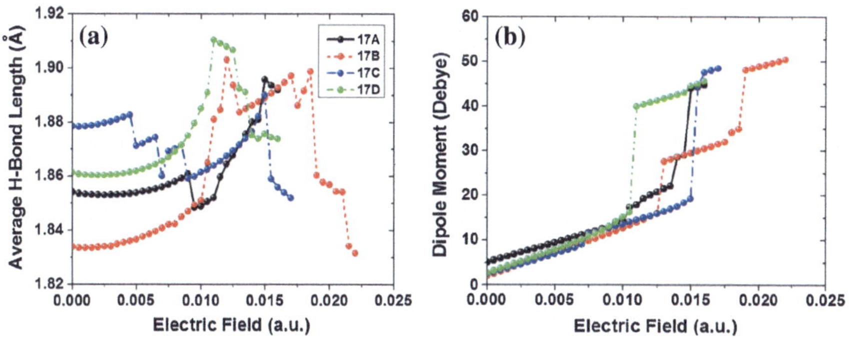


**Fig. 15.9** **a** Average O:H length as function of the applied electric field (in atomic units) for the lowest energy conformers of $(H_2O)_{17}$ showing maxima (*peaks*) corresponding to opening up of the cluster followed by breakdown of the O:H network. **b** Electric dipole moment showing abrupt increments associated with reorientation of the molecules in the cluster and breaking of O:H nonbonds at characteristic field values (Reprinted with permission from [50].)

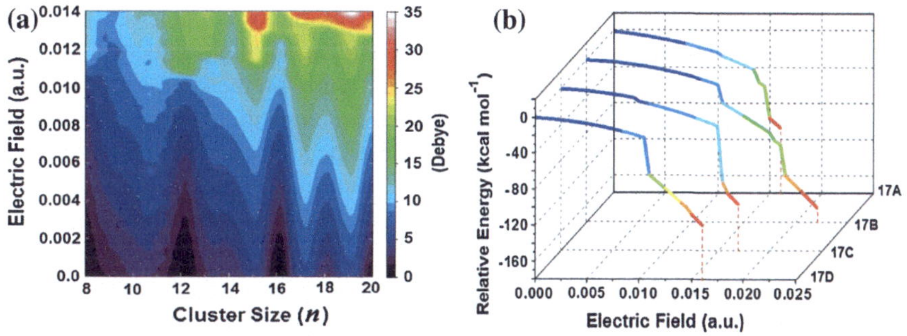


**Fig. 15.10** **a** Computational map for $(H_2O)_N$ cluster size and electric field dependence of the dipole moment. **b** Relative energy (with respect to the respective field free energy) of $(H_2O)_{17}$ conformers (17A–17D) during conformational transition brought about by the electric field applied along the permanent dipole moment (Reprinted with permission from [50].)

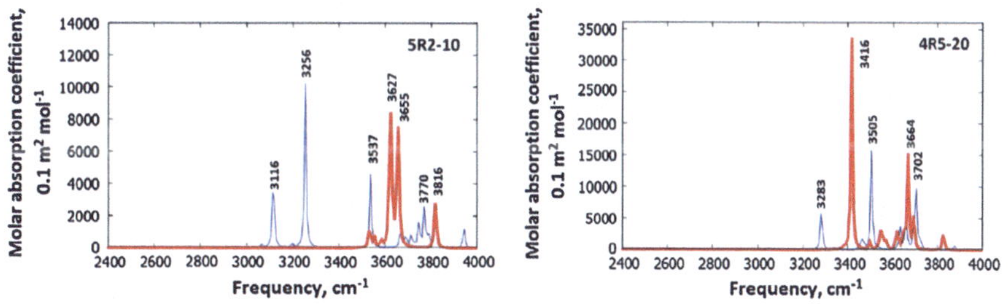


**Fig. 15.11** Typical IR frequencies for $H_2O$ cluster under zero (*thinner blue lines*) and the breakdown field (*thicker red lines*) (Reprinted with permission from [51].) (color online)

## 15.5.2 *Droplet Electrofreezing: Quasisolid Phase Dispersion*

### 15.5.2.1 Multifield Effect

The critical temperatures for melting $T_m$ and for homogeneous ice nucleation $T_N$ are different but they are often confused as identical. Water transits from liquid to quasisolid at 273 K and then from quasisolid to homogeneous ice at 258 K [52]. The quasisolid phase boundaries are symmetrically dispersible by applying stimuli such as mechanical compression or tension, molecular undercoordination [6] and electrification. The freezing temperature depression is always accompanied with melting point elevation. O:H–O bond electrification by any form of electric field also tunes the freezing and melting temperature of water, which is of topical importance in the living and inanimate worlds.

Both skin molecular undercoordination and charge field electrification disperse the quasisolid phase boundary outwardly. As a consequence of the H–O phonon stiffening and O:H phonon softening, the quasisolid phase boundary disperses outwardly, which raises the $T_m$ and lowers the $T_N$.

However, repulsive field created by like charges opposes the electrification, called anti-electrification. Because of the skin polarization that produces localized dipoles with excessive negative charge locked uniformly on the skin, water droplet has the intrinsic effect of self-anti-electrification, a competition of the undercoordination, artificial electrification, and self-antielectrification dictates the thermodynamic behavior of a droplet of neat water.

Therefore, the floating bridge is actually in the quasisolid phase under electrification and the bridge does not melt up to 60 °C temperature. However, if a repulsive field is applied across a droplet, the situation reverts, which lowers the $T_m$ and raises the $T_N$.

### 15.5.2.2 Substrate Charging Effect on the $T_m$

Without applying a bias, the skin of a droplet still possesses excessive, negative, strongly localized charge due to undercoordination induced molecular polarization (BOLS-NEP description). The freezing/melting behavior of the droplet will change with the quantity and sign of charge on the substrate that the droplet deposited [17].

The negatively charged substrate will establish a repulsive field between the substrate and the skin of the water droplet—anti-electrification, which disperses the quasisolid phase boundary inwardly, resulting in $T_N$ elevation and $T_m$ depression. Indeed, water droplets cooled down on a negatively charged $LiTaO_3$ and $SrTiO_3$ surfaces remain liquid at −11 °C because of $T_m$ depression. The preference of liquid-quasisolid transition at the air/water interface [17] indicates that the local repulsive field is weaker in the air/water interface.

An attractive field forms in the liquid between the positively charged substrates and the electrons in the droplet skin, being the same to inonic electrofication in aqueous solutions, which raises the $T_m$ for liquid-quasisolid transition and lowers the $T_N$ for quasisolid-ice transition. Exactly, the positively charged $LiTaO_3$ and $SrTiO_3$ raises the $T_m$ from −11 to at −8 °C, immediately. The preference liquid-quasisolid transition t the solid/water interface [17] indicates that the stronger attractive field at the solid/water interface.

The intrinsic $T_m$ between −8 and −11 °C for the droplet is lower than that of the bulk melting at 0 °C, which evidences the skin dipole antielectrification of the droplet.

#### 15.5.2.3 Capacitor Bias Effect on the $T_m$

Figures 15.12 and 15.13 show the electrofreezing of deionized water droplet on different substrates with controlled current leaking [16]. Bias is applied between the upper skin of the droplet and the dielectric substrates with and without pinholes for limiting the current flow. The electric field is focused on the solid skin by charging

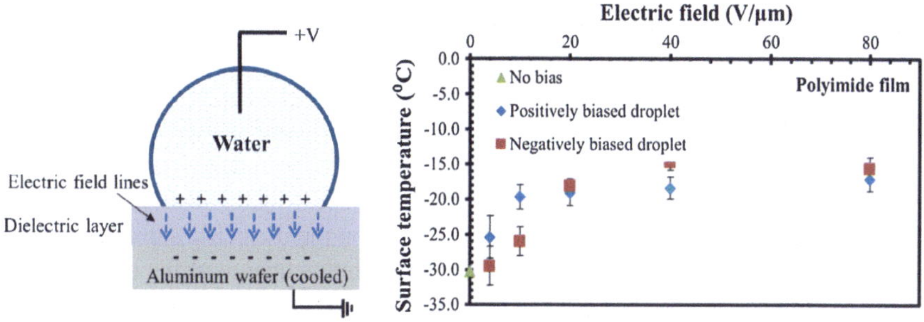


**Fig. 15.12** Bias dependence of the skin $T_N$ elevation of water droplet deposited on dielectric polymide substrate. Droplet positive bias creates a relatively weaker field that depresses the $T_N$ than stronger field (Reprinted with permission from [16].)

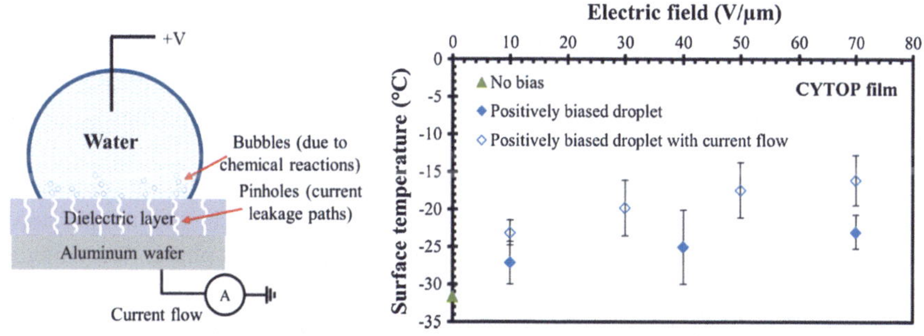


**Fig. 15.13** Bias dependence of the skin $T_N$ of water droplet deposited on dielectric CYTOP substrate with and without and current flow. The less depression of the $T_N$ by current flow indicates the weaker inner field of the droplet (Reprinted with permission from [16].)

in opposite sign accumulated on the droplet bottom, or on the surface of the dielectric layer. The resultant field of the applied and the substrate polarization-created electrify water molecules.

Observations in Fig. 15.12 revealed that the resultant field elevates the freezing temperatures from −30 to −15 °C, in the absence of current flow, the freezing starts from the triple phase contact line rather than from the solid-liquid interface or the air/water interface. The $T_C$ saturates at $4 \times 10^7$ V/m field. The polarity of the interfacial charge influence slightly the electrofreezing. Below saturation, the positive bias raises the $T_C$ but after saturation the negative raises the $T_C$.

However, a close examination of observations revealed that the negative bias creates positive charge on the substrate to form an attraction field within the droplet, which raises the $T_m$ but lowers the $T_N$. Therefore, the critical temperature corresponds to the quasisolid-ice transition $T_N$. The $T_N$—E profiles in Fig. 15.12 indicates that the positively biased droplet creates a weaker field inside than the negatively biased before saturation. The weaker field depresses the $T_N$ less than a stronger electric field.

In contrast, pinhole formation in the dielectric film allows current flow and bubble creation when water molecules penetrate into the holes, which isolates the effect of current flow from the effect of electric field. Current flow reduces the $T_N$ by 6.5 °C or so compared with that without current flow.

According to the present premise of the electric field superposition, pinhole current flow brought in two factors that depress the $T_N$: one is the bubble creation and the other is the internal field weakening. Undercoordinated molecules in the bubble skin lowers locally the $T_N$ because of its supersolidity nature; the current flow reduces the number of electrons accumulated on the dielectric substrate, which weakens the field between the substrate and the negative charge of the droplet skin. Therefore, the current leaking weakens the quasisolid nature of the droplet and lowers its $T_N$ less that that without current flow.

#### 15.5.2.4 Quantum Depression of Electrofreezing

Figure 15.14 shows a typical example for the E field dependence of the in-plane diffusion coefficient of 0.79 nm thick monolayer water molecules. The $T_C$ of the film decreases from 325 to 278 K when the external electric field is increased from 0 to $10^9$ V/m [18]. This observation agrees with the present prediction that $T_N$ drops with the increase of electric field strength.

Figure 15.15 shows the electrification effect on the in-plane diffusion coefficient for a bilayer of 0.95 nm across at 240 and 300 K [19]. Within this temperature range, the highly undercoordinated specimen is in the quasisolid phase. Assigning a charge of ±q to the O atoms located diagonally in neighboring hexagons across the film creates a perpendicular electric field across.

For the 300 K specimen, the coefficient approaches to zero when the q increases to 0.7 e, which indicates that quasisolid turns into ice by electrification that lengthens the O:H bond and softens its phonon. For the 240 K specimen,

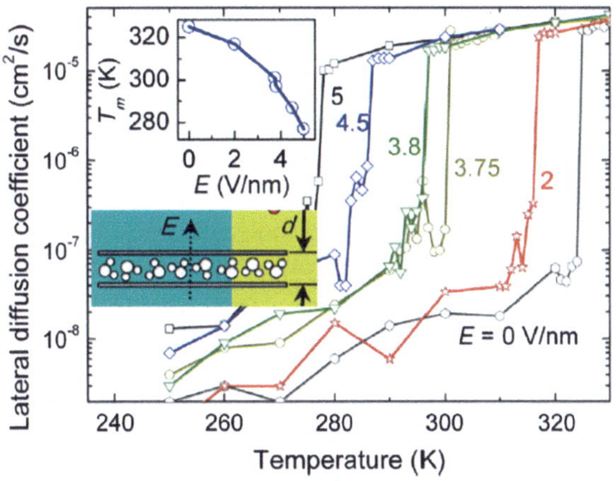


**Fig. 15.14** Temperature and electric field intensity dependence of the in-plane diffusion coefficient of the confined monolayer water of d = 0.79 nm thick (*inset*). The *insets* show the $T_N$ (instead of $T_m$ according to the present premise) as a function of E and the setup of the monolayer molecules and the direction of the E field (Reprinted with permission from [18].)

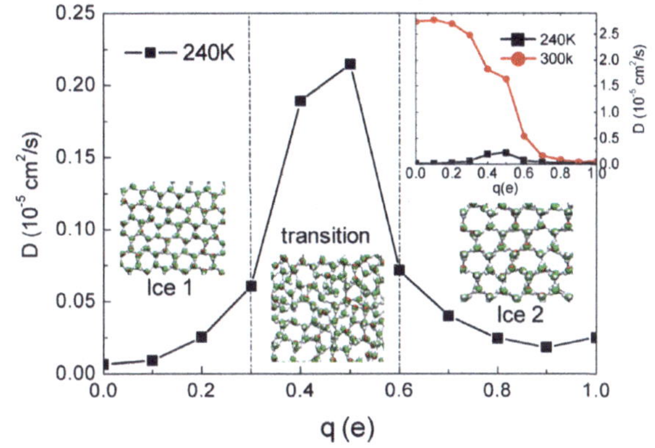


**Fig. 15.15** In-plane diffusion coefficient of water at T = 240 K and at T = 300 K (the *inset* shows the comparison with the maximum for 300 K at $2.7 \times 10^{-5}$ $cm^2/s$) as a function of charge ±q assigned to O atoms located diagonally in neighboring *hexagons* across the film (Reprinted with permission from [19].)

electrification by field due to q = 0.3−0.6 e transit from ice into a state of nonzero diffusivity because of disruption of the hexagonal arrangement in ice 1. The diffusion coefficient in the quasisolid phase is about 0.2 and in the ice it is below 0.05 unit. Further q increase induces phase transition that stabilize ice molecules. The density maximum of the bilayer molecules shifts from 0.34 to 0.29 nm away from the wall plates, which indicates the electrification stretching of water molecules, which agrees the mirror charge electric field as discussed in Sect. 3.4.5.

#### 15.5.2.5 Melting Point $T_m$ Elevation

The "freezing (should be melting instead) temperature" of water is also altered by electrical fields in narrow cracks at the hydrophobic faces of alpha-amino acid single crystals [53]. Water vapor cooled to freezing between the hydrophobic faces of single crystals of a series of pairs of racemic and chiral-resolved hydrophobic alpha-amino acids. The morphologies and molecular packing arrangements within each crystal pair are similar but only one of each pair exhibits a polar axis, parallel to the hydrophobic face exposed to water. Those crystals that have a polar axis raises the freezing (melting) point by 4–5 °C compared with crystals that do not have a polar axis.

Liquid water transits into ice (should be quasisolid instead) inside a gap of nanometer spacing under the control of electric fields and gap distance. Choi and co-workers [54] trapped the water in a nanometer wide gap between the gold-plated tip of an STM and a gold surface. The water frozes in a much weaker electric field of just $10^6$ volts per meter. The STM bias raised $T_m$ of liquid water evidences the H–O bond contraction by electrification, which is the same to the Water Bridge. The sudden, reversible phase transition occurs in electric fields of $10^6$ V $m^{-1}$ at room temperature [54], which agrees with observations that the rate of ice nucleation (quasisolid formation) from the vapor phase substantially increases in electric fields above $10^4$ $Vm^{-1}$ with respect to the normal growth rate [55].

Observations [53, 55] evidence the effect of electrification and molecular undercoordination raise the temperature transiting liquid into the quasisolid state–melting point elvation that is associated with freezing and evaporating point depression.

### 15.5.3 *Soap Film Electrification*

An electric field can drive fluid through a narrow pipe by pulling on ions in the solution. The ions are normally an equal mixture of positive and negative charges, but an excess of one type typically collects along the surface of the channel because the walls are charged. When these surface ions are pulled by an electric field, they drag fluid along with them, including all of the fluid from the interior (or "bulk") of the channel. This flow—called electroosmosis—is dominant in micro- and nanochannels, where the surface-to-volume ratio is high.

Bonhomme et al. [56, 57] demonstrated in Fig. 15.16 that an electric field can pull fluid through a soap film of a 100 nm thick, which is essentially a narrow channel with deformable walls. They found that the film thickness, the flow rate, and the conductance relaxation time increase with the strength of the electric field applied across the electrodes.

According to the present notion for the Armstrong water floating bridge formation, the increase of wall thickness, flow rate, and the conductance relaxation

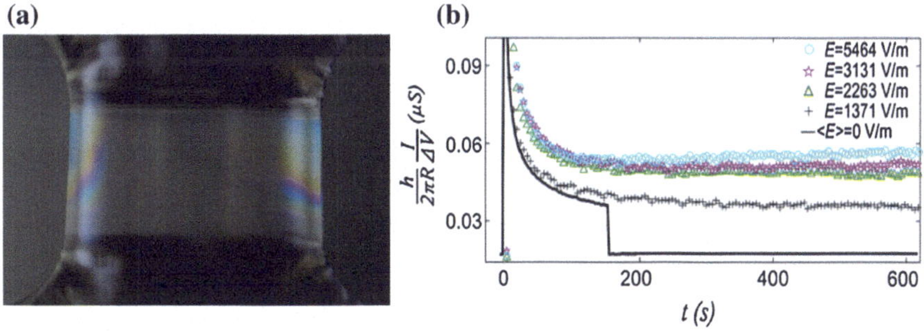


**Fig. 15.16** **a** Voltage applied across this cylindrical soap film causes fluid to flow up against gravity in the roughly 100 nm thick film. **b** Geometrically normalized conductance G versus time and for different applied electric fields. Applying a voltage thickens the film, lengthens the relaxation time of conductance, and raises the flow rate of the solution (Reprinted with permission from [56, 57].)

time indicate that the electrification of water molecules increases the viscosity and slow the molecular fluctuation dynamics, which raise the tension, viscosity and structure order of water molecules in quasisolid soap film. Electrification disperses the quasisolid phase boundaries outwardly so the soap film is in the quasisolid phase.

### *15.5.4 Phonon Spectrometric Evidence*

However, the Raman residual phonon spectrum in Fig. 15.1b reveals that the bridge formation raises the $\omega_H$ from 3100 to 3500 $cm^{-1}$, slightly higher than the characteristic peak of the supersolid skin at 3450 $cm^{-1}$. The preliminary Raman DPS for 298 K deionized water shown in Fig. 15.17 is in agreement with Fig. 15.1b. Therefore, polarization and viscoelasticity due to skin supersolidity may dominate in holding the bridge. The $\omega_H$ offsets from 3450 to 3500 $cm^{-1}$ may indicate that the electric effect enhances skin supersolidity by promoting molecular polarization that shortens and stiffens the H–O bond further. Therefore, the skin supersolidity, long-range electric force, and gravitational force will contribute to the bridge.

## 15.6 Soil Wetting by Aqueous Solution

Soil particles also produce strong electric field by the surface charges, which profoundly influence water movement by diffusion. Studies [58] suggested that the electric field influences water transport in soil by affecting: (i) the electrostatic forces between the electric field and water molecular dipoles; and (ii) the soil water

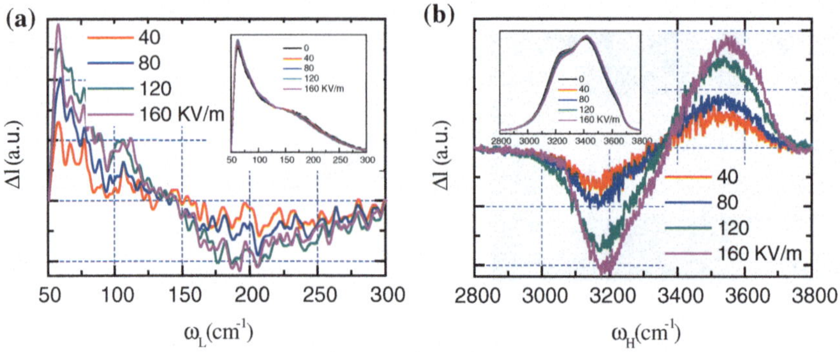


**Fig. 15.17** Raman DPS as a function of the electric field (3.8 mm electrode separation) of deionized water at 298 K. The H–O bond is insensitive to the applied potential below 1000 V

conductivity through controlling stability of soil aggregates. There exists a critical surface potential of 153 mV. Higher electric field above this critical value breaks soil aggregation and enhances the field-dipole interaction. Conversely, lower electric field stabilizes the soil aggregation. This observation suggests means to regulate soil water diffusion by adjusting soil electric field using electrolytes. Figure 15.18 shows the time dependent water transition depth as a function of electrolyte concentration [58].

According to the present notions of solute electrification and particle skin charge induction, both soil particles and aqueous solutes have the same effect of electrification that stretches and polarizes the molecules to enhance the viscosity that is

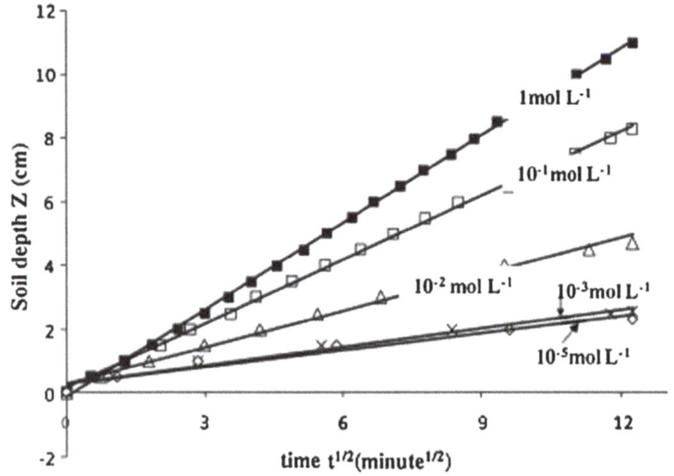


**Fig. 15.18** $NaNO_3$ concentration dependence of the wetting front positions in soil as a function of time (Reprinted with permission from [58].)

detrimental to water transportation within pore of cylindrical soil columns. However, soil particles and solute hydration clusters are at different size scales. Water molecular dipoles and solute dipoles orient in such a manner that cancel the field of the soil article, which much like the solute ions destabilize the water bridge. Therefore, the resultant electrification effect is weaker than under the soil particle or the solute point polarizer potential alone. The electrolyte contained fluid becomes thinner in the soil pores.

## 15.7 Summary

The combination of skin supersolidity induced by molecular undercoordination and the quasisolidity enhanced by ordered electrification is responsible for the formation, plasticity, and stability of the Water Bridge and the electrofreezing phenomena in various circumstances:

(1) Electrification by the capacitor's field raises the long-range order of water molecules, which aligns, stretches, and polarizes the O:H–O bond. The O:H–O elongation results in H–O contraction and O:H elongation associated with $\omega_H$ phonon stiffening and $\omega_L$ softening as a consequence of O–O repulsion.
(2) Electric field due to the supersolid skin dipoles plays its determining the droplet thermodynamics.
(3) Phonon relaxation offsets the Debye temperatures and disperses the boundary of the quasisolid phase, which raises the $T_m$ for liquid/quasisolid transition and depresses the $T_N$ for quasisolid/ice transition. The Bridge is in the quasisolid phase that is plastic and thermally stable.
(4) Electrification of O:H–O bond by field of like charges disperses inwardly but field of unlike charges disperses outwardly the boundary of the quasisolid phase, and hence, electrofreezing and electromelting proceed in contrast manners.
(5) Solute ions add additional electric fields that oppose the field of the capacitor so ions destabilize the bridges.
(6) Water molecules and the O:H–O bond are extremely sensitive to the electric perturbation in a long-range ordering manner.

## References

1. W. Armstrong, Electrical phenomena the newcastle literary and philosophical society. Electr. Eng. **10**, 153 (1893)
2. R.M. Namin, S.A. Lindi, A. Amjadi, N. Jafari, P. Irajizad, Experimental investigation of the stability of the floating water bridge. Phys. Rev. E **88**(3), 033019 (2013)
3. J. Woisetschlaeger, A.D. Wexler, G. Holler, M. Eisenhut, K. Gatterer, E.C. Fuchs, Horizontal bridges in polar dielectric liquids. Exp. Fluids **52**(1), 193–205 (2012)

4. F. Saija, F. Aliotta, M.E. Fontanella, M. Pochylski, G. Salvato, C. Vasi, R.C. Ponterio, Communication: an extended model of liquid bridging. J. Chem. Phys. **133**(8), 081104 (2010)
5. R.C. Ponterio, M. Pochylski, F. Aliotta, C. Vasi, M.E. Fontanella, F. Saija, Raman scattering measurements on a floating water bridge. J. Phys. D: Appl. Phys. **43**(17), 175405 (2010)
6. Y. Huang, X. Zhang, Z. Ma, Y. Zhou, W. Zheng, J. Zhou, C.Q. Sun, Hydrogen-bond relaxation dynamics: resolving mysteries of water ice. Coord. Chem. Rev. **285**, 109–165 (2015)
7. X. Zhang, P. Sun, Y. Huang, Z. Ma, X. Liu, J. Zhou, W. Zheng, C.Q. Sun, Water nanodroplet thermodynamics: quasi-solid phase-boundary dispersivity. J. Phys. Chem. B **119**(16), 5265–5269 (2015)
8. E.C. Fuchs, A.D. Wexler, A.H. Paulitsch-Fuchs, L.L.F. Agostinho, D. Yntema, J. Woisetschlager, The Armstrong experiment revisited. Eur. Phys. J. Spec. Top. **223**(5), 959–977 (2014)
9. E.C. Fuchs, Can a century old experiment reveal hidden properties of water? Water **2**(3), 381–410 (2010)
10. C.T.R. Wilson, G.I. Taylor, The bursting of soap-bubbles in a uniform electric field. Math. Proc. Camb. Philos. Soc. **22**(05), 728–730 (1925)
11. W.A. Macky, Some investigations on the deformation and breaking of water drops in strong electric fields. Proc. R. Soc. Lond. A: Math. Phys. Eng. Sci. **133**(822), 565–587 (1931)
12. G. Taylor, Disintegration of water drops in an electric field. in *Proceedings of the Royal Society of London A: Mathematical, Physical and Engineering Sciences*. The Royal Society (1964)
13. D.B. Victor, Tylore cone-jet-spray (2013), https://www.youtube.com/watch?v=r6TGvG7RUyo
14. F. Liu, C.-H. Chen, Electrohydrodynamic cone-jet bridges: stability diagram and operating modes. J. Electrostat. **72**(4), 330–335 (2014)
15. M. Gunji, M. Washizu, Self-propulsion of a water droplet in an electric field. J. Phys. D: Appl. Phys. **38**(14), 2417–2423 (2005)
16. K. Carpenter, V. Bahadur, Electrofreezing of Water Droplets Under Electrowetting Fields (Langmuir, 2015)
17. D. Ehre, E. Lavert, M. Lahav, I. Lubomirsky, Water freezes differently on positively and negatively charged surfaces of pyroelectric materials. Science **327**(5966), 672–675 (2010)
18. H. Qiu, W. Guo, Electromelting of confined monolayer ice. Phys. Rev. Lett. **110**(19), 195701 (2013)
19. F. Mei, X. Zhou, J. Kou, F. Wu, C. Wang, H. Lu, A transition between bistable ice when coupling electric field and nanoconfinement. J. Chem. Phys. **142**(13), 134704 (2015)
20. L. Dufour, Uber das gefrieren des wassers und uber die bildung des hagels. Poggendorfs Ann. Physik **114**, 530–554 (1861)
21. W. Rau, Eiskeimbildung durch dielektrische Polarisation. Zeitschrift für Naturforschung A **6** (11), 649–657 (1951)
22. H. Pruppacher, Electrofreezing of supercooled water. Pure Appl. Geophys., **104**(1), 623–634 (1973)
23. H.R. Pruppacher, The effects of electric fields on cloud physical processes. Zeitschrift für angewandte Mathematik und Physik ZAMP **14**(5), 590–599 (1963)
24. P. Wilson, K. Osterday, A. Haymet, The effects of electric field on ice nucleation may be masked by the inherent stochastic nature of nucleation. CryoLetters **30**(2), 96–99 (2009)
25. J. Doolittle, G. Vali, Heterogeneous freezing nucleation in electric fields. J. Atmos. Sci. **32**(2), 375–379 (1975)
26. T. Hozumi, A. Saito, S. Okawa, K. Watanabe, Effects of electrode materials on freezing of supercooled water in electric freeze control. Int. J. Refrig. **26**(5), 537–542 (2003)
27. E.C. Fuchs, A. Cherukupally, A.H. Paulitsch-Fuchs, L.L.F. Agostinho, A.D. Wexler, J. Woisetschläger, F.T. Freund, Investigation of the mid-infrared emission of a floating water bridge. J. Phys. D: Appl. Phys. **45**(47), 475401 (2012)

28. E.C. Fuchs, P. Baroni, B. Bitschnau, L. Noirez, Two-dimensional neutron scattering in a floating heavy water bridge. J. Phys. D: Appl. Phys. **43**(10), 105502 (2010)
29. E.C. Fuchs, B. Bitschnau, J. Woisetschläger, E. Maier, B. Beuneu, J. Teixeira, Neutron scattering of a floating heavy water bridge. J. Phys. D: Appl. Phys. **42**(6), 065502 (2009)
30. E.C. Fuchs, J. Woisetschlager, K. Gatterer, E. Maier, R. Pecnik, G. Holler, H. Eisenkolbl, The floating water bridge. J. Phys. D: Appl. Phys. **40**(19), 6112–6114 (2007)
31. A. Widom, J. Swain, J. Silverberg, S. Sivasubramanian, Y.N. Srivastava, Theory of the Maxwell pressure tensor and the tension in a water bridge. Phys. Rev. E **80**(1), 016301 (2009)
32. L. Piatkowski, A.D. Wexler, E.C. Fuchs, H. Schoenmaker, H.J. Bakker, Ultrafast vibrational energy relaxation of the water bridge. Phys. Chem. Chem. Phys. **14**(18), 6160–6164 (2012)
33. E.C. Fuchs, K. Gatterer, G. Holler, J. Woisetschläger, Dynamics of the floating water bridge. J. Phys. D: Appl. Phys. **41**(18), 185502 (2008)
34. A.D. Wexler, M. Lopez Saenz, O. Schreer, J. Woisetschlager, E.C. Fuchs, The preparation of electrohydrodynamic bridges from polar dielectric liquids. J. Vis. Exp. (91), e51819 (2014)
35. A.H. Paulitsch-Fuchs, E.C. Fuchs, A.D. Wexler, F.T. Freund, L.J. Rothschild, A. Cherukupally, G.J. Euverink, Prokaryotic transport in electrohydrodynamic structures. Phys. Biol. **9**(2), 026006 (2012)
36. R.J. Raco, Electrically supported column of liquid. Science **160**(38–25), 311 (1968)
37. A.G. Marin, D. Lohse, Building water bridges in air: Electrohydrodynamics of the floating water bridge. Phys. Fluids **22**(12), 122104 (2010)
38. J. Woisetschlager, K. Gatterer, E.C. Fuchs, Experiments in a floating water bridge. Exp. Fluids **48**(1), 121–131 (2010)
39. L.B. Skinner, C.J. Benmore, B. Shyam, J. Weber, J.B. Parise, Structure of the floating water bridge and water in an electric field. Proc. Natl. Acad. Sci. **109**(41), 16463–16468 (2012)
40. J. Woisetschläger, A.D. Wexler, G. Holler, M. Eisenhut, K. Gatterer, E.C. Fuchs, Horizontal bridges in polar dielectric liquids. Exp. Fluids **52**(1), 193–205 (2012)
41. O. Teschke, D. Mendez Soares, J.F. Valente Filho, Floating liquid bridge tensile behavior: electric-field-induced young's modulus measurements. Appl. Phys. Lett. **103**(25), 251608 (2013)
42. D. Rai, A.D. Kulkarni, S.P. Gejji, R.K. Pathak, Water clusters $(H_2O)n$, n = 6–8, in external electric fields. J. Chem. Phys. **128**(3), 034310 (2008)
43. H. Nishiumi, F. Honda, Effects of electrolyte on floating water bridge. Adv. Phys. Chem. **2009** (2009)
44. S.T. van der Post, C.S. Hsieh, M. Okuno, Y. Nagata, H.J. Bakker, M. Bonn, J. Hunger, Strong frequency dependence of vibrational relaxation in bulk and surface water reveals sub-picosecond structural heterogeneity. Nat. Commun. **6**, 8384 (2015)
45. O. Teschke, W. Gomes, J. Roberto de Castro, J.F.V. Filho, Relaxation oscillations in floating liquid bridges. Chem. Sci. J. **6**, 99 (2015)
46. Y.C. Choi, C. Pak, K.S. Kim, Electric field effects on water clusters (n = 3–5): systematic ab initio study of structures, energetics, and transition states. J. Chem. Phys. **124**(9), 094308 (2006)
47. A.A. Aerov, Why the water bridge does not collapse. Phys. Rev. E **84**(3), 036314 (2011)
48. G.H. Pollack, *The Fourth Phase of Water: Beyond Solid, Liquid, and Vapor* (Ebner and Sons Seattle, USA, 2013)
49. D. Rai, A.D. Kulkarni, S.P. Gejji, R.K. Pathak, Water clusters $(H(_2)O)(n)$, n = 6–8, in external electric fields. J. Chem. Phys. **128**(3) (2008)
50. D. Rai, A.D. Kulkarni, S.P. Gejji, L.J. Bartolotti, R.K. Pathak, Exploring electric field induced structural evolution of water clusters, $(H_2O)n$ [n = 9−20]: density functional approach. J. Chem. Phys. **138**(4), 044304 (2013)
51. L.J. Bartolotti, D. Rai, A.D. Kulkarni, S.P. Gejji, R.K. Pathak, Water clusters $(H_2O)n$ [n = 9–20] in external electric fields: exotic OH stretching frequencies near breakdown. Comput. Theor. Chem. **1044**, 66–73 (2014)

52. C.Q. Sun, X. Zhang, X. Fu, W. Zheng, J.-L. Kuo, Y. Zhou, Z. Shen, J. Zhou, Density and phonon-stiffness anomalies of water and ice in the full temperature range. J. Phys. Chem. Lett. **4**, 3238–3244 (2013)
53. M. Gavish, J.L. Wang, M. Eisenstein, M. Lahav, L. Leiserowitz, The role of crystal polarity in alpha-amino-acid crystals for induced nucleation of ice. Science **256**(5058), 815–818 (1992)
54. E.-M. Choi, Y.-H. Yoon, S. Lee, H. Kang, Freezing transition of interfacial water at room temperature under electric fields. Phys. Rev. Lett. **95**(8), 085701 (2005)
55. J.T. Bartlett, A.P. Vandenheuvel, B.J. Mason, Growth of ice crystals in an electric field. Z. Angew. Math. Phys. **14**(5), 599 (1963)
56. M. Schirber, Focus: tuning the flow through a soap film. Physics **6**, 12 (2013)
57. O. Bonhomme, O. Liot, A.-L. Biance, L. Bocquet, Soft nanofluidic transport in a soap film. Phys. Rev. Lett. **110**(5), 054502 (2013)
58. Z. Yu, H. Li, X. Liu, C. Xu, H. Xiong, Influence of soil electric field on water movement in soil. Soil Tillage Res. **155**, 263–270 (2016)

# Chapter 16
# Miscellaneous Issues

- *Multiple stimuli relax the O:H–O bond in a superposition manner.*
- *Charge induction and energy absorption proceed in a long-range order.*
- *O:H–O bond cooperativity is general to systems with nonbonding lone pairs being involved.*
- *Discriminative specific heats due to the asymmetrical, coupled, short-range interactions stem the negative thermal expansion of other substances.*

**Abstract** Furnished with the rules established for O:H–O bond relaxation and polarization, one can readily gain consistent insight into the common mechanism behind the unusual behavior of water and ice under excitations by multiple fields such as isotope addition, charge induction, electromagnetic radiation, ac field electrification. The O:H–O relaxation and polarization promises the resolution to general situations of X:H–Y interaction, negative thermal expansion, dielectric relaxation, and clarify the quasisolidity of the "polywater". Dominated by the asymmetrical, short-range, and coupled interactions, the O:H–O bond responds to excitation in a manner of long-range order.

## 16.1 Multifield Coupling

### *16.1.1 Undercoordination Versus Heating*

Water droplet size reduction raises its curvature and the fraction of the undercoordinated molecules. The curvature increase also lowers the CN of molecules in the skin. Thus, the supersolidity of the highly curved skin is stronger than a less curved skin. The H–O bond there becomes even shorter, polarization is even stronger, molecular motion becomes slower, the skin is stiffer, the melting point becomes higher and the freezing and evapirating point turns to be lower when the skin is highly curved.

Undercoordination effects the same of liquid and solid heating on O:H–O relaxation but undercoordination enhances the polarization. Medcraft et al. [1] confirmed this trend by examining the joint effects of size reduction and thermal excitation on the $\omega_H$ frequencies of ice nanoparticles of 3–200 nm in diameter over a temperature range of 5–209 K. They observed that reducing particle size below 5 nm stiffens the $\omega_H$ by some 40 cm$^{-1}$. The size effect is not apparent for particles larger than 8 nm. They also found that the peak $\omega_H$ shifts up by 35 cm$^{-1}$ from

 
C.Q. Sun and Y. Sun, *The Attribute of Water*, Springer Series
in Chemical Physics 113, DOI 10.1007/978-981-10-0180-2_16

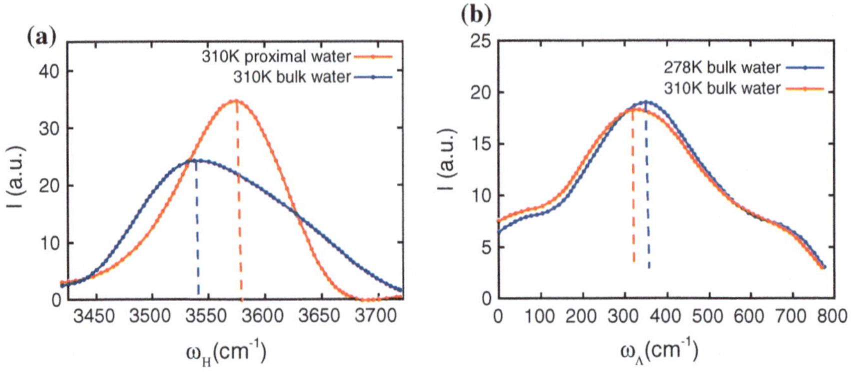


**Fig. 16.1** Cluster size reduction (proximal) (**a**) stiffens the $\omega_{\mathrm{H}}$ at the same temperature; and (**b**) heating softens the $\omega_{\mathrm{L}}$ of bulk water because both molecular undercoordination and heating shorten and stiffen the H–O bond, and lengthen and soften the O:H nonbond [5] (Reprinted with permission from [2].)

3218 $\mathrm{cm}^{-1}$ at 30 K to 3253 $\mathrm{cm}^{-1}$ at 209 K, demonstrating the superposition of the heating and undercoordination effects.

MD calculations [2] shown in Fig. 16.1 also confirm this size and heating coupling effect on the phonon frequency shift. The $\omega_{\mathrm{H}}$ of molecules at the polymer proxy undergoes a further 35 $\mathrm{cm}^{-1}$ blueshift at 310 K compared to bulk water at the same temperature; instead, the $\omega_{\mathrm{L}}$ of bulk water undergoes a redshift upon heating. In fact, cluster size reduction lengthens the O:H bond and softens the $\omega_{\mathrm{L}}$; heating enhances this size trend on $\omega_{\mathrm{L}}$ softening. The joint effect results in the MD—derived trends in Fig. 16.1. These observations confirm that size reduction and heating have the same effect on $\omega_{\mathrm{H}}$ stiffening and $\omega_{\mathrm{L}}$ softening.

However, at $T < 60$ K(varies with cluster size), the $\omega_x$ shows almost no change [1, 3], except for a slight increase in volume [4] if the particle size remains unchanged. Although O:H–O angle stretching increases the volume, the specific heat $\eta_x \approx 0$ relaxes neither the length nor the stiffness of the O:H or the H–O bond, conserving the $\omega_x$.

### 16.1.2 Compression Compensating Undercoordination

Figure 16.2 shows the joint effect of undercoordination and compression on the O 1s energy shift $\Delta E_{1s}$ and the valence band shift for water clusters of different sizes [6]. Except for the O 1s peak at 539.7 eV for gaseous molecules, size growth and mechanical compression jointly shift the O 1s energy from 539.7 to 538.2 eV towards the component centered at 538.1 eV for the skin of bulk water [7, 8]. UPS

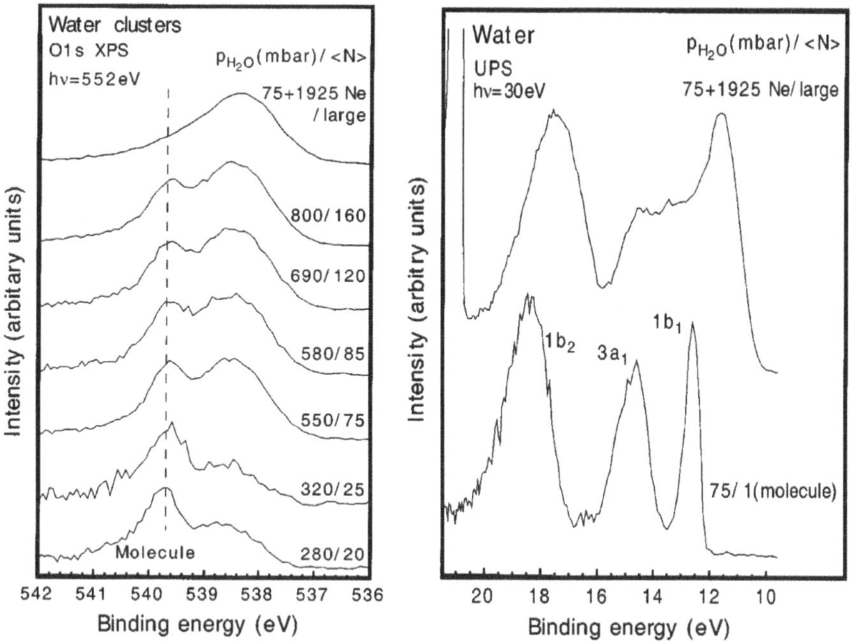


**Fig. 16.2** Joint effect of mechanical compression and molecular undercoordination on the $\Delta E_{1s}$ (*left*) and the valence band (*right*) for water clusters (10 mbar = 1 kPa). The vertical broken line denotes the $E_{1s}$ at 539.7 eV for gaseous molecules. There also show the $1b_1$, $1b_2$, and $3a_1$ orbitals of water molecular monomers. Compression and cluster size growth effect the same on raising the $E_{1s}$ and the valence band by H–O softening as the energy shift is proportional to the H–O bond energy [9] (Reprinted with permission from [6].)

reveals that the entire valence band of a molecule subjected to 7.5 kPa pressure shifts up and expands in width when the cluster size grows [6].

However, both compression and molecular undercoordination promote polarization of the nonbonding electrons [5]. Solvated electrons require 1.6 eV to escape from the bound of the supersolid skin but they need 3.3 eV from the liquid bulk water. These energies reduce further when the droplet size is reduced as a result of skin supersolidity enhancement. Compression widens the band gap via a polarization mechanism instead of the bond energy gain held in normal semiconductors.

The effect of compression and size growth coupling on the O 1s and valence band follows the BOLS notion—that is, the amount of energy shift is proportional to the $E_H$. Compression softens but molecular undercoordination stiffens the H–O bond. Therefore, size growth enhances the effect of compression on the binding energy shift in all energy bands.

Systematic studies [10, 11] have also revealed that cooling enhances the effect of compression on the structure phase transition and dipole moment of ice. In the solid phase, both cooling and compression shorten the O:H nonbond and lengthen the H–O bond [5].

### 16.1.3 Minimal Compressibility–Polarization

Compression shortens the O–O distance but lengthens the H–O bond, which results in the low compressibility of ice compared to the normal materials [12]. The low compressibility of water leads to its often being assumed as incompressible. Compression at sufficiently low temperature can raise the diffusivity but depress the viscosity of water as compression elongates and softens the H–O bond, which is also responsible for ice regelation—melting point depression. In contrast, compression of most other liquids leads to a progressive loss of fluidity as molecules are squeezed closer together [13].

The compressibility of liquid water is lower than that of ice. As shown in Fig. 16.3, at 0 °C and zero pressure, the compressibility is $5.1 \times 10^{-10}$ $Pa^{-1}$. At the zero-pressure limit, the compressibility reaches a minimum of $4.4 \times 10^{-10}$ $Pa^{-1}$ around 45 °C before increasing again with increasing temperature. As the pressure is increased, the compressibility decreases, being $3.9 \times 10^{-10}$ $Pa^{-1}$ at 0 °C and 100 MPa [14]. The low compressibility of water means that even in the deep oceans at 4 km depth, where pressures are 40 MPa, there is only a 1.8 % decrease in volume and 0.6 % compression of the O:H–O bond.

The temperature dependence of compressibility display the joint effect of compression and thermal expansion and depolarization. Compression shortens the O:H–O bond and enhances the polarization but heating does the opposite. The inset in Fig. 16.3 shows the heating depolarization of neat and salted water. Salting enhances but heating depresses the polarization that provides repulsive force between oxygen ions. For instance, the contact angle drops from 47 to 30° when the droplet is heated from 20 to 80 °C. The weakening of interoxygen repulsion and thermal depolarization eases the recovery of the compressibility as temperature rises. Therefore, opposing compression, polarization takes the responsibility for the compressibility minimum around 50 °C [16]. When heated, depolarization proceeds, which raises the compressibility slightly. When cooled below the least

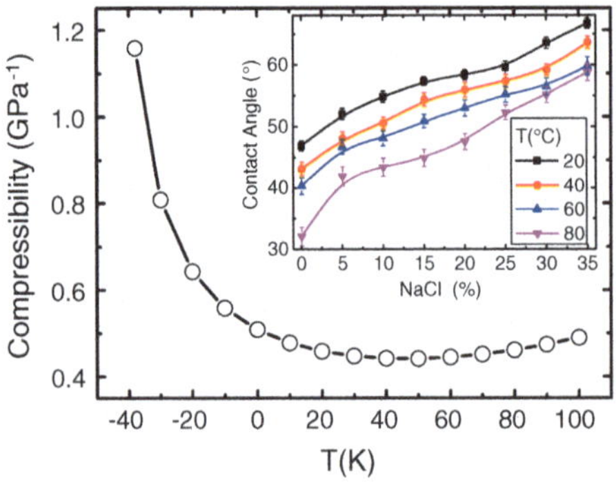


**Fig. 16.3** Temperature dependence of the compressibility of water and ice shows the anomalous minimum around 50 °C. The inset suggests that heating depolarization recovers the compressibility as temperature rises (Reprinted with permission from [15].)

compressibility temperature, mechanical compression and cooling contraction of the O:H nonbond work cooperatively, elevation the compressibility.

### *16.1.4 Electrification Enhancing Undercoordination*

Electric fields associated with aqueous solutes or capacitors align, stretch, and polarize the O:H–O bond throughout the entire volume of the liquid, which disperses outwardly the quasisolid phase boundaries. Molecular undercoordination not only disperses outwardly the quasisolid phase boundaries but also creates the supersolid phase at sites surrounding the undercoordinated molecular sites.

Molecular electrification and electrification enhance each other on stretching and polarizing the O:H–O bond, but particularly capacitor long-range volumetric electrification tends to align the O:H–O bond.

## 16.2 Isotope Effect on Phonon Frequency Shift

An addition of isotope D (Deuterium, $^2$H) into water has insignificant effect on the length and energy of the O:H–O bond but the vibration frequencies by altering the effective mass.

Figure 16.4 shows that the isotope effects on the IR spectrum of ordinary $H_2O$ [17]. Two features are apparent: one is the general intensity attenuation of all peaks and the overall red shift because of the rise of the reduced mass [17]. However, the $\omega_x$ maintains the heating trend—$\omega_H$ stiffening and $\omega_L$ softening. The intensity of a specific peak varies with its frequency in the phonon spectroscopy. If the peak

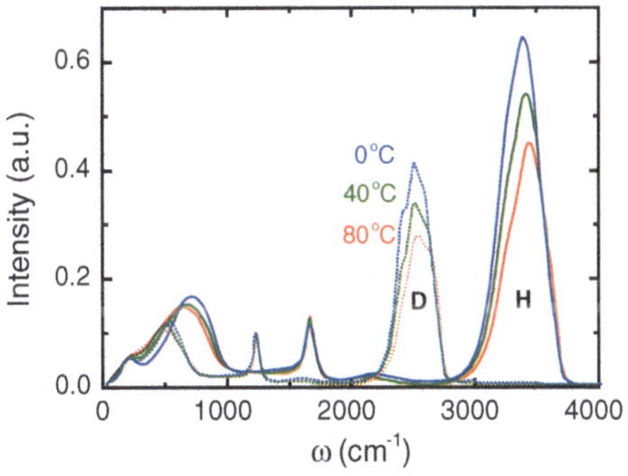


**Fig. 16.4** IR spectra of 1 μm—thick ordinary water (*H*) and heavy water (*D*) reveal that the isotope *D* attenuates the intensity and softens all phonons (*redshift*) of the $H_2O$ in general, although the trend of thermal $\omega_H$ stiffening and $\omega_L$ thermal softening remain (Reprinted with permission from [17].) (color online)

shift to higher frequency its intensity increases and vice versa, which is the artefact of the spectroscopy.

The effective mass reduction due to isotope dictates the phonon relaxation. The isotope contributes only to reducing the $\mu(m_1, m_2) = m_1 m_2/(m_1 + m_2)$ in the $\omega_x \alpha (E_x/\mu_x)^{1/2}/d_x$ expression. Considering the mass difference, both vibration modes shift their relative frequencies compared to ordinary water in the following manner:

$$\frac{\Delta\omega_{xH}}{\Delta\omega_{xD}} \cong \left(\frac{\mu_D(m_1, m_2)}{\mu_H}\right)^{1/2} = \begin{cases} \left[\frac{\mu_D(2,16)}{\mu_H(1,16)}\right]^{1/2} = (17/9)^{1/2} = 1.374 & (\omega_H) \\ \left[\frac{\mu_D(20,20)}{\mu_H(16,16)}\right]^{1/2} = (5/4)^{1/2} = 1.180 & (\omega_L) \end{cases}, \tag{16.1}$$

where, for intramolecular H–O vibration, $m_1$ is the mass of H (1 atomic unit) or D (2 atomic units); and $m_2$ is the mass of O (16 units); and for intermolecular ($H_2O$):($H_2O$) vibration, $m_1 = m_2$ is the mass of 2H + O (18 units) or 2D + O (20 units). Measurements shown in Fig. 16.4 yield the following:

$$\frac{\Delta\omega_{xH}}{\Delta\omega_{xD}} \approx \begin{cases} 3400/2500 = 1.36 & (\omega_H) \\ 1620/1200 = 1.35 & (\omega_B) \\ 750/500 = 1.50 & (\omega_L) \end{cases}. \tag{16.2}$$

The difference between the numerical derivatives in (16.1) and measurements in (16.2) arises mainly from Coulomb coupling, particularly for $\omega_L$. Such a first-order approximation is effective for describing the isotopic effect on the phonon relaxation dynamics of $\omega_x$. Therefore, the addition of the isotope softens all the phonons by mediating the effective mass of the coupled oscillators in addition to the quantum effect that may play a certain role of higher order. The peak intensities in the isotope are also lower because low-frequency vibrations enhance phonon scattering.

Figure 16.5 shows that the O 1s K-edge absorption spectra as a function of temperatures for $D_2O$ and $H_2O$ [18, 19]. D addition shifts the spectral features slightly resulting from the energy level shift of the O 1s band and the valence band. Bedsides the slight shift due to isotope effect, the spectra resolve hardly the thermal effect on the electronic binding energy shift dominated by the H–O bond.

## 16.3 Energy Exchange: Long-Range Perturbation

### *16.3.1 Ice Flake Formation under Perturbation*

Water absorbs all sorts of energy in a long-range manner. Besides electromagnetic field radiation that penetrates into water and ice throughout the entire body,

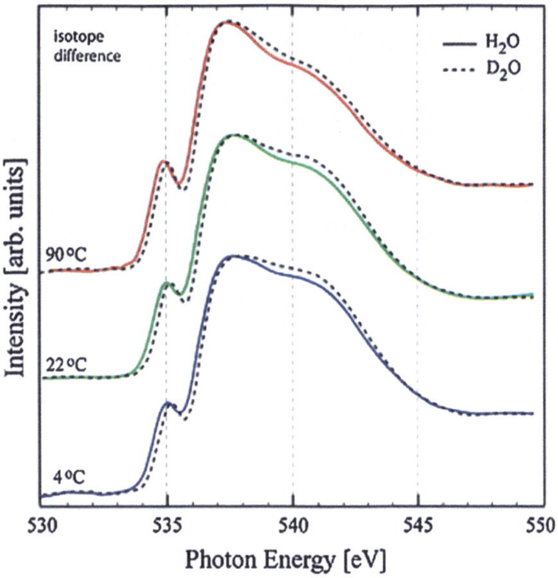


**Fig. 16.5** O K-edge XRS spectra of $D_2O$ and $H_2O$ at 4, 22, and 90 °C showing that the O 1s energy is insignificant to the isotope or the temperature effect (Reprinted with permission from [18].)

perturbation by the local perturbation affects the alignment and relaxation of the O:H–O bond in a long-range manner, resembling the Domino effect because of the weak O:H interaction.

For instance, the ∠O:H–O containing angle, the ∠O–H–O angle, the segmental length, and the cohesive energy determine the geometries of snowflakes and ice crystals. Any perturbation by bioelectronic signals like emiotion [20], mechanical excitation like sound tones and frequencies, and thermal pulse or fluctuational signals will mediate the growth modes—shape and geometry of ice flakes. Water is highly sensitive to a perturbation because the extremely weak O:H nonbond with binding energy in the order of ~$10^1$ meV that is compatible to stimulus perturbation. Room temperature at 300 K corresponds to 25 meV and 1 K temperature rise corresponds to $8 \times 10^{-5}$ eV energy for absorption.

Because of the extremely high sensitivity of water to the thermal and experimental conditions, any perturbation can change the patterns of ice crystal [20] and the cooling rate of water freezing [21]. Masaru Emoto, a Japanese scientist, examined the crystal patterns from water samples subject to simple words like love, thank you, war and hate. He observed the ice crystals under the microscope (see Fig. 16.6 for instance). The samples subjected to love and thank you form brilliant crystal shapes. The hate and war samples formed ugly, amorphous shapes.

Geometries of snowflakes that have fascinated many eminent scientists and philosophers such as René Descartes, Johannes Kepler, and Robert Hooke, but the man who literally devoted his entire life to showing us the diversity and beauty of snowflakes is American Wilson A. Bentley (1865–1931).

It is not surprising that harmonious classical music like that of Bhimsen Joshi, Pandit Ravi Shankar, Ali Akbar Khan, Mozart and Beethoven or of harmonious

**Fig. 16.6** Geometries of snowflakes (*Credit* Masaru Emoto [20] and Timeea Vinerean 2011, Public domain.)

new-age music, and natural sounds of the sea, whales, etc., have a benevolent influence on the patterns of the ice crystal. The opposite has been the case with disharmonious music like heavy metal, sounds like traffic noise and words like "I hate you". Any sound waves or bioclcctronic signals of thinking a demotion at different tones or frequencies affect the growth manner of ice crystals.

As shown in Fig. 16.6, snowflakes are all built on the hexagonal shape that occurs as water droplets freeze into crystals. The basic patterns include hexagonal plates, simple columns, thin columns, needles, and stellar, crystal or branched shapes. These shapes can build on each other to create complex, hybrid constructions that reflect the temperature and humidity conditions during formation. Warm, humid conditions are especially conducive to large, complex flakes. Any point defect will alter the pattern of growth. Although these experimental are highly reproduciple and even some people thought somewhat superstitious, but they do show the extraordinary sensitivity of the O:H–O bond to a tiny perturbation.

### *16.3.2 Solution Precipitation under Compression*

Figure 16.7 shows the optical images of 3 M NaI solution forming ice under compression and 298 K. No ice could form until 2.39 GPa (a). After holding 35 h of this hydrostatic pressure, ice crystal forms participially but the rest remains as liquid due to salt concentration become nonuniform—low concentration liquid forms ice preferentially (b). Raising the pressure to 2.90 GPa and hold for another 2 h, liquid remains but ice transforms into ice VII and the gauge pressure drops to 2.61 GPa (c). Raising the pressure to 3.0 GPa and holding for 2 h, the liquid part of solution turns to be ice lastly.

Figure 16.8 shows the Raman $\omega_x$ spectra of the liquid part as a function of pressure in contrast to the spectra for the solid part presented in Sect. 13 Fig. 16.8. The experimental conditions are consistently uniform but relaxation of the detecting time leads to the concentration precipitation during aging. The solution evolves into

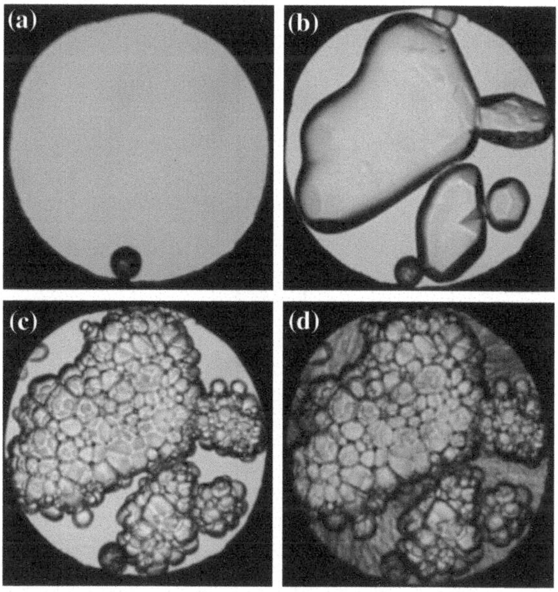


**Fig. 16.7** Solute precipitation of 3 M NaI solution during ice formation under compression. **a** Liquid state at 2.39 GPa and below. **b** Keeping 35 h under 2.39 GPa hydrostatic pressure, partial ice forms due to solution precipitation. Low concentration solute forms ice preferentially. **c** Raising pressure up to 2.90 GPa and holds the cell volume steadily, the pressure drops to 2.61 GPa. The liquid remains but the ice turns from VI to VIII phase. **d** Increase the pressure to 3.00 GPa and hold it for 2 h, part of the liquid forms ice (Reprinted with permission from [22].)

the lower and the higher concentration parts after long period holding. The part of lower concentration forms ice preferentially and the part of higher concentration retains its liquid form till the critical pressure it needs for ice formation.

### *16.3.3 Icing Solute Exclusion*

In order to examine the concept of ice exclusion of salt, one can measure the $\omega_H$ spectrum of the mother solution at a certain concentration first, and then put the solution in a fridge for ice formation. The fridge temperature and freezing time depend on the solute concentration. Once some of the solution is frozen, then separate the ice from the solution and warm them up to the room temperature. If ice exclusion occurs, the $\omega_H$ of the molten will not shift at all or shift less than the mother solution and even less than the remnant solution that undergoes even further freezing temperature depression.

Figure 16.9 compares the Raman spectra of the 3 M NaI solution before and after solution freezing precipitation and their DPS with respect to neat water measured under the ambient pressure and 298 K. Results conformed the expectations about the

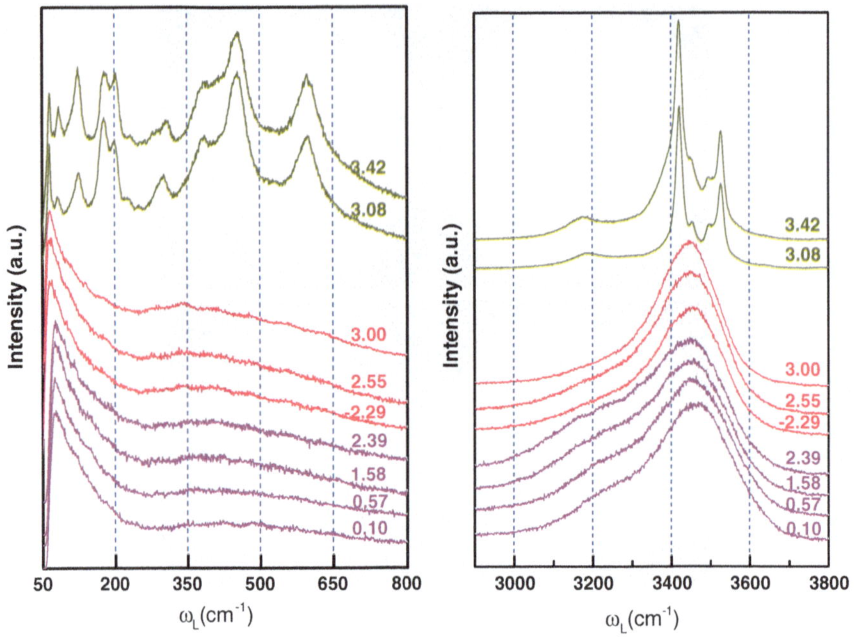


**Fig. 16.8** Pressure dependence of the Raman spectra for the high-concentration liquid part in the 3 M NaI solution in contrast to the spectra for the solid part formed the VI phase in the second state (*red lines*) (Reprinted with permission from [22].) (color online)

effect of concentration precipitation up on ice formation. Water is indeed sensitive to this tiny change of local chemical environment and excludes solutes. Results show that the ice exclusion is more significant in the bulk than it is in the skin for the 3 M NaI solution. The diluted concentration in the skin shift the $\omega_H$ more towards high frequency as the 3 M concentration passes the saturation and recovers the phonon frequency slightly, as well as the preferential occupancy of I anions. Nevertheless, ice excludes indeed solutes and further investigation is in progress. Table 16.1 summarizes information gained from the spectra decomposition.

This observation may suggest a means for sea water desalting. Ice formation in cold winter can exclude partially harmful organic and inorganic impurities, which ease the distillation of sea water for drinking and watering when the ice melts.

## 16.4 Induction and Polarization

### *16.4.1 Kelvin Water Dropper Battery*

The Kelvin water dropper, invented by British scientist William Thomson (Lord Kelvin) in 1867, is a type of electrostatic generator. The apparatus is widely called

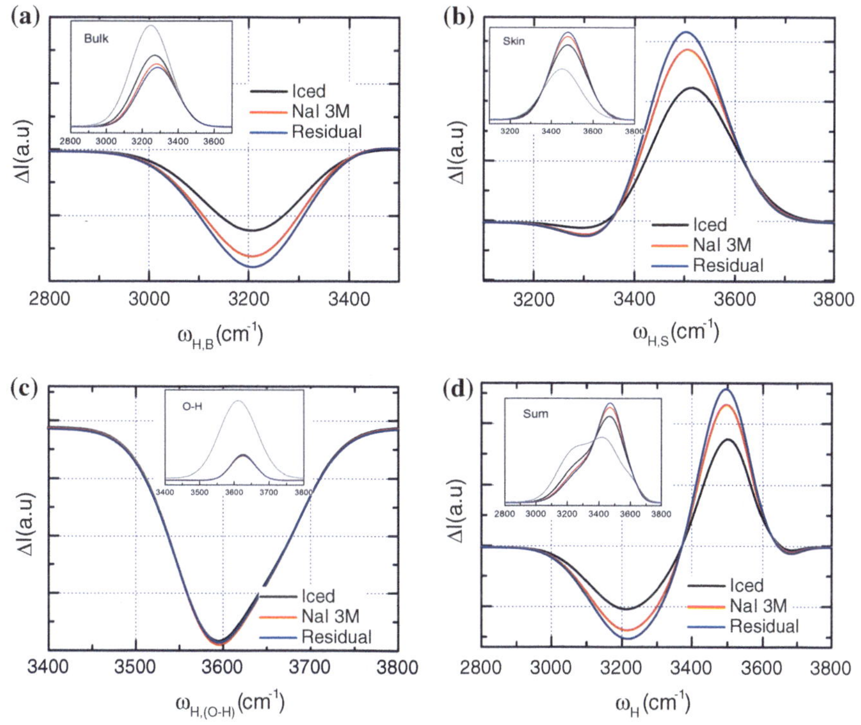


**Fig. 16.9** The **a** bulk, **b** skin, **c** H–O radical, and the **d** resultant Raman DPS for 3 M NaI mother solution, iced solution, and the residual solution measured under the ambient pressure and 298 K. The sum of spectral area loss equals the sum of gain. The H–O blueshift and its integral loss indicates the skin preferential occupancy of the $I^-$ anions. The substantial gain of the skin component includes both skin and the hydration shells of the ions

the Kelvin hydroelectric generator. The device uses falling water to generate voltage differences by electrostatic induction occurring between interconnected, oppositely charged systems.

Figure 16.10 shows the typical setup for the dropper battery [23]. A reservoir of conducting liquid (water or otherwise) has holes or piping that releases two falling streams. Each stream passes without touching through a conducting ring, and lands in one of two containers. The containers must be electrically insulated from each other and from electrical ground. Similarly, the rings must be electrically insulated from each other and their environment. The left ring is electrically connected with (wired to) the right container and the right ring is wired to the left container. It is necessary for the streams to break into separate droplets before reaching the containers. Typically, the containers are conductors, such as metal buckets.

Instead of buckets, after falling through the charging electrodes the drops fall into metal funnels which collect the charge but let the water through. This machine is equipped with Leyden jar capacitors *(cylindrical objects at base)* which store the

**Table 16.1** Decomposition of the neat water, 3 M NaI mother solution, deiced solution and the residual solution at 298 K

| | | Bulk | Skin-hydration shells | H–O | Notes |
|---|---|---|---|---|---|
| $\omega_H$ ($cm^{-1}$) | $H_2O$ | 3246.88 | 3451.05 | 3611.77 | Phonon stiffness |
| | Iced solution | 3269.11 | 3477.50 | 3625.64 | |
| | 3 M NaI | 3281.01 | 3479.60 | 3626.02 | |
| | Residual solution | 3285.72 | 3479.64 | 3625.57 | |
| FWHM ($cm^{-1}$) | $H_2O$ | 215.78 | 175.65 | 108.06 | Fluctuation order |
| | Iced solution | 214.69 | 177.76 | 68.79 | |
| | 3 M NaI | 213.70 | 169.36 | 69.23 | |
| | Residual solution | 212.12 | 166.12 | 70.18 | |
| Component integral (a.u.) | $H_2O$ | 4869 | 4197 | 0685 | Abundance of the phonons |
| | Iced solution | 3431 | 6297 | 0141 | |
| | 3 M NaI | 3023 | 6719 | 0136 | |
| | Residual solution | 2828 | 6921 | 0141 | |

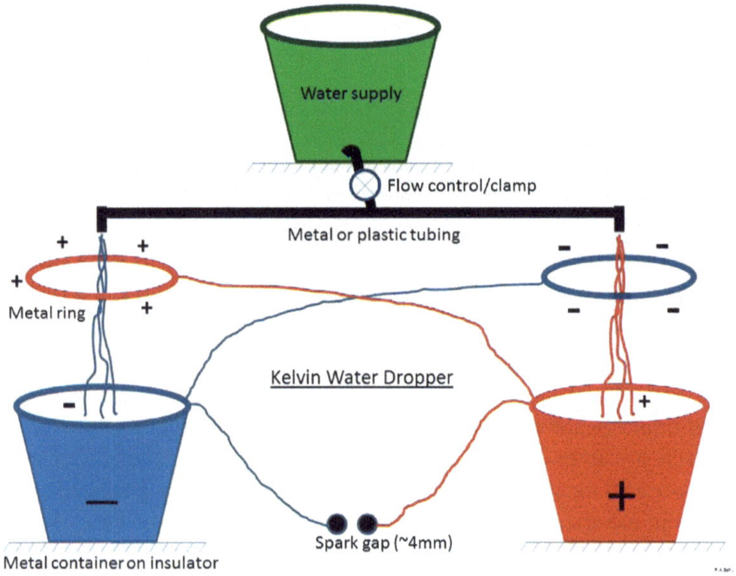


**Fig. 16.10** A typical setup for the Kevin dropper battery (Reprinted with permission from [23].)

charge. Any small charge on either of the two buckets suffices to begin the charging process. For example, the right bucket has a small positive charge. Now the left ring also has some positive charge since it is connected to the bucket. The charge on the left ring will attract negative charges in the water ions into the left-hand stream by electrostatic attraction. When a drop breaks off the end of the left-hand stream, the drop carries negative charge with it. When the negatively charged water drop falls into its bucket (the left one), it gives that bucket and the attached ring (the right one) a negative charge.

Once the right ring has a negative charge, it attracts positive charge into the right-hand stream. When drops break off the end of that stream, they carry positive charge to the positively charged bucket, making that bucket even more positively charged. So positive charges are attracted to the right-hand stream by the ring, and positive charge drips into the positively charged right bucket. Negative charges are attracted to the left-hand stream and negative charge drips into the negatively charged left bucket. The positive feedback of this process makes each bucket and ring increasingly charged. The higher the charge, the more effective this process of electrostatic induction is. During the induction process, there is an electric current that flows in the form of positive or negative ions in the water of the supply lines. This is separate from the bulk flow of water that falls through the rings and breaks into droplets on the way to the containers. For example, as water approaches the negatively charged ring on the right, any free electrons in the water can easily flee toward the left, against the flow of water.

Eventually, when both buckets have been highly charged with different effects, an electric pulse creates spark arc between the two buckets or rings, decreasing the charge on each bucket. If there is a steady stream of water through the rings, and if the streams are not perfectly centered in the rings, one can observe the deflection of the streams prior to each spark due to the electrostatic attraction of opposite charges. As charging increases, a smooth and steady stream may fan out due to self-repulsion of the net charges in the stream. The voltages reached by this device can be in the range of kilovolts, but the amounts of charge are small, so there is no more danger to persons than that of static electrical discharges produced by shuffling feet on a carpet, for example.

The charge separation and build-up of electrical energy ultimately comes from the gravitational potential energy released when the water falls. The charged falling water does electrical work against the like-charged containers, converting gravitational potential energy into electrical potential energy, plus motional kinetic energy.

### *16.4.2 Clouds and Fogs: Intrinsic Polarization*

Clouds and fogs consist of clusters of a few water molecules. They neither separate nor agglomerate but float in the air about the ground though the cluster's mass density is higher than that of air. Why does this happen?

Water clusters are subject to the skin supersolidity with strong polarization because of the undercoordination of water molecules. The skin locally pinned dipoles of the clusters repel them one another to prevent them from being forming bigger drops or falling to the ground that is negatively charged. In the nature

**Fig. 16.11** Fogs at the Nicaragua waterfalls and clouds in the sky consist of molecular clusters that are subject to the molecular undercoordination induced skin polarization and supersolidity. The skin polarization and supersolidity make each cluster an isotropic dipole pointing outwardly. Equilibrium of skin-skin Coulomb repulsion and skin-core attraction prevents clusters from being agglomerated or falling on the ground that is generally negatively charged

environment, electrostatic force forms easily by charge induction though rubbing and cluster size reduction. The skin polarization of the cluster due to H–O bond contraction leaves its core being positively charged in equivalent. Equilibrium of the neighboring skin-skin Coulomb repulsion and the core-skin attraction make the fogs and clouds as they stay like giant atoms, see Fig. 16.11. This giant atom model is general to molecular clusters like dust droplets or particles in the air. The droplet is covered by a dipole layer because of the skin molecular undercoordination that shortens the intramolecular bond and lengthens the intermolecular nonbond with strong polarization.

This intrinsic charging by dipole formation or extrinsic charging by induction is common to all substance on this planet, particularly at the nanometer scale. The skin of any substance consists of undercoordinated atoms is subject to the global bond contraction and the associated quantum entrapment and subjective polarization of nonbonding electrons. If nonbonding electron exists, intrinsic polarization pertains, otherwise, the entrapment of the core electrons make the skin positive like but the skin will be attached by negative charge from the earth that serves as the reservoir of electrons.

Nanoparticle interactions in solution affect their binding to biomolecules, their electronic properties, and their packing into larger crystals. However, theories that describe larger colloidal particles fail for nanoparticles, because the interactions do not add together linearly. Nanoparticles also have complex shapes and are closer in size to the solvent molecules. The skin polarization and charge induction make crystalline and amorphous particles as well as dispersions of inorganic, organic, and biological nanomaterials nonadditivity compared with larger droplets [24].

## 16.5 Electromagnetic Radiation and AC Electrification

### *16.5.1 Dipoles Moving in the Lorentz Force Field*

Electromagnetic radiation exerts its power mainly through the electrical effect [25]. Similar to the effect of electric field alone, electromagnetic radiation can also reshape water structure in totally opposite manner under different conditions. As a result, there are some divergences on the effect of electromagnetic field on the O:H–O bond network. The weak electromagnetic field also weakens the lower energy part of absorption spectrum [26]. An extremely low frequency electromagnetic radiation of 0.15 T intensity can raise the dielectric constant by 3.7 % than the control over the frequency range of 1–10 GHz of 45 μT filed [27], which indicates that low-frequency high-intensity radiation promotes polarization and association of water molecules.

Electromagnetic radiation has been widely used in the microwave heating with industrial and domestic purpose. The oscillating electric field is able to force water dipole moments to reorient [25], the frictional interaction between reorienting molecule and unaffected or lag molecule leads to energy loss by heating. This process is largely dependent on the strength of HB; qualitatively, this effect could be invoked at GHz frequencies (microwaves) for ambient liquid water, whilst the frequency descends down to KHz (long electromagnetic waves) for solid ice.

Additionally, electromagnetic field has biological effects as well. An electromagnetically conditioned water had both beneficial and adverse effects on the growth of higher organisms at optimum and higher levels of conditioning, respectively [28], so this effect can be applied in stimulating the growth of beneficial organisms or inhibiting the unwanted microorganisms.

The magnetic field influences on water structure reflected by thermodynamic properties, spectrum, etc. Increasing magnetic field intensity can change the internal energy, heat capacity and the radial distribution function of water [29]. Magnetic fields also modulate water freezing in a much more irregular manner. MD calculations reveal that a 10 T magnetic field can raise the freezing temperature of a hydrophobically confined water nanodroplet up to 340 K [30]. Increasing the magnetic field up to 10 T, the surface tension of regular water at 298 K increases from 71.7 to 73.3 mN/m and that of $D_2O$ increases to 74.0 mN/m [31] in a $B^2$ manner. $^1$H—NMR measurements revealed that a 0.01–1.0 T magnetic field reduces the surface tension but raises the viscosity of water [32]. However, a 60 × 10 T d/c magnetic field lowers the freezing point of the ambient water to −7 °C [33]. Confirmation of the magnetic field effect on water freezing would be very interesting.

One can observe the so called diamagnetic property of water by putting a container on top of a magnet that generate inhomogeneous magnetic field, and applying a bias across the center and the edge of the container filled with liquid water, see Fig. 16.12. When an electrical charge travels through a magnetic field, in a direction perpendicular to the field, it is deflected in a third dimension

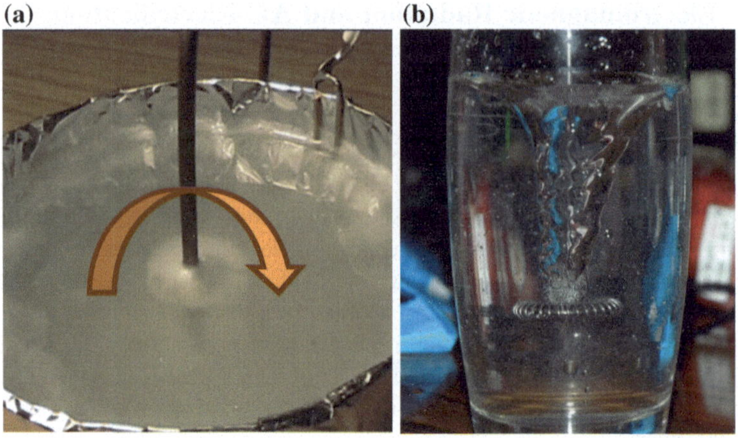


**Fig. 16.12** **a** Water rotates under magnetic field when current flows between the central and the circumference of a container [34]; **b** Whirlpool formation of flowing water under magnetic field [35]

perpendicular to both the direction of its own movement and the direction of the *B* field, rotating and transiting water molecule. The same principle is used in the cathode ray tube used in old television sets. There the charges move through vacuum, whereas here they move through water dipole. The term of diamagnetic water may be improper because the Lorentz force mediates the angular and transitional motion of water dipoles. On top of a magnetic field, flowing water can form whirlpool due to the Lorentz forcing, shown in Fig. 16.12b.

The magnetic field activated angular and translational motion of molecular dipoles may improve the microscopic circulation in the blood tissue, which could ease hypertension to a certain extent, for instance.

The surface-wave resonance measurements [31, 36] revealed that the tension of water–air interface increases linearly with the square of the magnetic field by $1.57 \pm 0.25$ % under the magnetic field up to 10 T (see Fig. 16.13). This observation suggests that O:H–O bond relaxation and polarization also occurs in the magnetic field.

## *16.5.2 AC Electric Wetting*

In dynamic electrical field experiments, Wang and Zhao [37, 38] trapped tiny water droplets and caused them to 'dance', by combining the effects of surface tension, elastic force and electrical force to manipulate a flexible thin film and encapsulate or release a tiny droplet in a controlled and reversible manner. The film—supported droplet vibrated at twice the frequency of the input a/c signal. During this action, the droplet was observed to lie flat on the surface at the maximum applied voltage, and bent upward as the voltage was reduced (see Fig. 16.14).

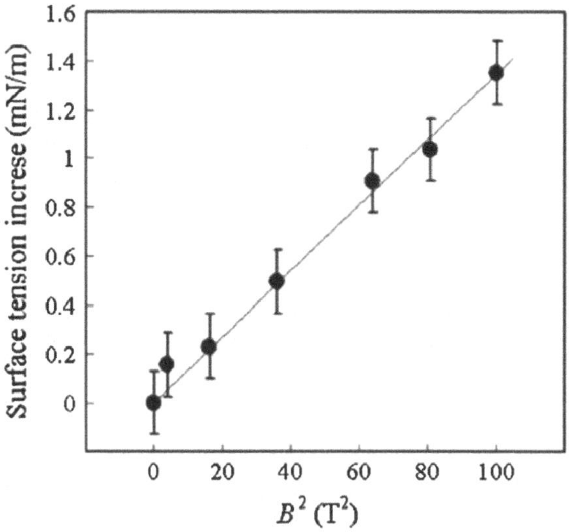


**Fig. 16.13** Magnetic field intensity dependence of water surface tension (Reprinted with permission from [31].)

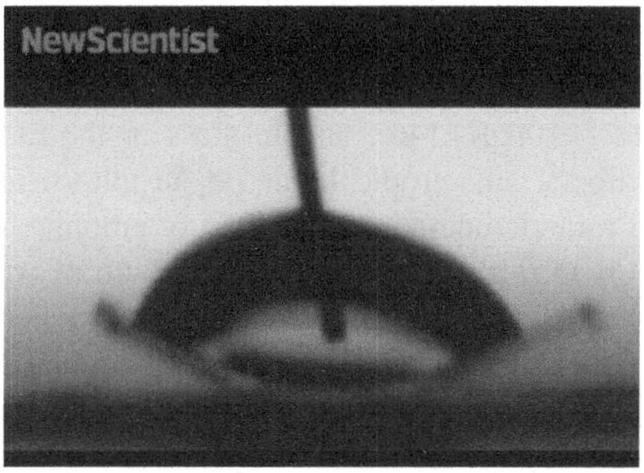


**Fig. 16.14** A droplet is wrapped in a thin film and when a changing voltage is applied, the duo starts to move up and down. The vibration of the droplet makes it look like it is tap dancing while the flexible film seems like its dancing skirt. During the dance, the droplet lies flat on the surface when the voltage reaches its peak, allowing it to be unwrapped in a controllable and reversible way (Reprinted with permission from [37, 38].)

Dynamic wetting of water droplets on either a hydrophobic or a hydrophilic surface of different roughness and at various temperatures is an interesting issue, as it is related to liquid–solid interactions. Yuan and Zhao [39, 40] reported the development of a mechanism of multiple—scale dynamics of a moving contact line on lyophilic pillars with the scaling relation $R \approx t^x$ ($x = 1/3$ for a rough surface and 1/7 for a smooth surface), where R is the spreading radius and t is time. The spreading of a liquid drop on a hydrophilic, flexible pillared surface followed the same scaling relationship. The flexible pillars accelerated the liquid when the liquid approached, and trapping the liquid as it passed. The liquid deformed the pillars, resulting in energy dissipation at the moving contact line. The joint effect of the surface topology, intrinsic wettability and elasticity of a solid influenced the flow pattern and the flow field of the droplet on the pillar-arrayed surface.

Water-hydrophilic substance interaction is dominated by O:H-O bond like formation. The substrate must be positively charged or have quanum entrapment dominance without charge for poalrization. The positively charged skin will attract the negatively charged water skin molecules to form such bond through elecsteatic attraction but no exchange interaction, like water contunuing its topology at the interface, which has been confirmed by phonon spectrometrics. Water-hydrophobic interaction proceeds by electron repulsion with an air gap presenting bwtween the droplet and the substrate, as addressed previously.

## 16.6 Negative Thermal Expansion

The vast majority of materials have a positive coefficient of thermal expansion ($\alpha(\theta) > 0$) and their volume increases on heating. There are also another very large number of materials that display the opposite behavior: their volume contracts on heating, that is, they have a negative thermal expansion (NTE) coefficient [41–44]. A typical specimen is the cubic $ZrW_2O_8$ that contracts over a temperature range exceeding 1000 K [45]. NTE also appears in diamond, silicon and germanium at very low temperatures (<100 K) [46], and in glass in the titania–silicate family, Kevlar fiber, carbon fibers, anisotropic Invar Fe–Ni alloys, and certain kinds of molecular networks at room temperature. The NTE of graphite [47], graphene oxide paper [48], and $ZrWO_3$ [45] all share the NTE attribute of water at freezing, see Fig. 16.15. NTE materials may be combined with other materials with a positive thermal expansion coefficient to fabricate composites having an overall zero thermal expansion (ZTE). ZTE materials are useful because they do not undergo thermal shock on rapid heating or cooling.

The typical model that explain the NTE effect suggests that NTE arises from the transverse thermal vibrations of the bridging oxygen in the M–O–M linkages inside $ZrW_2O_8$, $HfW_2O_8$, $SC_2W_3O_{12}$, $AlPO_{4-17}$, and faujasite-$SiO_2$ [49, 50]. The phonon modes (centered around 30 meV or 3200 $cm^{-1}$) [51] can propagate without distorting the $WO_4$ tetrahedron or the $ZrO_6$ octahedron, termed the 'rigid-unit mode'. The rigid-unit mode also accounts for the structural phase transition of $ZrW_2O_8$ and $ZrV_2O_7$ [52].

Cubic scandium trifluoride ($ScF_3$) has a large negative thermal expansion over a wide range of temperatures [54]. Inelastic neutron scattering experiments probed the phonon DOS as a function of temperature varying from 7 to 750 K revealed a large anharmonic contribution with a thermal stiffening of modes around 25 meV. Phonon calculations with first-principles methods identified the individual modes in the densities of states, and frozen phonon calculations showed that some of the modes with motions of F atoms transverse to their bond direction behave as quantum quartic oscillators. The quartic potential originates from harmonic

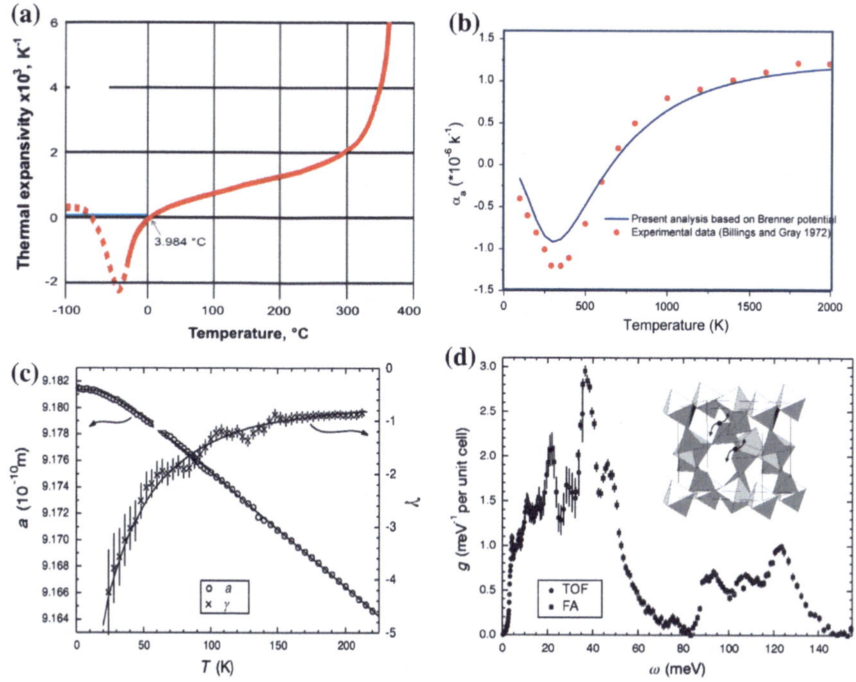


**Fig. 16.15** The NTE of **a** $H_2O$; **b** graphite; and **c** $ZrW_2O_8$ with thermal expansion coefficient $\alpha$ (*open circles*) and Grüneisen parameter $\gamma = 3\alpha B/C_v$ (*crosses*), where $B$ is the bulk modulus and $C_v$ is the specific heat at constant volume; **d** shows the associated phonon spectrum measured at $T = 300$ K. The inset illustrates the 'rigid rotation model' model (Reprinted with permission from [47, 51, 53].) These NTEs share the same behavior as water freezing albiet temperature ranges, which is evidence of the essentiality of two types of coupled short-range interactions with specific-heat disparity to these materials

interatomic forces in the DOS structure of $ScF_3$, and accounts for phonon stiffening with the temperature and a significant part of the negative thermal expansion.

Extending the NTE mechanism for water in the quasisolid phase provides a bonding point view on the NTE in general. The NTE results from the involvement of at least two kinds of coupled, short-range interactions and the associated specific-heat disparity. In the instance of graphite, the (0001) intralayer covalent bond and the interlayer vdW interactions play their roles, in much the same way as the O:H–O bond does in water. O, N and F all create lone pairs of electrons upon reaction, which create the weaker short-range nonbonding interaction. Phonon spectroscopes have the capability to monitor the relaxation process easily and directly, as they do for water. Figures 16.15 and 16.16 evidence clearly the presence of the phonon DOS below 25 meV for $ScF_3$ and $ZrW_2O_8$.

It is expected the layered $WX_2$ (W = Mo, W; X = S, Se) substances also have the NTE effect becaues of the short-range interactions. A monolayer $WX_2$ consists of three atomic shells in $X^{-\delta}$-$W^{+\delta}$-$X^{-\delta}$ configuration ($\delta$ is net charge), which forms a

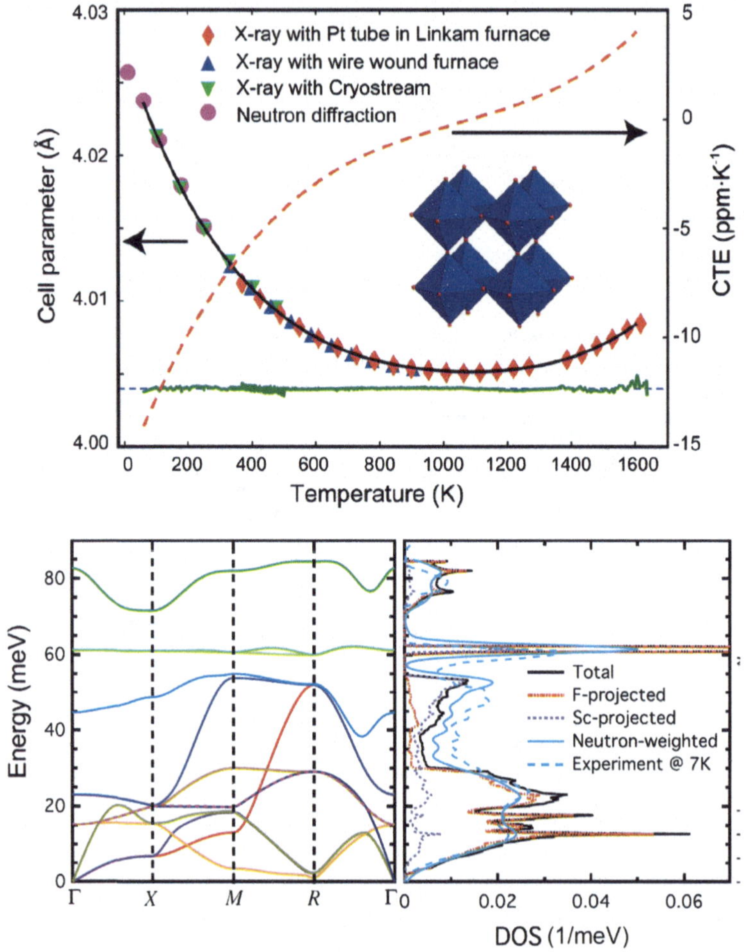


**Fig. 16.16** Measured NTE for $ScF_3$ crystals and the total and partial phonon DOS curves at 0 K from first-principles calculation, neutron-weighted phonon DOS with instrument broadening at 120 meV added, and experimental neutron-weighted phonon DOS at 7 K (Reprinted with permission from [54,55].)

two-dimentional diant atom skined with negative charge and centered positive. The interaction bewtween such giant atoms mimics water clusters of fogs an clouds. The mixture of short- and long-range interactions derives different specific heats. The superposition of theses specific heats will result in the same to water having different specific ratios at different temperature regions and thus NTE could happen.

## 16.7 Dielectric Relaxation: Polarization

The transition of electrons from the valence band to the conduction band determines the relaxation of the static dielectric constant of a semiconductor [56]. Processes of bandgap $E_G$ expansion, lattice relaxation, and electron–phonon coupling contribute to the dielectric relaxation. The dielectric permittivity ($\chi = \varepsilon_r - 1$) of a semiconductor is approximately proportional to the inverse square of its bandgap $E_G$ [57–59] that is proportional to the bond energy or the inverse of atomic distance [56, 58]. The refractive index, $n = \varepsilon_r^{1/2} = (\chi + 1)^{1/2}$, drops accordingly when the specimen is compressed or cooled, since $E_G$ increases under such conditions [60–62]. The dielectric constants of semiconductors also drop in the skin region, and drops with solid size [56].

However, the refractive index of liquid water at room temperature increases with pressure in the same trend as density (see Fig. 16.17). The dielectric constant also increases with the drop of temperature. Both compression and cooling stiffen the O:H nonbond and soften the H–O bond but they increase the dielectric constant. This is controversy to the dielectric behavior of 'normal' materials.

The following possible mechanisms may be responsible for the dielectric enhancement of water when compressed or cooled. These stimuli shorten the O:H bond and lengthen the H–O bond, producing a gain in mass density yet a loss in $E_H$. Some or all of the following parameters may raise the dielectric constant of water but polarization could be the dominant factor:

$$\varepsilon_r \propto \begin{cases} \rho(d_{OO}) & (Mass\ density) \\ E_H^{-2} & (H - O\ energy) \\ P & (Polarization) \end{cases}$$

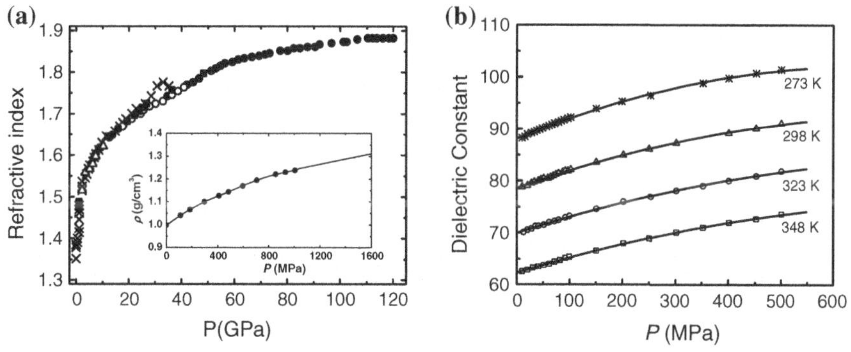


**Fig. 16.17** **a** Compression and **b** cooling enhance the refractive index of water. The *inset* in (**a**) shows the pressure-dependence of the mass density of water at 298 K (Reprinted with permission from [63, 64].)

Water molecular polarization happens under the stimuli of compression, cooling, molecular undercoordination, electrification, and magnetization. However, mass density gains only under compression and cooling excitation but density losses in undercoordination, electrification, and magnetization. The dielectric enhancement proceeds in these situations, which suggest that molecular polarization dictates the dielectrics of water and ice.

According to Gregory et al. [40], the dipole moment of small water cluster, $n$ = 1–8, climbs up to ~2.76 D from 1.865 D for a monomer as $n$ increases. Non-overlapping configuration of positive and negative charge center of an isolated water monomer is responsible for the comparatively large dipole moment of 1.855 D and anomalous behaviors of bulk water. Further, this value in condensed phase evolves to 2.4–2.6 D [41] because the size growth weakens the undercoordination effect [5]. Formulation of the dielctrics of water and ice under various stimuli remains a challenge.

## 16.8 Hydrophilic Interface: EZ Water

Professor Gerald Pollack [23] at University Washington defined in his book the fourth phase of water associated with hydrophilic interfaces. This IVth phase named an exclusion zone (EZ), as described in Fig. 16.18. The different colors of the IR absorption image for a triangle Nafion embedded in water indicate different IR absorption magnitudes. Far from the Nafion, the uniform blue color indicates a uniformly low level of absorption. The Nafion blends a hydrophobic backbone of tetrafluoroethylene with side chains containing exceptionally hydrophilic sulfonic acid groups. The color change closer to the Nafion (green) indicates that the EZ absorption differs from bulk water absorption. The magnetic resonance image in Fig. 16.18b separates the gel-like EZ water zone from the liquid water. The UV-vis absorption spectra in Fig. 16.18c show absorption intensity at 270 nm (4.59 eV) increases closer to the interface. Figure 16.18d shows the site-resolved viscosity of the EZ water. The inset in Fig. 16.18d illustrates the $H_3O_2$ structure model. Arrow and the circles show the offset of the neighboring layers of molecules. The question remains is that how the monovalent H atom can bond to two neighboring O atoms simultaneously.

The EZ water was observably more stable, more viscous, and more ordered planner hexagonal structures of $H_3O_2$ packing with alternative shift along a certain axis to let positive change on the top of a negative.

Compared with bulk water, molecular motions in the EZ water are more restricted; and it has a 10 % higher refractive index. The EZ water excludes almost anything suspended or dissolved in the water. The hydrophilic surface could induce nearby water molecules to line up as they would in a liquid crystal in a shelled order. As the ordered zone grew, it would push out solutes in the same way that a growing glacier pushes out rocks.

MD calculations suggest that when approaching the wall of a neutral pores, or lowering the temperature of observation, water molecules become less mobile but

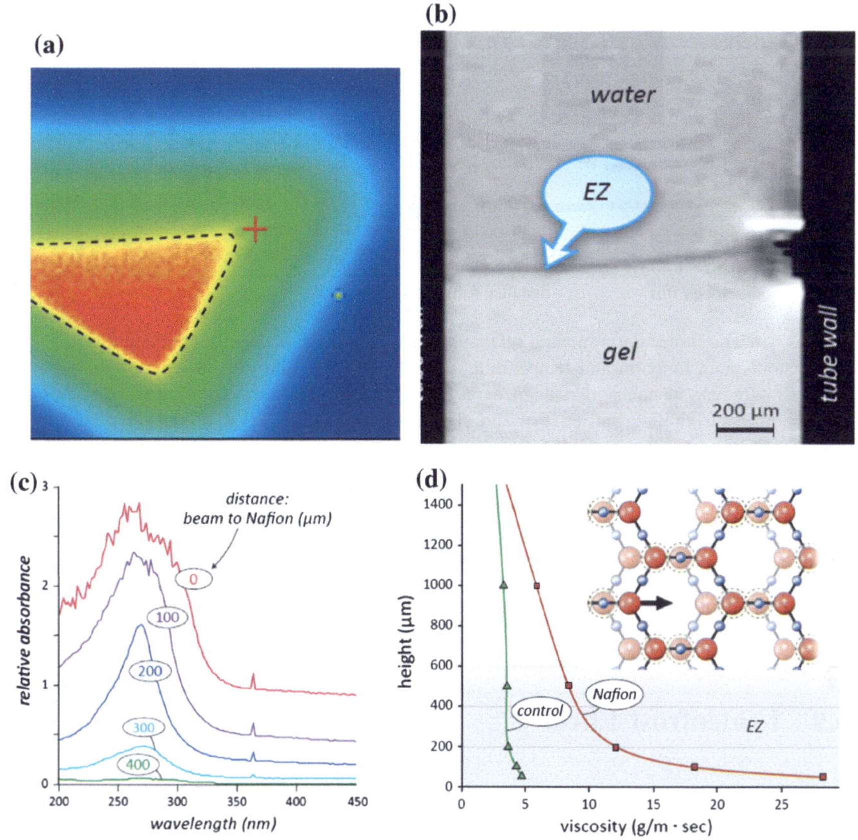


**Fig. 16.18** **a** IR absorption image of triangle Nafion embedded in water showing the EZ in *green* color. **b** Magnetic resonance image of the gel-like EZ water, **c** Distance-resolved UV absorption spectrum, and (**d**) the viscosity of the EZ water. The *inset* in **d** illustrates the $H_3O_2$ structure model. *Arrow* and the *circles* show the offset of the neighboring layers of molecules (Reprinted with permission from [23].) (color online)

the amplitude of O oscillation increases, as shown in Fig. 16.19 [65]. As we demonstrated that the $r^2$ increase is associated with $\omega_L$ frequency depression and the $\tau_\alpha$ increase corresponds to the $\omega_H$ blueshift and viscosity enhancement [66], which results from molecular undercoordination induced O:H–O elongation and non-bonding electron polarization and cooling in the quasisolid phase [5].

Under certain conditions, water could become unexpectedly stable: it became difficult to freeze and equally difficult to vaporize but easy to melt because the O:H–O bond relaxation disperses the quasisolid phase boundaries. The behavior of the EZ water is quite similar to the supersolid phase associated with the free skin of water or at its hydrophobic contact. The supersolid phase is viscoelastic, highly polarized, less dense, mechanically stiffer, molecular less mobile, and thermally

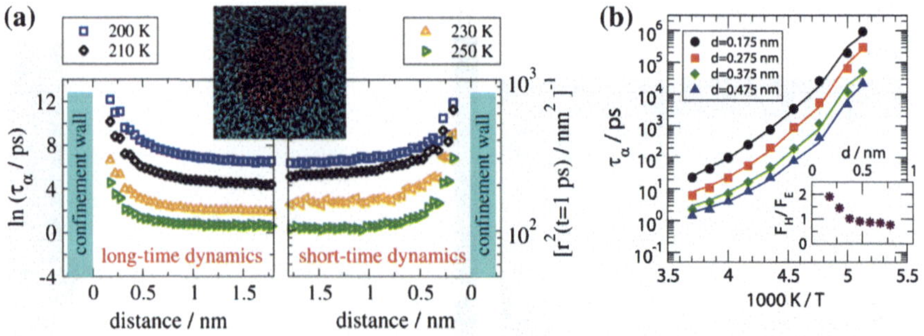


**Fig. 16.19** **a** The molecular structural relaxation time $\tau_\alpha$ and the means-square displacement $r^2$ of oxygen atoms as a function and temperature and distance from the wall of neutral confinement. **b** Temperature and distance dependence of the relaxation time at various distances *d*. *Inset* Resulting ratio of hopping barrier ($F_H$) and elastic barrier ($F_E$) at 210 K in various pore regions (Reprinted with permission from [65].)

more stable. Particularly, the 4.59 eV absorption energy corresponds to the dissociation energy for the undercoordinated H–O bond at a stretching frequency of 3450 $cm^{-1}$ compared with the H–O bond energy 3.97 eV and phonon frequency 3200 $cm^{-1}$ for bulk water or the 5.10 eV and 3700 $cm^{-1}$ for a monomer [5].

## 16.9 Leidenfrost Effect

The Leidenfrost effect denotes the skittering of water when it hits a hotplate. When a liquid hits something that is very hot—about double the liquid's boiling point—it never comes directly in contact with its surface. The action is similar to that of liquid nitrogen poured on the floor. The liquid droplets each produce an insulating vapor layer that prevents the liquid from boiling rapidly, and acts as a barrier that appears to levitate the droplet. Water droplets will also 'climb' up a steep incline, the steepness increasing with the surface roughness, and the direction of the droplets' movement depends on the thermal field of the surface [67].

The dynamics of molecular evaporation and the momentum of the ejected molecules may help in understanding this phenomenon. A water molecule evaporates more readily at higher temperatures and less-saturated vapour pressures. Both heating and unsaturated vapour pressure lengthen and soften the O:H bond with the memory effect, as they do in the Hofmeister effect [16]. Therefore, the liquid–vapor phase transition at the contact ejects molecules with considerable momentum, applying a reaction impulse to the droplet. The direction of the impulse depends on the contact conditions: the component of the impulse parallel to an incline pushes the droplet up the incline; and the component normal to the horizontal hotplate surface separates the droplet from the hotplate. Theoretical formulation of the Leidenfrost effect, in particular the upward movement of the droplet, from the perspective of O:H–O bond memory and evaporation impulse could be of interest.

## 16.10 Polywater—Electrification and Undercoordination

### *16.10.1 The Story—Pathological Science*

Polywater is a hypothesized polymerized form of water that was the subject of much scientific controversy during the late 1960s. By 1969 the popular press had taken notice and sparked fears of a "polywater gap" in the USA. The Pentagon was bankrolling efforts to push this country's polywater technology ahead [68]. Increased press attention also brought with it increased scientific attention, and as early as 1970 doubts about its authenticity were being circulated, turning eventually out the pathological science.

The Soviet physicist Nikolai Fedyakin, working at a lab in Kostroma, Russia, had performed measurements on the properties of water that had been condensed in, or repeatedly forced through, narrow quartz ($SiO_2$) capillary tubes. Some of these experiments resulted in what was seemingly a new form of water with a higher boiling point, lower freezing point, and much higher viscosity than ordinary water. Boris Derjaguin, director for surface physics laboratory at the Institute for Physical Chemistry in Moscow, improved on the method to produce the new water, and though he still produced very small quantities of this mysterious material, he did so substantially faster than Fedyakin did. Investigations of the material properties showed a substantially lower freezing point of −40 °C or less, a boiling point of 150 °C or greater, a density of approx. 1.1–1.2 g/cm$^3$, and increased expansion with temperature. They published these results in Soviet science journals, and short summaries in *Chemical Abstracts* in English without being noticed by Western scientists.

In 1966, Derjaguin travelled to England, presented the work again, in the "Faraday Discussions" in Nottingham. English scientists then started researching the effect as well, and by 1968 it was also under study in the United States. By 1969 the concept had spread to newspapers and magazines. The New York Times [69] featured that: *Water is so essential, so abundant, so simple in composition, and so intensively studied over the centuries that it seems a most unlikely substance to provide a major scientific surprise. Nevertheless, this is precisely what has recently occurred. American chemists have confirmed that there is a form of water with properties quite different from that of the fluid everyone takes for granted. Polywater as this substance has been named is an organized aggregate or polymer of ordinary water molecules but it has very different properties.*

There was fear by the United States military that there was a polywater gap with the Soviet Union [68]. "The U.S. has apparently closed the polywater gap and the Pentagon is bankrolling efforts to push this country's polywater technology ahead."

However, a scientific furor followed shortly. Some experimentalists were able to reproduce Derjaguin's findings, while others failed. Several theories were advanced to explain the phenomenon. Some proposed that it was the cause for increasing resistance on trans-Atlantic phone cables, while others predicted that if polywater were to contact ordinary water, it would convert that water into polywater, causing a disaster or a war. By the 1970s, polywater was well known in the general population.

During this time several people questioned the authenticity of what had come to be known in the West as polywater. The main concern was contamination of the water. Denis Rousseau and Sergio Porto of Bell Labs [70] carried out infrared spectrum analysis which showed polywater was made mostly of chlorine and sodium, which has also the identical spectroscopic feature of sweat. Rousseau [71] then suggested that polywater was nothing more than water with small amounts of biological impurities.

Another wave of research followed, this time more tightly controlled. Invariably the polywater could no longer be made. Chemical analysis found that samples of polywater were contaminated with other substances, and examination of polywater via electron microscopy showed that it also contained small particles of various solids from silica to phospholipids, explaining its greater viscosity.

*Time* magazine [72] reported that: *Challenged by critics to let impartial scientists analyze his polywater, Derjaguin had turned over 25 tiny samples of the substance to investigators of the Soviet Academy of Sciences' Institute of Chemical Physics. The results, which were published in the journal, showed that Derjaguin's polywater was badly contaminated by organic compounds, including lipids and phospholipids, which are ingredients of human perspiration.*

When the experiments that had produced polywater were repeated with thoroughly cleaned glassware, the anomalous properties of the resulting water vanished, and even the scientists who had originally advanced the case for polywater agreed that it did not exist. This took a few years longer in the Soviet Union, where scientists still clung to the idea. In August, 1973, Derjaguin and Churaev published a letter in the journal *Nature* in which they wrote [73], "these anomalous properties should be attributed to impurities rather than to the existence of polymeric water molecules."

Today, polywater is best known as an example of pathological science [74].

### *16.10.2 Clarification—Density, Stability, and Viscosity*

Nevertheless, the story is over but the mechanism for the ionic effect remain open for discussion. The rounds of experimental outcomes indicate the extremely high sensitivity of water to even trace contamination. The unanswered question remains why organic and inorganic impurities even the sweat make the solution—polywater denser, more viscous, and thermally more stable than regular water?

The phenomenon is within the description of the present understanding:

The higher density arises from the masses of impurities and its viscosity results from the dispersion of the quasisolid phase boundaries—the polywater is in quasisolid state. Both molecular undercoordination between the hydrophobic impurities and liquid water and the ionic electrification of the solute impurities, which enhance each other on shortening the H–O bond and stiffening its phonons but the O:H nonbond responds to the perturbation oppositely. The phonon cooperative relaxation mediates the respective Debye temperature and hence raises the melting point and depresses the freezing temperature. Nothing more than the impurity electrification and molecular undercoordination effects play around the polywater.

## 16.11 $H_2O$–Cell and $H_2O$–DNA Interactions

### *16.11.1 Phonon Spectrometrics of $H_2O$–Cell Interaction*

Forming important component of the Hofmeister series, interactions between water molecules and cells, membranes, proteins and so on is one of the most challenging areas in biology [75]. For instances, solvation water around proteins is denser than bulk water [76], implying that the H–O bond becomes longer and softer, and that the interaction is hydrophilic - O:H-O bond like forms. Ice absorbs and entraps albumin protein in solution [77]. The geometry of the O:H–O network within the solvation layer differs from that in bulk water when interacting with a protein surface.

Figure 16.20 shows the Raman $\omega_H$ spectra in normal (non-cancerous) and cancerous breast tissue (infiltrating ductal cancer) with bulk pure water [78]. The spectral features distinguish the H–O features of cancerous tissue, healthy tissue and pure water, and reveal how the length and stiffness of the H–O bond change once water interacts with cancerous tissue. Besides the blueshift of the 2900 $cm^{-1}$ (two fold of the C-C bond frequency in methanol, ethanal, and glycerol) feature of healthy tissue, cancer tissue creates additional H–O features with 3200 and 3450 $cm^{-1}$ components—the cancer tissue is highly flooded with high fraction of undercoordinated $H_2O$ molecules.

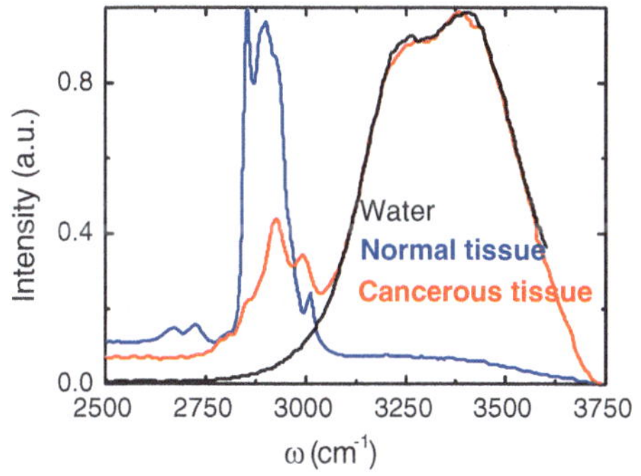


**Fig. 16.20** Raman $\omega_H$ spectra in normal (non-cancerous) and cancerous breast tissue compared with bulk pure water (Reprinted with permission from [78].)

### *16.11.2 Neutron Diffraction from $H_2O$–DNA*

DNA stores all information regarding the structure and function of every living cell. Interaction between DNA and water molecules determines largely its structure and function. However, little is known about the DNA hydration mechanism. Signal of light—scattering from DNA and from proteins in general is very weak and they are in a different energy transfer range. It is only possible to resolve vibrations that are largely due to water–water interactions while presence of biomolecules perturbs the water-water interactions [79]. The neutron diffraction spectrum (predominated by H motion) in the low-frequency region is mainly due to O:H–O bonding between water and DNA, or between water molecules. Inelastic incoherent neutron scattering provides information within the same energy ranges as IR and Raman spectra, but the intensity is sensitive to the phonon DOS without subjecting to the selection rule in Raman and IR spectroscopy.

The concentration-dependence of the neutron spectra of $H_2O$–DNA at 200 K shown in Fig. 16.21 revealed that lowering hydration levels shifts the $\omega_H$ from 400 meV to higher energies and the $\omega_L$ from ~5 meV to lower energies. The spectral peaks approach those of ice if the hydration level is sufficiently high. This effect is the same as for heating and salting, and undercordinating, but it is in an inverse order of DNA concentration.

Inelastic neutron scattering experiments and MD simulations [80] of hydration water in selected hydrophilic and hydrophobic biomolecules revealed that the plasticity of the hydrogen-bond network of hydration water molecules changes with the biomolecular site. At 200 K, the measured low frequency density of states of hydration water molecules of hydrophilic peptides is remarkably similar to that of

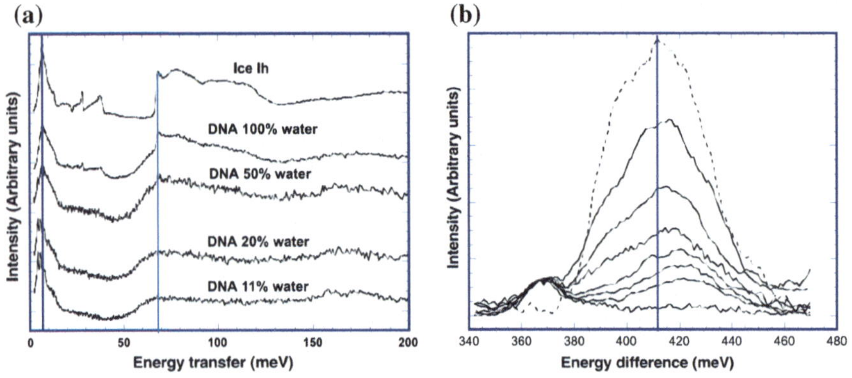


**Fig. 16.21** Neutron phonon spectra of $H_2O$–DNA at 200 K as a function of water concentration. **a** The *percentage* indicates grams of water per 100 g dry DNA. **b** Order of lyophilized DNA, upwards from the *bottom curve*, is 25, 50, 75, 100, 150, 200 g of water per 100 g DNA. *Vertical lines* denote the characteristic $\omega_x$ of neat water (*broken-line spectrum*) (Reprinted with permission from [79].)

high density amorphous ice, whereas, for hydrophobic biomolecules, it is comparable to that of low density amorphous ice behavior. In both hydrophilic and hydrophobic biomolecules, the high frequency modes show a blue shift of the libration mode as compared to the room temperature data. Therefore, it is suggested that the apparent local density of water is larger in a hydrophilic environment than hydrophobic. According to the present premise, H–O bond elongation and O:H nonbond contraction result in higher water density. However, results shown in Fig. 16.21 indicate the opposite—H–O bond contraction and O:H nonbond elongation, which indicate hydrophobic interaction dominance between water molecules and DNA.

## 16.12 X:H–O Bond

Asymmetrical, short-range O:H–O bond potentials are intrinsic to specimens containing F, O, and N element. The short-range interactions and Coulomb coupling is applicable to inter- and intra-molecular interactions of these materials. For instance, Raman measurements have revealed the coupled $\omega_L$ stiffening (110–290 $cm^{-1}$) and $\omega_H$ softening ($\approx$3000 $cm^{-1}$) in the O:H–N bonds in oxamide subjected to compression [81]. The pressure-trend of the Raman shifts of Oxamide ($CO(NH_2)_2$) [82] and Biurea ($C_2H_6N_4O_2$) [83] super molecules, shown in Fig. 16.22, exactly emulate the trend of compressed water ice [84].

Compression at pressures greater than 150 GPa also softens the phonons of hydrogen crystal ($\approx$4000 $cm^{-1}$) at various temperatures [85]. Computations reveal that compression symmetrizes the intra- and inter-$H_2$ molecular distance [86]. These observations may indicate that short-range inter- and intra-molecular interactions and the Coulomb coupling exist in hydrogen crystals.

Figure 16.23 shows the pressure dependence of the (a) $\omega_L$ spectra, (b) $\omega_L$ shift, (c) X:H–O length and, (d) $\omega_H$ for the F:H–O and Cl–H:O bond in $CuF_2(H_2O)_2$(3-chloropyridine) [87]. The change in line color denotes a new phase (or coexistence of phases). HP-I is the first high pressure phase, HP-II is the second high pressure phase, AFM is antiferromagnetic, and FM is ferromagnetic.

Likewise, all the X:H–O bonds share the same relaxation trend of segmental length and stiffness under pressure. Compression shortens the X:H length and stiffens its stretching phonon $\omega_L$ but lengthens the H–O distance slightly and soften its stretching phonon $\omega_H$, resulting in the X:H–O contraction or mass density gain. The La:F [88] and Fe:S [89] stretching phonons at $\omega_L \leq 500$ $cm^{-1}$ also undergo compression stiffening. However, in situ Raman spectroscopy and synchrotron XRD revealed that both high- and low-frequency phonons for the Cl–H:N in the solid hydrazinium monochloride undergoes blueshift under pressure [90], as the Coulomb repulsion between Cl and N ions could be insufficiently strong.

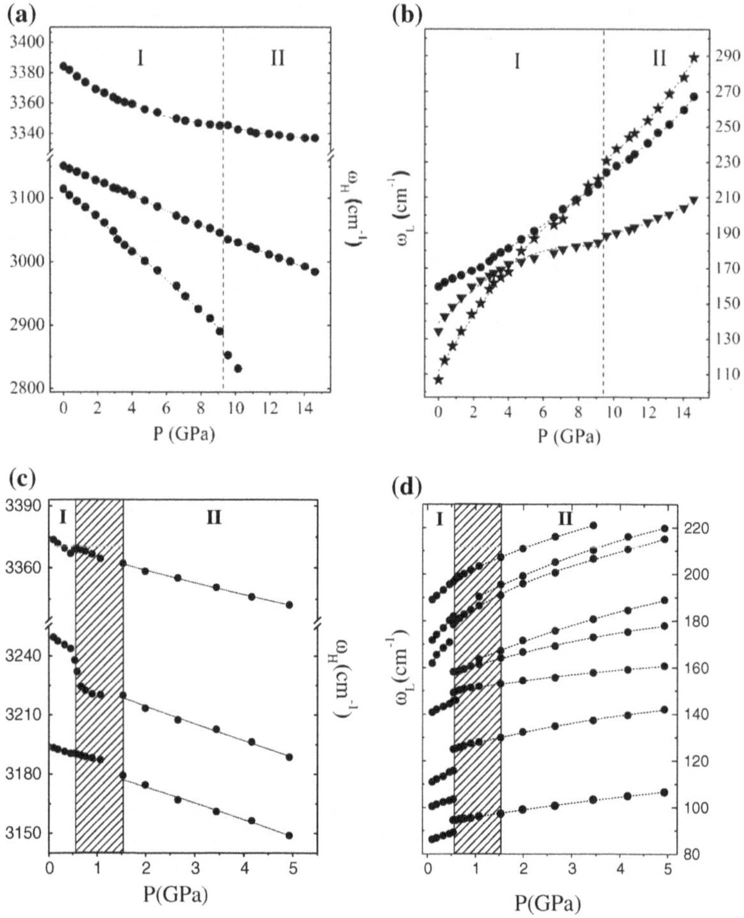


**Fig. 16.22** Compression stiffens the $\omega_L$ and softens the $\omega_H$ phonon of the O:H–N bond in **a**, **b** Oxamide ($CO(NH_2)_2$) and **c**, **d** Biurea ($C_2H_6N_4O_2$) molecular crystals at different phases I and II, indicating the presence of inter electron pair repulsion coupling the O:H–N bond (Reprinted with permission from [81, 83].)

Therefore, X:H–O bond exists in a wide range of materials—$H_2O$, $NH_3$, HF, $H_2$, oxides, nitrides and fluorides—with the presence of asymmetrical and short-range interactions. On the basis of the current notion of O:H–O bond cooperativity, it is expected that asymmetrical relaxation in length and stiffness of the O:H–O bond dictates the functionality of species with O:H–O bond-like involvement, including biomolecules, organic materials, H crystals, among others.

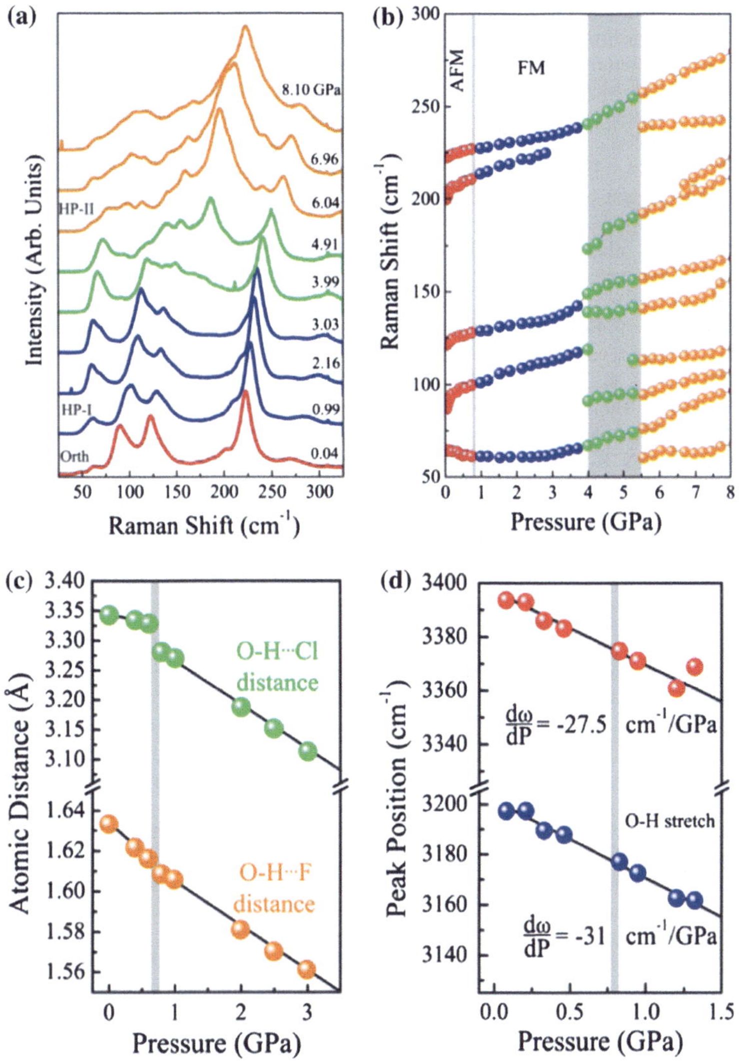


**Fig. 16.23** Pressure dependence of the **a** X:H stretching phonon $\omega_L$ spectra, **b** $\omega_L$ shift, **c** X:H–O length, and **d** H–O stretching phonon $\omega_H$ for $CuF_2(H_2O)_2$(3—chloropyridine) (Reprinted with permission from [87].)

## References


1. C. Medcraft, D. McNaughton, C.D. Thompson, D.R.T. Appadoo, S. Bauerecker, E.G. Robertson, Water ice nanoparticles: size and temperature effects on the mid-infrared spectrum. PCCP **15**(10), 3630–3639 (2013)

2. S.A. Deshmukh, S.K. Sankaranarayanan, D.C. Mancini, Vibrational spectra of proximal water in a thermo-sensitive polymer undergoing conformational transition across the lower critical solution temperature. J. Phys. Chem. B. **116**(18), 5501–5515 (2012)
3. C. Medcraft, D. McNaughton, C.D. Thompson, D. Appadoo, S. Bauerecker, E.G. Robertson, Size and temperature dependence in the far-Ir spectra of water ice particles. Astrophys. J. **758** (1), 17 (2012)
4. K. Rottger, A. Endriss, J. Ihringer, S. Doyle, W.F. Kuhs, Lattice-constants and thermal-expansion of $H_2O$ and $D_2O$ Ice ih between 10 and 265 K. Acta Crystallog. B **50**, 644–648 (1994)
5. Y. Huang, X. Zhang, Z. Ma, Y. Zhou, W. Zheng, J. Zhou, C.Q. Sun, Hydrogen-bond relaxation dynamics: resolving mysteries of water ice. Coord. Chem. Rev. **285**, 109–165 (2015)
6. O. Bjorneholm, F. Federmann, S. Kakar, T. Moller, Between vapor and ice: free water clusters studied by core level spectroscopy. J. Chem. Phys. **111**(2), 546–550 (1999)
7. M. Abu-Samha, K.J. Borve, M. Winkler, J. Harnes, L.J. Saethre, A. Lindblad, H. Bergersen, G. Ohrwall, O. Bjorneholm, S. Svensson, The local structure of small water clusters: imprints on the core-level photoelectron spectrum. J. Phys. B. **42**(5), 055201 (2009)
8. K. Nishizawa, N. Kurahashi, K. Sekiguchi, T. Mizuno, Y. Ogi, T. Horio, M. Oura, N. Kosugi, T. Suzuki, High-resolution soft X-ray photoelectron spectroscopy of liquid water. PCCP **13**, 413–417 (2011)
9. X.J. Liu, M.L. Bo, X. Zhang, L.T. Li, Y.G. Nie, H. TIan, Y. Sun, S. Xu, Y. Wang, W. Zheng, C.Q. Sun, Coordination-resolved electron spectrometrics. Chem. Rev. **115**(14), 6746–6810 (2015)
10. D.D. Kang, J.Y. Dai, J.M. Yuan, Changes of structure and dipole moment of water with temperature and pressure: a first principles study. J. Chem. Phys. **135**(2), 024505 (2011)
11. D.D. Kang, J. Dai, H. Sun, Y. Hou, J. Yuan, Quantum similation of thermally driven phase transition and O k-edge absorption of high-pressure ice. Sci. Rep. **3**, 3272 (2013)
12. G.M. Marion, S.D. Jakubowski, The compressibility of ice to 2.0 kbar. Cold Reg. Sci. Technol. **38**(2–3), 211–218 (2004)
13. P.G. Debenedetti, H.E. Stanley, Supercooled and glassy water. Phys. Today **56**(6), 40–46 (2003)
14. R.A. Fine, F.J. Millero, Compressibility of water as a function of temperature and pressure. J. Chem. Phys. **59**, 5529 (1973)
15. G.S. Kell, Density, thermal expansivity, and compressibility of liquid water from 0. deg. to 150. deg. Correlations and tables for atmospheric pressure and saturation reviewed and expressed on 1968 temperature scale. J. Chem. Eng. Data **20**(1), 97–105 (1975)
16. X. Zhang, T. Yan, Y. Huang, Z. Ma, X. Liu, B. Zou, C.Q. Sun, Mediating relaxation and polarization of hydrogen-bonds in water by NaCl salting and heating. PCCP **16**(45), 24666–24671 (2014)
17. Y. Marechal, The molecular structure of liquid water delivered by absorption spectroscopy in the whole IR region completed with thermodynamics data. J. Mol. Struct. **1004**(1–3), 146–155 (2011)
18. A. Nilsson, D. Nordlund, I. Waluyo, N. Huang, H. Ogasawara, S. Kaya, U. Bergmann, L.Å. Näslund, H. Öström, P. Wernet, K.J. Andersson, T. Schiros, L.G.M. Pettersson, X-ray absorption spectroscopy and X-ray Raman scattering of water and ice; an experimental view. J. Electron Spectrosc. Relat. Phenom. **177**(2–3), 99–129 (2010)
19. J. Meibohm, S. Schreck, P. Wernet, Temperature dependent soft X-ray absorption spectroscopy of liquids. Rev. Sci. Instrum. **85**(10), 103102 (2014)
20. M. Emoto, E. Puttick, *The Healing Power of Water* (Hay House, Incorporated, 2007)
21. J.D. Brownridge, When does hot water freeze faster then cold water? A search for the Mpemba effect. Am. J. Phys. **79**(1), 78 (2011)
22. Q. Zeng, Y. Zhou, W. Kai, B. Zou, C.Q. Sun, *Solution room temperature phase precipitation under compression.* Communicated (2015)

23. G.H. Pollack, *The Fourth Phase of Water: Beyond Solid, Liquid, and Vapor* (Ebner & Sons Seattle, USA, 2013)
24. C.A. Silvera Batista, R.G. Larson, N.A. Kotov, Nonadditivity of nanoparticle interactions. Science **350**(6257) (2015)
25. N.J. English, J. MacElroy, Molecular dynamics simulations of microwave heating of water. J. Chem. Phys. **118**(4), 1589–1592 (2003)
26. A. De Ninno, A.C. Castellano, On the effect of weak magnetic field on solutions of glutamic acid: the function of water, in *Journal of Physics: Conference Series* (IOP Publishing, 2011)
27. X. Shen, Increased dielectric constant in the water treated by extremely low frequency electromagnetic field and its possible biological implication, in *Journal of Physics: Conference Series* (IOP Publishing, 2011)
28. A. Goldsworthy, H. Whitney, E. Morris, Biological effects of physically conditioned water. Water Res. **33**(7), 1618–1626 (1999)
29. K. Zhou, G. Lu, Q. Zhou, J. Song, S. Jiang, H. Xia, Monte Carlo simulation of liquid water in a magnetic field. J. Appl. Phys. **88**(4), 1802–1805 (2000)
30. G. Zhang, W. Zhang, H. Dong, Magnetic freezing of confined water. J. Chem. Phys. **133**(13), 134703 (2010)
31. Y. Fujimura, M. Iino, The surface tension of water under high magnetic fields. J. Appl. Phys. **103**(12), 2940128 (2008)
32. R. Cai, H. Yang, J. He, W. Zhu, The effects of magnetic fields on water molecular hydrogen bonds. J. Mol. Struct. **938**(1–3), 15–19 (2009)
33. Z. Zhou, H. Zhao, J. Han, Supercooling and crystallization of water under DC magnetic fields. CIESC J. **63**(5), 1408–1410 (2012)
34. D. Taylor, *Standard YouTube License* (2013), https://www.youtube.com/watch?v=kt—n8N_kqto
35. P. domain, *Magnetic field whirpool*, https://www.wikiwand.com/en/Whirlpool
36. L. Chen, C.J. Li, Z.M. Ren, Variation of surface tension of water in high magnetic field. Adv. Mater. Res. **750–752**, 2279–2282 (2013)
37. Z. Wang, F.C. Wang, Y.P. Zhao, Tap dance of a water droplet. Proc. Roy. Soc. London A. **468** (2145), 2485–2495 (2012)
38. S. Ceurstemont, *Zapped droplets tap dance to the beat (Zhao Yapu)*. New Scientist: http://www.newscientist.com/blogs/nstv/2012/04/zapped-droplets-tap-dance-to-the-beat.html (2013)
39. Q. Yuan, Y.-P. Zhao, Multiscale dynamic wetting of a droplet on a lyophilic pillar-arrayed surface. J. Fluid Mech. **716**, 171–188 (2013)
40. Q.Z. Yuan, Y.P. Zhao, Wetting on flexible hydrophilic pillar-arrays. Sci. Rep. **3**, 1944 (2013)
41. S. Iikubo, K. Kodama, K. Takenaka, H. Takagi, M. Takigawa, S. Shamoto, Local lattice distortion in the giant negative thermal expansion material $Mn_3Cu_{1-x}Ge_xN$. Phys. Rev. Lett. **101**(20), 205901 (2008)
42. A.L. Goodwin, M. Calleja, M.J. Conterio, M.T. Dove, J.S.O. Evans, D.A. Keen, L. Peters, M. G. Tucker, Colossal positive and negative thermal expansion in the framework material $Ag_3$ Co(CN)(6). Science **319**(5864), 794–797 (2008)
43. A.C. McLaughlin, F. Sher, J.P. Attfield, Negative lattice expansion from the superconductivity-antiferromagnetism crossover in ruthenium copper oxides. Nature **436** (7052), 829–832 (2005)
44. J.S.O. Evans, Negative thermal expansion materials. J. Chem. Soc.-Dalton Trans. (19): 3317–3326 (1999)
45. T.A. Mary, J.S.O. Evans, T. Vogt, A.W. Sleight, Negative thermal expansion from 0.3 to 1050 Kelvin in $ZrW_2O_8$. Science **272**(5258), 90–92 (1996)
46. S. Stoupin, Y.V. Shvyd'ko, Thermal expansion of diamond at low temperatures. Phys. Rev. Lett. **104**(8), 085901 (2010)
47. Q.H. Tang, T.C. Wang, B.S. Shang, F. Liu, Thermodynamic properties and constitutive relations of crystals at finite temperature. Sci. China-Phys. Mech. Astron. G. **55**, 933 (2012)

48. Y.J. Su, H. Wei, R.G. Gao, Z. Yang, J. Zhang, Z.H. Zhong, Y.F. Zhang, Exceptional negative thermal expansion and viscoelastic properties of graphene oxide paper. Carbon **50**(8), 2804–2809 (2012)
49. A.W. Sleight, Compounds that contract on heating. Inorg. Chem. **37**(12), 2854–2860 (1998)
50. J.S.O. Evans, T.A. Mary, T. Vogt, M.A. Subramanian, A.W. Sleight, Negative thermal expansion in $ZrW_2O_8$ and $HfW_2O_8$. Chem. Mater. **8**(12), 2809–2823 (1996)
51. G. Ernst, C. Broholm, G.R. Kowach, A.P. Ramirez, Phonon density of states and negative thermal expansion in $ZrW_2O_8$. Nature **396**(6707), 147–149 (1998)
52. A.K.A. Pryde, K.D. Hammonds, M.T. Dove, V. Heine, J.D. Gale, M.C. Warren, Origin of the negative thermal expansion in $ZrW_2O_8$ and $ZrV_2O_7$. J. Phys.-Condens. Matter **8**(50), 10973–10982 (1996)
53. M. Chaplin. *Water structure and science*, http://www.lsbu.ac.uk/water/
54. C.W. Li, X. Tang, J.A. Muñoz, J.B. Keith, S.J. Tracy, D.L. Abernathy, B. Fultz, Structural relationship between negative thermal expansion and quartic anharmonicity of cubic $ScF_3$. Phys. Rev. Lett. **107**(19), 195504 (2011)
55. B.K. Greve, K.L. Martin, P.L. Lee, P.J. Chupas, K.W. Chapman, A.P. Wilkinson, Pronounced negative thermal expansion from a simple structure: cubic $ScF_3$. J. Am. Chem. Soc. **132**(44), 15496–15498 (2010)
56. L.K. Pan, S.Q. Xu, W. Qin, X.J. Liu, Z. Sun, W.T. Zheng, C.Q. Sun, Skin dominance of the dielectric-electronic-phononic-photonic attribute of nanostructured silicon. Surf. Sci. Rep. **68** (3–4), 418–455 (2013)
57. L.K. Pan, H.T. Huang, C.Q. Sun, Dielectric relaxation and transition of porous silicon. J. Appl. Phys. **94**(4), 2695–2700 (2003)
58. L.K. Pan, C.Q. Sun, T.P. Chen, S. Li, C.M. Li, B.K. Tay, Dielectric suppression of nanosolid silicon. Nanotechnology **15**(12), 1802–1806 (2004)
59. R. Tsu, D. Babic, Doping of a quantum-dot. Appl. Phys. Lett. **64**(14), 1806–1808 (1994)
60. J.W. Li, L.W. Yang, Z.F. Zhou, P.K. Chu, X.H. Wang, J. Zhou, L.T. Li, C.Q. Sun, Bandgap modulation in ZnO by size, pressure, and temperature. J. Phys. Chem. C **114**(31), 13370–13374 (2010)
61. L.K. Pan, Y.K. Ee, C.Q. Sun, G.Q. Yu, Q.Y. Zhang, B.K. Tay, Band-gap expansion, core-level shift, and dielectric suppression of porous silicon passivated by plasma fluorination. J. Vac. Sci. Technol. B **22**(2), 583–587 (2004)
62. G. Ouyang, C.Q. Sun, W.G. Zhu, Atomistic origin and pressure dependence of band gap variation in semiconductor nanocrystals. J. Phys. Chem. C. **113**(22), 9516–9519 (2009)
63. C.S. Zha, R.J. Hemley, S.A. Gramsch, H.K. Mao, W.A. Bassett, Optical study of H2O ice to 120 GPa: dielectric function, molecular polarizability, and equation of state. J. Chem. Phys. **126**(7), 074506 (2007)
64. W.B. Floriano, M.A.C. Nascimento, Dielectric constant and density of water as a function of pressure at constant temperature. Braz. J. Phys. **34**(1), 38–41 (2004)
65. F. Klameth, M. Vogel, Slow water dynamics near a glass transition or a solid interface: a common rationale. J. Phys. Chem. Lett., 4385–4389 (2015)
66. S.T. van der Post, C.S. Hsieh, M. Okuno, Y. Nagata, H.J. Bakker, M. Bonn, J. Hunger, Strong frequency dependence of vibrational relaxation in bulk and surface water reveals sub-picosecond structural heterogeneity. Nat. Commun. **6**, 8384 (2015)
67. E. Zolfagharifard. *Leidenfrost effect:* http://www.dailymail.co.uk/sciencetech/article-2442638/Leidenfrost-Effect-makes-high-temperature-water-travel-uphill.html (2013)
68. Editorial, U.S. Begins efforts to exceed the USSR in polywater science, in *Wall Street Journal* (1969)
69. Editorial, Polywater, in *New York Times* (1969)
70. D.L. Rousseau, S.P.S. Porto, Polywater: polymer or artifact? Science **167**(3926), 1715–1719 (1970)
71. D.L. Rousseau, "Polywater" and sweat: similarities between the infrared spectra. Science **171** (3967), 170–172 (1971)
72. Editorial, Science: doubts about polywater, in *Time* (1970)

73. B.V. Derjagui, N.V. Churaev, Nature of anomalous water. Nature **244**(5416), 430–431 (1973)
74. D. Eisenberg, A scientific gold rush. Science **213**(4512), 1104–1105 (1981)
75. G.H. Zuo, J. Hu, H.P. Fang, Effect of the ordered water on protein folding: An off-lattice Gõ-like model study. Phys. Rev. E **79**(3), 031925 (2009)
76. A. Kuffel, J. Zielkiewicz, Why the solvation water around proteins is more dense than bulk water. J. Phys. Chem. B (2012)
77. A. Twomey, R. Less, K. Kurata, H. Takamatsu, A. Aksan, In Situ spectroscopic quantification of protein-ice interactions. J. Phys. Chem. B **117**(26), 7889–7897 (2013)
78. I.V. Stiopkin, C. Weeraman, P.A. Pieniazek, F.Y. Shalhout, J.L. Skinner, A.V. Benderskii, Hydrogen bonding at the water surface revealed by isotopic dilution spectroscopy. Nature **474** (7350), 192–195 (2011)
79. I. Michalarias, I. Beta, R. Ford, S. Ruffle, J.C. Li, Inelastic neutron scattering studies of water in DNA. Appl. Phys. A Mater. Sci. Process. **74**, s1242–s1244 (2002)
80. D. Russo, J. Teixeira, L. Kneller, J.R.D. Copley, J. Ollivier, S. Perticaroli, E. Pellegrini, M.A. Gonzalez, Vibrational density of states of hydration water at biomolecular sites: hydrophobicity promotes low density amorphous ice behavior. J. Am. Chem. Soc. **133**(13), 4882–4888 (2011)
81. T. Yan, S. Li, K. Wang, X. Tan, Z. Jiang, K. Yang, B. Liu, G. Zou, B. Zou, Pressure-induced phase transition in N-H...O hydrogen-bonded molecular crystal oxamide. J. Phys. Chem. B **116**(32), 9796–9802 (2012)
82. K. Wang, D. Duan, R. Wang, A. Lin, Q. Cui, B. Liu, T. Cui, B. Zou, X. Zhang, J. Hu, G. Zou, H.K. Mao, Stability of hydrogen-bonded supramolecular architecture under high pressure conditions: pressure-induced amorphization in melamine-boric acid adduct. Langmuir **25**(8), 4787–4791 (2009)
83. T. Yan, K. Wang, X. Tan, K. Yang, B. Liu, B. Zou, Pressure-induced phase transition in N-H...O hydrogen-bonded molecular crystal biurea: combined raman scattering and X-ray diffraction study. J. Phys. Chem. C. **118**(28), 15162–15168 (2014)
84. Y. Yoshimura, S.T. Stewart, H.K. Mao, R.J. Hemley, In situ Raman spectroscopy of low-temperature/high-pressure transformations of $H_2O$. J. Chem. Phys. **126**(17), 174505 (2007)
85. C.-S. Zha, Z. Liu, R. Hemley, Synchrotron infrared measurements of dense hydrogen to 360 GPa. Phys. Rev. Lett. **108**(14), 146402 (2012)
86. H. Liu, H. Wang, Y. Ma, Quasi-molecular and atomic phases of dense solid hydrogen. J Phys. Chem. C. **116**(16), 9221–9226 (2012)
87. K.R. O'Neal, T.V. Brinzari, J.B. Wright, C. Ma, S. Giri, J.A. Schlueter, Q. Wang, P. Jena, Z. Liu, J.L. Musfeldt, Pressure-induced magnetic crossover driven by hydrogen bonding in $CuF_2(H_2O)2$(3-chloropyridine). Sci. Rep. **4**, 6054 (2014)
88. W.A. Crichton, P. Bouvier, B. Winkler, A. Grzechnik, The structural behaviour of LaF(3) at high pressures. Dalton Trans. **39**(18), 4302–4311 (2010)
89. A. Kleppe, A. Jephcoat, High-pressure Raman spectroscopic studies of $FeS_2$ pyrite. Mineral. Mag. **68**(3), 433–441 (2004)
90. S. Jiang, D. Duan, F. Li, X. Huang, X. Yang, W. Li, Y. Huang, K. Bao, Q. Zhou, B. Liu, T. Cui, The hydrogen-bond effect on the high pressure behavior of hydrazinium monochloride. J. Raman Spectrosc. **46**(2), 266–272 (2015)

# Chapter 17
# Approaching Strategies

- *Interplay of quantum computation and phonon and electron spectrometrics enables correlation and quantification of the bond–phonon–electron–property relaxation dynamics.*
- *Lagrangian oscillating mechanics transforms the known segmental length and vibration frequency into the respective force constant and bond energy and hence enabled mapping the potential paths of O:H–O bond at relaxation.*
- *Fourier thermal fluid transport dynamics confirm skin supersolidity and O:H–O bond memory by reproducing the Mpemba paradox.*
- *Phonon spectrometrics provides comprehensive and quantitative information on the O:H–O bond cooperative relaxation, charge polarization, viscoelasticity, etc.*

**Abstract** Diffraction crystallography, scanning tunneling microscopy, molecular-site-resolved spectrometrics of electronic binding energy and phonon frequencies are essential techniques for probing atomistic, dynamic, multifield bond–electron–phonon relaxation that determine the observable physical properties. However, these techniques possess each advantages and limitations in dealing with water and ice because of the O:H–O bond cooperativity and segmental disparity. The asymmetrical, ultra-short-range interactions, atomistic anisotropy and localization, and the involvement of polarization also hindered quantum computations and conventional statistic thermodynamics from describing accurately and consistently the true situations in water and ice. A combination of the electron and phonon spectrometrics, the Lagrangian oscillating mechanics, and the Fourier thermo-fluid transport dynamics are necessary to complementing the quantum and statistic thermodynamic approaches.

## 17.1 Numerical Approaches

### *17.1.1 Quantum Computations*

#### 17.1.1.1 Quantities of Concern

Density functional theory (DFT) and molecular dynamics (MD) computations aid verification of hypotheses and expectations on the framework of O:H–O bond cooperativity. Major concerns include the stimulated relaxation of the following quantities:



C.Q. Sun and Y. Sun, *The Attribute of Water*, Springer Series in Chemical Physics 113, DOI 10.1007/978-981-10-0180-2_17

(1) Geometrical configuration of molecular clusters
(2) O:H–O bond containing angle and segmental $d_x$ and $E_x$
(3) The power spectra for phonon frequency $\omega_x$
(4) Electron and phonon DOS distribution and skin charge accumulation
(5) Electron binding energy $E_{Bind}$ and skin stress and viscosity
(6) Molecular diffusivity as an indication of liquid/quasisolid phase transition

Calculation derivatives are used as reference for concept verification as they are subject to deviation from experimental observations in occasions. Inclusion of strong localization and fluctuation, and proper representation of the asymmetrical and short-range interactions remain challenge. The inclusion of dispersion correction by introducing O:H interaction in the DFT package still limits the accuracy of the segmental length and vibration frequencies with respect to measurements. Nevertheless, quantum computations form the powerful means for proving the hypothesis, and importantly, one should be focused more on the physical origin and the trend of O:H–O bond segmental cooperative relaxation in length and energy than on the absolute values of concern.

#### 17.1.1.2 Molecular Dynamics

Having taken the flexibility, polarizability, and nuclear quantum effects of water into consideration, the ab initio optimized MD force field, the Condensed-phase Optimized Molecular Potentials for Atomistic Simulation Studies (COMPASS'27) [1], is an elegant approach. This package derives the phonon (power) spectra in terms of Fourier transformation of molecular velocity autocorrelation function, Cor (t), $I(\omega) = 2\int_0^\infty \mathrm{Cor}(v(t)) \cos \omega t \, dt$ [2]. A 360-molecule supercell of proton-disordered ice $I_h$ was simulated, in which the proton was sufficiently optimised to avoid a net dipole moment or to minimise a net quadruple moment [3]. The unit cell was relaxed in the isothermal–isobaric ensemble (NPT) at atmospheric pressure and different temperatures. Andersen's thermostat and barostat approach maintained the temperature and pressure of the closed system [4]. The relaxation time is extended to 120 ps, in order to ensure the stability of the single-phase system in terms of temperature, density and energy. A 15 nm vacuum slab was inserted into the supercell, shown in Fig. 17.1, to represent effect of the skin, which was relaxed in the canonical ensemble (NVT) at 200 K for 100 ps to obtain equilibrium in a 0.5 fs time interval. The Nosé-Hoover thermostat algorithm with a Q ratio of 0.01 was adopted to control the temperature. The ice structure was relaxed in the NPT ensemble for 30 ps in 0.5 fs steps to converge for T, P, and energy.

Deviation may exist between numerical derivatives and experimental observations because of the dominance of asymmetrical, short-range interactions. This deviation can however be corrected with respect to spectrometric derivatives.

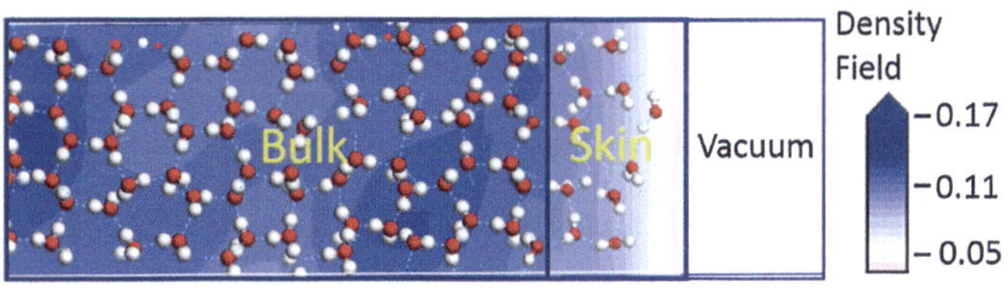


**Fig. 17.1** Schematic illustration of the water supercell with an insertion of a vacuum slab representing the supersolid skin of ice at 200 K. This comprised three regions, l. to r.: the bulk, the skin, and the vacuum. The skin contains undercoordinated molecules and free H–O dangling bond radicals. The colors along the horizontal axis indicate the MD-derived mass density distribution in the unit cell. This unit cell also applies to the one-side shell of a nanobubble (Reprinted with permission from [5].) (color online)

#### 17.1.1.3 DFT and MD Calculation Procedures

The structural optimization and the bond angle–length–stiffness relaxation of $(H_2O)_N$ clusters and ice skin were conducted using the DFT Dmol$^3$ code of Perdew and Wang (PW) [6] and the dispersion-corrected PW code of Ortmann-Bechstedt-Schmidt (OBS) [7]. The latter includes the O:H nonbond (referred as hydrogen bond) interactions. The all-electron method was used to approximate the wave functions with a double numeric and polarization basis sets. The self-consistency threshold of total energy was set at $10^{-6}$ Hartree. In the structural optimization, the tolerance limits for the energy, force and displacement were set at $10^{-5}$ Hartree, 0.002 Hartree/Å, and 0.005 Å, respectively. Harmonic vibrational frequencies were computed by diagonalizing the mass-weighted Hessian matrix [8].

The CASTEP code [9] was used within the Perdew–Burke–Ernzerhof (PBE) [10] functional parameterization to calculate the compressed ice-VIII. The norm–conserving pseudopotential (NCP) scheme was adopted, in which the H $1s^1$ and the O $2s^2p^4$ are treated as valence electrons. The use of the plane-wave kinetic energy cutoff of 500 eV converged in total energies. Ice-VIII consisting of two interpenetrating cubic ice lattices with eight molecules in each unit cell, was examined.

The MD calculations were performed to examine the evolution of the H–O and O:H distances in a 2 × 2 supercell of ice-VIII containing 32 molecules, subjected to a pressure changing from 1 to 20 GPa. The structure was dynamically relaxed in the isoenthalpic–isobaric ensemble for 30 ps and showed sufficiently stable convergence [11]. The average H–O and O:H lengths were taken over the final 10 ps (20,000 steps).

DFT calculations optimize quantities of concern without capable of including thermal excitation. Furthermore, involvement of O:H dispersion correction in computation gives rise to trend of O:H–O bond cooperativity despite slight difference in absolute values for water clusters compared with the derivatives of the non-dispersed package.

### 17.1.2 Skin Stress and Viscosity

Calculations using the following method quantify the stress and viscosity of water skin. The difference between the stress components in directions parallel and perpendicular to the interface defines the skin tension $\gamma$ [12, 13]:

$$\gamma = \frac{1}{2}\left(\frac{\sigma_{xx} + \sigma_{yy}}{2} - \sigma_{zz}\right) \cdot L_z, \tag{17.1}$$

where $\sigma_{xx}$, $\sigma_{yy}$, and $\sigma_{zz}$ are the stress tensor elements and $L_z$ is the length of the supercell. The skin shear viscosity $\eta_s$ is correlated to the volume stress $\sigma$ as [14, 15]:

$$\eta_s = \frac{V}{kT}\int_0^{\infty} \langle \sigma_{\alpha\beta}(0)\sigma_{\alpha\beta}(t)\rangle dt, \tag{17.2}$$

where $\sigma_{\alpha\beta}$ denotes the three equivalent off-diagonal elements of the stress tensors. The volume viscosity $\eta_v$ depends on the decay of fluctuations in thc diagonal elements of the stress tensor:

$$\begin{aligned} \eta_v &= \frac{V}{kT}\int_0^{\infty} \langle \delta\sigma(0)\delta\sigma(t)\rangle dt \\ \delta\sigma &= \sigma - \langle\sigma\rangle. \end{aligned} \tag{17.3}$$

On basis of these notations, $\gamma$ was first optimized using the MD calculations to derive the stress tensors. The auto-correlation functions of the stress tensors were also used to calculate $\eta_s$ and $\eta_v$ in accordance with (17.2) and (17.3).

### 17.1.3 Lagrangian Mapping O:H–O Bond Potentials

Lagrangian mechanics is an ideal approach dealing with the oscillating dynamics of the coupled O:H–O bond oscillator pair. A linear approximation of the O:H–O bond is convenient because the ∠O:H–O angle in water ice is greater than 160° [16]. The angle relaxation contributes only to density and has little effect on the segmental energy or vibration frequency.

The reduced mass of the $H_2O$:$H_2O$ oscillator is $m_L = 18 \times 18/(18 + 18)m_0 = 9m_0$. For the H–O oscillator, it is $m_H = 1 \times 16/(1 + 16)m_0 = 16/17m_0$, where $m_0$ is the unit mass of $1.66 \times 10^{-27}$ kg. Inclusion of isotope or chemical complex will change the value of the reduced mass, which varies the oscillator frequency. The motion of the coupled O:H–O oscillators follows the Lagrangian equation [17]:

$$\frac{\mathrm{d}}{\mathrm{d}t}\left(\frac{\partial L}{\partial(\mathrm{d}q_x/\mathrm{d}t)}\right) - \frac{\partial L}{\partial q_x} = Q_x, \tag{17.4}$$

where L = T − V, in which T is the total kinetic energy and V is the total potential energy. $Q_x$ is the generalized non-conservative forces. The time-dependent $q_x(t)$ here is $u_L$ and $u_H$, representing the generalized coordinates of O atoms from their equilibrium positions in the L and H springs. The kinetic energy T consists of two terms:

$$T = \frac{1}{2}\left[m_\mathrm{L}\left(\frac{\mathrm{d}u_\mathrm{L}}{\mathrm{d}t}\right)^2 + m_\mathrm{H}\left(\frac{\mathrm{d}u_\mathrm{H}}{\mathrm{d}t}\right)^2\right] \tag{17.5}$$

The potential energy V contains three terms that approximate the coupled oscillator pair:

$$2V = k_L(\Delta u_L)^2 + k_H(\Delta u_H)^2 + k_C[\Delta(u_L - u_C)]^2 + 0(\ldots), \tag{17.6}$$

where $k_x$ is the force constant; $\Delta u_x$ is the amplitude of vibration of the O:H nonbond and the H–O bond; and $k_C$ is the second derivative of the Coulomb potential. The first two terms are the isolated O:H and H–O oscillators and the last term is the O–O coupling. The $f_x = 0$, $f_x + f_c = 0$, and $f_x + f_c + f_{ext} = 0$ represent the equilibrium with or without the Coulomb interaction $f_c$ or the external non-conservative $f_{ext}$ interactions. These constraints define the parameters involved in the O:H–O bond potentials.

A Laplace transformation and then an inverse transformation of the Lagrangian equation can convert the measured segmental length and vibration frequency onto the respective force constant and the cohesive energy, which enables the mapping of the asymmetrical, short-range potential paths of the O:H–O bond relaxing under excitation.

### *17.1.4 Fourier Thermo–Fluid Transport Dynamics*

The Fourier thermal–fluid transport equation [18] with appropriate initial and boundary conditions best describes the process of thermo–fluid transportation (diffusion and convection) in the liquid water with the skin as an additional component, which resolves the Mpemba effect as a consequence of O:H–O bond memory in energy emission, skin supersolidity in heat transmission, and non-adiabatic heat dissipation between the source and drain.

The time-dependent temperature change at a site (x) in the container being cooled follows a step-function representing the ordinary bulk and the supersolid skin:

$$\frac{\partial\theta(x)}{\partial t} = \nabla \cdot (\alpha(\theta(x), x)\nabla\theta(x)) - v \cdot \nabla\theta(x), \tag{17.7}$$

where $\alpha$ is the thermal diffusion coefficient and v the convection velocity in the fluid. The interface between the bulk and the skin and the interface between the skin and the drain must satisfy certain conditions. Chapter 11 gives more details on the boundary conditions and calculation procedures. In order to examine all possible factors contributing to the Mpemba effect that integrates the heat "emission–conduction–dissipation" dynamics in the "source–path–drain" cycle system, this initial and boundary conditioned problem was solved numerically using the finite-element method [19].

## 17.2 Probing Strategies

### *17.2.1 X-ray and Neutron Diffraction*

#### 17.2.1.1 Principles and Derivatives

X-ray and neutron diffraction provides powerful means to determine the O–O distance and the order of geometry, which satisfies laws of Bragg diffraction and provides the relaxation of the structure factor S(q) and the O–O radial distribution function (RDF) $g_{oo}(r)$—density distribution of the first and the subsequent shells surrounding a certain oxygen atom. The q is the vector in the reciprocal space. Protons serve as the scatters in neutron diffraction and atoms as the scatters in X-ray diffraction. Therefore, neutron diffraction also provides $g_{OH}(r)$ and $g_{HH}(r)$ functions with improved accuracy. Besides, neutron diffraction probes the density-of-states for phonons over the full energy range for water and ice.

Figure 17.2 describes the sophisticated measurements of structure factor S(q) for micrometer-sized water droplets as a function of temperature [20], which is also used for water floating bridge formed under electric field [21].

#### 17.2.1.2 Advantage and Limitations

The experiments have reached a stage where data acquisition with good statistical accuracy is now in hand. However, it is the correction of difficult systematic errors in the data that have made for a history of uncertainty in the reported structural correlation functions. In essence, most of the systematic experimental error corrections are relatively straightforward to account for polarization, absorption, and other geometric corrections relevant to the details of a given scattering experiment.

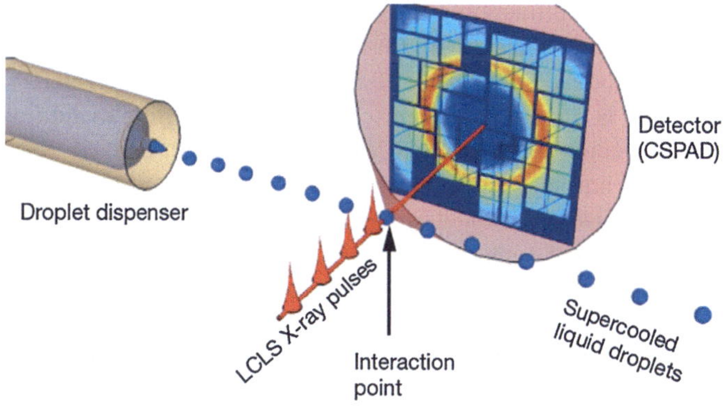


**Fig. 17.2** X-ray scattering from individual water micrometer-sized droplets with a single-shot selection scheme [20] or from water bridge formed between beakers under electric field [21]. Neutron diffraction can be realized by replacing the source of X-ray with the neutron beam in high vacuum [22] (Reprinted with permission from [20].)

Head-Gordon and Hura [23] have reviewed the scattering techniques of X-ray and neutron scattering, with special attention focused on the inherent difficulties in data analysis in each of these experiments to extract the structural content from the measured intensity. Furthermore, structure is only one of many properties of water, and other experiments that probe dynamics over many time scales that characterize the thermodynamic phase diagram of water or its properties in solution and in its role as a solvent are vital for full understanding of the fluid.

These very different but complementary experiments have converged on a virtually identical $g_{OO}(r)$ under ambient conditions. The question is whether these experimental measurements using better data analysis are becoming a reliable source of quantitative structural information for water. An optimistic view is that the new X-ray measurements have less uncertainty than previous measurements because they use a state-of-the-art synchrotron source and high-quality CCD detectors, have much better estimates of unwanted Compton scattering, and have more careful theoretical work to restore chemical bonding effects in the weighting of the partial structure factors.

With the advances in the construction and capabilities of diffractometers and low temperature facilities for data collection, highly precise structural information is now available, but the location and refinement of H atom positions often still remains at the limits of the technique. A more fundamental concern lies in the fact that in X-ray structure determinations, distance between the H atom and the bonded heavier atom (X–H) is on the average 0.1–0.2 Å shorter than the internuclear distances. This happens because X-rays are scattered by electrons and the atomic position derived for an H atom from an X-ray analysis approximates the centroid of the electron density. The latter is not centred at the H nucleus, but is displaced towards the atom X.

The use of neutron diffraction analysis avoids this problem since the scattering centres are the atomic nuclei themselves. The distances derived from neutron analysis therefore correspond nearly to the interatomic distances and, accordingly, neutron diffraction is a most important technique in the determination of accurate hydrogen bond parameters [24–26]. While the neutron-derived distances are accurate, this does not necessarily mean that they are the most chemically meaningful. This is because one cannot simply identify an atom with its nucleus, but rather consider it as being composed of nucleus and electrons. In any event, neutron distances have established themselves as benchmarks in hydrogen bond research.

However, X-ray diffraction probes the O–O separation under excitation but limited information on the O:H–O bond cooperative relaxation. In fact, the O:H and H–O cooperative relaxation in opposite directions by different amounts determines the O–O distance. Neutron diffraction probes the H–O and O:H distances but it needs attention to the O:H, H–O, and O–O correlation for comprehension of the O: H–O bond cooperative relaxation dynamics. The discriminative relaxation of the segmental length and energy dictating the performance of water and ice. It remains a question how the accuracy could be adequate for the strongly correlated and fluctuating water and aqueous solutions. Focusing on the nature of origin and the trend of change of bond-electron-phonon changing under peertubation could be more effcient.

### *17.2.2 Electron Spectrometrics*

There are four types of electron spectrometrics probing binding energies of electrons at different energy levels. These techniques include [27]:

(1) Scanning tunneling microcopy/spectroscopy (STM/S) probes the nonbonding lone pair states and the antibonding dipole states of a few electron volts across Fermi energy at extremely low temperatures and ultra-high vacuum, in the solid and quasisolid states [28, 29]. Atomic force microscopy using conditioned tips images the bonding and nonbonding states confirming absence of charge between molecules [30].
(2) Liquid-jet ultraviolet photoelectron spectroscopy (UPS) using two light sources for ionizing the H–O bond and exciting the solvent electrons [31], collects information of vertical binding energy and time delay for hydrated or adsorbed electrons in the skin and in the volume of the liquid, as well as molecular clusters.
(3) O 1s near edge X-ray absorption/emission fine stricture spectroscopy (NEXAFS/XES) that involves the O 1s core level and the upper valence band of water with dynamics of thermalization of the excited electrons. This technique is complicated as both of the involved levels are subject to relaxation when the sample is conditioned.
(4) X-ray photoelectron spectroscopy (XPS) probes directly the energy shift of the O 1s electrons under external stimulation under a certain vacuum condition.

Electron spectroscopies need to operate in vacuum conditions and the energy shift of the O 1s level that is dominated by the relaxation of the H–O covalent bond. However, XPS collects hardly information of the O:H nonbond as its cohesive energy is only 3 % or less of the H–O bond. The O:H relaxation and polarization dominates the upper occupied and unoccupied states and the H–O relaxation dictates the O 1s shift, which contribute to the NEXAFS in a complicated way.

#### 17.2.2.1 STM/S and Ultrafast Liquid-Jet UPS

Figure 17.3 illustrates the STM/S probing the unoccupied antibonding and occupied nonbonding energy states across the Fermi energy for $(H_2O)_N$ clusters deposited on NaCl substrate under low-temperature and high-vacuum [28, 29]. The bias between the tip and the $(H_2O)_N$ molecules determines the direction of electron flow. If the $(H_2O)_N$ is positively biased, the nonbonding electrons will flow from the $(H_2O)_N$ to the tip, otherwise, electron will flow from the tip to occupy the initially unoccupied antibonding states of the $(H_2O)_N$. Observations have confirmed that the $sp^3$-orbital hybridization of oxygen occurs at 5 K or even lower and that the molecular interaction discriminates the energy states between a $H_2O$ monomer and a $(H_2O)_4$ tetramer.

Figure 17.4 illustrates the liquid-jet UPS for detecting the lifetime and bound energy. The nozzle ejects the liquid beam of pure water or solutions in front of the time-of-flight (TOF) spectrometer. A laser beam of 267 nm wavelength (4.64 eV energy) ionizes water molecules to generate hydration electrons that can be trapped by proper solution ions. The extreme UV (EUV) beam of 38.7 eV excites the hydration electrons to escape from the bound with kinetic energy and then be caught by the TOF spectrometer. Time delay between the pump pulse of the laser or the EUV and the probe pulse give the dynamics and the lifetime of the hydration electrons.

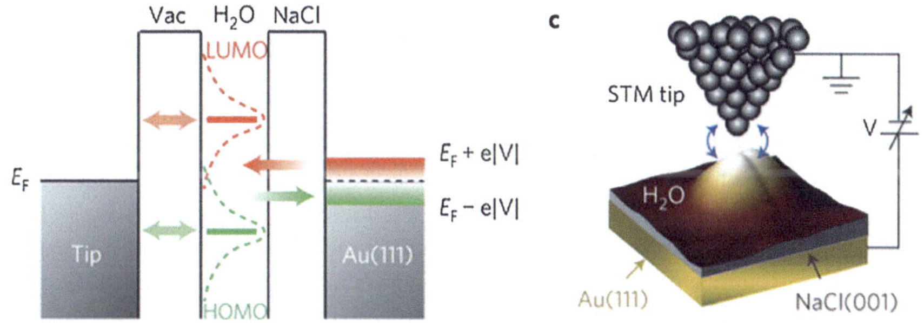


**Fig. 17.3** Principle of STM/S probing the energy states across Fermi level for $(H_2O)_N$ clusters deposited on NaCl substrate under high-vacuum and low-temperature (Reprinted with permission from [29].)

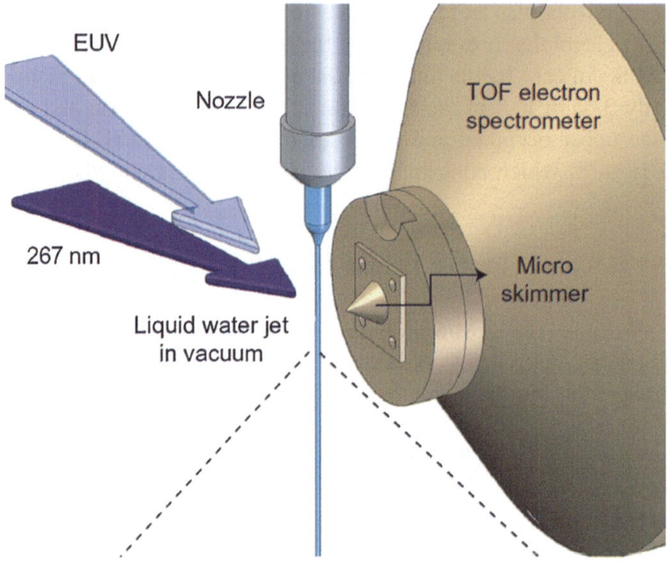


**Fig. 17.4** Ultrafast liquid-jet UPS detects the bound energy and life time of the hydrated electrons in the liquid jet. A laser beam of 267 nm wavelength creates hydrate electrons in the liquid jet and the highly harmonic light pulses in extreme UV (EUV, hν = 38.7 eV) excites the photoelectrons, the time-of-flight (TOF) spectrometer measures the kinetic energy and the pump-probe time delay of the ejected electrons (Reprinted with permission from [31].)

### 17.2.2.2 XPS, XAS, and XES

Figure 17.5 illustrates the band structure and the electronic processes involved in the NEXAFS, XPS, and NEXES. These techniques are based on the fact that atoms will absorb X-ray radiation to emit electrons from the O 1s level. The excited electrons that become free in XPS or fall on the upper unoccupied band, depending on the incident beam energy.

XPS measures the O 1s core level shift $\Delta E_{1s}$ of the liquid. The XPS spectrum shows a peak as a mixture of the skin and the bulk components. The PES monitors the consequence of crystal field change due to bond relaxation and bond nature alteration upon reaction. In the XPS and UPS process, the incident radiation of hν energy excites an electron from the O 1s level. The excited electrons will overcome the bound of work function ϕ and the O 1s level and then escape from the specimen becoming free with kinetic energy $E_K$. The electronic process of UPS and XPS follows the energy conservation of Einstein's photoelectron effect: $h\nu = E_{1s} + E_K$. The energy shift $\Delta E_{1s} = -\Delta E_K$ is proportional to the H–O bond energy at equilibrium. Contribution of the weak O:H nonbond energy is only 3 % or less that of the H–O bond energy of 4.0 eV.

In the NEXAFS and XES, the O 1s electron is excited to the uppermost unoccupied states leaving a hole behind [32]. All bands will relax and the excited electron undergoes a thermalization and then transiting from the unoccupied to the

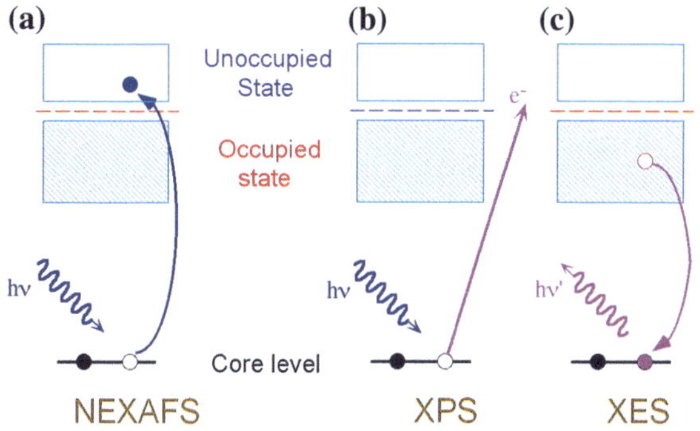


**Fig. 17.5** Principles for the electronic spectrometrics. O 1s electron absorbs energy in **a** the NEXAFS and **b** the XPS processes. The excited electron (**a**) transits to the upper unoccupied levels or (**b**) becomes free with kinetic energy. After thermalization to the bottom edge of the upper band, the excited electron in (**a**) transits to the O 1s level again and emits the XES energy (**c**). The XPS probes the energy shift of the O 1s level for an isolated atom and its bulk shift while the rest involves two levels that shift simultaneously under external stimulus [27]. The energy shift of the energy level from the isolated atom is proportional to the H–O bond energy only with negligible contribution of the O:H nonbond

high occupied states. The relaxed electron will revert to the initial 1s level and emit light. The difference between the absorption and the emission energies is a convolution of thermalization and electron-phonon coupling [27]. On the other hand, both the ground 1s level and the excited orbitals are subject to shift that is proportional to the H–O bond energy.

The EXAFS is sensitive to bond lengths where the NEXAFS is sensitive to both bond angles and bond lengths. EXAFS provides enough energy to eject an electron but NEXAFS has just enough energy to excite an electron to jump up to an unoccupied higher energy level-band edge. The steplike vertical rise in absorption intensity resides between the absorption edge and the EXAFS region, hence the "near-edge" designation. The energy at which this rise occurs differs according to the individual element, chemical bond, or molecular orientation. One can determine the surface molecular orientation of a liquid and its intramolecular bond lengths by tuning X-ray frequencies in the NEXAFS. The XEAFAS or the XES provide limited information on the binding energy shift as the processes involve multiple factors [33, 34].

Figure 17.6 shows the typical NEXAFS spectra for water and ice at different temperatures [34, 35]. At the O 1s energy level, a sudden increase in the absorption coefficient is observed. These energies, called absorption edges, correspond to the energy required to eject an electron from the atom. For an isolated atom, the sudden peak at the absorption edge occurs as the electron is ejected and then a gradual decrease in X rays absorbed occurs as the X-ray energy levels are increased. However, for atoms in a molecule or in a liquid or solid state, the closeness of other atoms around the absorbing atom causes oscillations in the amount of X-ray absorption just past the absorption edge. These "wiggles" detected in the absorption edge, called EXAFS oscillations, arise from the ejected electron backscattering off

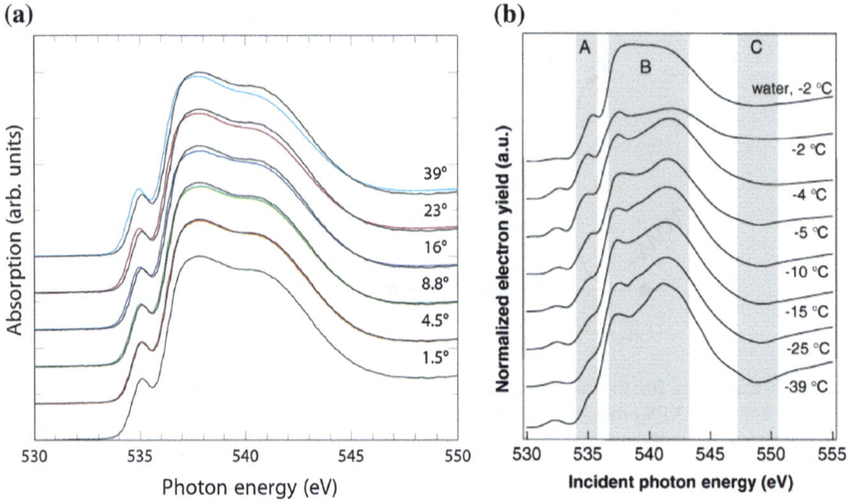


**Fig. 17.6** Temperature dependence of the O 1s NEXAFS spectra of **a** water and **b** ice (Reprinted with permission from [34, 35].)

neighboring atoms. The structure of the oscillations—that is, their frequency and amplitude-depends on the distance and number of neighboring atoms. The length of bonds between neighboring atoms—such as oxygen atoms or hydrogen atoms in water, for instance—can be determined by analyzing these EXAFS oscillations.

## *17.2.3 Phonon and Dielectric Spectrometrics*

### 17.2.3.1 IR, Raman, and SFG

The phonon resonance frequency of a substance is measured as $\omega = \omega_0 + \Delta\omega$, where $\omega_0$ is the reference point from which the shift $\Delta\omega$ proceeds under an applied stimulus. The referential $\omega_0$ may vary with the frequency of the incident radiation but the $\Delta\omega$ is independent of the incident beam energy. one can derive the vibration frequency of a harmonic system in the first order approximation by matching the second derivative of the interatomic potential u(r) in a Taylor series [36]:

$$\begin{aligned} u(r) &= \sum_{n=0} \left( \frac{\mathrm{d}^n u(r)}{n!\mathrm{d}r^n} \right)_{r=d_i} (r - d_i)^n \\ &= E_i + \frac{\mu\omega^2 (r - d_i)^2}{2} + o\left[ (r - d_i)^{n \geq 3} \right] \ldots \end{aligned} \tag{17.8}$$

The term with index n = 0 corresponds to the equilibrium binding energy $E_b$ that determines the energy shift of electrons in various levels. XPS probes the core level shift that is proportional to the $E_b$. The term n = 1 is the force at equilibrium; the term n = 2 corresponds to the force constant of harmonious vibration of a dimer oscillator, which dominates the frequency shift in phonon spectrometrics [37]. Terms n ≥ 3 corresponds to the nonlinear vibrations, which contribute to the transportation dynamics such as thermal expansion and thermal conductivity. Equaling the vibration energy of the dimer to the third term in the Taylor series of the interatomic potential yields:

$$\frac{1}{2}\mu(\Delta\omega)^2x^2 \cong \frac{1}{2}\frac{\partial u(r)}{\partial r^2}\bigg|_{r=d_i} x^2 \propto \frac{1}{2}\frac{E_i}{d_i^2}x^2 \tag{17.9}$$

The harmonic approximation is valid for bond vibrating near the equilibrium. Here we use the proportional relationship and dimension analysis as we consider the relative change of the frequency shift with respect to the bulk reference. This relation indicates that phonon frequency shift, $\Delta\omega(d_i, E_i)$, depends functionally on the bond length $d_i$, bond energy $E_i$, and the reduced mass $\mu = \frac{m_1 m_2}{m_1 + m_2}$ of the representative oscillator dimer,

$$\Delta\omega(d_i, E_i, \mu_i) \propto \frac{\sqrt{E_i/\mu_i}}{d_i} \propto \frac{\sqrt{Y_i d_i}}{\mu_i} \propto \sqrt{\frac{k_i + k_C}{\mu_i}} \tag{17.10}$$

The $k_i$ is the force constant of the bond or nonbond and $k_C$ is the force constant of the O–O Coulomb coupling, which are the second derivative of the respective short-range potentials. In fact, the phonon shift is proportional to the square root of the stiffness $Y_i d_i$ of the specific segment of the O:H–O bond. The $Y_i$ is the elastic modulus that is proportional to the local energy density $E_i/d_i^3$.

Spectrometrics of Raman scattering, IR absorption, and neutron diffraction probe the phonon frequencies of the O:H stretching vibration $\omega_L$, the H–O stretching vibration $\omega_H$, ∠O:H–O containing angle bending $\omega_{B1}$, and ∠H–O–H bending $\omega_{B1}$, in the full frequency range. There are three quantities of concern:

(1) The peak frequency shift from the reference corresponds to the variation of segmental stiffness
(2) The full width at maximum (FWHM) represents the degree of fluctuation in the stiffness of the particular segment
(3) The component integral (abundance) represents the richness of the specific segment of the same vibration frequency. Generally, the intensity increase with the frequency due to the increased reflexivity.

The IR absorption peak frequency is generally slightly higher than that of Raman reflection. Neutron diffraction gives more fine structure details than phonon spectrometrics as the neutron scattering collects information on the phonon density of states.

Sum frequency generation spectroscopy (SFG) is a technique used to analyze surfaces and interfaces. This nonlinear laser spectroscopy method was developed in 1987 [38] and rapidly applied to deduce the composition, orientation distributions, and structural information of molecules at gas–solid, gas–liquid and liquid–solid interfaces. SFG is monolayer surface sensitive and it is comparable to the second harmonic generation and infrared and Raman spectroscopy.

IR-visible SFG spectroscopy uses two laser beams that overlap at a surface of a material or the interface between two materials. An output beam is generated at a frequency of the sum of the two incident light. The two incident beams have to be able to access the surface, and the output beam needs to be able to leave the surface to be picked up by a detector. One of the beams is a visible wavelength laser held at a constant frequency and the other is a tunable infrared laser. By tuning the IR laser, the system can scan over resonances and obtain the vibrational spectrum of the interfacial region. The intensity of the SFG corresponds to the second order susceptibility depending on the interface molecular dipole orientation. It is also called dielectric spectroscopy.

The output beam is collected by a detector and its intensity $I$ is calculated using

$$I(\omega_3; \omega_1, \omega_2) \propto \left|\chi^{(2)}\right|^2 I_1(\omega_1) I_2(\omega_2)$$

where $\omega_1$ is the visible frequency, $\omega_2$ is the IR frequency and $\omega_3 = \omega_1 + \omega_2$ is the SFG frequency. The constant of proportionality varies across literature, many of them including the product of the square of the output frequency, $\omega_2$ and the squared secant of the reflection angle, $\sec^2 \beta$. Other factors include index of refractions for the three beams.

The second order susceptibility has two contributions

$$\chi = \chi_{nr} + \chi_r$$

where $\chi_{nr}$ is the non-resonating contribution and $\chi_r$ is the resonating contribution. The non-resonating contribution is assumed to be from electronic responses. Although this contribution has often been considered to be constant over the spectrum, because it is generated simultaneously with the resonant response, the two responses must compete for intensity. This competition shapes the nonresonant contribution in the presence of resonant features by resonant attenuation. Because it is not currently known how to adequately correct for nonresonant interferences, it is very important to experimentally isolate the resonant contributions from any nonresonant interference, often done using the technique of nonresonant suppression.

SFG spectroscopy probes the vibrational response of interfacial water. The second-order susceptibility $\chi^{(2)}$ determines the SFG intensity and the imaginary part of the second-order susceptibility, $\mathrm{Im}[\chi^{(2)}]$, can be determined using phase-resolved methods. $\mathrm{Im}[\chi^{(2)}]$ constitutes the surface equivalent of the bulk infrared absorption spectrum. Moreover, a positive (negative) band in the $\mathrm{Im}[\chi^{(2)}]$ spectrum indicates the net up (down)-orientation of the H–O stretch transition dipole moment at the

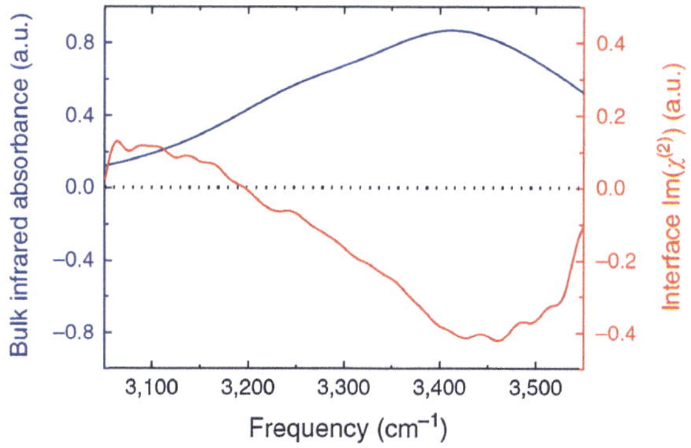


**Fig. 17.7** H–O stretching phonon band of $H_2O$. The infrared absorption spectrum of bulk water and the Im[$\chi^{(2)}$] spectrum of the air/water interface in the H–O stretching vibration frequencies (Reprinted with permission from [39].)

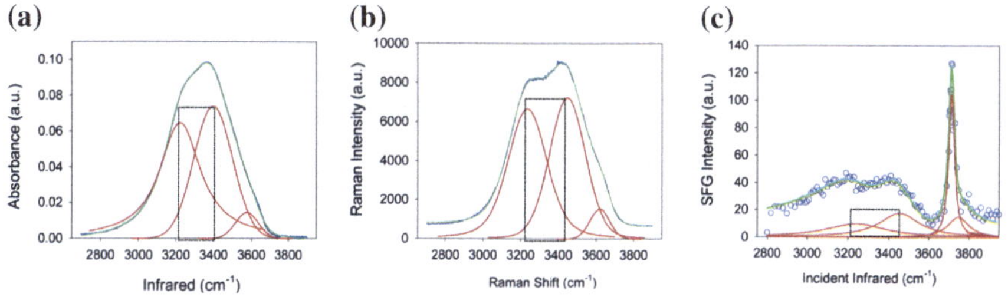


**Fig. 17.8** Deconvolution of the **a** attenuated total reflectance Fourier transform infrared (ATR-FTIR) spectrum and **b** Isotropic Raman spectrum of neat water, and **c** SSP-polarized SFG spectrum of the air- water interface (Reprinted with permission from [40]). Features centered at 3200 and 3450 cm$^{-1}$ correspond to the bulk and the skin and the feature at 3650 cm$^{-1}$ to H–O dangling bond radicals in Raman and IR spectra [41]. SFG collects the information from the monolayer skin and resolves the frequency of the H–O radicals at 3700 cm$^{-1}$

air/water interface. The spectral shape of the Im[$\chi^{(2)}$] spectrum of water at the air/water interface is similarly broad as the bulk infrared absorption spectrum of $H_2O$ (see Fig. 17.7).

Figure 17.8 compares the spectra of (a) Infrared absorption, (b) Raman reflection and (c) vapor-liquid interface SFG for neat water in the H–O stretching frequency regime ($\omega_H$). There are slight difference in the relative intensity and peak position for the respective bulk (~3200 cm$^{-1}$), skin (~3450 cm$^{-1}$), and the H–O dangling bond radicals (~3650 cm$^{-1}$) components.

As the signals at low frequencies are considerable weak compared to high frequencies, researchers are often focused more on the high frequency regime, which prevents information on the O:H relaxation ($\omega_L$) and O–O repulsion from being resolved.

#### 17.2.3.2 Multifield Phonon Spectrometrics

Phonon spectrometrics is most powerful as it collects quantitative information on the O:H–O bond angle and segmental cooperative relaxation in length and energy stimulated by various stimulus, which suits different experimental conditions without needing high vacuum. Ultrafast spectroscopy also gives information of molecular dynamics that is related to the order of fluctuation for molecular dynamics or the viscoelasticity of the specimen.

Raman and FTIR database can be established by probing over the full frequency range of water and ice during cooling, heating, compressing, salting, electrification, molecular clustering, and at different emission angles between the incident beam and the normal of the skin. The established database enables verification of predicted O:H–O bond cooperative relaxation and electron entrapment and polarization under various stimuli.

#### 17.2.3.3 DPS: Site Resolved O:H–O Bond Relaxation

Extension of the zone-resolved XPS strategy [27, 42] results in the differential phonon spectrometrics (DPS) that distills clearly the characteristic phonons due to the conditioning, by differencing the spectra collected before and after conditioning, or collected at different reflection angles from the same specimen upon the spectral peak areas being normalized and background-corrected. This DPS strategy also applies to discriminate the characteristic $d_x$ or $\omega_x$ peaks obtained under different conditions in calculations. For instance, it can distil the bond length and vibration frequency features for molecules with lowest CN from that of the bulk molecules with the standard nearest four CN in the bulk [43].

One can imagine what will happen to the outcome by differentiating two spectra collected under any of the following conditions from:

(1) Water or ice skin at different emission angles.
(2) Liquid water after and before conditioning such as salting at different concentration, heating to different temperature, compression and tension, droplet size reduction, electrification, magnetization, etc. under the same probing conditions.

Upon the standard processes of background correction and spectral peak area normalization, the DPS in (i) distills the spectral features due to the monolayer skin by filtering out the bulk information as the DPS collects more information from the surface at larger emission angles. Likewise, the DPS in (ii) purifies merely the spectral features due to conditioning. This DPS strategy can monitor the phonon relaxation both statically and dynamically with high sensitivity and accuracy without needing any approximation or assumption.

If the solute, such as alcohol and glycerol possess the features overlapping that of water, one can purify the solute effect on the phonon frequency shift by subtracting the composed from that of compound spectra in the form: $\Delta I = I_{x,solute} + I_x$

$_{-1,water}-(xI_{solute}+(1-x)I_{water})$, where $I_{solute}$ and $I_{water}$ are the spectra for the pure solute and pure water, respectively, collected under the same conditions. The x and (1−x) are the molar fractions of the constituent solute and water in the solution.

The DPS distils the phonon DOS gain as a component presenting above the horizontal axis and features the DOS loss as valley below the axis in the DPS spectrum. This process removes the commonly shared spectral feature that is out of concern.

Ideally, the resultant DPS components conserve as the spectral areas above and areas below the lateral axis are identical. Any improper background correction or spectral normalization may asymmetrize the spectral gain and loss compared with the ideal situation. With these criteria, one can readily gain quantitative information on the local bond length, bond energy, and charge entrapment and polarization, etc.

Conventionally, one needs to correct the spectral background using the standard Tougaard method [44–46] by employing Gaussian-, Lorentz-, or Doniach-Sunjic-type functions before decomposing the DPS profiles. However, DPS saves such tedious procedures of background correction, component specification, and peak energy fine-tuning. DPS gives directly the skin or conditioned component as emerging peaks and the bulk component as a valley.

The convoluted DPS applied to both O:H and H–O phonons gives the trend of phonon relaxation direction and an estimation of the overall $\omega_x$ phonon frequency shift of each segment.

One can obtain the molecular-site resolved DPS to discriminate the phonon relaxation dynamics in the bulk, the skin, and the H–O dangling bond radicals in the following procedures:

(1) Decompose the high-frequency H–O spectral peak of the reference water into the 3200 $cm^{-1}$ bulk component and the 3450 $cm^{-1}$ skin component with a 3610 $cm^{-1}$ feature for the dangling H–O bond radicals.
(2) Decompose the H–O peak of the conditioned spectrum into the respective components but the centers of these components are subject to shifting.
(3) Subtraction of the reference spectrum from the conditioned one upon being the overall area normalized and background corrected to resolve the site resolved O:H–O bond relaxation dynamics, elucidating the local O:H–O bond segmental length and energy and the situation of polarization or ions occupation tendency.

In the process of phonon spectrometrics, the central frequency, and the FWHM, and the integral area of the peak are of key concerns:

(1) The $\omega_H$ shift results from bond length and energy or the bond stiffness relaxation, which is proportional to the H–O lifetime $\tau_H$, molecular associativity (opposite to self-diffusivity), and water viscosity $\eta$ in a proportional way. The $\omega_H$ shift is generally coupled with the $\omega_L$ shift in the opposite direction.
(2) The FWHM decrease means lower order of fluctuation and self-diffusivity of $H_2O$ molecules.

(3) The peak area integral represents the abundance of the specific phonon mode in a particular molecular site.
(4) The peak area normalization before subtraction suggests that the sum of spectral area loss be identical to its gain of the three components, as the total number of phonons conserves.
(5) Molecular undercoordination resolves the $\omega_H$ for neat water into the bulk, skin, and H–O dangling bond components. However, the $\omega_H$ depends functionally on the H–O length and energy without discriminating the source of perturbation. Therefore, compression, electrification, and thermal excitation can change the frequencies in their respective ways.

## 17.3 Bond–Electron–Phonon–Property

### 17.3.1 The $E_H$–$d_H$–$\Delta E_{1s}$ Correlation

According to the BOLS-NEP notion [47], the cohesive energy of a specific segment denoted by subscript x is proportional to the inverse m power of its length under z stimulus [48]:

$$\frac{E_x(z)}{E_x(\infty)} = \left(\frac{d_x(z)}{d_{x0}(\infty)}\right)^{-m} = [C_x(z)]^{-m}, \tag{17.11}$$

where ∞ denotes the bulk standard, and $C_x(z)$ is the bond relaxation coefficient. However, the m holds no constant for water and ice.

The tight-binding theory [49] formulates the shift of the ν-th energy level of an isolated atom in the Hamiltonian:

$$H = \left[-\frac{\hbar^2\nabla^2}{2m} + V_{atom}(r)\right] + V_{cry}(r).$$

The intra-atomic potential $V_{atom}(r)$ determines the ν-th core level energy of an isolated atom, $E_\nu(0)$, and the crystal potential $V_{cry}(r)$ determines its shift $\Delta E_\nu(\infty)$. They follow the relations [49, 50]:

$$\begin{aligned} E_\nu(0) &= \langle \nu, i|V_{atom}(r)|\nu, i\rangle \\ \Delta E_\nu(\infty) &= \langle \nu, i|V_{cry}(r)|\nu, i\rangle\left[1 + \frac{z\langle \nu, i|V_{cry}(r)|\nu, j\rangle}{\langle \nu, i|V_{cry}(r)|\nu, i\rangle}\right] \\ &\cong E_b\left(1 + \left(\frac{\text{overlap integral}}{\text{exchange integral}} < 3\%\right)\right) = E_b, \end{aligned} \tag{17.12}$$

where $|v, i\rangle$ is the eigen wave function at the i-th atomic site, which satisfies the relationship $\langle v, i \mid v, j\rangle = \delta_{ij}$ because of the strong localization of the core electrons; and $E_b$ is the bond energy in ideal bulk. The $V_{cry}(r)$ sums the pairing potentials over all neighbors. The zeroth approximation of the Tylor series of the interatomic potential or the bond energy at equilibrium $E_b$ determines the energy shift of the νth energy level $\Delta E_v(\infty)$. External stimulus perturbs the crystal potential from $V_{cry}(r)$ to $V_{cry}(r)(1 + \Delta_H) \cong E_b(1 + \Delta_H)$ at equilibrium without modulating the wavefunction.

Because four lone pairs isolate a $H_2O$ molecule, the H–O bonding potential dominates the $V_{cry}(r)$. Therefore, the O 1s energy shift $\Delta E_{1s}(z)$ from that of an isolated oxygen atom $E_{1s}(0)$ is proportional to the H–O bond energy $E_H$ [51]. Any relaxation of the H–O bond shifts the $E_{1s}(z)$ away from the bulk reference $E_{1s}(\infty)$. The $\Delta E_{1s}(z)$ may be positive or negative, depending on the sources of perturbation, including bond length relaxation [52, 53], charge polarization [53], Coulomb coupling, etc. $\Delta E_{1s}(z)$ in water ice follows the relation [51]:

$$\frac{\Delta E_{1s}(z)}{\Delta E_{1s}(\infty)} = \frac{E_{1s}(z) - E_{1s}(0)}{E_{1s}(\infty) - E_{1s}(0)} = \frac{E_H(z)}{E_H(\infty)} = [C_H(z)]^{-m}. \tag{17.13}$$

As the O:H nonbond energy is around 3 % of the H–O bond energy of 4.0 eV, XPS collects no nonbond information, which limits the application of the XPS.

### 17.3.2 $Y_x$–$\omega_X$–$\Delta E_{1s}$ Correlation

From a dimensional viewpoint, the second-order derivative of the $u_x(r)$ at equilibrium is proportional to $E_x/d_x^2$ [54]. Approximating the vibration energy, $\mu_x\omega_x^2x^2/2$, of an oscillator to the second differential of the Taylor series of its potential, $u_x(r)$, yields the $\omega_x$ of the oscillator, where $\mu_x$ is the reduced mass of the oscillator and x is the amplitude of vibration. The following expressions also approximate the elastic modulus by definition, and $\omega_x$ is correlated with the elasticity in the following relationships [16, 55, 56]:

$$\begin{cases} \Delta\omega_x \propto \left( \left. \frac{\partial^2 u_x(r)}{\mu_x \partial r^2} \right|_{r=d_x} / \mu_x \right)^{1/2} \propto \sqrt{E_x/\mu_x} \Big/ d_x \\ Y_x \propto -V \left. \frac{\partial^2 u_x(r)}{\partial V^2} \right|_{r=d_x} \propto \frac{E_x}{d_x^3} \\ (\Delta\omega_x)^2 \cong Y_x d_x, \end{cases}$$

where $\omega_x(0)$ is the reference point from which the $\omega_x(z)$ shifts; V is the molecular volume; and $\Delta\omega_x(z)$ is a direct measure of the bond stiffness, which is the product of the Young's modulus $Y_x$ and the length $d_x$ of the bond. For liquid water, the elastic modulus is dominated by the weaker O:H nonbond (x = L), and polarization. Coulomb coupling contributes to the vibration by replacing the force constant $k_x$

with ($k_x + k_C$) in the form of $\Delta\omega \propto [(k_x + k_C)/\mu_x]^{1/2}$. The following describes their relative shifts:

$$\begin{cases} \frac{\Delta\omega_x(z)}{\Delta\omega_x(\infty)} = \frac{\omega_x(z)-\omega_x(0)}{\omega_x(\infty)-\omega_x(0)} = [C_x(z)]^{-(1+m_x/2)} \\ \frac{Y_x(z)}{Y_x(\infty)} = [C_x(z)]^{-(3+m_x)}. \end{cases} \quad (17.14)$$

Thus, $\Delta E_{1s}$ correlates with $\Delta\omega_H$ as follows:

$$\begin{cases} \Delta E_{1s} \propto E_H & (O_{1s}\ shift) \\ \Delta\omega_H \propto d_H^{-1}\sqrt{E_H} & (\omega_H\ shift) \\ (d_H\Delta\omega_H)^2 \cong \Delta E_{1s} & (Correlation). \end{cases} \quad (17.15)$$

This correlation means that $\Delta E_{1s}$ and the $\Delta\omega_H$ shift cooperatively in the same direction, but at different rates.

### 17.3.3 $T_C$ Versus $E_x$

For other 'normal' materials, $T_C$ is proportional to the atomic cohesive energy, $T_C \propto zE_z$, where z is the effective atomic CN and $E_z$ is the bond energy of the z-coordinated atom [48]. However, for water molecules, the $T_C$ for meting is proportional to $E_H$ only, because of the 'isolation' of the $H_2O$ molecule by its surrounding lone pairs. $E_L$ determines the $T_C$ for evaporation $T_V$, as this process dissociates the O:H nonbond:

$$\frac{T_C(z)}{T_C(\infty)} = \begin{cases} \frac{E_H(z)}{E_H(z_b)} = [C_H(z)]^{-m_H} & (T_C < T_V) \\ \frac{E_L(z)}{E_L(z_b)} = [C_L(z)]^{-m_L} & (T_C = T_V). \end{cases} \quad (17.16)$$

## 17.4 Summary

Interplay of the multiscale computations, diffraction crystallography, atomistic and electronic microscopy, electron and phonon spectrometrics are necessary to verify theory predictions on the performance of water and ice at different scales and under various constraints. Instrumental information gained using these approaches enables holistic insight into the O:H–O bonding and electronic dynamics, which determine uniquely the physical behaviour of water and ice.

The bond segmental length and energy and the associated charge polarization correlate and unify the macroscopic properties of water and ice. The response of bonding parameters to external stimuli stem the anomalies of water and ice. All detectable properties vary functionally with the relaxation of either the H–O bond or the O:H nonbond. The relaxation of the H–O bond shifts the O 1s energy $\Delta E_{1s}$, the critical temperature $T_C$ for phase transition (except for evaporation), and the H–O phonon frequency $\omega_H$. The relaxation of the O:H nonbond contributes to polarization, $\omega_L$ frequency shift, elasticity, and molecular dissociation energy $E_L$.

Except for heating that is associated with depolarization, H–O bond contraction under other stimulation is always associated $\Delta\omega_H$ blueshift and $\Delta\omega_L$ redshift as a result of polarization. The polarization elevates the viscoelasticity that slows the molecular dynamics and hence prolong the relaxation times in all electronic, phononic, rotational dynamics. The slowed dynamics corresponds to the supersolidity and quasisolidity. Therefore, all the detectable and imaginable properties of water and ice are correlated. One supposed to know all in the cycle if any one change of the quantities is well measured.

## References


1. H. Sun, COMPASS: an ab initio force-field optimized for condensed-phase applications overview with details on alkane and benzene compounds. J. Phys. Chem. B **102**(38), 7338–7364 (1998)
2. J. Wang, Q.H. Qin, Y.L. Kang, X.Q. Li, Q.Q. Rong, Viscoelastic adhesive interfacial model and experimental characterization for interfacial parameters. Mech. Mater. **42**(5), 537–547 (2010)
3. J.A. Hayward, J.R. Reimers, Unit cells for the simulation of hexagonal ice. J. Chem. Phys. **106** (4), 1518–1529 (1997)
4. H.C. Andersen, Molecular-dynamics simulation at constant pressure and-or tempertaure. J. Chem. Phys. **72**(4), 2384–2393 (1980)
5. X. Zhang, Y. Huang, Z. Ma, J. Zhou, Y. Zhou, W. Zheng, C.Q. Sun, A common superslid skin covering both water and ice. PCCP (2014). doi:10.1039/C4CP02516D
6. J.P. Perdew, Y. Wang, Accurate and simple analytic representation of the electron-gas correlation-energy. Phys. Rev. B **45**(23), 13244–13249 (1992)
7. F. Ortmann, F. Bechstedt, W.G. Schmidt, Semiempirical van der Waals correction to the density functional description of solids and molecular structures. Phys. Rev. B **73**(20), 205101 (2006)
8. E.B. Wilson, J.C. Decius, P.C. Cross, *Molecular Vibrations* (Dover, New York, 1980)
9. S.J. Clark, M.D. Segall, C.J. Pickard, P.J. Hasnip, M.J. Probert, K. Refson, M.C. Payne, First principles methods using CASTEP. Zeitschrift Fur Kristallographie **220**(5–6), 567–570 (2005)
10. J.P. Perdew, K. Burke, M. Ernzerhof, Generalized gradient approximation made simple. Phys. Rev. Lett. **78**(7), 1396 (1997)
11. X. Su, Z.J. Zhang, M.M. Zhu, Melting and optical properties of ZnO nanorods. Appl. Phys. Lett. **88**(6), 061913-3 (2006)
12. L.X. Dang, T.-M. Chang, Molecular dynamics study of water clusters, liquid, and liquid–vapor interface of water with many-body potentials. J. Chem. Phys. **106**(19), 8149–8159 (1997)
13. M.J.P. Nijmeijer, A.F. Bakker, C. Bruin, J.H. Sikkenk, A molecular dynamics simulation of the Lennard-Jones liquid–vapor interface. J. Chem. Phys. **89**(6), 3789–3792 (1988)
14. M.S. Green, Markoff random processes and the statistical mechanics of time-dependent phenomena. J. Chem. Phys. **20**, 1281 (1952)
15. R. Kubo, Statistical mechanical theory of irreversible processes. I. general theory and simple applications to magnetic and conduction problems. J. Phys. Soc. Jpn. **12**, 570 (1957)

16. C.Q. Sun, X. Zhang, X. Fu, W. Zheng, J.-L. Kuo, Y. Zhou, Z. Shen, J. Zhou, Density and phonon-stiffness anomalies of water and ice in the full temperature range. J. Phys. Chem. Lett. **4**, 3238–3244 (2013)
17. L.N. Hand, J.D. Finch, *Analytical Mechanics* (Cambridge University Press, 2008)
18. J. Fourier, *The Analytical Theory of Heat* (Dover Publications, New York, 1955)
19. X. Zhang, Y. Huang, Z. Ma, C.Q. Sun, *O:H-O bond anomalous relaxation resolving Mpemba paradox*. http://arxiv.org/abs/1310.6514
20. J.A. Sellberg, C. Huang, T.A. McQueen, N.D. Loh, H. Laksmono, D. Schlesinger, R.G. Sierra, D. Nordlund, C.Y. Hampton, D. Starodub, D.P. DePonte, M. Beye, C. Chen, A.V. Martin, A. Barty, K.T. Wikfeldt, T.M. Weiss, C. Caronna, J. Feldkamp, L.B. Skinner, M.M. Seibert, M. Messerschmidt, G.J. Williams, S. Boutet, L.G. Pettersson, M.J. Bogan, A. Nilsson, Ultrafast X-ray probing of water structure below the homogeneous ice nucleation temperature. Nature **510**(7505), 381–384 (2014)
21. L.B. Skinner, C.J. Benmore, B. Shyam, J. Weber, J.B. Parise, Structure of the floating water bridge and water in an electric field. Proc. Natl. Acad. Sci. **109**(41), 16463–16468 (2012)
22. E.C. Fuchs, P. Baroni, B. Bitschnau, L. Noirez, Two-dimensional neutron scattering in a floating heavy water bridge. J. Phys. D **43**(10), 105502 (2010)
23. T. Head-Gordon, G. Hura, Water structure from scattering experiments and simulation. Chem. Rev. **102**(8), 2651–2669 (2002)
24. J. Li, Inelastic neutron scattering studies of hydrogen bonding in ices. J. Chem. Phys. **105**(16), 6733–6755 (1996)
25. A. Kolesnikov, J. Li, S. Parker, R. Eccleston, C.-K. Loong, Vibrational dynamics of amorphous ice. Phys. Rev. B **59**(5), 3569 (1999)
26. J. Li, D. Ross, Evidence for two kinds of hydrogen bond in ice. Nature **365**, 327–329 (1993)
27. X.J. Liu, M.L. Bo, X. Zhang, L.T. Li, Y.G. Nie, H. Tian, Y. Sun, S. Xu, Y. Wang, W. Zheng, C.Q. Sun, Coordination-resolved electron spectrometrics. Chem. Rev. **115**(14), 6746–6810 (2015)
28. X. Meng, J. Guo, J. Peng, J. Chen, Z. Wang, J.-R. Shi, X.-Z. Li, E.-G. Wang, Y. Jiang, Direct visualization of concerted proton tunnelling in a water nanocluster. Nat. Phys. **11**(3), 235–239 (2015)
29. J. Guo, X. Meng, J. Chen, J. Peng, J. Sheng, X.-Z. Li, L. Xu, J.-R. Shi, E. Wang, Y. Jiang, Real-space imaging of interfacial water with submolecular resolution. Nat. Mater. **13**, 184–189 (2014)
30. J. Zhang, P. Chen, B. Yuan, W. Ji, Z. Cheng, X. Qiu, Real-space identification of intermolecular bonding with atomic force microscopy. Science **342**(6158), 611–614 (2013)
31. K.R. Siefermann, Y. Liu, E. Lugovoy, O. Link, M. Faubel, U. Buck, B. Winter, B. Abel, Binding energies, lifetimes and implications of bulk and interface solvated electrons in water. Nat. Chem. **2**, 274–279 (2010)
32. J. Stöhr, *NEXAFS Spectroscopy*, vol. 25 (Springer Science & Business Media, 2013)
33. A. Nilsson, D. Nordlund, I. Waluyo, N. Huang, H. Ogasawara, S. Kaya, U. Bergmann, L.Å. Näslund, H. Öström, P. Wernet, K.J. Andersson, T. Schiros, L.G.M. Pettersson, X-ray absorption spectroscopy and X-ray Raman scattering of water and ice; an experimental view. J. Electron Spectrosc. Relat. Phenom. **177**(2–3), 99–129 (2010)
34. J. Meibohm, S. Schreck, P. Wernet, Temperature dependent soft x-ray absorption spectroscopy of liquids. Rev. Sci. Instrum. **85**(10), 103102 (2014)
35. H. Bluhm, D.F. Ogletree, C.S. Fadley, Z. Hussain, M. Salmeron, The premelting of ice studied with photoelectron spectroscopy. J. Phys.: Condens. Matter **14**(8), L227 (2002)
36. G. White, Solid state physics, in *Physics in Australia: A Review by the National Committee for Physics* (1981)
37. C.Q. Sun, Relaxation of the chemical bond. Springer Ser. Chem. Phys. **108**, 550 (2014)
38. Y.R. Shen, Basic theory of surface sum-frequency generation. J. Phys. Chem. C **116**, 15505–15509 (2012)

39. S.T. van der Post, C.S. Hsieh, M. Okuno, Y. Nagata, H.J. Bakker, M. Bonn, J. Hunger, Strong frequency dependence of vibrational relaxation in bulk and surface water reveals sub-picosecond structural heterogeneity. Nat. Commun. **6**, 8384 (2015)
40. L.M. Levering, M.R. Sierra-Hernández, H.C. Allen, Observation of hydronium ions at the air—aqueous acid interface: vibrational spectroscopic studies of aqueous HCl, HBr, and HI. J. Phys. Chem. C **111**(25), 8814–8826 (2007)
41. Y. Huang, X. Zhang, Z. Ma, Y. Zhou, W. Zheng, J. Zhou, C.Q. Sun, Hydrogen-bond relaxation dynamics: resolving mysteries of water ice. Coord. Chem. Rev. **285**, 109–165 (2015)
42. C.Q. Sun, *Atomic scale purification of electron spectroscopic information (Patent publication: US20130090865)*. (USA, 18 June 2010)
43. T.F. Kahan, J.P. Reid, D.J. Donaldson, Spectroscopic probes of the quasi-liquid layer on ice. J. Phys. Chem. A **111**(43), 11006–11012 (2007)
44. S. Hajati, S. Coultas, C. Blomfield, S. Tougaard, XPS imaging of depth profiles and amount of substance based on Tougaard's algorithm. Surf. Sci. **600**(15), 3015–3021 (2006)
45. M.P. Seah, I.S. Gilmore, S.J. Spencer, Background subtraction—II. General behaviour of REELS and the Tougaard universal cross section in the removal of backgrounds in AES and XPS. Surf. Sci. **461**(1–3), 1–15 (2000)
46. X.B. Zhou, J.L. Erskine, Surface core-level shifts at vicinal tungsten surfaces. Phys. Rev. B **79** (15), 155422 (2009)
47. C.Q. Sun, Relaxation of the chemical bond. Springer Ser. Chem. Phys. **108**, 807 (2014)
48. C.Q. Sun, Size dependence of nanostructures: impact of bond order deficiency. Prog. Solid State Chem. **35**(1), 1–159 (2007)
49. M.A. Omar, *Elementary Solid State Physics: Principles and Applications* (Addison-Wesley, New York, 1993)
50. C.Q. Sun, Y. Sun, Y.G. Nie, Y. Wang, J.S. Pan, G. Ouyang, L.K. Pan, Z. Sun, Coordination-resolved C-C bond length and the C 1s binding energy of carbon allotropes and the effective atomic coordination of the few-layer graphene. J. Phys. Chem. C **113**(37), 16464–16467 (2009)
51. C.Q. Sun, Surface and nanosolid core-level shift: Impact of atomic coordination-number imperfection. Phys. Rev. B **69**(4), 045105 (2004)
52. C.Q. Sun, Y. Nie, J. Pan, X. Zhang, S.Z. Ma, Y. Wang, W. Zheng, Zone-selective photoelectronic measurements of the local bonding and electronic dynamics associated with the monolayer skin and point defects of graphite. RSC Adv. **2**(6), 2377–2383 (2012)
53. C.Q. Sun, Y. Wang, Y.G. Nie, Y. Sun, J.S. Pan, L.K. Pan, Z. Sun, Adatoms-induced local bond contraction, quantum trap depression, and charge polarization at Pt and Rh surfaces. J. Phys. Chem. C **113**(52), 21889–21894 (2009)
54. W.T. Zheng, C.Q. Sun, Underneath the fascinations of carbon nanotubes and graphene nanoribbons. Energy Environ. Sci. **4**(3), 627–655 (2011)
55. C.Q. Sun, X. Zhang, J. Zhou, Y. Huang, Y. Zhou, W. Zheng, Density, elasticity, and stability anomalies of water molecules with fewer than four neighbors. J. Phys. Chem. Lett. **4**, 2565–2570 (2013)
56. G. Ouyang, G.W. Yang, C.Q. Sun, W.G. Zhu, Nanoporous structures: smaller is stronger. Small **4**(9), 1359–1362 (2008)

# Chapter 18
# Laws for Water

- *Water prefers the four-coordinated geometry because of the $sp^3$-orbital hybridization of oxygen.*
- *O:H–O bond disparity and O–O Coulomb repulsion dictates its cooperativity, adaptivity, recoverability, and sensitivity when respond to stimulation.*
- *The O:H–O bond relaxation in length and energy and the associated charge polarization determine the unusual behavior of water and ice.*
- *O:H–O bond absorbs all sorts of energy and responds to perturbation in a long-range manner.*

**Abstract** Sixty basic rules govern the O:H–O bond relaxation and electron polarization and their consequences on detectable properties of water and ice such as the phonon frequencies, O 1s binding energy shift, crystal geometry, H–O phonon lifetime, mass density, elasticity, hydrophobicity, fluidity, lubricity, supersolidity, quasisolidity, viscosity, skin stress, skin solubility, molecular dynamics, degree of fluctuation, thermal stability and their interdependence.

## 18.1 Key Thrusts and General Rules

Table 18.1 summarizes the key thrusts that have enabled the present progress and will drive future endeavor in the fields. In addition to the tabulated premises that complement the main stream convention, an introduction of the quasisolid phase between liquid and solid and the supersolid phase due to molecular undercoordination for water skin, nanodroplets, nanobubbles, and hydration shells and their dispersion to the quasisolid phase are crucial to comprehension of water and ice. Introduction of super-HB quatum compression, anti-HB fragmentation, and ionic polarization to aqueous and impurified solutions could improve the understanding of water interaction with other substances like drug and biomolecules. Furthermore, all detectable properties depend functionally on the O:H–O bond segmental length and energy and the associated charge entrapment and polarization.

Table 18.2 features the O:H–O bond identities under typical conditions. O:H–O bond cooperative relaxation under the examined stimuli determine the physical properties of water and ice. Table 18.3 summarizes the respond of the O:H–O bond and the associated properties to applied stimuli. Understanding these relaxation trends should be of use in controlling O:H–O bond relaxation dynamics so as to harnessing the performance of water ice in a predictive way.



C.Q. Sun and Y. Sun, *The Attribute of Water*, Springer Series in Chemical Physics 113, DOI 10.1007/978-981-10-0180-2_18

**Table 18.1** Perspective transition in the present volume dealing with the structure, potential, and the physical anomalies of ice and water

| | Conventional premise | Complementary thrust |
|---|---|---|
| 1 | Single snapshots for an individual parameter | Statistic mean of a complete set of correlated parameters |
| 2 | Randomly domained mixed-phase or mono-phase fluctuating structural order | Core-shell structured two-phase order of the same tetrahedral geometry of different bond lengths |
| 3 | Proton "two-in two-out" fluctuation or "transitional tunneling transition" | Universality of O:H–O bond cooperative relaxation in length and energy |
| 4 | Interaction between rigid or non-rigid dipoles | Asymmetric, coupled O:H–O bond oscillator pair |
| 5 | Asymmetrical potential for proton tunneling transition or symmetrical potential for proton fluctuation | Asymmetrical, short-range, and strongly coupled potentials |
| 6 | Classical statistical thermodynamics | Segmental specific heat disparity and specific heat ratios |
| 7 | Supercooling and superheating at skin and nanoscale | Dispersion of the quasisolid phase boundary |
| 8 | Surface toughness and slipperiness | Skin and hydration shell supersolidity-bond stiffness, electronic repulsivity, phonon elasticity, order of fluctuation, and thermal stability |
| 9 | Electron and phonon spectroscopy, crystallography | Coordination-resolved multifield spectrometrics of electrons and phonons |
| 10 | Quantum computation | Quantum computation, Lagrangian oscillation dynamics, Fourier thermo-fluid transport dynamics |

## 18.2 Water Structure and Phase Diagram

(1) Preference of the $sp^3$-orbital hybridization of oxygen laid the foundation for the structure of water and ice despite its strong fluctuation and relaxation in $H_2O$ molecular size and molecular separation under excitation.
(2) Water prefers the four-coordinated quasi-tetrahedron geometry with the regular bulk covered with the supersolid skin of the same geometry but subject to undercoordination induced O:H–O bond elongation.
(3) The strong H–O bond energy of 4.0 eV prevents the $(H_2O)_2 \rightarrow OH_3 + OH$ proton tunneling transition or frustration from happening unless under extreme mechanical and thermal activation or viewed macroscopically or statistically.
(4) The tetrahedron geometry and the detectable mass density unify the size and separation of molecules packed in water and ice. The H–O and O:H projecting length at 1.0004 and 1.6946 Å at 4 °C.

**Table 18.2** Definition of the basic parameters for the segmented O:H–O bond and their reference values for the bulk water and ice, and the supersolid skin[a]

| Quantities | | | | 4 °C bulk | ±20 °C skin | −20 °C ice |
|---|---|---|---|---|---|---|
| O:H–O bond attributes | H–O Bond | Length (Å) | $d_H$ | 1.0004 | 0.84 | 0.9713 |
| | | Energy (eV) | $E_H$ | 3.97 | 4.66 | 3.97 |
| | | Frequency (cm$^{-1}$) | $\omega_H$ | 3200 | 3450 | 3150 |
| | | Debye temperature (K) | $\Theta_{DH}$ | 3200 | 3400 | |
| | | Melting point (K) | $T_m$ | 273 | 315 | 273 |
| | | O1s energy shift (eV) | $\Delta E_{1s}$ | 536.6 | 538.1 | – |
| | | $\omega_H$ life time (ps) | $\tau_H$ | 0.25 ± 0.90 | – | – |
| | O:H Nonbond | Length (Å) | $d_L$ | 1.6946 | 2.18 | 1.7985 |
| | | Energy (eV) | $E_L$ | 0.1 | 0.095 | 0.1 |
| | | Frequency (cm$^{-1}$) | $\omega_L$ | 175 | 75 | |
| | | Debye Temperature (K) | $\Theta_{DL}$ | 198 | 100 | 198 |
| | | Freezing point (K) | $T_N$ | 258 | 240 | – |
| | | Evaporation/dew point (K) | $T_V$ | 373 | 330 | – |
| | | O:H related molecular lifetime (fs) | $\tau_{H_2O}$ angle (°) | 200 | 700 | 300 |
| | | $\angle O:H{-}O$ angle (°) | $\theta$ | 160 | – | 165 |
| | | Mass density (g/cm$^3$) | $\rho$ | 1 | 0.75 | 0.92 |

[a]H–O phonon frequency $\omega_H$ *and life time* $\tau_H$, *molecular associativity (opposite to fluctuation degree and self-diffusivity* $\mu$), *skin stress* $\gamma$, *water viscosity* $\eta$ *are proportionally correlated*

(5) Discriminative relaxation of the O:H–O bond angle and segmental lengths determines the $T_C - P_C$ slopes of phase boundaries that fall into four categories in the phase diagram.

(6) H–O bond compression–elongation dictates the Liquid–Quasisolid and the VII–VIII phase boundaries; O:H nonbond elongation determines the Liquid-Vapor phase boundary; O:H–O frozen dictates the XI–$I_c$ boundary of constant $T_C$, and identical H–O and H:O relaxation dictates the X-(VII, VIII) boundary of constant pressure. Matching the negative $T_C - P_C$ slopes boundary results in 3.97 eV H–O bond energy for water and ice, and reproduction of the positive slope liquid-vapor boundary turns out the pressure dependence of the O:H length.

**Table 18.3** O:H–O bond relaxation and its consequences on the physical properties of water and ice

| | | Stimulus |
|---|---|---|
| O:H–O bond elongation | $\Delta d_{H} < 0$<br>$\Delta E_{H} > 0$<br>$\Delta \omega_{H} > 0$<br>$\Delta \Theta_{DH} > 0$ | Thermal excitation<br>• Quasisolid cooling (ice floating)<br>• Liquid heating<br>• Ice $I_{c+h}$ heating |
| | $\Delta T_{m} > 0$<br>$\Delta \tau_{H} > 0^{*}$ | Mechanical activation<br>• Tension |
| | $\Delta E_{1s} > 0$<br>$\Delta d_{L} > 0$<br>$\Delta E_{L} < 0$<br>$\Delta \omega_{L} < 0$<br>$\Delta \Theta_{DL} < 0$<br>$\Delta T_{N} < 0$<br>$\Delta T_{V} < 0$<br>$\Delta \rho < 0$<br>$\Delta Y > 0$ | Molecular undercoordination—supersolidity<br>• Water and ice skins (hydrophobicity and slipperiness)<br>• Nanodroplet and nanobubble (supercooling/superheating)<br>• Molecular clusters<br>• Hydration shell<br>• Hydrophobic capillary confinement (polywater), etc.<br>• Superhydrophobicity<br>• Superfluidity<br>• Supersolidity |
| | $\Delta \eta > 0$<br>$\Delta \tau > 0$<br>$\Delta \varepsilon_{r} > 0$<br>$\Delta \gamma > 0$ | Electrification<br>• Solute ionic short-range electrification (Hofmeister effect, superlubricity)<br>• Capacitor long-range electrification (Armstrong Water Bridge)<br>• Field of unlike mirror charges ± q |
| | $\Delta P > 0$ | |
| O:H–O compression (opposite effect) | | Thermal excitation<br>• Quasisolid heating<br>• Liquid cooling<br>• Ice $I_{c+h}$ cooling |
| | | Mechanical activation<br>• Compression<br>• Base solution quantum compression |
| | | Electrification<br>• Field of like charges<br>• Superposition of oppositely directed fields |

*$\tau_{H} > 0$ is associated with lower molecular fluctuation dynamics and higher water viscosity due to $d_{H}$ contraction induced polarization, excluding liquid heating that shortens $d_{H}$ but depolarizes nonbonding electrons and enhances the fluctuation

*Acid solution polarization and quantum fragilation lenghtens the O:H-O bond but destructs surface stress

## 18.3 O:H–O Bond Cooperativity

(7) Presence of the electron lone pars laid the foundation for the O:H–O bond in water and ice, which is general to X:B-Y with X and Y being atoms of other electronegative elements and B the electropositive element.
(8) In place of the dipole-dipole interaction convention, the O:H–O bond bridging $O^{2-}$ anions is universal to all phases of water and ice even in the $OH_3^+$:$OH^-$ superionic state albeit the segmental length and bending angle relaxation.
(9) O:H–O bond forms the basic building block that mimic an asymmetrical, H-bridged oscillator pair coupled with O–O Coulomb repulsion.
(10) The "master-slave" rule drives the O:H–O bond segmental cooperative relaxation; if one part contracts, the other part expands, irrespective the nature of excitation.
(11) $O^{2-}$ ions dislocate in the same direction along the O:H–O bond by different amounts with respect to the $H^+$ proton at coordination origin; the O:H always relaxes more than the H–O bond.
(12) O:H–O angle distortion results in phases of different geometries like the 2D ice, $(H_2O)_N$ clusters, and the XVI hollow cages but the O:H–O attribute retains.
(13) Depending on the segmental cohesive energy and length, phonon frequency shift fingerprints the segmental stiffness, irrespective of the type or nature of perturbation.
(14) The segmental phonon becomes stiffer when the segment becomes shorter, and vice versa.
(15) O:H–O bond segmental disparity and O–O repulsion defines the adaptivity, cooperativity, sensitivity, and recoverability of water and ice when subject to stimulation.
(16) O:H–O bond segmental relaxation in length and energy and the charge entrapment and polarization determine discriminatively all detectable properties such as phonon frequencies, Debye temperatures, degree of fluctuation, dipole moment, lifetimes, mass density, O 1s energy shift, phase transition dynamics, stiffness, solubility, thermal stability, and viscoelasticity.

## 18.4 O:H–O Bond Potentials and Specific Heats

(17) Short-range vdW-like potential, Morse exchange interaction, and the O–O Coulomb coupling repulsion describe adequately O:H–O bond interactions with other common long-range or quantum nuclear entanglement interactions being discarded as general background.
(18) Lagrangian oscillating mechanics transform the known segmental length and phonon frequency into the respective force constant and cohesive energy and

their respond to stimulation, which enables mapping the potential paths for the O:H–O bond at relaxation.

(19) The O:H nonbond and the H–O polar-covalent bond has respective specific heat of Debye approximation; the segmental Debye temperature defines the shape and the cohesive energy determines the thermal integral of the specific heat.

(20) The superposition of the specific heat curves defines two intersecting temperatures that correspond to the extreme densities at temperatures close to the melting point and freezing temperature for homogeneous ice formation; the intersecting temperatures also divide the full temperature range into the liquid, quasisolid, solid $I_{c+h}$, and ice XI phases with different specific heat ratios $\eta_L/\eta_H$.

(21) The segment having a relatively lower specific heat follows the regular rule of thermal expansion but the other segment serving as slave responds oppositely to thermal excitation because of O–O coupling.

(22) O:H–O phonon stiffness relaxation mediates the critical temperatures of freezing and melting in opposite directions by dispersing the quasisolid phase boundaries through the Eisenstein relation: $\Theta_D \propto \omega$.

## 18.5 Thermal Excitation at the Ambient Pressure

(23) The relatively lower specific heat of the O:H nonbond in the liquid and the solid $I_{c+h}$ phases entitles the O:H nonbond cooling contraction and driving the H–O bond slight expansion, which result in the seemingly normal thermal expansion of the liquid water and the $I_{c+h}$ ice at different rates.

(24) The relatively lower specific heat of the H–O bond in the quasisolid phase undergoes cooling contraction and the O:H nonbond expands more significantly, resulting in O:H–O cooling elongation, and therefore ice floats. This mechanism extends to other solids of negative thermal expansion with asymmetric, short-range, coupling interactions.

(25) The $\angle O:H{-}O$ angle remains constant at $\sim 160°$ in liquid but cooling stretches it in the quasisolid and in the solid phase up to 174°, which contributes positively to quasisolid cooling expansion but negatively to solid cooling densification.

(26) In the phase XI or at 100 K or lower, O:H–O bond length and stiffness remain constant because of its zero specific heat. The containing angle cooling stretching expands the volume slightly.

(27) The strong H–O bond dictates the extremely high specific heat of water and ice, which has little to do with the O:H liquid matrix.

(28) Thermal fluctuation softens the phonon frequency of the H–O dangling bond but raises its probability of phonon probing in reflection.

## 18.6 Mechanical Compression

(29) Compression shortens the O:H nonbond and lengthens the H–O bond via O–O repulsion, which results in the low compressibility of water and ice and the O:H–O length symmetrization toward phase X.

(30) Compression widens the bandgap of ice by polarization rather than H–O bond energy gain as happened to other semiconductors.

(31) Compression disperses the quasisolid phase boundary inwardly but tension does it outwardly, which results in the regelation—ice melts under compression and freezes again when the pressure is relieved. Tendency of $sp^3$-orbital hybridization and coordination environment restorage entitle the extraordinary ability of the O:H–O bond recovering deformation and dissociation.

(32) O:H–O bond thermal depolarization and O:H thermal expansion recovers its compressibility being lowest at 45 °C.

## 18.7 Molecular Undercoordination

(33) Bubbles, droplets, hydration shells, defects, and skins of water and ice share the same attribute of molecular undercoordination that shortens the H–O bond and stiffens its phonon spontaneously, which lengthens the O:H nonbond and polarize the nonbonding electrons in two rounds: core electron entrapment induced lone-pair polarization and the interoxygen electron pair repulsion.

(34) Molecular undercoordination not only disperses the quasisolid phase outwardly but also creates a supersolid skin. Quasisolid phase dispersion is often regarded as "supercooling" at freezing and "superheating" at melting. The supersolid phase is hydrophobic, ¼ less dense, thermally more stable, and viscoelastic being characterized with a common H–O phonon frequency 3450 $cm^{-1}$ for ±20 °C water and ice.

(35) Curvature elevation enhances the droplet supersolidity, which makes the highly curved skins of droplets and nanobubbles even more elastic, ice like, polarized, stiffer, and thermally more stable.

(36) Nanodroplets and nanobubbles prefer the two-phase structure in the core-skin manner.

(37) The skin supersolid makes long-lived nanobubbles mechanically stronger, chemically more active, less mobile, and thermally more stable.

## 18.8 Interfaces Contact: Friction and Wetting

(38) Electrostatic repulsivity (high density of dipoles) and O:H phononic elasticity (low frequency but high magnitude vibration) drive the friction and wetting phenomena at contacting interfaces, performing like maglev or hovercraft.

(39) Atomic undercoordination induced local quantum entrapment and subjective polarization stem the Wenzel-Cassie-Baster's notions for superhydrophobicity (polarization dominance) and superhydrophilicity (entrapment dominance) enhancement by nanofabrication.

(40) An air gap of 0.5–1.0 nm exists at the hydrophobic interface because of the water skin supersolidity; supersolid nanofluid flows freely in microchanels than in macrochanels because of the phonon elasticity and electrostatic repulsivity, making the supersolid-skinned quasisolid droplet contactless motion in microchanells.

(41) A supersolid skin covers both water and ice that makes ice slippery and water skin tough and hydrophobic.

## 18.9 Salt Solutions: Hofmeister Effect

(42) The electric field of an ionic solute in salt aligns, stretches, and polarizes its surrounding molecular dipoles, which effects the same to molecular undercoordination on the O:H–O bond relaxation and polarization and the associated physical properties of liquid water—polarizability, quasisolidity, viscoelasticity, and the fluctuation degree of molecular dynamics.

(43) The degree of O:H–O bond electrification is molecular site, solute concentration, solute type, and temperature dependent.

(44) Preferential skin occupancy of heavy anions stiffens the phonon frequency of the H–O dangling bond but weakens its probability of observation because the anions not only enhance the local electric field but also screen the spectral signal.

(45) Solute electrification raises the critical pressures for transiting liquid water into ice VI and then into ice VII at room temperature because it needs excessive energy to recover the electrification-deformed O:H–O bond; gelation time and critical temperatures are sensitive to electrification that mediates the phase boundaries.

(46) Different from changing solute type, raising solute concentration effects similarly to heating on the critical pressures for liquid water transiting into ice VI and then into ice VII. Salts also raise the critical pressure for VII-X phase transtion, stiffen the $\omega_H$ and lenghtn the O-O distance through out the course of VII-X transition.

## 18.10 Acid and Base Solutions: Quantum Specification

(47) The HX-type acids dissolve into the $X^-$ and the $H^+$ that binds to a $H_2O$ to form the hydronium $H_3O^+$ interacting with one of its four $H_2O$ neighbors through the O–H↔H–O anti-HB. The anti-HB breaks the HB network of the acid solution in a point-by-point manner, which makes the solution "fragile", capable of destructing skin stress and diluting human blood.

(48) The YOH-type bases dissolves into the $Y^+$ and the hydroxide $HO^-$ interacting with one $H_2O$ neighbor through the O:↔:O super-HB. The super-HB serves as a point compressor to lengthen and soften the H–O bond. Burning heat is released when the reaction, or the H–O bond elongation takes place.

(49) The $Y^+$ and $X^-$ ions serves as the point polarizer to the form the supersolid hydration shells that thicken the human blood, promoting hypertension.

## 18.11 Electrofreezing and Water Bridging

(50) O:H–O bond long-range electrification enhances water's quasisolidity. Combination of the skin supersolidity and quasisolidity maintain Armstrong's water bridge floating at the ambient condition.

(51) Aqueous solutes destabilize the bridge by their electric field opposing that of the capacitor; ion solutes promote soil wetting as the overlapping of the opposite electric fields of the aqueous solutes and soil particles, which reduces the viscosity of the liquid water from that under either field alone.

(52) Opposing to unlike charges, field of separated, like charges raises the freezing temperature by inwardly dispersing the quasisolid phase boundary.

## 18.12 Magnetification: Dipoles Moving in the Lorentz Force Field

(53) The Lorentz force of the electromagnetic field creates not only an additional angular component to the transitional motion but also a transitional component to the circuler motion, which not only enhances the molecular dynamics but also creates a whirpool in the flowing water. further induction of electric current creates magnetic field opposing the source magnetic field, making water antiferromagnetic.

(54) Magnetification enhances slightly the quasisolidity and skin supersolidity, which elevates the skin stress and viscosity only slightly compared with electrification.

## 18.13 Energy Absorption, Emission, Conduction, and Dissipation

(55) O:H–O bond absorbs all sorts of energies in a long-range manner by exciting the molecular fluctuation, O:H–O bond bending, O:H and H–O segmental relaxation and dissociation.

(56) Mpemba effect integrates the processes of O:H–O bond energy "emission-conduction-dissipation" in a "source-path-drain" cycling system.

(57) O:H–O bond memory, or extraordinary recoverability, entitles its rate of energy emission proportional to its initial storage.

(58) Heating and the supersolid promote skin thermal diffusivity mainly by density depression, favoring an outward heat flow.

(59) Mpemba effect occurs only in the strictly non-adiabatic source-drain interface. Other factors such as convection, evaporation, impurity, and supercooling play roles of insignificance.

## 18.14 Probing Strategies: Advantages and Limitations

(60) Phonon spectrometrics monitors the molecular-site-resolved O:H–O relaxation in terms of bond stiffness, order of molecular fluctuation, and phonon abundance.

(61) The micro-jet UPS and XPS collects molecular site-resolved information of electron polarization and the O 1s electron energy quantum entrapment.

(62) O:H–O segmental cooperative relaxation is beyond the scope of diffraction and the O:H nonbond energetic contribution is beyond the capacity of O k-edge X-ray absorption/emission fine structure spectroscopies that collect information involving the O 1s and the valence band energy shift.

(63) Fourier thermo-fluid transport dynamics describes adequately the heat conduction dynamics with capability of skin discrimination. Lagrangian mechanics is hitherto the most efficient approach dealing with the asymmetrical, short-range, and coupled interactions of the O:H–O oscillator pair, in terms of segmental length, cohesive energy, force constant, and vibration frequency.

# Erratum to: The Attribute of Water

**Chang Q. Sun and Yi Sun**

## Erratum to:
## C.Q. Sun and Y. Sun, *The Attribute of Water*, Springer Series in Chemical Physics 113, DOI 10.1007/978-981-10-0180-2

The original version of this book was inadvertently published with incorrect spelling in the introductory quote "O:H–bond segmental disparity……". The text should read as "O:H–O bond segmental disparity……"

---

The updated original online version for this book frontmatter can be found at DOI 10.1007/978-981-10-0180-2

---

C.Q. Sun (✉)
School of Electrical and Electronic Engineering, Nanyang Technological University, Singapore, Singapore
e-mail: ecqsun@gmail.com

Y. Sun
Xiangtan University, Changsha, China

Y. Sun
China Jiliang University, Hangzhou, China



C.Q. Sun and Y. Sun, *The Attribute of Water*, Springer Series in Chemical Physics 113, DOI 10.1007/978-981-10-0180-2_19

# Index




C.Q. Sun and Y. Sun, *The Attribute of Water*, Springer Series in Chemical Physics 113, DOI 10.1007/978-981-10-0180-2

**Q**



**R**



**S**

The manufacturer's authorised representative in the EU is Springer Nature Customer Service Centre GmbH, Europaplatz 3, 69115 Heidelberg, Germany. If you have any concerns regarding our products, please contact ProductSafety@springernature.com

Printed and bound by CPI Group (UK) Ltd, Croydon, CR0 4YY
30/09/2026
02225443-0019